Where to watch birds in
WORLD CITIES

Paul Milne

YALE UNIVERSITY PRESS
NEW HAVEN AND LONDON

Published 2006 in the United Kingdom by Christopher Helm,
an imprint of A & C Black Publishers Ltd., and in the United
States by Yale University Press.

Text copyright © 2006 by Paul Milne
Illustrations copyright © 2006 by Chris Shields

ISBN-13: 978-0-300-11691-5
ISBN-10: 0-300-11691-8
Library of Congress Control Number: 2006933444
Printed and bound in China via Compass Press

A catalogue record for this book is available from the
British Library.

The paper in this book meets the guidelines for permanence and
durability of the Committee on Production Guidelines for Book
Longevity of the Council on Library Resources.

10 9 8 7 6 5 4 3 2 1

Contents

Acknowledgements

Writing this book could only be possible with the cooperation and willingness of hundreds of birders to document their observations in all the cities mentioned. I have been overwhelmed with the generosity of many birders around the world who have provided me with information and assistance about their home cities and other cities that they have visited. They have given generously of their time and their knowledge to provide site information and checklists, and have proofread the many drafts that it took to get the city accounts as accurate as possible. Inevitably, I may have omitted some information or made some errors in the accounts and for these I take full responsibility.

So many birders have contributed to this book that it would be impossible to list everyone and, for this, I apologise in advance, but I have to make special mention of the following: Keith Betton, who rescued the project by writing the accounts for the Australian and New Zealand cities and for Honolulu; Bill Harvey, who wrote the account of New Delhi; and Chris Shields and Nick Moran who, between them, covered Shanghai. Chris Shields drew the highly original and unique illustrations. Brian Southern rendered the maps with great care and attention to detail, despite my confused and often contradictory instructions. Ernest Garcia edited the final copy.

Additionally, I received assistance in the form of information, advice, critical input and proofreading from: Joe Adamson, Nilo Arribas, Simon Ball, Rob Batchelder, Giff Beaton, Scott Carpenter, Hugh Cotter, Fabrice Ducordeau, Gunnar Engblom, Luiz Fernando, Ricardo Gagliardi, Urs Geiser, Charles Harper, Wayne Hsu, Mikkel Kure Jakobsen, Ottavio Janni, Johnny Kamugisha, Claudio Köller, Waldemar Krasowski, Martin Kühn, Mark Kuiper, Rachel Lawson, Igor Lebedev, Dominic Mitchell, Dr Saeed A. Mohamed, Christophe Moning, Pete Morris, John Muddeman, Dermot O'Mahoney, Brian Rapoza, Judy Raskin, Martin Riesing, Jonathan Rossouw, Cyril Schönbächler, Eric Secker, Lim Kim Seng, Jevgeni Shergalin, Andy Smith, Lloyd Spitalnik, Ulrik Svane, Don Turner, Paul Walser and Chip Weseloh.

The vital task of refereeing the city accounts fell to: Simon Aspinall, Margaret Bain, Greg Butcher, Geoff Carey, Callan Cohen, Eduardo de Juana, Joe Doolan, Brian Finch, Tim Fisher, Gerard Gorman, Ricard Gutiérrez, Guy Kirwan, Guilhem Lessaffre, Etienne Marais, Ger Meesters, Jeremy Minns, Sunjoy Monga, Klaus Malling Olsen, Hal Opperman, Simon Papps, Sebastian Patti, Mark Pearman, Wayne Petersen, Derek Pomeroy, Andreas Ranner, Laurent Raty, Nigel Redman, Richard W. Roberts, Philip Round, Chris Sharpe, Adam Supriatna, Ted Shimba, Mano Tharmalingam, Uthai Treesucon, Negussie Toye, David Winter and Rob Williams.

I would also like to extend heartfelt thanks to the staff of A&C Black whose goodwill and encouragement were vital ingredients in getting this project started and brought to fruition. In particular I thank Mike Unwin for helping me put a coherent shape on the project and for guiding my thoughts on so many occasions; Sophie Page for helping to keep me on track during my more forgetful moments; Sharmila Logathas for her help and assistance during the early stages of the project; and, most especially, Nigel Redman, for taking me seriously when I first suggested this book and for maintaining a guiding hand right through the project. Indeed, his experience, knowledge and address book proved to be essential components in the successful completion of this book.

Finally, I would like to acknowledge the support and encouragement of my long-suffering wife, Patti, and children, Claire and Declan, who have accompanied me (reluctantly) on so many birding diversions during family holidays, for their patience and understanding during the many lost evenings and weekends while I worked on this book, and for never letting me forget what is most important in life.

Introduction

The idea for this book germinated one hot July day in Tokyo 2002. I was staring out of my hotel window at a pair of Large-billed Crows flying about and indulging in their aerobatics between the glass and steel canyons of the downtown Tokyo skyscrapers. As they flew off into the distance, I wondered where and how far they were going and whether there was a good birding site where the crows and other birds might congregate. As a fairly frequent business traveller, I often visited foreign cities for a day or two with a vague knowledge of the local avifauna but absolutely no opportunity of seeing any of it. On occasion, I found myself delayed in an airport or, due to cancellation of a meeting, stuck in the city for a half-day and looking for some way to kill time. This book is intended to assist the birder to maximise his or her time in precisely these circumstances, so that they may sample the amazing birdlife which often exists just a short drive from the city centre.

As more and more of the world's birding hotspots are documented by an increasing number of excellent birding guides, it seems that the world's cities need to catch up. This book can in no way compete with such comprehensive guides that are aimed at those on a full-blown birding trip, which may last anything from a week to six months or more. Rather, it is aimed at the birder who is visiting a city for business, pleasure, family or some other non-birding reason but who may find a few hours or perhaps a weekend to spare, and would like to find out what birding possibilities exist. Such trips are often characterised by no planning, no binoculars, no weatherproof clothing and no local knowledge of what birds can be found.

I have selected the 61 cities in this book on the basis of their popularity as business, tourist or transit destinations and, inevitably, this has created a bias in favour of the developed world. Their selection was in no way influenced by the richness of their birdlife as, by definition, the reason for your visit in the first place was for non-birding purposes. Nor were the choices made according to the city's size or whether it is the capital city of a country. In some cases, the city in question may be the starting point for a bird tour to the interior of the country and an unscheduled delay at the start or end of the trip will provide the opportunity to do some birding in the city itself. Generally, the guide assumes that the birder is visiting the city principally for purposes other than birding, and it aims to allow the best possible use to be made of any birding opportunities which may arise.

In writing this guide I have attempted to share the benefit of my own (often hapless) experiences and the generous cooperation of many friends and acquaintances. My intention is to give a flavour of the kind of birding that can be experienced at any time of year within an 80km radius of the city. Only for very compelling cases have I broken the 80km rule, and I have usually selected sites that are accessible by public transport. The treatment of each site has been more general than would be the case in the country-specific guides and gives the birder an overview of the site's birdlife but, where possible, I have attempted to provide a full picture of what birds are likely at all seasons of the year, not just during the optimum periods.

I hope that every travelling birder will keep this guide in his or her briefcase or travel bag to be used whenever a few spare hours present themselves on a city trip, and that the experience gained will prompt a return trip to the city and country in question. I also hope that it will stimulate others to share their experiences: constructive comments and corrections to assist in updating this guide for future editions will be gratefully received. Please send your suggestions to the author, care of the publishers.

Urban Birding

On initial consideration, no one would think that a city offered any potential as a birding location. However, the physical factors which favoured the location of a city — a river crossing, sheltered harbour, valley or proximity to an agricultural area — can also provide a rich habitat for birds and, very often, the development of the city itself can create habitats which are exploited by birds. The heightened awareness and increased resources now made available to nature conservation have given a measure of protection to birds and their habitats in many urban areas but, on the other hand, the relentless expansion of urbanisation has placed enormous pressures on many habitats. It is often the destruction of a buffer zone, rather than of the site itself, which spells disaster for the wildlife. It is hoped that the ability of a city's birding sites to interest and attract visiting birders will assist in raising awareness and improving the chances of their conservation.

URBAN HABITATS

On arriving in an unfamiliar city there are a couple of factors to consider when deciding on a birding strategy. Coastal cities have an obvious advantage, as the shoreline will provide opportunities to see seabirds and other coastal species while the hinterland will support the typical landbirds of the area. However, even the presence of an adjacent lake or large river will also provide the mosaic of habitats which is essential to supporting a diversity of species. Elevation is also an important factor, and a city which has access to a range of altitudes will often have a variety of habitats and ecological niches on its doorstep. Always look for physical features in a city which may concentrate birds, such as waterways that can support aquatic species, 'greenbelt' areas which penetrate into the heart of the city and, of course, any coastal zones.

HOTELS

If you have a choice, you should choose a hotel that has its own extensive grounds or backs onto a park or golf course. Your hotel gardens will provide the most immediate opportunity for birding, particularly if they are extensively irrigated and contain plenty of dense cover. They may not support any specialist species but will probably provide enough habitat to support all the typical species of the city area, and a quick chat with the hotel gardener or grounds keeper may reveal the presence of a roosting owl or other speciality which might otherwise go unnoticed. If you are visiting another continent for the first time, the hotel grounds will also provide a 'gentle' introduction to the typical birds of the region and enable you to familiarise yourself with the common bird families before venturing into more productive habitats. This experience can be particularly valuable in the tropics, where it is very easy to become completely overwhelmed when venturing out into the prime hotspots for the first time.

TALL BUILDINGS

There are many examples of tall buildings providing habitats for normally cliff-dwelling species and nearly every city has a special bird or two. For swifts, martins, falcons and other cliff-dwellers, tall structures have become prime urban breeding sites. In many cities that are famous for their historic buildings and antiquities, these popular tourist sites are often themselves good for birds and may provide a nest-site for a cliff specialist. Very often it is their popularity as a tourist site that has drawn a birder there at some point to record the birds which can be seen on a typical visit.

PUBLIC PARKS

Every city has public parks and recreation areas which act as green lungs for the city and as much-needed amenities for its citizens. The wooded areas act as patches of forest for breeding and migrant forest species, while most parks will have a lake or other waterway which attracts wildfowl in winter and sometimes to breed. On the downside, many parks get crowded, particularly at weekends, making birding impossible. Others may have restricted opening hours which preclude an early morning visit, which is always the best time. Birding in a park requires careful selection to ensure the best possibilities. Always look for a park with a lake or a river running through it. Apart from the obvious attraction for waterfowl, the littoral or riparian habitats also support species which would not be found in woodland alone. Parks that are contiguous with more extensive forest, usually on the outskirts of the city, will support a larger range of species. A feature of most European capital cities are remnant areas of forest that were originally preserved as hunting estates for the ruling monarchies. The monarchies may now be gone, or at least may be no more than shadows of their former splendour, but the estates provide a wonderful habitat for the wild inhabitants of the forest, safe from hunting pressures and with many ecological niches still intact. The provision of nest-boxes and other aspects of wildlife management have enhanced these areas as wildlife refuges.

ZOOS AND BOTANICAL GARDENS

In dry climates, and in hot climates which experience seasonal rains, the year-round irrigation provided in botanical gardens and in zoos promotes vigorous plant growth and often provides an oasis of green vegetation in an otherwise parched environment. The attendant plant and insect life can make these sites the best birding habitats in the city. The birdlife is often well documented and there may even be nature trails and perhaps a bird feeding station, attracting finches, sparrows, sunbirds or hummingbirds. The fact that you must pay to gain entrance means that such places are likely to be less crowded than the public parks but the downside is that opening hours are restricted and they may be popular with noisy school parties mid-week, when one would normally expect some respite from the crowds.

AIRPORTS

The main requisite for building an airport is a large expanse of flat open terrain close to a city. In many cases, this has meant locating airports on reclaimed marshes or estuaries or other 'low-value' locations rather than on relatively expensive farmland. Unfortunately the environmental price paid by these decisions was rarely taken into consideration. However, in some cases, the perimeter of the airport still retains some remnant marshlands or coastal wetlands which, due to their proximity to the airport, make them worthwhile sites to visit. Enlightened airport development may sometimes ensure that such sites are retained and managed for wildlife conservation. In other cases, the airport itself provides habitat for grassland species which intensive agricultural techniques may have destroyed elsewhere. However, heightened security around airports has rendered many such sites out of bounds, and local enquiries should always be made prior to visiting.

GOLF COURSES

The pursuit of golf has preserved some of the world's most expensive urban property from being built on, and where other recreational facilities have been sacrificed for development, golf seems to have its own system of self-preservation. Moreover, the maintenance of golf courses requires a regime of irrigation and landscaping which offers habitats for birds and other animals, and the periphery of golf courses provides cover in the form of shrubbery, small woods and often dense scrub. This kind of habitat can be scarce in arid zones or in dense urban environments, and a golf course can provide an oasis of habitat for migrating and breeding birds. Unfortunately, many golf clubs jealously guard their privacy and exclusivity but it is always worth asking permission from the groundsman or other official for permission to go birding on the course, even if it does mean having to ignore the scowls of the snootier members and dodging the odd golf ball!

REFUSE TIPS AND SEWAGE LAGOONS

The ideologically pure environmentalist will probably want to boycott such sites as a protest, but the visiting birder needs to make the most of what the city offers. Refuse tips and sewage ponds are probably man's greatest gift to urban birds. Refuse dumped in open landfill attracts huge flocks of gulls and crows in the temperate latitudes, and kites, storks and vultures in warmer countries. Those who have the forbearance to withstand the stench for periods of time can obtain some of the best possible prolonged views of feeding birds at close range.

The technology of pumping sewage into large shallow lagoons and allowing it to fester over time provides a very basic solution to a universal problem; it also has great appeal for birds and has provided some of the best wetland sites for ducks, herons and waders in any urban environment. As water levels are raised and lowered as part of the treatment process, a variety of freshwater marginal habitats are created and result in a mosaic of different habitat types supporting a wide diversity of species. The rich concentration of organic nutrients promotes abundant plant and insect life which are, in turn, exploited by breeding, migrating and wintering birds. The turbulence produced by the discharge of effluent into an adjacent river or lake helps to prevent icing-over in cold climates where sewage outfalls are often the only stretch of ice-free open water available to gulls and waterfowl.

Open landfill and sewage treatment are falling out of favour, particularly in the developed world, as incineration, thermal treatment and other technologies take the place of traditional methods. But it is hoped that the resultant abandoned sites will be redeveloped as managed wildlife reserves, thus preserving wildlife diversity.

WETLANDS

Wetland habitats have probably suffered most as a result of urban expansion. Their flat terrain adjacent to a river or coast and their apparent uselessness for any other purpose have always made them prime candidates for drainage and infill. Once lost, they are extremely difficult to recreate. However, in some cities, the few remnant sites of once more extensive marshes have been conserved. These sites are usually valued components of a city's amenities and tend to be well managed and protected. Not only do they have a good infrastructure of hides, walkways and nature trails, but also they are often actively managed to attract birds and may have a higher species diversity than larger marshes, more remote from the city but neglected. Enclosed wetlands especially are in a state of succession from open water to dry land and unless they are constantly dredged, the vegetation cut back and water levels regulated, the mosaic of different habitats cannot be maintained and overall species diversity suffers.

PIERS AND JETTIES

In cities which are adjacent to the coast, a lake or other large waterway, locating a pier which extends out into deeper water should be a key objective. A harbour pier will get you closer to any waterfowl which require deep water as well as to seabirds or to waders on mudflats. To get full benefit of this access you would probably need a telescope, but there are still opportunities with binoculars to see certain species which just cannot be identified satisfactorily from shore. If the harbour is frequented by fishing boats, then good numbers of gulls can usually be seen. Even inland cities are almost always built beside a river and if it is a particularly broad one then the city bridges can be a surprisingly valuable vantage point for scanning upstream and downstream.

FERRIES

Certain coastal cities and, indeed, lakeside ones, use commuter ferries to provide public transport to populated islands and outlying communities. These ferries can often provide a worthwhile watching platform for the birder as they are usually quite slow-moving, with plenty of deck space and a handrail at elbow height on which you can rest your binoculars. Even harbour cruises or other tourist-type vessels can give you access to open waters or may pass near enough to island breeding colonies to make it worthwhile. Whale-watching trips are nearly always good for birds, as the feeding activity of whales draws seabirds from a wide area.

SKI RESORTS

Temperate zone cities that are fortunate enough to have high mountains nearby will usually have a ski resort within commuting distance. A ski resort offers several benefits to a birder. Its existence will usually mean that there is good road access and public transport up to high elevations and with all amenities present. The ski lifts can transport you to the higher levels necessary to get to the specific habitats of the high altitude species, while the presence of cross-country ski runs means that there is a network of trails through the forests and open slopes. In summer, most ski resorts encourage hiking and backpacking, and so remain open, although some ski lifts are shut down for maintenance at this time. In winter, many montane species move to lower elevations but the areas around the ski resorts often provide feeding for finches and other species.

HIGH GROUND

It is well known that migrating birds will make use of physical features of a landscape as an aid to navigation. Moreover, migrating raptors, storks and other soaring birds use the upcurrents associated with high ground. Ridges running along a north/south axis, or isolated hills in the midst of flat land, not only provide good vantage points from which to scan the area, but during migration may be productive sites from which to watch for visible passage. Well-known migration watchpoints near cities usually attract groups of birders during favourable weather conditions and offer a chance to meet local enthusiasts and share information.

Useful Tips

Arriving in a foreign city always brings home how much we take for granted about our own familiar locations. Such routine information about how far the airport is, which bus routes serve a location, how to navigate the city, finding parking and general orientation become a major challenge, and if you do not speak the language these challenges are compounded.

For the birder on a tight schedule, finding the time to slot in some birding in between non-birding business, accommodating the wishes of a non-birding partner and getting from hotel to airport become a major logistical exercise. Happily, there are a number of ways in which you can quickly take advantage of the good fortune of being in a foreign country, with possibly several new birds close at hand.

LOCAL BIRDERS

Birding is such an infectious hobby that most birders cannot help sharing their enthusiasm. Even though you may know no one in a strange city, most birders are usually delighted to show an out-of-towner around their local patch. If you have not been able to make contact with anyone before your trip, you could always try the excellent Birding Pal website to locate a birder in the city you are visiting. There is no substitute for local knowledge and expertise when it comes to birding in a new locality, and you will inevitably see much more if accompanied by a local birder. If you do not have any birding gear with you, another birder will probably be able to provide a spare pair of binoculars and a weatherproof jacket.

TOUR GUIDES

If you cannot make contact with a local birder you may, alternatively, be able to contact a local bird tour company, particularly if the city is close to a well-known birding hotspot or the country is a 'hot' birding destination. Most tour guides carefully prepare their birding trips well in advance and all the participants are pre-booked. However, they may be able to allow an extra birder to join the group for a day or two, or, if they are not leading a trip at the time, may be prepared to put together an impromptu 'mini-trip' for a day or a weekend. The cost of a one- or two-day trip would be quite modest and when you consider the efficiency with which a knowledgeable local guide will get you the birds, you will probably 'clean-up' in half the time you would take to do it on your own, thus leaving you with an extra day or two to spend with family or on other activities.

INTERNET

There is now a vast amount of information available online which can help you get orientated and out birding in a foreign city. Unfortunately, sifting through all this information in order to access the really useful material can be very time-consuming and may ultimately be fruitless. Wherever possible, this guide provides a list of useful websites for each city, as well as contact information. In addition there are several very good websites which provide general birding information on a worldwide basis and which act as portals to more specialist ornithological websites. Among the most helpful sites are:

www.fatbirder.com
One of the best birding sites available, packed with information about birding on a worldwide scale and with many useful local links.

www.birdguides.com
Another excellent birding site, providing a comprehensive up-to-the-minute information service for the UK, and also for further afield.

www.surfbirds.com
A UK birding site with a large amount of content pertaining to world birding.

www.camacdonald.com/birding/
Tina MacDonald's website – 'Birding Hotspots Around the World'.

www3.ns.sympatico.ca/maybank/main.htm
Birding the Americas – a comprehensive set of trip reports for North and South America.

www.eurobirding.com
Trip lists from just about every country in the world.

http://birdingonthe.net
A very useful portal site with links to almost all of the mailing-list archives in North America and seven-day archives for everywhere else.

www.birdlife.org
A good source of information about the conservation status of the country you are visiting. It also contains brief site notes.

www.birdingpal.org
A website based on an excellent idea of putting visiting birders in touch with local birders.

www.mapquest.com
Electronic maps of varying quality for any location.

www.myforecast.com
Weather forecasts, snow reports and tidal conditions for just about anywhere in the world.

http://easytide.ukho.gov.uk/
Not terribly easy to navigate, but all the tidal information is there with tidal predictions for up to seven days in advance.

www.busstation.net
Portal site for locating public bus services anywhere in the world.

www.rentingcars.com
One of the best portal sites for getting good car rental deals at short notice.

MAPS

The first thing any birder should do on arriving in a foreign city is to purchase two maps: a city map and a regional map. Often the best place to do this is in the airport terminal as the best selection is likely to be available there. City maps are good for finding the main transport hubs within the city and perhaps some of the birding sites, but you will most probably also need a map of the city's hinterland to locate the adjacent sites and it may be difficult to get the required scale to give sufficient detail. Whatever you do, purchase the best available: don't skimp on your map purchases as there is nothing more frustrating than trying to locate a site with the aid of an imprecise map. The price differential is usually very little, when compared to the wasted effort.

RENTING CARS

Think very carefully before renting a car if you are on a short-term visit to a city. Although birders are conditioned to the speed, mobility and flexibility of using a car to cover large distances in the course of a day in order to maximise their birding opportunities, it is not always so practical on a short trip. Most birders finding some spare time in a foreign city are not looking for one particular species and are not aiming to visit any particular site. It is probably better to visit just one or two sites and see a good cross-section of local birds. The cost of renting a car for just one or two days can be quite expensive and when you weigh up the 'bang for your buck' you may find that it is not worth it. Timing can work against you, as a 'day' rental of a car is usually based around office hours, so it means not picking up the car until rather late (for birding) in the morning and having to drop it back rather early (for birding) in the evening if you are catching a flight later. Finally, do not underestimate the challenges of driving through and navigating an unfamiliar city, often with signposts in a foreign language or even an indecipherable script. Carefully consider the alternatives in terms of public transport, which can be cheap, reliable and, in some cities, very comprehensive.

If you wish to rent a car, there are a few things worth noting. Try to arrange a pick-up and a drop-off that is convenient for you. Generally, a city centre office is the least convenient as you will have to allow time to pick up, drop off and then get to your final destination, probably the airport. Be wary of 'advice' that you need a 4WD or need to take out extra insurance because you are going birding, as this counsel is usually for the benefit of the rental agency not yours. In some developing countries, carefully weigh up the cost of paying extra for an accompanied driver. The incremental cost may be quite small and the added benefits of a driver who is familiar with the roads, speaks the language and has a much better grasp of local practices will very often far outweigh the expense.

GETTING KITTED OUT

If you arrive in a city for a business meeting, wearing a suit and carrying a briefcase, going birding will probably be the last thing on your mind. However, if you experience an unscheduled delay do not pass up the chance to do some city birding. Seasoned birders who regularly travel on business, or for other non-birding purposes, will probably always carry some light binoculars and will have a suitable compact waterproof garment in their luggage. Otherwise buy a cheap pair of binoculars and some cheap weatherproof gear: after all, you are in a city so they should not be difficult to obtain. It does not have to be durable or even all that good: you will only use it once and you can leave it behind if you wish. It never ceases to amaze me how many retail outlets sell binoculars, especially the cheap, low-quality glasses which no birder would normally buy. Camera stores, sporting goods stores, large pharmacies and even flea markets sell cheap binoculars: even opera glasses are better than nothing.

NATURAL PESTS

Most birders have a casual attitude towards the possibility of problems with insect and other animal pests when birding on their own patch. Familiarity breeds a fairly healthy disregard of these possibilities. However, when on unfamiliar territory, there are probably valid grounds for some precautions. What is surely one of the great mysteries of the world is the propensity for temperate zone males immediately to bare their legs when birding in the tropics. Shorts are probably the most ill-advised type of clothing to wear. Apart from the risk of sunburnt thighs and calves you are simply asking to be attacked by every biting and stinging insect within range. Long, loose-fitting trousers are much more suitable and practical.

Despite popular misconception, it has been my experience that biting and stinging insects are less of a problem in the tropics than would be thought. Indeed, the worst insect pests can be the plagues of mosquitoes that are a feature of temperate zone forests in eastern North America and eastern Europe. Probably of much greater concern are deer ticks and other parasites, which are vectors for Lyme's Disease and other infections. Again, wear loose-fitting clothing which covers your legs and spray your ankles with insect repellent. A further consideration is that the short-term visitor to tropical cities may not have received the recommended immunisations nor have arranged malaria prophylaxis, and this may need to be borne in mind if an unexpected opportunity to travel into the countryside presents itself.

Leeches may also be a problem, particularly in tropical forests. Make sure that you tuck your trouser legs into your socks and check your ankles periodically: otherwise you will experience the squelching sound of your own blood sloshing about inside your shoes.

In many years of birding in the tropics, I have rarely come across a snake of any description, and it is probably fair to say that these creatures do not represent much of a threat to a birder. Needless to say, the practical advice of keeping to well-marked trails is the best preventer of trouble from this quarter of the animal kingdom.

HUMAN PESTS

I am happy to say that in all of my birding life, I have never felt even remotely threatened by anyone I have encountered while birding. Generally, in the developed world you will be ignored by passers-by or be the subject of some polite enquiry, whereas in the developing world the curiosity may be a little more exuberant but is always good-humoured and non-threatening. That is not to say that some threats do not exist, and normal precautions should be exercised. Single women, in particular, need to be extra vigilant not only for the obvious dangers of secluded areas but also when birding alone in cultures where lone women may invite hostile comment.

Perhaps it is a sign of the growing popularity of birding that in many well-known bird sites in the developing world, self-appointed 'bird guides' will offer their services for a fee. It is really a question of judgement whether there is any benefit in being shown the birds in this way, or if it is desirable to encourage local people to gain monetarily from the birdlife. Bear in mind that your bird guide probably originally honed his skill as a bird trapper and is now putting that skill to more wholesome use.

There are a number of websites, such as those of the CIA and FCO, which advise on personal safety in individual countries. However, it is likely that their travel advice will tend to exaggerate the risks somewhat. The best way to utilise such information is to check what they are saying about your own country and use that as a benchmark from which to gauge the validity of the advice.

How to Use This Book

Stealing a few hours from a busy schedule in order to go birding in a new city requires a certain amount of initiative and a willingness to take a few risks in order to make the most of an unexpected opportunity, but you will be rewarded with a wonderful experience and perhaps a good story to tell your birding friends when you return. As much as we all enjoy seeing good birds, it is sharing our birding experiences with our friends that adds to our enjoyment of birding. I hope that this book will encourage you to take advantage of the many opportunities that exist in the world's cities to see some amazing birds. Most of the sites described are limited to those that lie within an 80-kilometre (50-mile) radius of the city centre – many are much closer of course. The exceptions are a small number of truly exceptional sites which lie just a little bit further away but which offer such a unique birding experience that it is worth making the effort to travel the extra distance.

This book follows a standard format for ease of use. Each city account features an **Introduction** to the city and its birdlife, including lists of **Typical Species**, which give those species which are found commonly in the area where the city is located and which can be seen without too much difficulty in the right habitat. There follows a section dealing with **Getting Around** the city and making use of public transport wherever possible.

The **Locations** section offers a number of sites selected on the basis of their convenience to the city centre or airport, the richness of their birdlife and the opportunity presented to experience all of the habitats that lie within easy reach of the city. Each city account features a map, showing the relative position of the selected sites. The sites are numbered, roughly from the city centre outwards. Sites that are beyond the city limits, or ones that are a bit further afield are sometimes given in a separate section entitled **Further Away**.

Each site description contains information on **distance from the city centre** (in miles or kilometres depending on local usage), **nearest public transport facility**, a brief description of the site and its habitats, directions to access the site, a listing of the key species which can be expected at each relevant season of the year and, in certain cases, a map of the location. The **Key Species** listing includes those species, in addition to the Typical Species of the city itself, which are likely to occur regularly at the site. Sometimes 'typical species' are repeated in the site accounts where they are characteristic, but often they are not, for reasons of space. Certain nocturnal species, such as owls, have been left off the list of key species as, although they may occur regularly, the visiting birder has little chance of seeing them without local assistance. Species whose names are followed by **(I)** are **introduced species** which have established feral populations in the area. The Feral Pigeon, which is ubiquitous and abundant in urban areas throughout the world, is deliberately excluded from all accounts as its presence can be assumed and it will be largely ignored by most birders anyway.

Maps have been selected to assist the reader where the site is large or difficult to find. Most are drawn from memory or from larger scale maps, but it is hoped they will be adequate for the purposes of finding the site. Any city birder will need a good city map and in most cases a map of the adjacent counties or provinces.

All urban locations are subject to development pressures – much more so than birding sites in more remote areas of a country. Inevitably, by the time this book goes to print, some of the sites mentioned may be radically altered so, where possible, always try to get some local information before visiting the site in question.

NOMENCLATURE

A book which spans the world across all zoogeographical regions with the exception of Antarctica presents a number of challenges in terms of species taxonomy and nomenclature. Species limits and names used in this book generally follow that of two standard lists: *The British Birds List of Birds of the Western Palearctic* for cities within the Western Palearctic and *Birds of the World: A Checklist* by James F. Clements (5th edition) for everywhere else. Although the use of these two lists leads to some inconsistencies between names favoured on either side of the Atlantic, they do ensure that the species lists reflect, as closely as possible, the vernacular names within the region, and we hope readers will be prepared to be lenient in this respect. Whilst 'Clements' is the world list of choice for North Americans and is particularly strong in respect of New World names, there are a handful of Old World names that Clements uses which are not used in local field guides outside the Western Palearctic and which may be unfamiliar to many people. We have therefore taken the liberty in a few cases to disagree with Clements and use the more widespread name. These few discrepancies are listed below for the sake of completeness.

The use of hyphens in bird names is a constant source of disagreement. They are used abundantly in Clements, and tend to be more favoured in North America than in Europe. However, we have chosen to follow Howard & Moore (Dickinson 2003) and the new IOC-endorsed list of Recommended English Names (Gill and Wright 2006) with regard to the hyphenation of English names. Both these works explain the rationale for their use of hyphens, but generally they take the minimalist approach, avoiding hyphens unless it is considered essential to use them. Where our names differ from Clements only in the usage of hyphens (and the capitalisation of a letter following a hyphen), they are not listed as alternatives below. Similarly, we have chosen to use the English spellings of words such as 'grey' for Old World cities, and these are not listed below either.

NAME DIFFERENCES FROM CLEMENTS

As we have elected to follow the *British Birds* list for the Western Palearctic, the problem of unfamiliar Clements names only affects Africa and Asia (there are no such problems with New World and Australasian names). The list below details those cases where we have used a name that is not the Clements name. In a few cases, this is because Clements has not recognised a recent split, and the scientific name of the elevated subspecies is given after the Clements name.

In southern Africa, the names used in local field guides can differ markedly from Clements. Our policy of using more familiar names will alleviate some of the more extreme differences. Other names differ from Clements in more minor (and recognisable) ways. For example, 'widow' is used for 'widowbird' in southern Africa, but we follow Clements and use 'widowbird' here throughout. Furthermore, there are often significant name differences between East and South Africa. We have tried to use the most appropriate name for each region, but for some widespread species we have had to make the choice of a single name for the sake of consistency.

List of non-Clements names used in this book

Name used in World Cities	Name used in Clements	Name used in World Cities	Name used in Clements
Black-necked Grebe (OW)	Eared Grebe	Little Grey Greenbul	Gray Greenbul
Black Ibis	Red-naped Ibis	Cape Rockjumper	Rufous Rockjumper
Goosander (OW)	Common Merganser	Lynes's Cisticola	Wailing Cisticola
Great Sparrowhawk	Black Goshawk		(C. l. distinctus)
Common Buzzard	Eurasian Buzzard	Little Rush Warbler	African Bush Warbler
African Crowned Eagle	Crowned Hawk-Eagle	Desert Lesser Whitethroat	Small Whitethroat
Common Kestrel	Eurasian Kestrel	Chestnut-vented Tit-babbler	Rufous-vented Warbler
Common Pheasant	Ring-necked Pheasant	Brown Parisoma	Brown Warbler
White-quilled Korhaan	White-quilled Bustard	Pallas's Warbler	Lemon-rumped Warbler
Lesser Sand Plover	Mongolian Plover	Yellow-browed Warbler	Inornate Warbler
Kentish Plover (OW)	Snowy Plover	Dark-sided Flycatcher	Siberian Flycatcher
Grey Plover (OW)	Black-bellied Plover	Lead-coloured Flycatcher	Gray Tit-Flycatcher
Arctic Skua (OW)	Parasitic Jaeger	Red-throated Flycatcher	Taiga Flycatcher
Pomarine Skua (OW)	Pomarine Jaeger	Brown Rock Chat	Indian Chat
Pallas's Gull	Great Black-headed Gull	Rufous-tailed Wheatear	Red-tailed Wheatear
Vega Gull	East Siberian Gull		(O. x. chrysopygia)
African Olive Pigeon	Rameron Pigeon	Northern Black Tit	White-winged Black Tit
White-eared Brown Dove	White-eared Dove	Montane White-eye	Broad-ringed White-eye
Grey-bellied Cuckoo	Plaintive Cuckoo	Abyssinian Oriole	Dark-headed Oriole
	(C. m. passerinus)	Montane Oriole	Black-tailed Oriole
Linchi Swiftlet	Cave Swiftlet	Crimson-breasted Shrike	Crimson-breasted Gonolek
Red-throated Wryneck	Rufous-necked Wryneck	Taiwan Magpie	Formosan Magpie
Grey-headed Woodpecker	Gray-faced Woodpecker	Vitelline Masked Weaver	African Masked-Weaver
Japanese Green Woodpecker	Japanese Woodpecker	Black-breasted Weaver	Bengal Weaver
Japanese Pygmy Woodpecker	Pygmy Woodpecker	Scaly-breasted Munia	Nutmeg Mannikin
Sand Martin (OW)	Bank Swallow	Black-headed Siskin	Abyssinian Siskin
Grassland Pipit	African Pipit	White-capped Bunting	Chestnut-breasted Bunting
Buff-bellied Pipit	American Pipit		

OW = in Old World only

Glossary of Geographical Features

Alpine meadow: mountain grassland specific to higher elevations above the treeline.

Babul (India): a species of acacia native to India, which favours rivers and waterways.

Bog (Eurasia): acidic, waterlogged soil where the decomposition of dead organic matter proceeds at a slower than normal pace and accumulates as peat.

Bukit (Malaysia): Malay for hill.

Cerrado (South America): dry grassland with scattered trees.

Campo (South America): higher elevation grassland with some scrub but no trees.

Cerro (South and Central America): mountain peak.

Cloud forest: montane forest which is often enveloped in damp cloud although situated in an otherwise dry area.

Dam (Africa): a small artificial lake or reservoir, created by damming a river or stream.

Dehesa (Spain): grazing woodland, a habitat which is unique to the Iberian Peninsula comprising extensive pastures with scattered cork oaks and holm oaks.

Fen (Eurasia): alkaline peat bog which receives relatively high amounts of mineral salts and water from the ground.

Fynbos (South Africa): shrub habitat of the Cape Province, which contains more plant species per square metre than any other habitat in the world.

Gallery forest: waterside forest and woodland where forested areas merge into more open areas.

Garrigue (Mediterranean): low, often thorny, open scrub which usually results from overgrazing of maquis (qv).

Gunung (Indonesia): Bahasa Indonesian for mountain.

Hammock (Florida): forest habitat that is typically higher in elevation than surrounding areas and characterised by hardwood forest of broadleaved evergreens.

Heath (Eurasia): land with vegetation dominated by evergreen dwarf shrubs which usually grow on well-drained, acidic, gravelly or sandy soils in lowlands near the Atlantic. Sometimes distinguished from moorland (qv) in having an average annual total precipitation of less than 1,000mm.

Jheel (India): specifically an oxbow lake but the term is more generally applied to any lake.

Karoo (South Africa): semi-desert shrub and grassland habitat.

Kebun Raya (Indonesia): Bahasa Indonesian for Botanical Gardens.

Khor (Arabia): marine bay or inlet.

Lowland forest: tropical forest at elevations from sea-level up to 1,000m.

Mallee (Australia): a semi-arid area of open country interspersed with thickets of eucalyptus trees.

Mangrove: tropical or subtropical woodland typical of tidal swamps.

Manguezal (Brazil): mangrove swamp-forest consisting of a limited number of tree species specific to the southeast coast of Brazil.

Maquis (Macchia) (Mediterranean): tall, mostly evergreen, dense scrub, up to four metres high, usually including species such as gorse, heather, mastic, rosemary, sage and broom and typically found on Mediterranean slopes.

Montane forest: forest typical of higher elevations in tropical zones characterised by broken canopy encouraging lush undergrowth and the absence of conifers.

Moorland (Eurasia): acidic peatland in northern European uplands dominated by grasses, heath-like plants or sedges. Sometimes distinguished from heath (qv) in having an average annual total precipitation of over 1,000mm.

Pampas (South America): grassland (dry or wet).

Pan (Africa): a natural depression, usually found in arid areas, which forms shallow lakes at times of high rainfall.

Pedregal (Mexico): a solidified bed of lava rock which supports stunted oak forest and other vegetation.

Polder (Netherlands): a low-lying area of farmland which has been reclaimed from the sea.

Prairie (North America): a flat or gently undulating plain that is grassy and generally treeless – dominated by tall grasses.

Pulau (Indonesia): Bahasa Indonesian for island.

Puszta (Hungary): natural and semi-natural grasslands interspersed with seasonal and permanent wetlands, as well as cultivation.

Ras (Arabia): a point or headland on the coast.

Restinga (Brazil): sandy strip of land that often separates a shallow lagoon from the sea.

Riparian woodland: woodland community found primarily along the banks of rivers and streams and beside lakes, ponds, marshes and earthen reservoirs.

Saltmarsh: a marsh which is periodically flooded by the sea or which contains water which is rich in salt for some other reason.

Savanna (Africa): extensive area of grassland interspersed with tree or shrub cover to a greater or lesser extent (*Acacia* being dominant).

Secondary forest: forested areas which have been cleared but have since partly regenerated.

Shan (China): mountain or mountain range.

Slough (USA): salt or freshwater inlet with deep mudbanks.

Steppe (Eurasia): extensive, usually treeless, plains where the vegetation is dominated by short grasses or dwarf shrubs.

Veld (South Africa): large, generally flat expanse of grassland or low scrub.

    Lowveld: veld below 1,500m.

    Highveld: veld above 1,500m.

    Bushveld: veld with more extensive bush cover.

    Thornveld: bushveld where *Acacia* spp. are the dominant tree species.

Wadi (Egypt and Arabia): river course in desert area which is dry for most of the year.

The Hammerkop builds one of the largest nests of any bird

TYPICAL SPECIES

Resident
Great Cormorant, Egyptian Goose, Cattle Egret, Hamerkop, Wattled Ibis, Hadada Ibis, Sacred Ibis, Marabou Stork, Black Kite, Black-shouldered Kite, Hooded Vulture, White-backed Vulture, Tawny Eagle, Augur Buzzard, Dark Chanting Goshawk, Helmeted Guineafowl, Spur-winged Plover, Speckled Pigeon, Dusky Turtle Dove, Laughing Dove, African Mourning Dove, Ring-necked Dove, Red-eyed Dove, Speckled Mousebird, Lilac-breasted Roller, Red-billed Hornbill, Eastern Yellow-billed Hornbill, Rock Martin, African Paradise Flycatcher, Fork-tailed Drongo, Cape Crow, Pied Crow, Fan-tailed Raven, Common Fiscal, Greater Blue-eared Glossy Starling, Lesser Blue-eared Glossy Starling, Superb Starling, Common Bulbul, Montane White-eye, Grey-backed Camaroptera, Swainson's Sparrow, Baglafecht Weaver, Village Indigobird, Red-cheeked Cordon-bleu, Village Weaver, Red-billed Quelea.

Oct–Mar
Western Marsh Harrier, Common Buzzard (*B. b. vulpinus*), Booted Eagle, Common Kestrel, Common Sandpiper, European Bee-eater, Northern Carmine Bee-eater, Common Swift, Yellow Wagtail, Barn Swallow, Sand Martin, Blackcap, Willow Warbler.

Addis Ababa

Sooner or later, any serious student of African birds will make a comprehensive trip to Ethiopia, as apart from having one of the highest birdlists in Africa, it also is home to almost 30 endemic species. However the short-term visitor to Addis Ababa still has the opportunity to sample this wonderful avifauna within one or two hours' drive of the city. The city itself has a rich birdlife with many beautiful tropical birds inhabiting the parks, gardens and hotel grounds while vultures, kites and Marabou Storks circle overhead, as would gulls or pigeons in any temperate zone city.

Addis Ababa is located in the Western Highlands at the centre of the country and at an altitude of 2,400m. Consequently its climate is quite temperate: in fact in winter the nights can be rather cold. To the north and west lie the Entoto Mountains and to the south and east lies the escarpment on the edge of the Great Rift Valley of east Africa. Because of its altitude, the surrounding countryside consists of high elevation savanna grassland interspersed with pockets of acacia woodland, and juniper forest higher up. However, much of the natural vegetation has been cleared for agriculture and for plantations of eucalyptus trees. Eucalyptus is a fast-growing, non-native evergreen which is very suitable for providing fuel, but its fast growth crowds out native species and also draws much of the moisture from the soil, resulting in severe habitat degradation in places.

The most popular time to visit Ethiopia is during winter when the weather is at its driest and when the local bird populations are augmented by many Palearctic migrants. There are a number of wetland sites close to the city which are magnets for migrating waders during spring and autumn and are also important sites for wintering wildfowl.

So whether you are passing through Addis at the start of a major birding expedition, or just have a day or two to spend, there is much to be seen in this high altitude city, located on the Roof of Africa.

GETTING AROUND

Getting around Addis Ababa really requires the use of a car as, apart from the hotel gardens, you will probably want to visit one or more of the sites outside the city. Many of the roads are in a bad state of repair and they tend to be crowded with slow-moving traffic and also livestock, so expect an average speed of no more than 40km per hour and plan your journeys accordingly. For those sites which are popular with tourists you could also use a shared taxi: these are distinctive blue minibuses which are a cheap

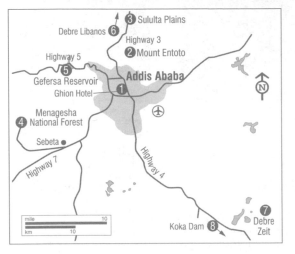

alternative to renting a car. Most of them depart from the railway station and although they will only set off when full, they provide quite a frequent service between Addis and the bigger towns. The only problem is that they are usually overcrowded. For other sites further afield such as Debre Zeit and Debre Libanos, you can also get a bus from the railway station terminal.

Bear in mind that Addis Ababa is situated at an elevation of 2,400m, so give yourself some time to acclimatise before exerting yourself. On the other hand, the climate is a lot cooler than other cities at the same latitude making all-day birding a very pleasant experience. None of the birding sites mentioned in this guide are national parks or reserves so there are no entrance fees and, unlike some other African birding sites, you are free to leave your car and hike on foot: there are no dangerous animals at large! Good walking shoes are all that is required as you will mostly make short walks from the roadside and there is no need to undertake long hikes through the bush.

The only widely available map covering the area is the Michelin No. 954 – Northeast Africa & Arabia (1:4,000,000). However, it will at least enable you to orient yourself on the main roads and give you some sense of distance.

Locations

1 Ghion Hotel (City Centre)

If you are fortunate enough to be staying here, then you have some of Addis Ababa's best urban birding just outside your window. Most of the expensive hotels have well-cultivated gardens which are good for birds, but the Ghion also has a stream running through it which attracts some aquatic species as well as passerines. One of Ethiopia's endemics, the Wattled Ibis, can be seen here.

The Ghion also has a bookshop which stocks some guidebooks and maps of Ethiopia.

The hotel is located centrally on Ras Desta Damtew Avenue, just opposite the Sports Stadium and close to the railway station.

KEY SPECIES

Resident
Wattled Ibis, White-collared Pigeon, Banded Barbet, Abyssinian Woodpecker, Grey Woodpecker, Black-winged Lovebird, Olive Thrush, Ruppell's Robin Chat, Brown Parisoma, Abyssinian Slaty Flycatcher, African Dusky Flycatcher, White-backed Black Tit, Tacazze Sunbird, Scarlet-chested Sunbird, Brown-rumped Seedeater, Streaky Seedeater

2 Mount Entoto (6km)

The Entoto Mountains, reaching an altitude of 3,200m, are located just north of Addis Ababa forming a spectacular semi-circular ring around the northern perimeter of the city and separating it from the Sululta Plains further north. Some of the original juniper forests still remain although, like many forests in Ethiopia, large areas have been cleared for agriculture and any subsequent reforestation has been carried out using the quick-growing but non-native eucalyptus. This area tends to be neglected by visiting birders as they quickly make their way to the birding hotspots further north. However there is a good selection of Afrotropical montane species to be found here and the proximity to the city centre makes it a very convenient location for birders on a short visit. Access to the forest is provided in the Entoto Natural Park which is run by the Ethiopian Heritage Trust.

Take the road north from the city centre (**Entoto Avenue**) to **Entoto Mariam Church**, a well-known tourist site on the outskirts of Addis Ababa.

KEY SPECIES

Resident
Rüppell's Vulture, Egyptian Vulture, Lammergeier, Wahlberg's Eagle, Mountain Buzzard, Rufous-chested Sparrowhawk, Moorland Francolin, Erckel's Francolin, White-collared Pigeon, Yellow-fronted Parrot, Black-winged Lovebird, White-cheeked Turaco, Narina's Trogon, Nyanza Swift, Abyssinian Woodpecker, Cardinal Woodpecker, Black-billed Barbet, Banded Barbet, Abyssinian Ground Thrush, Rüppell's Robin Chat, Moorland Chat, White-winged Cliff Chat, Abyssinian Catbird, Brown Parisoma, Brown Woodland Warbler, Abyssinian Slaty Flycatcher, White-backed Black Tit, Tropical Boubou, Slender-billed Starling, White-billed Starling, Sharpe's Starling, Abyssinian Oriole, Thick-billed Raven, Tacazze Sunbird, African Citril, Abyssinian Crimsonwing, Black-headed Siskin, Brown-rumped Seedeater, Streaky Seedeater

3 Sululta Plains (30km)

Driving north from Addis Ababa provides access to the high altitude grassland habitat for which the Ethiopian Highlands are famous. The most well-known and productive area is the Sululta Plains located on either side of the highway north of the city. These are surrounded by mountains, from which streams flow down and feed shallow water pans which flood during the rainy season. There is a mixture of habitats with grain cultivation, grazing, uncultivated grassland and small marshes. A number of 'special' birds occur: the rare and local White-winged Flufftail inhabits the wet meadows; the Red-chested Swallow is more widespread but requires careful observation to distinguish it from the Barn Swallows, and the endemic Spot-breasted Lapwing can be seen during the dry season.

Take the main highway north (**Dejazmach Belay Zeleke Street**) towards **Bahir Dar (Highway 3)**. After about 20km, this road passes the town of **Sululta**. For the next 30km between Sululta and **Muka Turi**, the road passes through the plains. Drive slowly (which is easy to do, given the road conditions!) and stop and scan often for raptors, wildfowl and plovers, and also at any water pans.

KEY SPECIES

Resident
Blue-winged Goose, Grey Heron, Black-headed Heron, Little Egret, Great Egret, Glossy Ibis, Yellow-billed Stork, Yellow-billed Duck, Secretarybird, Egyptian Vulture, African Harrier-Hawk, Greater Kestrel, Lanner Falcon, Senegal Thick-knee, Black-winged Stilt, Black-winged Lapwing, Three-banded Plover, African Snipe, White-collared Pigeon, Namaqua Dove, Abyssinian Roller, Abyssinian Ground Hornbill, Red-winged Lark, Erlanger's Lark, Chestnut-backed Sparrow-Lark,

Grassland Pipit, Abyssinian Longclaw, Red-chested Swallow, Grey-rumped Swallow, Mosque Swallow, Lesser Striped Swallow, Plain Martin, Red-breasted Wheatear, Mourning Wheatear, Moorland Chat, African Stonechat, Groundscraper Thrush, African Dusky Flycatcher, Red-billed Oxpecker, Pectoral-patch Cisticola, Winding Cisticola, Tacazze Sunbird, Yellow Bishop, Red-billed Firefinch, African Quailfinch, African Silverbill, Cut-throat Finch, Village Indigobird, Black-headed Siskin, Pin-tailed Whydah, Fan-tailed

Widowbird, Streaky Seedeater

Oct–Mar
Black Stork, Steppe Eagle, Long-legged Buzzard, Pallid Harrier, Common Kestrel, Common Crane, Spot-breasted Lapwing, Little Ringed Plover, Temminck's Stint, Ruff, Wood Sandpiper, Green Sandpiper, Marsh Sandpiper, White Wagtail, Red-throated Pipit, Red-rumped Swallow, Ortolan Bunting

Jul–Sep
White-winged Flufftail

4 Menagesha National Forest (37 km)

Addis Ababa is overlooked to the west by Mt Wechecha, an extinct volcano over 3,300m high. Its southwest slopes are covered in dense juniper forest with some clearings given over to cultivation. Unfortunately, due to its proximity to Addis Ababa, the forested area is under continual threat from overexploitation for timber and fuel. However, there is still sufficient good habitat to support many of the key species typical of the Afrotropical Highlands biome, including a few Ethiopian endemics.

Take the main road west from Addis Ababa (**Smuts Street**) towards **Jimma** (**Highway 7**). Turn north at Sebeta towards Mt Wechecha and the Menagesha Forestry Training Centre. Just before the training centre turn right into the park headquarters. This road ascends the mountain through the forest where there is a campsite and trails.

KEY SPECIES

Resident
Lammergeier, African Crowned Eagle, Mountain Buzzard, Rufous-chested Sparrowhawk, African Goshawk, White-collared Pigeon, Yellow-fronted Parrot, Black-winged Lovebird, White-cheeked Turaco, Narina's Trogon, Nyanza Swift, Abyssinian Woodpecker, Cardinal Woodpecker, Nubian Woodpecker, Black-billed

Barbet, Banded Barbet, Abyssinian Ground Thrush, Rüppell's Robin Chat, Banded Wattle-eye, Abyssinian Hill Babbler, Abyssinian Catbird, Brown Woodland Warbler, Abyssinian Slaty Flycatcher, White-backed Black Tit, Abyssinian White-eye, Grey-backed Fiscal, Tropical Boubou, Slender-billed Starling, Sharpe's Starling, Red-winged Starling, Long-tailed Glossy

Starling, Abyssinian Crimsonwing, Abyssinian Oriole, Thick-billed Raven, Variable Sunbird, Tacazze Sunbird, African Citril, Yellow-bellied Waxbill, Brown-rumped Seedeater, Streaky Seedeater

Oct–Mar
Alpine Swift, Mottled Swift

5 Gefersa Reservoir (18km)

This reservoir and the surrounding area comprise one of the best birding sites close to the city. Several of the Ethiopian endemics can easily be seen here as well as a wealth of wildfowl in winter and Palearctic waders during migration. As water levels fluctuate considerably, this has affected the growth of aquatic vegetation along the shoreline and the only rank vegetation occurs where feeder streams flow into the reservoir. This leaves substantial areas of exposed shoreline which provide feeding for herons and waders. The surrounding terrain consists of a mosaic of grasslands, cultivation and, where grazing is fenced off, some montane woodland. There are also substantial eucalyptus plantations. These areas support a rich birdlife including raptors and many seed-eating passerines.

The reservoir is located about an hour's drive west of Addis Ababa on the **Nekempte** Road (**Highway 5**). You can access the northern shore of the reservoir from the road.

KEY SPECIES

Resident
Little Grebe, Great Crested Grebe, Darter, Blue-winged Goose, Black-headed Heron, Great Egret, Hadada Ibis, Glossy Ibis, African Spoonbill, Yellow-billed Stork, African Black Duck, Yellow-billed Duck, African Fish Eagle, Rouget's Rail, Red-knobbed Coot, Black-winged Stilt, Three-banded Plover, African Snipe, White-collared Pigeon, Grey Woodpecker, Red-throated Wryneck, Grassland

Pipit, Abyssinian Longclaw, Grey-rumped Swallow, Plain Martin, Red-breasted Wheatear, Moorland Chat, Groundscraper Thrush, Abyssinian Catbird, Brown Woodland Warbler, Abyssinian Slaty Flycatcher, African Dusky Flycatcher, White-backed Black Tit, Red-billed Oxpecker, Winding Cisticola, Tacazze Sunbird, Black-headed Siskin, African Citril, Yellow-bellied Waxbill, Common Waxbill, Yellow-crowned Bishop, Yellow

Bishop, Pin-tailed Whydah, Brown-rumped Seedeater, Streaky Seedeater

Oct–Mar
Northern Pintail, Northern Shoveler, Eurasian Teal, Pallid Harrier, Common Ringed Plover, Little Ringed Plover, Temminck's Stint, Little Stint, Ruff, Green Sandpiper, Marsh Sandpiper, White Wagtail, Red-throated Pipit, Red-rumped Swallow, Ortolan Bunting

Further Away

6 **Debre Libanos** (105km)

Several tributaries of the Blue Nile cut through the mountains north of Addis Ababa forming spectacular gorges. These rivers have provided pathways for a number of species more typical of the lowlands to the west to penetrate deep into the highlands, resulting in an avifauna which is an interesting mix of lowland and highland species. One of the most impressive gorges and that nearest to the city is the Jemma River gorge at Debre Libanos, which has a famous monastery located at the edge of the canyon. The olive and juniper forest that covers the hills above the monastery has been preserved and supports a range of specialist montane forest species. The rocky slopes of the gorge provide good vantage points to observe soaring raptors as well as several passerine species endemic to the Ethiopian highlands. Another well-known Ethiopian endemic which is found here is the troop of Gelada Baboons.

The floor of the valley is about 700m below and supports a different variety of birds including the endangered, endemic Harwood's Francolin. However it is not accessible from this point, requiring a drive of almost 100km upstream in order to descend to a crossing point.

Take the main highway north (**Dejazmach Belay Zeleke Street**) towards **Fiche**. About 10km before reaching Fiche turn east at the sign for Debre Libanos. The monastery and gorge are located about 5km along this road. There is a small entrance fee into the grounds of the monastery.

KEY SPECIES

Resident
Rüppell's Vulture, Lappet-faced Vulture, Egyptian Vulture, Lammergeier, Verreaux's Eagle,

African Hawk-Eagle, Lanner Falcon, Peregrine Falcon, Common Kestrel, Erckel's Francolin, Senegal Thick-knee, Vinaceous Dove, Namaqua

Dove, Lemon Dove, Bruce's Green Pigeon, White-cheeked Turaco, Blue-breasted Bee-eater, Abyssinian Roller, Black-billed Wood Hoopoe, Nyanza

Swift, Hemprich's Hornbill, Nubian Woodpecker, Abyssinian Woodpecker, Cardinal Woodpecker, Black-billed Barbet, Banded Barbet, Plain-backed Pipit, Mountain Wagtail, Wire-tailed Swallow, Black Saw-wing, Little Rock Thrush, White-browed Robin-Chat, Rüppell's Robin-Chat, Rüppell's Black Chat, White-winged Cliff Chat, Mocking Cliff Chat, Mourning Wheatear, Singing Cisticola, Stout Cisticola, Brown-throated Wattle-eye, Abyssinian Slaty Flycatcher, White-backed Black Tit, Abyssinian White-eye, Abyssinian Catbird, Northern Puffback, Grey-backed Fiscal, Black-crowned Tchagra, Tropical Boubou, Red-winged Starling, Slender-billed Starling, White-billed Starling, Abyssinian Oriole, Variable Sunbird, Scarlet-chested Sunbird, Bush Petronia, Speckle-fronted Weaver, Vitelline Masked Weaver, Black-headed Weaver, Northern Red Bishop, Red-collared Widowbird, Yellow-rumped Serin, Red-billed Firefinch, Cut-throat Finch, Yellow-fronted Canary, Cinnamon-breasted Bunting

7 Debre Zeit (45km)

The town of Debre Zeit, southeast of Addis Ababa, is surrounded by a number of crater lakes. In the absence of any outlet, water levels and salinity are influenced solely by precipitation and evaporation, and with steep sloping sides there is relatively little margin to support waders. However, these are important wetlands for wintering wildfowl, the most significant being Lake Hora and Lake Bishoftu. The acacia scrub along the rim of the lakes is good habitat for many passerines as are the gardens of the Hora Ras Hotel on the shores of Lake Hora.

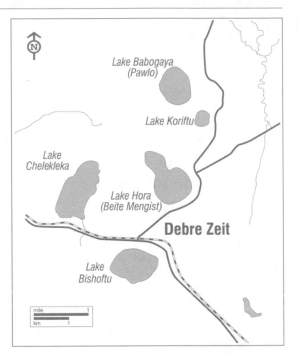

Lake Chelekleka to the southwest of the town, on the other hand, is a shallow water pan which floods during the rainy seasons. The area which floods varies from year to year but as flood waters recede, the resultant marshy areas are good for herons and waders.

Go south on **Ras Biru Avenue** which leads to the main road (**Highway 4**) southeast towards Debre Zeit. Just before entering the town, Lake Chelekleka is on the left-hand (north) side of the road while Lake Bishoftu is on the right-hand (south) side. Lake Hora is in the town centre.

KEY SPECIES

Resident
Little Grebe, Darter, Pink-backed Pelican, Lesser Flamingo, Black-headed Heron, Great Egret, Glossy Ibis, African Spoonbill, Spur-winged Goose, African Pygmy-goose, White-backed Duck, Hottentot Teal, Red-billed Teal, Southern Pochard, Rüppell's Vulture, Lesser Spotted Eagle, African Fish Eagle, Black Crowned Crane, Red-knobbed Coot, Common Moorhen, Black-winged Stilt, Crowned Lapwing, Three-banded Plover, African Snipe, Grey-headed Gull, Bruce's Green Pigeon, Great Spotted Cuckoo, Malachite Kingfisher, Pied Kingfisher, Blue-breasted Bee-eater, Black-winged

Lovebird, Black-billed Barbet, Plain Martin, White-winged Cliff Chat, Mocking Cliff Chat, Northern Black Flycatcher, White-rumped Babbler, Northern Black Tit, Rüppell's Glossy Starling, African Black-headed Oriole, African Silverbill, Chestnut Sparrow,

African Citril, Crimson-rumped Waxbill, Lesser Masked Weaver, Black-throated Canary, Streaky Seedeater

Oct–Mar
Garganey, Northern Pintail, Northern

Shoveler, Eurasian Teal, Ferruginous Duck, Tufted Duck, Pallid Harrier, Common Crane, Pied Avocet, Ruff, Green Sandpiper, Marsh Sandpiper, Black-headed Gull, Gull-billed Tern, White-winged Black Tern, Whinchat

8 Koka Dam (90km)

Koka Dam is on the Awash River, southeast of Addis Ababa, and forms a lake which holds good numbers of Palearctic wildfowl in winter. The surrounding cultivated land is an important winter site for Common Cranes. It is the first in a series of lakes which stretch further south along the Great Rift Valley all the way to Tanzania and therefore lies along an important migration flyway. For anyone making their way south to the birding hotspots of the Rift Valley and the Bale Mountains, this site is on the route and well worth a stop.

Take the main road (**Highway 4**) south and at Mojo turn south on **Highway 6** towards the Rift Valley. After about 20km, you pass the town of **Koka**, after which the road crosses the Awash River and there are extensive wetlands on either side of the road. The reservoir, also known as Lake Gelila, is on the left-hand side of the bridge.

KEY SPECIES

Resident
Little Grebe, Darter, Pink-backed Pelican, Great White Pelican, Goliath Heron, Black-headed Heron, Great Egret, Glossy Ibis, Yellow-billed Stork, African Spoonbill, Spur-winged Goose, African Pygmy-goose, Knob-billed Duck, Red-billed Teal, Southern Pochard, Rüppell's Vulture, African Fish Eagle, African Hawk-Eagle, Black-breasted Snake Eagle, Long-crested Eagle, Black Crowned Crane, Clapperton's Francolin, Red-knobbed Coot, Common Moorhen, African

Jacana, Lesser Jacana, Collared Pratincole, Black-winged Stilt, Senegal Thick-knee, Three-banded Plover, Common Snipe, Grey-headed Gull, Bruce's Green Pigeon, Abyssinian Ground Hornbill, Malachite Kingfisher, Pied Kingfisher, Blue-breasted Bee-eater, Black-billed Barbet, Blue-naped Mousebird, Red-fronted Tinkerbird, Chestnut-backed Sparrow-Lark, Wire-tailed Swallow, Lesser Striped Swallow, Winding Cisticola, Boran Cisticola, Lesser Swamp Warbler, Mariqua Sunbird, Beautiful Sunbird,

Chestnut Sparrow, Spectacled Weaver, Red-headed Weaver, Black-throated Canary

Oct–Mar
Garganey, Northern Pintail, Northern Shoveler, Montagu's Harrier, Pallid Harrier, Lesser Kestrel, Common Crane, Pied Avocet, Common Ringed Plover, Little Ringed Plover, Ruff, Green Sandpiper, Marsh Sandpiper, Black-headed Gull, Gull-billed Tern, White-winged Black Tern, Whiskered Tern, White Wagtail, Sedge Warbler, Basra Reed Warbler

Useful contacts and websites

Ethiopian Wildlife and Natural History Society
www.members.lycos.nl/Tigrai/ewnhs.htm
P.O.Box 13303, Addis Ababa, Ethiopia.
Tel: (+251) 1 614838.
BirdLife International Partner for Ethiopia; publishers of the scientific journal *Walia*.

African Bird Club www.africanbirdclub.org
African Bird Club, c/o BirdLife International, Wellbrook Court, Girton Road, Cambridge CB3 0NA, UK.

Recorder for Ethiopia: Mengistu Wondafrash, c/o Ethiopian Wildlife and Natural History Society, PO Box 13303, Addis Ababa, Ethiopia.

Nurgi Birding (Ethiopia)
www.nurgibirdingethiopia.com
P.O.Box 16201, Bole Road, Addis Ababa.
Tel: (+251) 911 623171.
Email: info@nurgibirdingethiopia.com
Local tour company offering guided birding tours of Ethiopia.

Books and publications

Important Bird Areas of Ethiopia
Ethiopian Wildlife and Natural History Society (1996) Addis Ababa

A Checklist of the Birds of Ethiopia
Emil K. Urban and Leslie H. Brown (1971) Haile Selassie I University Press, Addis Ababa.

Ethiopia: In search of endemic birds
Julian Francis & Hadoram Shirihai (1999)
ISBN 0953476201

A Guide to Endemic Birds of Ethiopia and Eritrea Jose Luis Vivero Pol (2001)
ISBN 1931253137

Where to Watch Birds in Africa
Nigel Wheatley (1995) Christopher Helm
ISBN 0713640138

Field Guide to the Birds of East Africa
Terry Stevenson & John Fanshawe (2001)
Christopher Helm
ISBN 0713673478

Illustrated Checklist: Birds of Eastern Africa
Ber Van Perlo (1995) HarperCollins
ISBN 0002199378

The striking Hoopoe Lark is a true desert bird but is easy to find at Giza

TYPICAL SPECIES

Resident
Grey Heron, Little Egret, Cattle Egret, Black-crowned Night Heron, Squacco Heron, Mallard, Black Kite (*M. m aegyptius*), Marsh Harrier, Common Kestrel, Moorhen, Senegal Thick-knee, Spur-winged Plover, Black-winged Stilt, Laughing Dove, Collared Dove, Pallid Swift, Hoopoe, Pied Kingfisher, Crested Lark, White Wagtail, Barn Swallow (*H. r. savignii*), Common Bulbul, Blackbird, Hooded Crow, Goldfinch, House Sparrow

Winter
Northern Lapwing, Black-headed Gull, Black Redstart, Blackcap, Lesser Whitethroat, Sardinian Warbler, Common Chiffchaff

Passage
White Stork, Turtle Dove, European Bee-eater, Common Swift, Barn Swallow, House Martin, Sand Martin, Willow Warbler

Cairo

Although just 150km from the Mediterranean Sea, the birder in Cairo will be immediately struck by the real African 'feel' to its avifauna. Kites soar overhead and egrets and herons stalk through any wet patches of open space, while bulbuls flit noisily through the treetops in the hotel gardens. Cairo lies on the River Nile with the great expanse of the Sahara Desert to either side. The vast Nile Delta, with its mosaic of canals, reedbeds and cultivation, is just to the north. To the south, the fertile Nile Valley stretches all the way to the heart of tropical Africa. These areas all present distinct options for the birder as each zone has species which are unique to the habitat and location. As Egypt lies at the southeast corner of the Western Palearctic region, the ranges of some African and Asian species overlap in the area.

The most popular periods for tourists to visit Egypt, early spring and late autumn, also coincide with the best birding seasons. The Nile is an important migration flyway for many Palearctic breeding birds and Cairo lies right in the path of this great wave of migrants. There are excellent sites on the Mediterranean and Red Sea coasts, the Sinai Peninsula and in the Upper Nile Valley which are too distant to be covered by this account of Cairo. However, for the birder with more than a couple of days to spare, there are several possibilities for field trips into these areas, which would yield many more species. For those who wish to remain closer to the city and also wish to visit the fascinating antiquities of Ancient Egypt, it is possible to combine both birding and archaeology as many of the historical sites are also good places to find birds.

GETTING AROUND

Cairo is the largest city in Africa and it is located in the most densely populated corner of Egypt, so getting around presents some challenges to the visitor. If you plan to visit sites outside the city, you would be best advised to rent a car, for a number of reasons. Public transport is available but, as neither timetables nor routes are published in English anywhere, it requires local knowledge to be able to use it. Also a car is very useful in

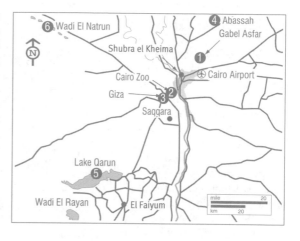

the field as distances are long, the areas to be covered at the sites are quite extensive and the vehicle can provide some welcome shade from the blistering heat in the middle of the day, which can be experienced even in spring and autumn. There is no ring road around Cairo and all highways radiate out from the centre, so if you are planning a trip which requires you to exit the city on the far side of town from where you are staying, you might consider hiring a car with a driver, as you will experience a lot of difficulty driving across the city. Also, your driver will speak Arabic and will be able to ask directions if you get lost or are unable to find the site and, at least, he will be able to find the way back to your hotel!

However, if you are determined to use public transport, you will find that the city area and Giza are very well served by Africa's only Metro. This, coupled with the ubiquitous city taxis, will be adequate for all the sites within Cairo. The train service to the outlying sites is regular but infrequent and usually crowded so you will probably need to book ahead and this may not be possible if you are on a short unplanned trip to the city. The other option is the long-distance bus service, which is very comprehensive but also very informal. There are several bus terminals within the city, all handling a variety of bus lines and levels of service, but most destinations are served by air-conditioned buses operating out of Turgoman Terminal, which is on the east bank of the Nile just west of the main railway station (Ramses Station).

Once you are out of the city, the main highways are quite good and you can make brisk time getting to the sites, most of which are well signposted. Needless to say, you will need to take adequate precautions from the intense glare and heat from the sun. A telescope would be very useful if birding around the lakes at Wadi El Natrun or in El Faiyum, but without a car it is an awkward burden to carry on foot, particularly since you will also have to carry sufficient water for a day in the field. Also bear in mind that you may have to pay an extra photographic admission charge to the Giza Pyramids unless you can convince the officials that your tripod and telescope are not a camera!

The only widely available map which covers the area is the Michelin No. 954 – North East Africa & Arabia (1:4,000,000). However, it will at least enable you to orient yourself on the main roads and give you some sense of distance.

Locations

1 Gabel Asfar (18km)

The sewage ponds at Gabel Asfar are one of the best wetland sites in the Cairo area, although apparently no longer as good as they were in the 1990s. However, this is still one of the most reliable sites close to Cairo for Senegal Coucals and White-breasted Kingfishers. Like all artificial sites, fluctuations in water levels can have a big impact on the amount of available habitat and numbers of birds. However any wetland in an arid area like

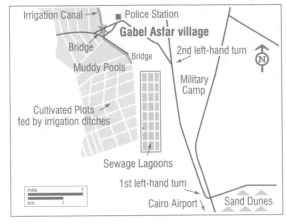

northeast Egypt will attract wildlife. Apart from the sewage lagoons, the waste water is used to irrigate farmland in the vicinity, so there is a large area of orchards, crop fields and irrigation canals, all of which support a good selection of birds. This is also a regular site for both Streaked Weavers and Red Avadavats, two introduced species which have become established in the Nile Delta.

Gabal Asfar is about 7km northeast of Cairo Airport and is a convenient site to visit if you have a few hours after arriving or before departing. Go east from the airport on the **Ismalia Road** for about 2.5km and turn north at the sign for **Zagazig** (you will need to make a U-turn at the next opportunity to get to the right side of the carriageway). Turn left at the first fork (first left-hand turn) in the road opposite some sand dunes and go north for about 5km, passing a military camp on the right, and take the left-hand fork (second left-hand turn), heading west on **Abu Zabel/El Khanka Road**. Continue along this road for about 3km, passing a Police Station on the right-hand side, until you reach an irrigation canal. There is a track going north along the east bank of the canal along which you can walk. If you return to the Police Station and take the road south for about 1.5km, passing the village, you reach another stretch of the canal, where there is a footpath going south and also some marshy pools beyond the west bank of the canal. There is a possibility of finding one of Egypt's most sought-after birds, the Painted Snipe, at these pools.

KEY SPECIES

Resident
Senegal Coucal, White-breasted Kingfisher, Little Green Bee-eater, Blue-cheeked Bee-eater, Yellow Wagtail (*M. f. pygmaea*), Zitting Cisticola, Graceful Prinia, Clamorous Reed Warbler

2 Cairo Zoo (5km)

Although zoos are not normally considered the most obvious places to go birding, if your time is limited this is a very productive spot to get a good flavour of the birdlife of Cairo. It acts as one of Cairo's green lungs in this extremely congested city and, together with the grounds of Cairo University which is just beside the zoo, it can attract some interesting species during migration periods. The freshwater ponds are good for ducks and aquatic species. The drawback, of course, is that it can be very crowded at weekends, and even during weekdays when noisy parties of schoolchildren may visit. An early morning visit is best: the zoo opens at 08.00.

KEY SPECIES

Resident
Little Bittern, Common Coot, Little Green Bee-eater, Zitting Cisticola, Graceful Prinia, Clamorous Reed Warbler

Summer
Sooty Falcon, Blue-cheeked Bee-eater, Eastern Olivaceous Warbler

Winter
Common Sandpiper, Common Kingfisher, Grey Wagtail, Nile Valley Sunbird

3 Giza (7km)

It is unlikely that any visitor to Cairo will pass up the opportunity to visit the site of the Pyramids at Giza. Luckily for the birder, this site is also quite good for birds and, despite the crowds, it provides an opportunity to go birding away from the congested areas. It is also one of the closest sites to Cairo which offers a chance to experience the true desert habitat as there is no cultivation or irrigation in the area. Many of the desert species are just at the edge of their preferred habitat and although numbers are small there are some interesting species. In spring, it would also be worth driving about 17km further south to Saqqara as the Step Pyramid at this site is a known stake-out for Pharaoh Eagle Owls.

Giza is accessed from the west bank of the Nile and can be easily reached by taxi, Metro or on a guided tour.

KEY SPECIES

Resident
Cream-coloured Courser, Little Green Bee-eater, Hoopoe Lark, Desert Lark, Short-toed Lark, Rock Martin, Mourning Wheatear, Hooded

Wheatear, White-crowned Black Wheatear, Brown-necked Raven

Winter
Red-tailed Wheatear, Isabelline

Wheatear, Desert Wheatear, Blue Rock Thrush, Common Stonechat, Spanish Sparrow

Further Away

4 Abassah (50km)

The area between Bilbeis and Abassah is typical of the Eastern Nile Delta: cultivated plots, flooded rice-fields, fishponds and fields of alfalfa, criss-crossed by irrigation canals. Many of the birds characteristic of this habitat can be seen along the canal which runs northeast from the town. The newly flooded rice-fields are also worth checking for Greater Painted-snipe.

Take the road north from Cairo to **Ismailiyah** via **Bilbeis**: this is not the main Ismailiyah Road which lies further to the south. Bilbeis is approximately 50km along this road. The stretch of road for about 13km west of Bilbeis to **Abassah**, which runs parallel to the irrigation canal, is most productive.

KEY SPECIES

Resident Little Bittern, Purple Heron, Glossy Ibis, Black-shouldered Kite, Common Coot, Purple Swamphen, Kentish	Plover, Greater Painted-snipe, Senegal Coucal, White-breasted Kingfisher, Little Green Bee-eater, Crested Lark, Zitting Cisticola,	Graceful Prinia, Clamorous Reed Warbler, Penduline Tit **Winter** White-tailed Plover, Spanish Sparrow

5 Lake Qarun (85km)

Lake Qarun is a large saline lake which lies in El Fayoum depression southwest of Cairo. El Fayoum Depression reaches a low point of about 40m below sea-level and the surrounding agricultural land drains into the lake. The lake has no outlet so its salinity has increased by evaporation to a level where saltwater fish can be farmed. Ultimately the lake will become as salty and as sterile as the Dead Sea in Israel but, for the moment, this is a very important wetland for both breeding and wintering aquatic birds. Although it is rather distant from Cairo, it is relatively close to the antiquities of the Nile Valley as well as being a tourist site in its own right, and it could be visited as part of a day trip out of the city.

Take the main road (**Highway 20**) to **El Fayoum** southwest from Giza. After about 70km turn west towards **Shakshuk** on **Route 204**. The road runs along the south shore of the lake all the way into Shakshuk and there are fishponds and mudflats along the lakeshore which support small numbers of waders. The salinity of the water prevents much aquatic vegetation from growing along the shore, but the small stands of reeds wherever there are drainage ditches feeding into the lake should be checked for herons, rails and passerines.

Also worth checking are two other salt lakes about 25km southwest of Shakshuk at Wadi El-Rayan. Unlike Lake Qarun, these lakes are artificial and have been caused by run-off from irrigated land. They support a similar selection of waterfowl to Lake Qarun, but the drive through the desert between the two sites offers an opportunity to look for some the more thinly-distributed Saharan species.

KEY SPECIES

Resident Little Bittern, Purple Heron, Glossy Ibis, Greater Flamingo, Common Coot, Purple Swamphen, Kentish Plover, Greater Painted-snipe, Senegal Coucal, Little Green Bee-eater, Crested Lark, Zitting Cisticola, Graceful Prinia, Clamorous Reed Warbler **Summer** Collared Pratincole, Slender-billed	Gull, Little Tern, Blue-cheeked Bee-eater **Winter** Black-necked Grebe, Great Crested Grebe, Eurasian Bittern, Great White Egret, Eurasian Teal, Northern Shoveler, Northern Pintail, Eurasian Wigeon, Common Pochard, Tufted Duck, Ferruginous Duck, Grey Plover, Ringed Plover, Little Stint, Temminck's Stint, Dunlin, Curlew	Sandpiper, Ruff, Common Redshank, Spotted Redshank, Greenshank, Marsh Sandpiper, Wood Sandpiper, Common Sandpiper, Pallas's Gull, Yellow-legged Gull, Whiskered Tern, White-winged Black Tern, Red-throated Pipit, Water Pipit, Great Grey Shrike

6 Wadi El Natrun (90km)

Wadi El Natrun is a set of nine brackish lakes in a depression northwest of Cairo. This wetland site is also quite distant from the city but the Coptic Monasteries beside the lakes put it firmly on the tourist map and your itinerary may take you here anyway. Apart from that, this is the most reliable site in the Western Palearctic for Kittlitz's Plovers, and during migration the lake shores are a haven for many migrant waders. The lakes are fed from freshwater springs which promote a luxuriant growth of vegetation at certain points along the shorelines. This growth is kept in check through grazing, resulting in a water meadow-like habitat which is very attractive to waders and herons.

Take the main road (**Highway 11**) northwest from **Giza** through the eastern Sahara Desert to **Alexandria** and after about 85km take the (only) turn left at **Bi'r Hooker,** which is a village lying between the third and fourth lakes. There is a regular bus service from Cairo to Wadi El Natrun departing every hour.

KEY SPECIES

Resident
Little Bittern, Purple Heron, Glossy Ibis, Black-shouldered Kite, Common Coot, Purple Swamphen, Kentish Plover, Kittlitz's Plover, Greater Painted-snipe, Senegal Coucal, Little Green Bee-eater, Desert Lark, Crested Lark, Yellow Wagtail (*M. f. pygmaea*), Zitting Cisticola, Graceful Prinia, Clamorous Reed Warbler

Summer
Slender-billed Gull, Little Tern, Blue-cheeked Bee-eater, Rufous Bush Robin

Winter
Eurasian Teal, Common Shelduck, Avocet, Grey Plover, Ringed Plover, Little Stint, Dunlin, Curlew Sandpiper, Ruff, Common Redshank, Spotted Redshank, Greenshank, Eurasian Curlew, Common Snipe, Richard's Pipit, Red-throated Pipit, Common Stonechat

Passage
Northern Shoveler, Common Crane, Little Crake, Temminck's Stint, Marsh Sandpiper, Wood Sandpiper, Black-headed Gull, Pallas's Gull, Yellow-legged Gull, Gull-billed Tern, Whiskered Tern, White-winged Black Tern

Useful contacts and websites

Birding Egypt www.birdingegypt.com
Website of the Egyptian Birding Community with lots of information about birding near Cairo.

Ornithological Society of the Middle East
www.osme.org
Ornithological Society of the Middle East,
c/o The Lodge, Sandy, Bedfordshire SG19 2DL, UK.

African Bird Club www.africanbirdclub.org
African Bird Club, c/o BirdLife International,
Wellbrook Court, Girton Road, Cambridge
CB3 0NA, UK.

Egypt Nature Adventures
www.birdingegypt.com/travel_assistance.htm
Bird tour guide for Cairo and rest of Egypt.
3 Abdalla El Katib St. Apt. 3, Dokki, Cairo, Egypt.
Tel & Fax. 202 7608160.
Email: info@birdingegypt.com

Egypt Bird Recorder
www.birdingegypt.com/rarities_reporting.htm
Sherif Baha El Din, 3 Abdalla El Katib St.,
Apt 3 Dokki, Cairo, Egypt.
Tel & Fax. 202 7608160.
E-mail: info@birdingegypt.com

Books and publications

Common Birds of Egypt
B. Brunn and S. Baha El Din (1994)
The American University in Cairo Press
ISBN 9774242394

Where to Watch Birds in Africa
Nigel Wheatley (1995) Christopher Helm
ISBN 0713640138

Directory of Important Bird Areas in Egypt
S. Baha El Din (1999) The Palm Press, Egypt

The Birds of Egypt Steven M. Goodman and
Peter L. Meininger (1989) Oxford University Press
ISBN 0198576447

**A Photographic Guide to Birds of Egypt
and the Middle East** Richard Porter and David
Cottridge (2001) American University in Cairo Press
ISBN 1859745121

**Pharaoh's Birds: A guide to ancient and
present-day birds of Egypt** J. Miles (1998)
ISBN 9774244907

Finding Birds in Egypt
Dave Gosney (1993) Gostours Publications
ISBN 1898110034

Birds of Middle East and North Africa
P. Hollom *et al.* (1988) T & AD Poyser
ISBN 085661047X

TYPICAL SPECIES

Resident
Little Grebe, Cape Gannet, Great Cormorant, Sacred Ibis, Little Egret, Cattle Egret, Egyptian Goose, Cape Francolin, Helmeted Guineafowl, Blacksmith Plover, Kelp Gull, Hartlaub's Gull, Ring-necked Dove, Little Swift, Speckled Mousebird, Eurasian Hoopoe (*U. e. africana*), Cape Wagtail, Pied Crow, Cape Crow, Sombre Greenbul, Cape Robin-Chat, Cape White-eye, Red-winged Starling, European Starling, Lesser Double-collared Sunbird, House Sparrow, Cape Sparrow

Summer
Common Buzzard (*B. b. vulpinus*), Barn Swallow

The Jackass Penguin colony at Boulders Beach is only a short drive from Cape Town

Cape Town

Apart from its beautiful setting and climate, any birder visiting Cape Town can experience the thrill of birding in superb habitats with spectacular scenery as a backdrop. Cape Town has a real 'open' feel and you never have the impression, as with other conurbations, that the city is closing in on the bird sites.

There is a belief that the Western Cape Province of South Africa is great for seabirds but rather poor in landbirds. However, bear in mind that this is only relative to the bird-rich hinterland of southern Africa. Most visiting birders will be overwhelmed by the variety of birdlife and will have more than enough to occupy them both on land and at sea. The dominant flora in the area is the fynbos, a shrub vegetation unique to the Western Cape and extremely rich in species diversity. This provides habitats for a number of specialist endemic bird species. However, the birds are thinly distributed and don't exactly give themselves up. In fact, when first viewed these areas look comparatively birdless and require intensive searching to yield results. It is always best to seek areas where the fynbos flora is in bloom, as this attracts sunbirds and other passerines. For many birders there will be an imperative to search for the endemics exclusively, but there are many other birds to be seen and a considerable number of these, although not endemic, are difficult to see elsewhere.

Any time of year is productive, although summer produces additional visitors from the Northern Hemisphere, the most interesting of which are waders. Winter, on the other hand, is better for pelagic birding as large numbers of Antarctic seabirds can be found offshore.

Although this guide is restricted to the areas immediately surrounding Cape Town, the visiting birder with some time to spare might consider venturing further afield. The Karoo is a semi-desert plateau that commences about 120km northeast of Cape Town and is home to many species, including a bewildering variety of larks. A pelagic trip is also well worth the effort, even for those prone to seasickness: the variety of seabirds encountered is so great that it will probably preclude the need to undertake a pelagic trip ever again!

GETTING AROUND

Cape Town is a delightful city with a well-developed eco-tourist infrastructure. There are several well-regarded bird-tour companies based in the city who can organise trips to local bird sites at fairly short notice, and this may be the best option if your time is limited and you have a specific target list of species to see. For those who wish to make their own way, the options are rather limited

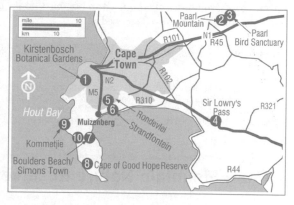

unless you rent a car. Only a small number of sites are accessible by public transport and the large size of these areas make them difficult to cover on foot.

If you are staying in the city the taxis don't operate much beyond the urban areas and you would be ill-advised, for safety reasons, to take a train or a bus, particularly to the Cape Flats wetland sites. So you really need to drive to make the most of your time in Cape Town. If you pick up a car at the airport, you are well-positioned on the N2 Freeway to get to the Cape Peninsula and to access the inland sites, without having to go into the city.

The dry climate in summer means that mosquitoes are not a problem and winter temperatures are fairly cool, so birding at all times of the year is very pleasant, although the mountain-tops and coast are subject to high winds. Be prepared to spend a lot of time clambering over rocky outcrops and boulders, so good hiking boots are recommended. Unless you plan to spend most of your time seawatching, a telescope is not essential as most sites permit fairly close approach to the birds using your car as a hide. All the nature reserves charge a small entrance fee for cars and are generally open from 08.00.

Locations

1 **Kirstenbosch Botanical Gardens** (13 km)

These famous botanical gardens are in the foothills of Table Mountain, surrounded by woodland and fynbos habitat. The flowering plants are excellent for attracting birds, and several Cape endemics can be seen here with ease. From the Visitors' Centre, there are several trails which lead through the gardens, but the best to follow is the Braille Trail which loops through an area of forest at the base of Table Mountain and returns back to the centre through fynbos and the cultivated gardens. There are several trails which also lead up through the ravines to the top of Table Mountain. Alternatively you could take the cable car to the top of the mountain and climb down to Kirstenbosch via the Contour Trail.

Kirstenbosch is in the western suburbs of Cape Town and can be reached by following the main freeway south to **False Bay** and turning west onto **Rhodes Drive** after about 7km, following signs for Kirstenbosch Botanical Gardens. If you don't have a car, this is probably the easiest site to get to as there is a safe and reliable bus service to Kirstenbosch during weekdays which departs from Adderley Street, and city taxis also operate to Kirstenbosch. If you need to buy bird books or maps, the gift shop located in the Visitors' Centre is well stocked.

KEY SPECIES

Resident		
Rufous-chested Sparrowhawk, Great Sparrowhawk, African Goshawk, Common Kestrel, African Olive Pigeon, Lemon Dove, Red-chested	Cuckoo, African Swift, Ground Woodpecker, Black Saw-wing, Olive Thrush, Knysna Scrub Warbler, Cape Grassbird, Tinkling Cisticola, African Paradise Flycatcher, African Dusky	Flycatcher, Cape Batis, Cape Sugarbird, Orange-breasted Sunbird, Common Waxbill, Brimstone Canary, Forest Canary, Cape Siskin

2 Paarl Mountain Nature Reserve (60km)

If you are planning to tour the wine-growing district of Cape Province, then it is worth visiting the Paarl Mountain Nature Reserve on the western outskirts of the town. This mountain habitat supports a good selection of species typical of mountain fynbos vegetation and the well-wooded slopes are a favoured habitat of the Protea Canary, a Cape endemic for which this area is well known. The Wild Flower (Veldblom) Garden within the reserve attracts sunbirds and other passerines to the flowering shrubs.

To get to the Paarl Mountain Nature Reserve, take the main freeway (**N1**) north from Cape Town to **Paarl** and take the first exit on the left (**R45**) towards the town centre. Then turn left at the **Afrikaans Language Monument** and then right onto a gravel road immediately before the monument. The gravel road leads to the reserve entrance.

KEY SPECIES

Resident		
Verreaux's Eagle, Jackal Buzzard, African Harrier-Hawk, Great Sparrowhawk, African Goshawk, African Olive Pigeon, Klaas's Cuckoo, African Swift, Pied Barbet, Ground	Woodpecker, Black Saw-wing, Cape Bulbul, Olive Thrush, Karoo Scrub Robin, Bar-throated Apalis, Cape Grassbird, Tinkling Cisticola, Karoo Prinia, African Dusky Flycatcher, Fiscal Flycatcher, Cape Batis, Cape	Sugarbird, Malachite Sunbird, Orange-breasted Sunbird, Brimstone Canary, Protea Canary, Streaky-headed Canary, Cape Siskin, Common Waxbill

3 Paarl Bird Sanctuary (64km)

This is a sewage farm in the town of Paarl, adjacent to the Berg River, which has been developed to provide birding facilities as well as performing its primary function. The site attracts a rich selection of aquatic birds, some of which are rare or difficult to find closer to Cape Town. There are gravel driveways between the lagoons and well-positioned bird hides which provide good viewing of some of the more secretive species.

From the **N1**, take the second exit on the left-hand side towards **Paarl Town Centre**. Continue through the centre on this road (**R303**) for about 2 or 3km until you come to a sign directing you to the Bird Sanctuary, which is a turn-off to the left.

KEY SPECIES

Resident
Black-necked Grebe, Darter, Little Bittern, Black-crowned Night Heron, Greater Flamingo, African Black Duck, Maccoa Duck, Southern Pochard, Cape Teal, Cape Shoveler, African Fish Eagle, African Goshawk, Black Crake, African Rail, Purple Swamphen, Water Thick-knee, Three-banded Plover, African Snipe, Grey-headed Gull, Malachite Kingfisher, Giant Kingfisher, Plain Martin, Cape Bulbul, Tinkling Cisticola, Little Rush Warbler, Lesser Swamp Warbler, Cape Weaver, Common Waxbill.

Summer
Little Stint, Curlew Sandpiper, Ruff, Common Greenshank, Marsh Sandpiper, Common Sandpiper, Wood Sandpiper, White-winged Black Tern, White-rumped Swift, Greater Striped Swallow, White-throated Swallow, African Paradise Flycatcher, African Reed Warbler

4 Sir Lowry's Pass (50km)

The Hottentot Holland mountain range lies about 50km east of Cape Town in a gentle arc sweeping right down to the ocean. Two of the 'must-see' Cape endemics, Cape Rockjumper and Victorin's Scrub Warbler, inhabit the rocky slopes of these mountains and Sir Lowry's Pass has become a famous site for these and for other mountain specialities. Unless you particularly want to see these birds, there is probably little point in the short-term visitor to Cape Town travelling this far out of town, but this would be a worthwhile stop for anyone who is touring the south coast Garden Route, and a chance to catch a breathtaking view of the Cape Flats, Table Mountain and the ocean beyond.

Take the **N2** freeway southwest for about 50km, passing **Somerset West** on the way. As you ascend the mountains, there is a lay-by at Sir Lowry's Pass, where you can pull in and cross the road to the track leading up to the summit. The rocky slopes leading up to the summit are the habitat of the Cape Rockjumpers, and wherever there are areas of dense thickets, you can listen and look for Victorin's Scrub Warblers. There are gravel paths leading up to the summit but you may need to clamber over the boulders when searching for these birds.

KEY SPECIES

Resident
Verreaux's Eagle, Common Kestrel, Peregrine Falcon, Ground Woodpecker, Cape Rock Thrush, Familiar Chat, Common Stonechat, Cape Rockjumper, Victorin's Scrub Warbler, Cape Grassbird, Karoo Prinia, Red-headed Cisticola, Piping Cisticola, Plain-backed Pipit, Cape Sugarbird, Orange-breasted Sunbird, Yellow Bishop, Cape Siskin, Cape Bunting

5 Rondevlei Nature Reserve (20km)

This freshwater lagoon is an excellent site for the birding newcomer to Africa to become familiar with the local birdlife and also to see Hippopotamuses, which have been introduced here. The reserve is very well-landscaped with trails and bird observation hides and possesses a variety of habitats including reedbeds, open water, riparian vegetation and exposed mud when water levels are low. The adjacent scrub is an example of coastal fynbos, which supports a limited number of passerines.

Take the main freeway (**M5**) south from Cape Town to **Muizenberg** and turn left (east) after about 11km onto **Victoria Road**. After about 2km turn right onto **Fisherman's Walk** which leads to the reserve entrance.

KEY SPECIES

Resident
Darter, Great White Pelican, Black-headed Heron, Little Bittern, Purple Heron, African Spoonbill, Greater Flamingo, Cape Teal, Yellow-billed Duck, Cape Shoveler, Red-billed Duck, African Fish Eagle, African Marsh Harrier, Purple Swamphen, Pied Avocet, Black-winged Stilt, African Snipe, Caspian Tern, White-backed Mousebird, Pied Barbet, Plain Martin, Lesser Swamp Warbler, Yellow Bishop, Cape Canary, Pin-tailed Whydah

Summer
Little Stint, Common Sandpiper, Wood Sandpiper, Common Greenshank, White-winged Black Tern, White-throated Swallow, African Reed Warbler

6 Strandfontein Sewage Works (22km)

This is located on the shore of False Bay approximately 25km south of Cape Town. Taken together with nearby Rondevlei Nature Reserve and the lake at Zeekoevlei, this area constitutes a large contiguous wetland which is one of the most important wildfowl sites in South Africa. The sewage farm is a large facility with 34 settling lagoons and, although changes in the sewage treatment process have reduced the organic content of the lagoons to some extent, it is still an excellent site for waders, wildfowl, herons and other aquatic species. Many Palearctic migrants can be found here in summer and there is a fine selection of African breeding species. The numbers and variety of birds in each of the lagoons vary depending on water levels but a visit is well worthwhile at any time of the year. It is also favoured by local birders as a vagrant hotspot. A network of tarmac and unsurfaced roads runs between the lagoons, along which you may drive.

If you are visiting Rondevlei Nature Reserve, you can get to Strandfontein Sewage Works by continuing east on **Fisherman's Walk** to **Strandfontein Road**. If travelling from Cape Town, take the **M5** freeway south towards **Muizenburg** and exit east at **Ottery** towards Strandfontein Road. Turn right (south) on Strandfontein Road for 4km and right again at the sign for **Zeekoevlei**. This road leads to the sewage farm entrance.

KEY SPECIES

Resident		**Summer**
Great Crested Grebe, Black-necked Grebe, Cape Cormorant, Great White Pelican, Black-crowned Night Heron, Glossy Ibis, Hadada Ibis, Greater Flamingo, Lesser Flamingo, Spur-winged Goose, South African Shelduck, Hottentot Teal, Maccoa Duck, Southern Pochard, Cape Teal,	Cape Shoveler, Yellow-billed Duck, Red-billed Duck, African Fish Eagle, African Marsh Harrier, Purple Swamphen, Common Moorhen, Red-knobbed Coot, African Oystercatcher, Black-winged Stilt, Plain Martin, Cape Bulbul, Tinkling Cisticola, Lesser Swamp Warbler, Little Rush Warbler	Little Stint, Ruff, Wood Sandpiper, Common Greenshank, White-winged Black Tern, Sandwich Tern, White-throated Swallow, African Reed Warbler

7 Boulders Beach (30km)   *(Simon's Town Train Station)*

Boulders Beach lies about 1km south of touristy Simon's Town. This is the site of the famous Jackass Penguin colony. This and Robben Island in Table Bay are the only sites near Cape Town where you can see the penguins so if you are visiting Simon's Town or the Cape of Good Hope, you should visit this beach. It is also one of the few sites near Cape Town which is served by public transport. The train service from Cape Town to Simon's Town is regular and has fewer problems with personal safety than the other routes.

Take the **M3** south and continue southwards on the **M4** following the east side of the **Cape Peninsula** as far as **Simon's Town**. Continue south past Simon's Town for about 1km and the nature park entrance is on the left-hand side. The rocks offshore should be scanned for cormorants and other seabirds, and you have a good chance to see Southern Right Whales from the shore. The coastal scrub and vegetation support some passerines.

KEY SPECIES

Resident		
Jackass Penguin, Bank Cormorant, Cape Cormorant, Crowned Cormorant, Great Crested Tern,	African Oystercatcher, White-fronted Plover, White-backed Mousebird, Pied Kingfisher, Cape Bulbul, Cape Grassbird, Southern Boubou, Cape	Sugarbird, Malachite Sunbird, Brimstone Canary

8 Cape of Good Hope Nature Reserve (40km)

This rather windswept area about 40km south of Cape Town is worth a visit not just for its historical and geographical significance but because several rather special birds are found here. However the Ostriches here, as everywhere else in the Western Cape, are introduced feral stock. There are extensive tracts of mountain fynbos and thicket and many kilometres of hiking trails across the reserve: you can obtain a useful trail map at the entrance gate. The moorland area southwest of Sirkelsvlei pond at the northern end of the reserve is the best place to look for Black-rumped Buttonquail. However, unless you have a large team of birders searching the moor systematically you are unlikely to flush one of these highly secretive birds. Even then it is likely that only some of the team will catch a glimpse of it, so it is probably not worth the frustration!

The Cape of Good Hope and nearby Cape Point are revered by seawatchers for the quality of seawatching that can be experienced here in westerly wind conditions. The possibilities are unrivalled anywhere else in the Southern Hemisphere. Winter brings large numbers of migrant seabirds from the Antarctic while summer is good for local breeding species as well as Northern Hemisphere and tropical migrants.

Take the **M5** south and continue southwards following the signs for Cape Peninsula National Park. There is a car park near the Cape of Good Hope but you will need to walk along the rocky path round to Cape Point.

KEY SPECIES

Resident		Summer
Cape Cormorant, Crowned Cormorant, Jackal Buzzard, Grey-winged Francolin, Black-rumped Buttonquail, Black Oystercatcher, White-fronted Plover, Speckled Pigeon, African Swift, Ground Woodpecker, Clapper Lark, Cape Bulbul, Familiar Chat, Cape Rock	Thrush, Sentinel Rock Thrush, Cape Grassbird, Cloud Cisticola, Red-headed Cisticola, Karoo Prinia, Fiscal Flycatcher, Plain-backed Pipit, Orange-throated Longclaw, Southern Boubou, Bokmakierie, Cape Sugarbird, Malachite Sunbird, Orange-breasted Sunbird, Cape Siskin, Cape Bunting	Cory's Shearwater, Sooty Shearwater, Ruddy Turnstone, Whimbrel, Arctic Skua, Sabine's Gull, Common Tern, Sandwich Tern
		Winter
		Black-browed Albatross, Shy Albatross, White-chinned Petrel, Brown Skua, Great Crested Tern

9 Kommetjie (25km)

This is a coastal town on the west coast of the Cape Peninsula and it is the most reliable site to see all of the endemic coastal seabirds as well as Antarctic Terns (in winter). The Benguela Current offshore enables a rich marine life which in turn supports large populations of seabirds, some of which are actually quite rare on the east side of the Cape Peninsula. The rocky shore is also good for waders, including migrants from the Northern Hemisphere.

Take the **M3** south and branch off southwest on the **M63** following the signs for Kommetjie. Take the road southwest from Kommetjie out to the **Lighthouse** and scan the offshore rocks, particularly at low tide. The small natural inlet known as the **Kom** is best for waders and roosting terns. In winter, during onshore winds, this is a good seawatching point.

KEY SPECIES

Resident		Turnstone, Whimbrel, Arctic Skua, Common Tern, Sandwich Tern
Bank Cormorant, Cape Cormorant, Crowned Cormorant, Black-headed Heron, Grey Heron, Hadada Ibis, Cape Shoveler, Yellow-billed Duck, Common Moorhen, Red-knobbed	Coot, Black-winged Stilt, African Oystercatcher, White-fronted Plover, Crowned Lapwing, Pied Kingfisher	
		Winter
	Summer	Sooty Shearwater, Antarctic Tern, Great Crested Tern
	Curlew Sandpiper, Sanderling, Ruddy	

Further Away

10 Simon's Town Pelagic Trips

Pelagic seabird trips are now something of a small industry off Cape Town and anyone visiting the city should take one out to the edge of the continental shelf, about 25km offshore. The cold Benguela Current ensures abundant feeding at all times of the year and its strategic location at the intersection of two oceans, plus a clear run from the Antarctic, has resulted in a long list of vagrants recorded here. Winter is the best time as the resident birds are augmented by wintering populations from the Antarctic as well as non-breeding birds from the Southern Ocean. Summer is quieter although migrants from the North Atlantic occur. Most of the pelagic trips originate from Simon's Town on the east coast of the Cape Peninsula. Contact details for trip operators are given below.

KEY SPECIES

All Year	Summer	Winter
Black-browed Albatross, Shy Albatross, White-chinned Petrel, Sooty Shearwater, Wilson's Storm-petrel, Cape Cormorant, Crowned Cormorant, Bank Cormorant, Brown Skua	Cory's Shearwater, European Storm-petrel, Pomarine Skua, Arctic Skua, Sabine's Gull, Common Tern, Arctic Tern, Sandwich Tern	Southern Giant Petrel, Northern Giant Petrel, Pintado Petrel, Antarctic Prion, Antarctic Tern

Useful contacts and websites

Birding Africa www.birdingafrica.com
Local guides and short tours around Cape Town.

Cape Birding Route www.capebirdingroute.org
Local guides, accommodation, car hire and free information service and birding updates.

Cape Bird Club www.capebirdclub.org.za
Local branch of BirdLife South Africa.

Zest for Birds www.zestforbirds.co.za
Local birding website with information, checklists and trip reports.

Cape Town Pelagics
www.capetownpelagics.com
Book on any pelagic trip off Cape Town here. All proceeds to albatross conservation.

Southern African Birding
www.sabirding.co.za
Information about birding sites and bird tour operators for southern Africa.

Books and publications

Essential Birding: Western South Africa
C. Cohen, C. Spottiswoode (2000) Struik Publishers ISBN 1868725243
Very detailed guide to the sites and available online at www.capebirdingroute.org

Southern African BirdFinder: Where to find 1400 species in southern Africa and Madagascar C. Cohen, C. Spottiswoode & J. Rossouw (2006) Struik Publishers ISBN 1868727254.

SASOL Birds of Southern Africa
I. Sinclair, P. Hockey, W. Tarboton (2002, 3rd edition) Struik Publishers, ISBN 1868727211

Birds of Africa, South of the Sahara
I. Sinclair & P. Ryan (2003) Struik Publishers ISBN 1868728579

TYPICAL SPECIES

Resident
Little Grebe, Long-tailed Cormorant, Hadada Ibis, Little Egret, Cattle Egret, Spur-winged Goose, Egyptian Goose, Helmeted Guineafowl, Red-knobbed Coot, Common Moorhen, Purple Swamphen, Crowned Lapwing, Blacksmith Plover, Ring-necked Dove, Laughing Dove, Little Swift, Speckled Mousebird, Eurasian Hoopoe (*U. e. africana*), Red-throated Wryneck, Cardinal Woodpecker, Cape Wagtail, Cape Robin-Chat, Pied Crow, Cape White-eye, Common Fiscal, Cape Glossy Starling, Red-winged Starling, House Sparrow, Cape Sparrow, Cape Weaver, Masked Weaver

Summer
Common Buzzard (*B. b. vulpinus*), Black Kite, Barn Swallow, House Martin, Willow Warbler

Greater Double-collared Sunbird feeding on Jacaranda blossom

Johannesburg and Pretoria

It is easy to see that Johannesburg owes its existence to the discovery of gold in the Witwatersrand Ridge which runs through this urban area, as there are no physical features which would have led to the establishment of human settlement. Johannesburg is in the midst of the highveld of South Africa at an elevation of almost 1,700m. The characteristic vegetation is grassland with tree cover limited to alien copses in gullies and on slopes. To the north (towards Pretoria) and west the grassland gives way to bushveld which is mostly woodland with rather dense acacia thornbush. Within these two extremes there are considerable variations in the habitat, depending on altitude, topography, soil type, rainfall and grazing regimes, and each type provides niches for distinct communities of birds, some of which include endemic species. Although the highveld is rather arid with few large watercourses, seasonal pans and depressions attract large numbers of waterfowl, waders and other waterbirds. In addition mining activities and agriculture have created numerous dams and reservoirs which have substantially increased the availability of wetland habitats for wildlife and which provide stretches of permanent water during the dry winters. A number of the better wetland sites are within an hour's drive of Johannesburg. The urban parks and gardens constitute another man-made habitat and these support a range of species which have benefited from the irrigation and the forest cover. The greatest diversity of birds is present during the summer, when the local populations are augmented by migrants from the Northern Hemisphere and Central Africa.

A birder visiting Johannesburg might be tempted to use the opportunity to make a trip to the renowned Kruger National Park, but unless you have more than four days to spare, the distance involved (about 400km) and the sheer size of Kruger militate against a casual visit. In fact, the highveld actually has more endemic bird species than Kruger, and Pretoria, a mere 50km north of Johannesburg, is indeed one of the richest birding areas on the subcontinent with over 400 species, including representatives of many typical African families such as hornbills and rollers. As Pretoria may well form part of your itinerary, some sites close to and within Pretoria are included here to give some options for birding there.

GETTING AROUND

Although Johannesburg forms the hub of a very densely populated and expansive urban area, it is very difficult to get around unless you have a car. Public transport is heavily dependent on the use of informal communal minibus taxis and unless you are familiar with the city and the routing of these taxis they are not suited to the visitor. Similarly, trains and buses are not advisable for visitors for reasons of personal safety, the one exception being the train service

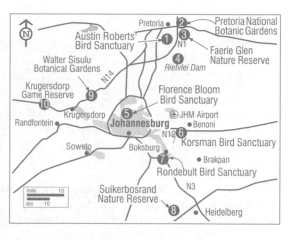

between Johannesburg and Pretoria. Driving through Johannesburg is made easy by a network of expressways which go through the city rather than around it, so that you are never too far from a slip-road, and they are by far the best way to get to the birding sites.

Almost all the birding sites are nature reserves operated by municipal or private authorities, and there is almost always an entrance charge. Consequently the sites are not heavily used by the local public and there is a degree of safety and security when birding there. In any of the nature reserves which contain dangerous game you will only be allowed access by car and there will be restrictions on getting out of your vehicle. This is not as limiting as it might seem, as driving enables you to traverse much greater distances than on foot, and the car can act as a portable hide to enable closer approach to both birds and mammals. There are nearly always trails at campsites and picnic sites where you can walk safely and these are usually near areas of good habitat for birds.

Summer on the highveld is rather mild, and the birdlife is much more diverse and conspicuous then, when many Palearctic and intra-african migrants are also present. Winter is drier, but bear in mind that the high elevation of Johannesburg means that birding in early morning can be extremely cold, with temperatures near or below freezing, requiring the kind of warm clothing not immediately associated with Africa. As some of the sites are a considerable distance from Johannesburg, you will need a map which covers Gauteng Province.

Locations

1 Austin Roberts' Bird Sanctuary (5km from Pretoria)

This bird reserve is just south of Pretoria city centre and is well worth a visit. There is a rather upmarket restaurant (aptly-called The Blue Crane) within the sanctuary, so if you are visiting Pretoria with a reluctant non-birding spouse or partner, it makes a good compromise excursion. It is named after Austin Roberts, author of the seminal work on the birds of South Africa, and it contains a really good selection of typical species which can be seen in a compact and easily walked park. There is some good woodland within the sanctuary as well as two lakes, and a hide has been erected at one of the lake shores. Many of the ducks here are pinioned so it is not much more than a bird park for these birds, but a nice spot nonetheless. The Cranes are not wild either.

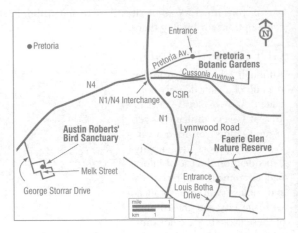

From the centre of **Pretoria**, go south on **Ben Schoeman Highway** and exit east onto **George Storrar Drive**. Continue east on George Storrar Drive to **Melk Street** and turn left (north) to get to the entrance.

KEY SPECIES

Resident
Black-crowned Night Heron, Little Bittern, Glossy Ibis, Blue Crane, White-browed Coucal, Red-faced Mousebird, Pied Kingfisher, Brown-hooded Kingfisher, Green Wood Hoopoe, Black-collared Barbet, Crested Barbet, Olive Thrush (*T. o. smithi*), Little Rush Warbler, Lesser Swamp Warbler, Grosbeak Weaver, Southern Red Bishop, Common Waxbill, Bronze Mannikin, Streaky-headed Seedeater, Southern Yellow-rumped Seedeater

---

2 Pretoria National Botanic Gardens (7km from Pretoria)

If your visit takes you to Pretoria, which is just about 50km north of Johannesburg, there are a number of sites near the city centre which are worth a look. For anyone new to Africa, a visit to Pretoria Botanic Gardens is an introduction to the avifauna of the region in very easy conditions. The area is small (75ha) and can be done in a morning or afternoon but combines a number of habitats including woodland, savanna and highveld grassland along the ridge which runs through the middle of the gardens.

The gardens are on Cussonia Avenue, just east of the N1/N4 interchange in the east of Pretoria. Take the **N4 east** and exit north at **Watermeyer Street**, followed by a left turn into **Cussonia Avenue**.

KEY SPECIES

Resident
Wattled Plover, Spotted Thick-knee, Wahlberg's Honeyguide, Pied Barbet, Golden-tailed Woodpecker, Olive Thrush, Kurrichane Thrush, Bar-throated Apalis, Piping Cisticola, African Paradise Flycatcher, Crimson-breasted Shrike, Southern Boubou, Black-backed Puffback, Brown-crowned Tchagra, Black-crowned Tchagra, Amethyst Sunbird, White-bellied Sunbird, Grosbeak Weaver, Bronze Mannikin

3 Faerie Glen Nature Reserve (12km from Pretoria)

The terrain around Pretoria is broken up by ridges and escarpments which have not been built over and south of the city centre many of these ridges have been preserved as nature reserves. One of these is Fairie Glen, southeast of the city centre, which provides habitat for highveld grassland species on the flat land on either side of the River Moreletta, and for woodland species in the acacia thornbush and woodland on the ridge slopes. The reserve is rather small, and a network of trails ensures that you can cover most of the important habitats in half a day or so. The high ridge along the northern flank of the reserve reaches almost 1,500m but there is no need to go to the top, except to see the view, and most birders confine their attention to the slopes.

From Pretoria, take the main freeway (N1) south to Johannesburg. After about 5km, exit east on **Lynnwood Road** and after about 2km, turn south on **Louis Botha Drive**. The reserve entrance is on the left-hand side of this road after about 1km.

KEY SPECIES

Resident
African Black Duck, Black-shouldered Kite, Wahlberg's Honeyguide, Pied Barbet, Golden-tailed Woodpecker, African Palm Swift, Rock Martin, Black Cuckoo-shrike, Arrow-marked Babbler, White-throated Robin-Chat, Red-backed Scrub Robin, Lesser Swamp Warbler, Cape Grassbird, Cape Crombec, Chestnut-vented Tit-babbler, Piping

Cisticola, Rattling Cisticola, Tinkling Cisticola, Tawny-flanked Prinia, Black-chested Prinia, Southern Boubou, Black-crowned Tchagra, Brown-crowned Tchagra, Bokmakierie, Sulphur-breasted Bush Shrike, White-bellied Sunbird, Amethyst Sunbird, Southern Red Bishop, White-winged Widowbird, Red-collared Widowbird, Streaky-headed Seedeater, Southern Yellow-rumped

Seedeater, Yellow-fronted Canary
Summer
Red-chested Cuckoo, Black Cuckoo, Striped Cuckoo, Dideric Cuckoo, White-rumped Swift, European Bee-eater, White-throated Swallow, Greater Striped Swallow, Lesser Striped Swallow, Marsh Warbler, Spotted Flycatcher, African Paradise Flycatcher, Violet-backed Starling

4 Rietvlei Dam (17km from Johannesburg)

Rietvlei Dam lies between Johannesburg and Pretoria and combines a number of wetland habitats which are excellent for waterfowl with good facilities, such as hides, at some spots. The surrounding grasslands are also rich in wildlife other than birds, and antelopes, zebras and rhinos can be seen. Additionally, the dense bush alongside the Sesmylspruit River, which feeds the reservoir is good for woodland species. The whole area is within the Rietvlei Dam Nature Reserve. You can only enter the reserve by car and your movements are restricted to the road network. However there are various picnic sites and other designated stopping points where you can explore on foot.

From the **R21** which connects Johannesburg, JIA Airport and Pretoria, exit east at **Olifantsfontein** where the reserve entrance is signposted.

KEY SPECIES

Resident
Great Crested Grebe, Darter, Little Bittern, Yellow-billed Duck, African Black Duck, Cape Shoveler, White-quilled Korhaan, Spotted Thick-knee, Grey-headed Gull, Palm Swift, Giant Kingfisher, Half-collared Kingfisher,

Greater Honeyguide, Lesser Honeyguide, Wahlberg's Honeyguide, Black-collared Barbet, Rufous-naped Lark, Spike-heeled Lark, Plain Martin, Rock Martin, Southern Anteater Chat, Capped Wheatear, Olive Thrush, Lesser Swamp Warbler, Desert

Cisticola, Tinkling Cisticola, Tawny-flanked Prinia, Buffy Pipit, Orange-throated Longclaw, Thick-billed Weaver, Long-tailed Widowbird, Zebra Waxbill, Streaky-headed Seedeater
Summer
Eurasian Hobby, Amur Falcon,

Common Sandpiper, White-rumped Swift, European Bee-eater, Dideric Cuckoo, White-throated Swallow, Lesser Striped Swallow, Banded Martin, African Yellow Warbler, Sedge Warbler, African Reed Warbler, Marsh Warbler, Garden Warbler, African Paradise Flycatcher

5 Florence Bloom Bird Sanctuary (10km from Johannesburg)

This is an area of reedbeds and woodland around two dams which has been set aside within the Delta Park in north Johannesburg as a bird sanctuary. Although the sanctuary area was planted fairly recently (in the 1980s) it attracts a good selection of landbirds and is augmented by bird populations from the more mature woodland in the surrounding park. There are two hides from which to scan the open waters and reedbeds. There is a well-known resident pair of Spotted Eagle Owls near the Environmental Centre, and the staff will be able to assist you in seeing them.

Take the **Jan Smuts Avenue** north from the city centre to **Blairgowrie**. Delta Park entrance is on the left-hand side of the road.

KEY SPECIES

Resident
Darter, Little Bittern, Black-crowned Night Heron, Greater Flamingo, Lesser Flamingo, African Black Duck, Southern Pochard, Cape Teal, Cape Shoveler, African Fish Eagle, Little Sparrowhawk, Ovambo Sparrowhawk, Black Crake, African Rail, Three-banded Plover, African Snipe, Grey-headed Gull, Spotted Eagle Owl, Malachite Kingfisher, Giant Kingfisher, Plain Martin, Tinkling Cisticola, Little Rush Warbler, Lesser Swamp Warbler, Common Waxbill

Summer
European Honey-buzzard, Little Stint, Curlew Sandpiper, Ruff, Common Greenshank, Marsh Sandpiper, Common Sandpiper, Wood Sandpiper, White-winged Black Tern, White-rumped Swift, Greater Striped Swallow, White-throated Swallow, African Paradise Flycatcher, African Reed Warbler

6 Korsman Bird Sanctuary (25km from Johannesburg)

Originally, the highveld of South Africa was dotted with both permanent and temporary freshwater pans, which were rich habitats for wildlife as well as valuable sources of water during dry conditions. Although much reduced and to some extent degraded by urbanisation, the area east of Johannesburg still contains a large number of these pans. One of these, Westdene Pan near Benoni, is protected as the Korsman Bird Sanctuary. Although the

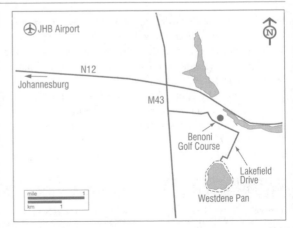

numbers and variety of birds can vary according to water levels and the conditions in neighbouring pans, this is one of the best sites in the Johannesburg area for aquatic birds. It is also about 15 minutes drive from Johannesburg Airport, so it is worth a visit on your way to or from there.

Take the **N12 east** from Johannesburg and after about 20km exit south towards **Benoni** on the **M43**. Take the next turn left (**Lakefield Drive**), followed by the next right at the **Golf Course**. This road leads to a circular drive which encircles the pan, enabling you to scan from the road.

KEY SPECIES

Resident
Great Crested Grebe, Darter, Black-headed Heron, Squacco Heron, Little Bittern, Glossy Ibis, African Spoonbill, Greater Flamingo, Lesser Flamingo, White-faced Whistling Duck, Fulvous Whistling Duck, Yellow-billed Duck, Red-billed Duck, Hottentot Teal, Cape Shoveler, Southern Pochard, Black

Crake, Pied Avocet, Black-winged Stilt, Kittlitz's Plover, Three-banded Plover, Wattled Plover, African Snipe, Grey-headed Gull, Marsh Owl, Giant Kingfisher

Summer
Goliath Heron, Purple Heron, Black Egret, Intermediate Egret, African

Black Duck, Maccoa Duck, Little Stint, Curlew Sandpiper, Ruff, Common Sandpiper, Wood Sandpiper, Common Greenshank, Whiskered Tern, White-winged Black Tern, Malachite Kingfisher

---

7 **Rondebult Bird Sanctuary** (20km from Johannesburg)

This wetland site is a series of freshwater lagoons attached to a waste-water treatment plant which, over the years, has developed into a good site for waterfowl as well as species typical of the adjacent grassland and scrub areas. The lagoons are interconnected by footpaths and a number of well-located hides permit viewing of the open waters and the reedbeds. Depending on water levels, the muddy margins are excellent for waders in summer.

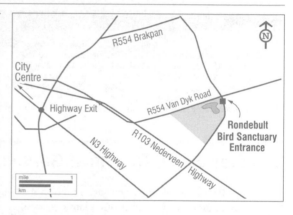

Take the main freeway (**N3**) southeast from Johannesburg and after about 10km exit east towards **Brakpan** on the **R554**. The reserve entrance is about 9km along this road on the right-hand (south) side.

KEY SPECIES

Resident
Striated Heron, Black-headed Heron, Little Bittern, Purple Heron, Glossy Ibis, Greater Flamingo, White-faced Whistling Duck, Fulvous Whistling Duck, Southern Pochard, Maccoa Duck, Yellow-billed Duck, Cape Shoveler, Red-billed Duck, Hottentot Teal, Black Crake, African Rail, Pied

Avocet, Black-winged Stilt, Wattled Plover, African Snipe, Little Rush Warbler, Lesser Swamp Warbler, Tinkling Cisticola, Zitting Cisticola, Tawny-flanked Prinia, Southern Boubou, Orange-throated Longclaw, Southern Red Bishop, Yellow-eyed Canary, Southern Yellow-rumped Seedeater, Zebra Waxbill

Summer
Black Egret, Little Stint, Curlew Sandpiper, Ruff, Common Sandpiper, Wood Sandpiper, Common Greenshank, Yellow Wagtail, Sedge Warbler, Great Reed Warbler, African Reed Warbler

8 Suikerbosrand Nature Reserve (50km from Johannesburg)

Suikerbosrand is a substantial game reserve (22,000ha) southeast of Johannesburg. It is at a higher altitude than the city and supports a number of distinctive highveld species, including many endemics. There is a network of hiking trails, but in order to sample as much of the reserve as possible you need a car to undertake the 60km loop of the reserve. The drive ascends from grassland and thorn thickets through several different habitats to the rocky summit of the Suikerbosrand Escarpment before descending again to acacia savanna. A variety of birds typical of each habitat can be seen as well as large mammals such as Eland, Red Hartebeest, zebra and the endemic Black Wildebeest.

Take the **N3** freeway southeast towards **Heidelberg** and after about 30km exit west onto the **R550** and drive about 6km before turning south at the signpost for the reserve. The reserve is open from 07.00 and it is best to be there early in the morning as birds and mammals will be most active then and the roads within the reserve less disturbed. The Visitors' Centre is well stocked with maps, trail guides and a bird checklist.

KEY SPECIES

Resident
Secretarybird, Black-shouldered Kite, Common Kestrel (*F. t. rupicolus*), Grey-winged Francolin, Red-winged Francolin, Orange River Francolin, White-bellied Bustard, Eastern Long-billed Lark, Rufous-naped Lark, Red-capped Lark, Spike-heeled Lark, Ashy Tit, Red-eyed Bulbul, Cape Rock Thrush, Familiar Chat, Mountain Wheatear, Southern Anteater Chat, Mocking Cliff Chat, Capped Wheatear, Kalahari Scrub Robin, Chestnut-vented Tit-babbler, Zitting Cisticola, Desert Cisticola, Cloud Cisticola, Wing-snapping Cisticola, Long-billed Pipit, Striped Pipit, Orange-throated Longclaw, Bokmakierie, Yellow Canary, Southern Yellow-rumped Seedeater

9 Walter Sisulu Botanical Gardens (25km from Johannesburg)

These wonderful botanical gardens are much more than fancy gardens with immaculate cut lawns and foreign and exotic plants. As well as presenting a microcosm of the rich flora of South Africa, they also embrace a number of wild habitats which support an excellent selection of birds and other wildlife in the midst of a built-up urban area. The variety of habitats includes bushveld, woodland and wetland areas as well as rocky outcrops along the Roodekrans Ridge. A network of trails fan out from the entrance area and the best birding areas lie in the northeast and northwest sectors. There is a hide on the shores of the Sasol Dam, while the resident pair of Verreaux's Eagles can be viewed easily at Witpoortjie Falls.

From the **N1** Motorway, exit east onto the **M47** towards Krugersdorp. After about 11km, where the **gardens** are signposted, turn south onto **Doreen Road**, and then right onto **Malcolm Road** which leads to the entrance gate.

KEY SPECIES

Resident
Verreaux's Eagle, Great Sparrowhawk, Common Kestrel, Red-winged Francolin, Wattled Plover, Spotted Thick-knee, African Olive Pigeon, Speckled Pigeon, Giant Kingfisher, Half-collared Kingfisher, Black-collared Barbet, Pied Barbet, Golden-tailed Woodpecker, Striped Pipit, Rock Martin, Black Cuckoo-shrike, Cape Rock Thrush, Olive Thrush (*T. o. olivaceus* and *T. o. smithi*), Bar-throated Apalis, Cape Grassbird, Zitting Cisticola, Neddicky, African Paradise Flycatcher, Southern Boubou, Black-backed Puffback, Black-crowned Tchagra, Malachite Sunbird, Greater Double-collared

Sunbird, Grey-headed Sparrow, Southern Red Bishop, Bronze Mannikin, Pin-tailed Whydah, Cape Bunting, Cinnamon-breasted Bunting

Summer
Red-chested Cuckoo, Black Cuckoo, Klaas's Cuckoo, African Swift, White-rumped Swift, European Bee-eater,

White-throated Swallow, Greater Striped Swallow, Garden Warbler

10 Krugersdorp Game Reserve (32km from Johannesburg)

Krugesdorp lies just a little over 30km west of Johannesburg and is the closest game reserve to the city where you have the opportunity to see some of the big game for which Africa is renowned. It is a rather small reserve (1,500ha) but it is not possible to walk anywhere within it due to the dangers of wild animals and so a car is needed to cruise around the roads. However, as most of the reserve is savanna grassland, a car offers the benefit of being able to cover large areas using your vehicle as a hide. There is a good selection of grassveld species in the reserve. The best places to look are around the dams and waterholes particularly in the mornings and evenings.

Exit west from the **N1** onto the **R24** towards **Krugersdorp**. Continue on the R24 past Krugersdorp: the entrance to the reserve is signposted on the right-hand (north) side of this road.

KEY SPECIES

Resident
Black-headed Heron, Glossy Ibis, White-faced Whistling Duck, Yellow-billed Duck, Small Buttonquail, White-quilled Korhaan, Rufous-naped Lark, Southern Anteater Chat, Cape Grassbird, Wing-snapping Cisticola, Cloud Cisticola, Desert Cisticola, Zitting Cisticola, Grassland Pipit, Southern Red Bishop, Red-collared Widowbird, White-winged Widowbird

Useful contacts and sites

BirdLife South Africa
www.birdlife.org.za
South African birding portal with links to local branches of BirdLife South Africa in Johannesburg area.

Pretoria Bird Club
www.birding.co.za
South African .

African Bird Club
www.africanbirdclub.org
African Bird Club, c/o BirdLife International, Wellbrook Court, Girton Road, Cambridge CB3 0NA, UK.

Indicator Birding www.birding.co.za
Local specialists with a section on birding the Johannesburg/Pretoria area on their website. Tel. 012 653 2030. Email info@birding.co.za

Southern African Birding
www.sabirding.co.za
Information about birding sites and bird tour operators for southern Africa.

Books and publications

SASOL Birds of Southern Africa
I. Sinclair, P. Hockey, W. Tarboton (2002, 3rd edition) Struik Publishers, ISBN 1868727211

Birds of Africa, South of the Sahara
I. Sinclair, P. Ryan (2003) Struik Publishers ISBN 1868728579

The Atlas of Southern African Birds
J. A.Harrison et al. (1997) BirdLife South Africa, Johannesburg, ISBN 0620207299

The Southern African Birdfinder
C. Cohen, C. Spottiswoode and J. Rossouw (2006) Struik Publishers, ISBN 1868727254

TYPICAL SPECIES

Resident

Long-tailed Cormorant, Black-headed Heron, Cattle Egret, Hamerkop, Openbill Stork, Marabou Stork, Hadada Ibis, Sacred Ibis, Black Kite, Hooded Vulture, African Goshawk, African Hobby, Helmeted Guineafowl, Grey Crowned Crane, Grey-headed Gull, Speckled Pigeon, Red-eyed Dove, Ring-necked Dove, Laughing Dove, Blue-spotted Wood Dove, Red-headed Lovebird, Grey Parrot, Meyer's Parrot, Great Blue Turaco, Ross's Turaco, Eastern Plantain-eater, African Palm Swift, Little Swift, Speckled Mousebird, Woodland Kingfisher, Pied Kingfisher, Broad-billed Roller, Black-and-white-casqued Hornbill, Double-toothed Barbet, Yellow-fronted Tinkerbird, Grey Woodpecker, Nubian Woodpecker, Angola Swallow, African Pied Wagtail, Common Bulbul, African Thrush, Tawny-flanked Prinia, Green-backed Camaroptera, Sooty Chat, Yellow White-eye, Scarlet-chested Sunbird, Black-headed Gonolek, Pied Crow, Rüppell's Glossy Starling, Splendid Glossy Starling, Grey-headed Sparrow, Slender-billed Weaver, Orange Weaver, Northern Brown-throated Weaver, Village Weaver, Vieillot's Weaver, Golden-backed Weaver, African Firefinch, Bronze Mannikin, Pin-tailed Whydah, Yellow-fronted Canary

Nov–Mar

Western Marsh Harrier, Common Buzzard (*B. b. vulpinus*), Eurasian Hobby, Common Tern, White-winged Black Tern, European Bee-eater, Common Swift, Yellow Wagtail, Barn Swallow, Sand Martin, Willow Warbler

Kampala and Entebbe

Any birder with the good fortune to have a few days to spare in Kampala is immediately faced with the wonderful dilemma of where to go birding in one of the most bird-rich countries in Africa. Although geographically situated well to the east on the continent of Africa, Uganda's location extending to one of the two great rift valley systems of Africa, and its climatic conditions, place it at the interface between the ranges of the East African and West African avifauna. Hence, despite its inland location, it has one of the longest bird lists in Africa, with a high number of regional endemic species. Located as it is just on the equator, Kampala experiences very dynamic changes in its avifauna associated with the seasonal rains and the seasonal changes of the Northern and Southern Hemispheres. The Lake Victoria Basin is an important wintering area for many Palearctic migrants, and also experiences considerable migration in spring and autumn.

Most of the large Ugandan National Parks are located around the periphery of the country, well away from Kampala in the centre (as also is the Albertine Rift Valley, an important Endemic Bird Area), and are well beyond the scope of this book. However its situation near the shore of Lake Victoria and the proximity of some excellent forest and wetland habitats make Kampala a birding destination in its own right. In fact, for many birders the starting point will be the grounds of their hotel as there is a profusion of secondary woodland habitat in the city and a rich selection of species.

GETTING AROUND

For the birder, the key route through Kampala is the Kampala/Entebbe Highway which runs along the spine of a peninsula which juts into Lake Victoria, for a distance of about 35km, with Entebbe and the airport at the tip. This

Black-and-white-casqued Hornbills can be seen in city parks in Kampala

highway provides access to several important birding sites between the two cities and to Entebbe itself, which also hosts some good birding sites. Kampala is a small city and, like most African cities, it functions by means of a thriving informal transport system. You will find that you are never short of a taxi to get anywhere, but for trips outside the city and along the Entebbe Highway the preferred method is by shared minibus taxi, or matatu as it is known locally.

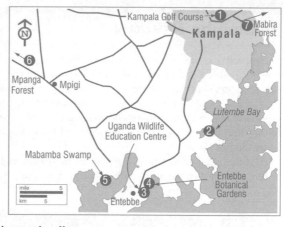

There are many car hire companies in Kampala and in Entebbe and rental rates are quite reasonable. If you do have a car, then driving is an easy option outside the congested areas of Kampala, with good paved main roads and reasonable secondary roads. Many of the access roads to birding sites and in remoter areas are unpaved murram roads, which are hard-packed gravel surfaces that are surprisingly smooth to drive on and, in fact, are much less susceptible to pot-holing than asphalt in the tropical climate, provided they are well maintained.

There is a thriving ecotourist industry in Uganda, which is quite strongly focused on birding, and there are many bird tour guides available to lead groups or individuals. These trips mainly need to be prebooked but it may be possible to make contact with a bird guide at short notice to take you around for a day or two in the Kampala area. There are a few good bookshops in Kampala which sell field guides and maps of Uganda and of Kampala. The best is probably the Aristoc which is in Plot 23, Kampala Road (opposite Steers Restaurant) in downtown Kampala.

Locations

1 Kampala Golf Course (City Centre)

If staying in Kampala, the quickest way to get to grips with the Ugandan avifauna is to spend half a day birding at the Kampala Golf Course. This is conveniently close to the hotel district in downtown Kampala and provides an area of open grassland and woodland which supports most of the typical species of the area. Even if you are planning a more extensive trip to the birding hotspots in the interior of the country, a visit here will help you to familiarise yourself with the different families of Ugandan birds, particularly if you are new to Africa. The golf course is northeast of the downtown area on Yusufu Lule Road. It is courteous to call at the club house to let them know that you would like to go birding around the edges of the course.

KEY SPECIES

Resident
Palm-nut Vulture, Lizard Buzzard, Long-crested Eagle, Bat Hawk, African Green Pigeon, Little Bee- eater, Crowned Hornbill, African Pied Hornbill, Speckle-breasted Woodpecker, Red-faced Cisticola, Grey-capped Warbler, Variable Sunbird, Tropical Boubou, Grey-backed Fiscal, Common Fiscal, Fork-tailed Drongo, Piapiac

2 Lutembe Bay (20km)

Lutembe Bay is one of the bays of Lake Victoria, about halfway between Kampala and Entebbe, which is fringed with papyrus swamps and large mats of water hyacinth. Many waders congregate here, particularly if water levels are low and large expanses of mud are exposed. It is also a very important site for other Palearctic migrants, especially marsh terns, during the Palearctic winter and during migration. Like most of the swamps in this area, it is best covered by boat and you can hire a boatman at the Lutembe Bay Resort.

Take the main road from Kampala to **Entebbe** and past **Kajansi** turn left (east) at the sign for **Lutembe Beach Resort**. Lutembe Bay is approximately 4km along this road.

KEY SPECIES

Resident
Long-tailed Cormorant, Great Cormorant, Pink-backed Pelican, Darter, Squacco Heron, Goliath Heron, Great Egret, Egyptian Goose, Spur-winged Goose, African Fish Eagle, Palm-nut Vulture, African Marsh Harrier, Long-crested Eagle, Grey Kestrel, Black Crake, Common Moorhen, African Jacana, Long-toed Lapwing, Water Thick-knee, African Skimmer, African Mourning Dove, African Green Pigeon, Dideric

Cuckoo, Blue-headed Coucal, White-rumped Swift, Malachite Kingfisher, Grey-headed Kingfisher, Little Bee-eater, White-throated Bee-eater, Crowned Hornbill, African Pied Hornbill, Wire-tailed Swallow, Lesser Striped Swallow, Mosque Swallow, Carruthers's Cisticola, Winding Cisticola, White-winged Warbler, Swamp Flycatcher, Brown-throated Wattle-eye, African Blue Flycatcher, Papyrus Gonolek, Red-chested Sunbird, Red-billed Firefinch, Grey-

headed Negrofinch, Fan-tailed Widowbird, Papyrus Canary

Nov–Mar
Little Stint, Curlew Sandpiper, Ruff, Marsh Sandpiper, Common Greenshank, Black-tailed Godwit, Common Sandpiper, Wood Sandpiper, Slender-billed Gull, Heuglin's Gull, Lesser Black-backed Gull, Gull-billed Tern, Whiskered Tern, Blue-cheeked Bee-eater

3 Uganda Wildlife Education Centre (32km)

The UWEC is basically a zoo located just outside Entebbe, but a zoo with a difference in that it has a self-contained nature trail within the grounds which provides some excellent birding. It also provides access to the shore of Lake Victoria so that a few hours wandering around can enable you to see a good range of woodland and aquatic species.

Take the main road from Kampala to **Entebbe** and after about 30km you will come to a sign indicating the turn-off to the left towards the **UWEC** entrance. The site is only 5km from Entebbe Airport, so if your flight is delayed you could easily spend a few hours here and at the adjacent Entebbe Botanical Gardens.

KEY SPECIES

Pink-backed Pelican, Long-tailed Cormorant, Great Cormorant, Egyptian Goose, Spur-winged Goose, African Fish Eagle, Palm-nut Vulture, Black-winged Stilt, African Green Pigeon, Little Bee-eater, Crowned Hornbill, African Pied Hornbill,

Malachite Kingfisher, Pygmy Kingfisher, Speckle-breasted Woodpecker, Little Greenbul, Yellow-throated Greenbul, Red-faced Cisticola, Winding Cisticola, Grey-capped Warbler, Black-and-white Shrike-flycatcher, African Paradise

Flycatcher, Variable Sunbird, Collared Sunbird, Olive-bellied Sunbird, Red-chested Sunbird, Tropical Boubou, Piapiac, Lesser Masked Weaver, Black-headed Weaver, African Citril

4 Entebbe Botanical Gardens (32km)

As most visitors will arrive in and depart from Uganda at Entebbe Airport, it is worth visiting Entebbe Botanical Gardens either at the beginning or end of your trip. In common with many botanical gardens in the tropics, the combination of a diversity of plant species and the practice of irrigation during the dry season produces a year-round supply of seeds and fruit, as well as an abundance of insect life, which in turn attracts the birds. These gardens are on the shore of Lake Victoria so that you can also scan the lakeshore for aquatic birds.

The gardens are close to the centre of Entebbe. On the main road from Kampala to **Entebbe Airport**, turn east at the south end of Entebbe town at the sign for **Botanical Beach Hotel**, and travel about half a kilometre on this road towards the hotel. The Botanical Gardens entrance is on this road just before the hotel entrance.

KEY SPECIES

Resident
Pink-backed Pelican, Long-tailed Cormorant, Great Cormorant, Egyptian Goose, Spur-winged Goose, African Fish Eagle, Palm-nut Vulture, Lizard Buzzard, Black-winged Stilt, African Green Pigeon, Little Bee-eater, Crowned Hornbill, African Pied Hornbill, Pygmy Kingfisher, Speckle-breasted Woodpecker, Little Greenbul, Red-faced Cisticola, Winding Cisticola, Grey-capped Warbler, Black-and-white Shrike-flycatcher, African Paradise Flycatcher, Variable Sunbird, Collared Sunbird, Olive-bellied Sunbird, Orange-tufted Sunbird, Mariqua Sunbird, Red-chested Sunbird, Superb Sunbird, Tropical Boubou, Piapiac, Purple Glossy Starling, African Citril

5 Mabamba Swamp (45km)

The north shore of Lake Victoria is lined with some extensive tracts of papyrus swamp. Mabamba Bay is one of the best areas for this habitat and supports a range of specialised papyrus swamp species, including the Shoebill which is regularly seen here. The papyrus is very dense and the only effective way to explore this area is by boat. At Mabamba you can hire boatmen to take you by canoe through the swamp and in so doing you will see vastly more birds than you could from the shore. Most of the boatmen are very knowledgeable of the birdlife and will add greatly to the birding experience.

Take the main road southwest from Kampala towards **Masaka** and after 30km turn south at **Mpigi** towards **Kasanje**. Drive through Kasanje and after 4km turn right for Mabamba, which is about 10km further along this road.

KEY SPECIES

Resident
Long-tailed Cormorant, Darter, Squacco Heron, Goliath Heron, Shoebill, African Pygmy-goose, White-faced Whistling Duck, Yellow-billed Duck, African Fish Eagle, Palm-nut Vulture, African Marsh Harrier, Lizard Buzzard, Martial Eagle, Long-crested Eagle, Grey Kestrel, Black Crake, Purple Swamphen, Common Moorhen, African Jacana, Long-toed Lapwing, Water Thick-knee, Gull-billed Tern, African Mourning Dove, African Green Pigeon, Dideric Cuckoo, Blue-headed Coucal, White-rumped Swift, Malachite Kingfisher, Grey-headed Kingfisher, Little Bee-eater, Blue-breasted Bee-eater, Crowned Hornbill, African Pied Hornbill, Yellow-rumped Tinkerbird, Yellow-fronted Tinkerbird, Spot-flanked Barbet, Blue Swallow, Wire-tailed Swallow, Lesser Striped Swallow, Mosque Swallow, Yellow-throated Longclaw, Carruthers's Cisticola, Winding Cisticola, White-winged Warbler, Swamp Flycatcher, Lead-coloured Flycatcher, Brown-throated Wattle-eye, African Blue Flycatcher, Northern Black Tit, Papyrus Gonolek, Red-billed Firefinch, Grey-headed Negrofinch, Fan-tailed Widowbird

6 Mpanga Forest (37km)

This small forest to the west of Kampala provides well-preserved lowland forest habitat. Like all tropical forests, it may seem almost birdless at first, but success in seeing the birds depends on slowly walking along a defined trail using all of your powers of observation: hearing as well as sight. A team of two observers is probably the optimum, as any more will probably create too much noise. If you are lucky enough to come across a feeding party of birds stay with the flock for as long as possible as its feeding activity attracts more birds. You will be amazed by the diversity of species contained within a small flock of, say, 30 birds. There are three trails through the forest of varying length and ease, and a slow methodical walk along each trail can reveal a surprising array of birds. The forest is popular with bird-tour groups so obviously you should avoid following directly behind a party as the birds will have already been disturbed. Instead try one of the alternative trails and return later. You can also engage the services of a guide at the forest entrance for a fairly modest fee, and for anyone who is unfamiliar with birding in tropical forests this is a worthwhile expense.

Take the main road southwest from Kampala towards **Masaka**. About 3km past **Mpigi** there is a sign indicating the forest entrance on the right-hand (north) side of the road.

KEY SPECIES

Resident
Palm-nut Vulture, African Harrier-Hawk, Long-crested Eagle, Red-chested Cuckoo, African Emerald Cuckoo, Black Cuckoo, Blue-breasted Kingfisher, Crowned Hornbill, African Pied Hornbill, Lesser Honeyguide, Brown-eared Woodpecker, White-throated Greenbul, Green Hylia, Buff-throated Apalis, African Paradise Flycatcher, Olive Sunbird, Weyns's Weaver, Brown Twinspot

Further Away

7 Mabira Forest (54km)

Mabira Forest Reserve is quite a large block of lowland forest, which is mainly secondary growth but also includes some primary forest. It supports a breathtaking number of species and is a site for the much sought-after Nahan's Francolin. The forest covers several low hills and valleys with papyrus swamps in the lower reaches. The reserve is very well appointed and caters as much for those who just have half a day to visit or those staying a whole week, as basic accommodation is provided. There are ten trails through the forest as well as a 25km cycle trail. For those who prefer to sit rather than walk there is a raised viewing platform where you can view the forest canopy without disturbing the wildlife. The clearing around the campsite and forest headquarters also provides an excellent birding spot as a result of the 'forest edge-effect' providing extra visibility as well as a more open canopy. Of the ten trails which penetrate the forest, one of the best is the Buwoola Pond Trail which leads through the forest to this pond which attracts many forest-edge birds. The primary forest is mainly across the road from the campsite and can be accessed via the Fig Junction Trail, while the Grassland Trail leads through several areas of clear-cut and more open canopy. One of the drawbacks of forest birding is the presence of armies of biting ants which will immediately launch an assault if you pause for a moment. However, these ant trails should be watched carefully as they are often followed by small flocks of the shyer ground-dwelling birds.

Like most reserves in Uganda the knowledgeable staff are available to act as bird guides and will also be able to advise on the whereabouts of fruiting trees, which can act as virtual bird-magnets.

Take the main road east from Kampala to **Jinja** and after 54km, as you come to the village of **Najjembe**, there is a sign for the forest entrance on the left-hand side (north) of the road. Turn north here and the entrance is about half a kilometre along this road.

KEY SPECIES

Resident
Palm-nut Vulture, African Harrier-Hawk, Long-crested Eagle, African Crowned Eagle, Ayres's Hawk-Eagle, Nahan's Francolin, White-spotted Flufftail, Afep Pigeon, Black-billed Turaco, Red-chested Cuckoo, Black Cuckoo, Dusky Long-tailed Cuckoo, African Emerald Cuckoo, Narina's Trogon, Blue-breasted Kingfisher, White-bellied Kingfisher, Shining Blue Kingfisher, Blue-throated Roller, Green Wood Hoopoe, Crowned Hornbill, African Pied Hornbill, Grey-throated Barbet, Hairy-breasted Barbet, Speckled Tinkerbird, Yellow-throated Tinkerbird, Lesser Honeyguide, Buff-spotted Woodpecker, Brown-eared Woodpecker, Purple-throated Cuckoo-shrike, Little Grey Greenbul, Red-tailed Greenbul, Slender-billed Greenbul, White-throated Greenbul, Green-tailed Bristlebill, Yellow-spotted Nicator, Forest Robin, Snowy-crowned Robin-Chat, Fire-crested Alethe, Green Hylia, Grey-throated Flycatcher, African Paradise Flycatcher, African Shrike-flycatcher, Jameson's Wattle-eye, Scaly-breasted Illadopsis, Tit-hylia, Blue-throated Brown Sunbird, Grey-headed Sunbird, Sooty Boubou, Chestnut-winged Starling, Red-headed Malimbe, White-breasted Negrofinch, Brown Twinspot

Useful contacts and websites

Nature Uganda
www.natureuganda.org
EANHS, P. O. Box 27034, Kampala, Uganda
Email: nature@natureuganda.org
BirdLife International Partner for Uganda.
Tel. (+256) 41 540719

African Bird Club
www.africanbirdclub.org
African Bird Club, c/o BirdLife International, Wellbrook Court, Girton Road, Cambridge CB3 0NA, UK.

Uganda Wildlife Authority
www.uwa.or.ug
Kira Road, Kololo, Nakasero PO Box 3530 Kampala, Uganda. Tel. (+256) 41 346287
Email: uwa@uwa.or.ug

Uganda Wildlife Society
www.uwa.or.ug
P.O. Box 7422, Kampala, Uganda. Tel. 041530891.
Email: uwa@uwa.or.ug

Birding In Paradise Safaris Ltd
Bird tours operator: www.birdinginparadise.com
P.O. Box 24015, Kampala, Uganda
Tel. (+256) 77 468521.

Uganda Bird Guides Club
www.ugandabirdguides.com
Network of bird guides in Uganda and information about birding throughout the country.

Enter Uganda Birding
www.enteruganda.com/travel/birds.php

Uganda Tourist Board
www.visituganda.com/birds.html

Uganda Wildlife Education Centre
www.uweczoo.org

Recorder for Uganda:
Prof. Derek Pomeroy.
Email: derek@imul.com
c/o Institute of Environment and Natural Resources, Makerere University, Box 7298 Kampala, Uganda.

Books and publications

Where to Watch Birds in Uganda
J. Rossouw & M. Sacchi (1998)
Uganda Tourist Board

Where to Watch Birds in Africa
Nigel Wheatley (1995) Christopher Helm
ISBN 0713640138

Field Guide to the Birds of East Africa
Terry Stevenson & John Fanshawe (2001)
Christopher Helm, ISBN 0713673478

The Bird Atlas of Uganda
M. Carswell, D. Pomeroy, J. Reynolds and H. Tushabe
British Ornithologists' Union and British Ornithologists' Club, ISBN 0952286648

Important Bird Areas in Uganda
A. Byaruhanga, P. Kasoma, D. Pomeroy (2001)
Nature Uganda, ISBN 9970714007

Nairobi

For many birders a trip to Nairobi is the starting point for a month-long odyssey through the legendary birding hotspots of Kenya, and they leave the city as soon as possible. However, doing this is to pass up an immense opportunity: Nairobi represents the pinnacle of city-based birding as more species have been recorded within its precincts than in any other urban area in the world – more than 600 species! For an inland city this may seem surprising but the key to Nairobi's species diversity, apart from its equatorial location, is its proximity to a range of different elevations with their associated climatic variations and vegetation zones. Its greatest attractions include the Great Rift Valley lakes, which are within an hour's drive, and Nairobi National Park, created by an inspired decision in 1945, which is just southeast of the city centre.

The maximum number of species occurs in the months corresponding with the Northern Hemisphere winter, when significant numbers of Palearctic migrants can be found in Kenya. These, though, can be quite unobtrusive as, apart from the waders, they don't form flocks or establish territories. In addition, this time of year corresponds with the dry season in East Africa and birding at Nairobi's moderate elevation (1,700m) can be conducted in very pleasant weather conditions. There are two wet seasons: the 'long rains' between March and June and the 'short rains' in October and November. A feature of these rainy seasons is the appearance of so-called seasonal wetlands when water pans form in depressions and previously dry rivers begin flowing. These sites are very attractive to birds and can provide spontaneous birding opportunities within normally arid terrain. The aftermath of the rainy season also brings many of the drab weavers into breeding condition and provides the opportunity to observe their elaborate courtship displays.

So, even if your visit does not permit you to venture far from Nairobi, there is a wealth of birdlife to be enjoyed in close proximity to the city. For those who are planning a more extensive trip to Kenya, a couple of days birding around Nairobi provides an excellent introduction to the rich birdlife of East Africa and a chance to get familiar with a variety

An Ostrich in Nairobi National park gets a great view of Nairobi's skyline

of tropical bird families without being overwhelmed. There will even be species in Nairobi which may be hard to find elsewhere on the usual safari circuit.

GETTING AROUND

Although rather large in terms of population (four million), the city centre area is quite small. It straddles the main highway (Uhuru Highway) which bisects the city and is the main route connecting the coast to the interior of the country. Consequently, you are never far from Uhuru Highway, and it provides a ready means of orientating yourself within the city and for travelling further afield.

Unless you are only planning to go birding for a day, or less, it is best to rent a car. As car rental, with or without a driver, is the main method for getting around Kenya for tourists, it is well catered for by numerous rental companies in Nairobi and at the airport. Taxis are relatively expensive, and the cost of a couple of taxi fares soon catches up with daily car rental rates. A car also enables you to cover the distances required within the national parks and, of course, acts as an excellent mobile hide with which to approach flocks of birds. Although people may try to persuade you otherwise, a conventional two-wheel drive car is perfectly adequate for all the sites mentioned here, even in the rainy season. Alternatively, the best way of getting around is to use the informal minibus taxis, or matatus, which are a popular means of public transport. There are some concerns about the safety of travelling by matatu, particularly at night when most accidents and hold-ups occur, but provided you travel during daylight and confine your route to the main roads, there should be no problems.

Most of the matatus operate from the central railway station and serve all of the outlying areas of Nairobi, as well as the provinces.

Maps of Kenya show the main roads but lack the detail necessary to navigate the minor road system so, if you are planning to explore any areas not mentioned in this guide, you should always double-check your directions with a knowledgeable local person.

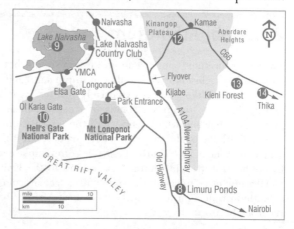

There are many safari companies in Nairobi which will offer one-day tours to the main birding spots such as Lake Naivasha or Nairobi National Park accompanied by skilled bird guides. This is probably the best option if you prefer to let someone else do the driving, or your time is limited and you don't mind spending extra. A visit to Thika, however, can easily be done on your own and can be combined with various other sites.

Many of the best sites are public or private properties of some kind and charge an entrance fee. It is a feature of access charges everywhere in Kenya that tourists pay much more than local residents. Rather like tipping in America, after a while this can seem like an irritating racket, but bear in mind that this money funds the enormous conservation effort that Kenya puts into preserving its fantastic natural heritage for all to enjoy.

Locations

1 National Museum (3km)

As well as having an extremely interesting section dealing with Kenyan ornithology, the National Museum is surrounded by splendid gardens and woodland which are teeming with birds. There is a small entrance fee which means fewer people are wandering about, making the birding a much more pleasant experience. On Wednesday mornings at 08:45 the Nature Kenya group meets and arranges bird walks through the gardens and also to other areas near the city. There are a number of other large city parks in Nairobi, notably Uhuru Park and Central Park, but they have much less woodland cover and support just a few common species, as well as being very crowded.

Take the **Uhuru Highway (A104)** northwest from the city centre and at the second large roundabout after the **Kenyatta Avenue** intersection, turn right (north) onto **Museum Hill Road**. The museum entrance is the first turning on the right.

KEY SPECIES

Resident
Hadada Ibis, Red-eyed Dove, Cinnamon-chested Bee-eater, White-headed Barbet, Lesser Honeyguide, Brown-backed Woodpecker, Cape Robin-Chat, Olive Thrush, Singing Cisticola, Green-backed Camaroptera, White-eyed Slaty Flycatcher, Montane White-eye, Tropical Boubou, Black-backed Puffback, African Black-headed Oriole, Violet-backed Starling, Amethyst Sunbird, Variable Sunbird, Bronze Sunbird, Parrot-billed Sparrow, Baglafecht Weaver, Holub's Golden Weaver, Red-billed Firefinch, Common Waxbill, Bronze Mannikin, Streaky Seedeater

2 Nairobi Arboretum (3km)

Nairobi Arboretum is not just an interesting site for botanists and plant enthusiasts: its undisturbed seclusion makes it also an important habitat for many birds in the urban environment. Although it is extensively planted with exotic trees from all over the world, there are significant stands of trees which are indigenous to the Aberdare mountain range and which support all the typical common species. The arboretum is popular with the public and with school groups, but it is well managed for wildlife and subject to none of the disturbance that is commonplace in the city parks. It is very close to the hotel district in central Nairobi and to the University of Nairobi, so this is a worthwhile place to visit for anyone with a free morning or afternoon and looking for a gentle introduction to the avifauna of East Africa. Over 180 bird species have been recorded here.

Go one block north on **Uhuru Highway** from the intersection with **Kenyatta Avenue**, and at the first roundabout turn west onto **State House Road**. The entrance to the arboretum is adjacent to the State House.

KEY SPECIES

Resident
Hadada Ibis, African Goshawk, Great Sparrowhawk, Red-eyed Dove, Dusky Turtle Dove, Cinnamon-chested Bee-eater, Silvery-cheeked Hornbill, Yellow-rumped Tinkerbird, White-headed Barbet, Eastern Honeybird, Cardinal Woodpecker, Brown-backed Woodpecker, Lesser Striped Swallow, Black Saw-wing, Yellow-whiskered

Greenbul, Cabanis's Greenbul, Rüppell's Robin-Chat, Olive Thrush, White-eyed Slaty Flycatcher, Singing Cisticola, Tawny-flanked Prinia, Green-backed Camaroptera, Yellow-breasted Apalis, African Paradise Flycatcher, Chinspot Batis, Montane White-eye, White-bellied Tit, Tropical Boubou, Black-backed Puffback, Black Cuckooshrike, African Black-headed

Oriole, Violet-backed Starling, Collared Sunbird, Amethyst Sunbird, Variable Sunbird, Northern Double-collared Sunbird, Bronze Sunbird, Parrot-billed Sparrow, Grosbeak Weaver, Baglafecht Weaver, Holub's Golden Weaver, Red-billed Firefinch, Bronze Mannikin, Streaky Seedeater

3 Nairobi National Park (7km)

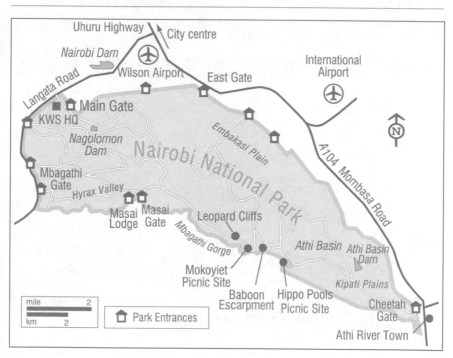

This is the single best site to visit if you have at least half a day to spare. At just over 100km² in size, Nairobi National Park is small by African standards but its value lies in its proximity to the city and its strategic location on the paths of the herds of zebra and wildebeest which undertake their migration during July and August. The park consists of a range of habitats which are representative of the East African savanna and over 516 species of bird have been recorded here, as well as many of the larger mammals such as Black Rhinoceroses, Lions, Cheetahs, Buffaloes and Giraffes. The terrain varies from open grassland with few trees to denser thickets of acacia bush, while the banks of the

Athi River, a permanent watercourse, support more luxuriant forest cover. Additionally, a number of man-made dams and permanent waterholes are attractive to mammals and birds, particularly during the dry season.

Like all national parks, access is only permitted by car. Take the **Uhuru Highway (A104/109)** southeast towards **Wilson Airport** and turn right (southwest) on **Langata Road**. The main entrance is on the left-hand side of the road.

You may only drive through the park on designated roads or tracks, so you will need to observe from your vehicle. This does have some advantages in that you can cover a much more extensive area and in the open grassland your car can work as a hide, allowing closer approach to birds and mammals. The fairly extensive network of tarmac and good quality gravel roads makes driving easy and safe. A worthwhile circular route is to start at the Nagolomon Dam, the largest permanent waterbody in the park, and drive along the southern flank of the park. The road runs parallel to the Mbagathi River through good stretches of riparian forest as far as the Hippo Pools Picnic Site. There is a Nature Trail at the picnic site and you may park here and walk the trail. This is excellent for woodland species and it is worth checking the river for African Finfoots which are occasionally seen here. From the Hippo Pools you can then loop back, first driving east and then north through the Kipati Plains and the Athi Basin for savanna species. If you are heading for Nairobi Airport, you may exit via the East Gate or else return to the Main Gate through the more densely wooded highlands.

KEY SPECIES

Resident
Ostrich, Striated Heron, Black-headed Heron, Hamerkop, Marabou Stork, White-faced Whistling Duck, African Black Duck, Yellow-billed Duck, Red-billed Duck, Secretarybird, Black-shouldered Kite, White-backed Vulture, Lappet-faced Vulture, Bateleur, Augur Buzzard, Tawny Eagle, Shelley's Francolin, Helmeted Guineafowl, Grey Crowned Crane, African Finfoot, White-bellied Bustard, Hartlaub's Bustard, Blacksmith Plover, Three-banded Plover, Yellow-throated Sandgrouse, Emerald-spotted Wood Dove, Red-eyed Dove, Ring-necked Dove, White-browed Coucal, Montane Nightjar, African Palm Swift, Striped Kingfisher, Malachite Kingfisher, Cinnamon-chested Bee-eater, White-headed Barbet, Grey Woodpecker, Rufous-naped Lark, Red-capped Lark, Fischer's Sparrow-Lark, Yellow-throated Longclaw, Banded Martin, Lesser Striped Swallow, Northern Pied Babbler, White-browed Scrub Robin, Southern Black Flycatcher, African Yellow Warbler, Winding Cisticola, Stout Cisticola, Rattling Cisticola, Siffling Cisticola, Pectoral-patch Cisticola, Tawny-flanked Prinia, Yellow-breasted Apalis, Red-faced Crombec, Buff-bellied Warbler, White-bellied Tit, Red-throated Tit, African Paradise Flycatcher, Chinspot Batis, Tropical Boubou, Black-backed Puffback, Long-tailed Fiscal, Common Fiscal, Sulphur-breasted Bush Shrike, Fork-tailed Drongo, Rüppell's Glossy Starling, Superb Starling, Red-billed Oxpecker, Yellow-billed Oxpecker, Scarlet-chested Sunbird, Variable Sunbird, Rufous Sparrow, Vitelline Masked Weaver, Holub's Golden Weaver, Red-billed Quelea, Yellow Bishop, White-winged Widowbird, Red-collared Widowbird, Red-billed Firefinch, Common Waxbill, Red-cheeked Cordon-bleu, Yellow-rumped Seedeater

4 Langata Giraffe Sanctuary (7km)

Rothschild's Giraffe is an endangered race which is indigenous to East Africa. Attempts are being made at the sanctuary to breed this subspecies for release into the wild. However, there are also two nature trails in the reserve area which traverse some good remnant forest habitat and this is a reliable site for Hartlaub's Turaco, an East African endemic species. There is a rather high entrance charge, by Kenyan standards, but if you only have a short amount of time available, it is worth it given the sanctuary's convenience and proximity to the city. If the alternative for the afternoon is killing time at the airport or in the hotel bar, then this is a much worthier cause on which to spend your money.

The Giraffe Sanctuary is not far from the main entrance to Nairobi National Park. Take the Langata Road going west from the National Park entrance and after about 1km turn south on Langata South Road. After 2km turn left into Koitobos Road and continue straight on into Gogo Falls Road where the main road makes a sharp swing to the right. Follow this road across several speed-bumps and, after several turns, the entrance to the sanctuary is on the right.

KEY SPECIES

Resident
African Goshawk, Little Sparrowhawk, Great Sparrowhawk, African Green Pigeon, Emerald-spotted Wood Dove, Red-chested Cuckoo, Dideric Cuckoo, Hartlaub's Turaco, Cinnamon-chested Bee-eater, Yellow-rumped Tinkerbird, White-headed Barbet, Lesser Honeyguide, Eastern Honeybird, Grey Woodpecker, Brown-backed Woodpecker, Black Saw-wing, Northern Pied Babbler, Cape Robin-Chat, Green-backed Camaroptera, White-eyed Slaty Flycatcher, Red-faced Crombec, Montane White-eye, White-bellied Tit, Chinspot Batis, African Paradise Flycatcher, Sulphur-breasted Bush Shrike, Tropical Boubou, Black Cuckoo-shrike, African Black-headed Oriole, Collared Sunbird, Amethyst Sunbird, Northern Double-collared Sunbird, Scarlet-chested Sunbird, Parrot-billed Sparrow, Spectacled Weaver, Common Waxbill, Bronze Mannikin, Village Indigobird, Streaky Seedeater

5 Ngong Road Forest Sanctuary (6km)

The last remnant of the original evergreen forest that existed in Nairobi is just south-west of the city centre, surrounded by the expanding city. After many years of neglect and piecemeal destruction the area has now been protected, and plans are under way to manage the reserve for the benefit of nature conservation and public amenity. Already a reforestation programme is in place and nature trails have been created through the forest. The Ngong Road Forest Sanctuary Trust organises bird walks through the forest at 09:00 on the first and third Saturday of every month, and this would be an excellent introduction to the local birdlife for birders who may be alone and new to Africa, as well as a chance to meet local birders and exchange information.

From central Nairobi take the Ngong Road southwest to Ngong Racecourse. You can access the main trail through the forest from the Racecourse Car Park.

KEY SPECIES

Resident
African Goshawk, Little Sparrowhawk, African Crowned Eagle, African Green Pigeon, Tambourine Dove, Emerald-spotted Wood-Dove, Red-chested Cuckoo, Dideric Cuckoo, Hartlaub's Turaco, Narina Trogon, Cinnamon-chested Bee-eater, Yellow-rumped Tinkerbird, White-headed Barbet, Lesser Honeyguide, Grey Woodpecker, Brown-backed Woodpecker, Black Saw-wing, Yellow-whiskered Greenbul, Brown-chested Alethe, White-starred Robin, Cape Robin-Chat, Green-backed Camaroptera, White-eyed Slaty Flycatcher, Red-faced Crombec, Montane White-eye, White-bellied Tit, Chinspot Batis, African Paradise Flycatcher, Sulphur-breasted Bush Shrike, Tropical Boubou, Black Cuckoo-shrike, African Black-headed Oriole, Collared Sunbird, Scarlet-chested Sunbird, Parrot-billed Sparrow, Spectacled Weaver, Common Waxbill, Bronze Mannikin, Streaky Seedeater

Further Away

6 Ngong Hills (20km)

The Ngong Hills are southwest of Nairobi just on the edge of the escarpment facing into the Great Rift Valley. The hills are actually a ridge formed by four extinct volcanoes and the highest point, Lamwia, is at an elevation of almost 2,500m. As you

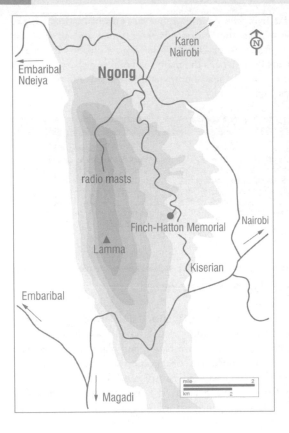

ascend the slopes the habitat is a mixture of cultivated plots, scattered woods and overgrown vegetation, and at the summit the terrain opens out into high altitude grassland with some scattered bush.

If driving, the best way to work the Ngong Hills is to drive to the town of **Ngong** and take the road west up to the **radio mast** and explore the general vicinity. You can then drive slowly back down the mountain stopping at likely spots. If you don't have a car and are prepared to hike, you can get a matatu to Ngong and walk up to the radio mast from there. You can then take a trail from the radio mast eastwards towards the **Nairobi-Magadi Road** from where you can get a matatu back into the city. Overall, this would involve a five-hour hike. You should be aware of personal safety in the Ngong Hills as this can be a bad area for muggings. You should only go there in groups and preferably with a local guide. It is not advisable to leave a car unattended here.

KEY SPECIES

Resident
African Goshawk, African Harrier-Hawk, Augur Buzzard, Namaqua Dove, Red-eyed Dove, Red-chested Cuckoo, African Emerald Cuckoo, Red-fronted Barbet, Cinnamon-chested Bee-eater, Little Bee-eater,

Silvery-cheeked Hornbill, Grassland Pipit, Rock Martin, Banded Martin, Northern Anteater Chat, African Stonechat, Rattling Cisticola, Pectoral-patch Cisticola, Montane White-eye, Black-backed Puffback, African Black-headed Oriole, White-necked Raven,

Cape Crow, Eastern Double-collared Sunbird, Bronze Sunbird, Parrot-billed Sparrow, Spectacled Weaver, Yellow Bishop, Common Waxbill, Bronze Mannikin, Yellow-crowned Canary, Streaky Seedeater

7 Magadi Road (10–130km)

Lake Magadi lies at the end of a 130km stretch of paved road, called Magadi Road. This lake, lying in baking lowlands at an altitude of 350m, often registers the highest temperatures in Kenya. It is the only location in the country where Chestnut-banded Plovers are resident, and they are in fact a common species here. The caustic nature of the lake restricts the number of species, but both species of flamingos and the soda-loving Cape Teal are a common sight. Avocets are often present in numbers and migrant waders come and go. However, it is not the lake itself that is the highlight of a trip to

Lake Magadi, but the countryside that you pass through to reach it. From Corner Baridi ('cold corner' in Swahili) at 2,100m, the journey down to the lake can provide the most varied and exciting dry-country birding that the continent has to offer.

To reach the Magadi Road, follow the directions for the main entrance of Nairobi National Park on Langata Road: go past this and after a couple of kilometres take the next paved road to the left. This is Magadi Road: stay on this road until it descends to the lake. The first settled area is Ongata Rongai, and the next is Kiserian. After this the road crosses highland grasslands and after a final long sweeping bend to the right, an impressive vista of the Great Rift Valley stretches to the horizon. To the right the land rises steeply to the summit of the Ngong Hills. The scrub on both sides of the road is home to the very local Lynes's Cisticola and a couple of pairs of Little Rock Thrushes, although the latter are secretive. The road commences its descent along the side of the Ngongs, shortly leaving the hillside in a sweep around to the left. In the acacia scrub there are Bush Pipits here seasonally. The small town of Kisamese soon appears and after this the road drops steeply into a narrow ravine. A couple of pairs of Schalow's Wheatears, a rift valley endemic, live along the rocky roadside. At the bottom of this hill the bird fauna will have changed very dramatically. In the mixture of acacia and commiphora scrub there will be a vast variety of dry-country forms, amongst the most interesting being Bare-eyed Thrushes, Tiny Cisticolas and Southern Grosbeak Canaries. After this there is another steep descent, the dense scrubby hillsides being home to such local species as Scaly Chatterers and Pringle's Puffbacks as well as more wide-spread species. For the next 30km any stops along the road could prove interesting. After the rains in June the countryside will have been transformed into a lush green park-like landscape, in contrast to the savage dry periods when not a leaf can be found: all four species of whydahs may be found at this time as well as several species of weavers including Vitelline Masked Weavers and Chestnut Weavers, the latter sometimes breeding in great numbers. There is a lot of vocal activity in the early mornings and birds are generally abundant. After the last descent, the seasonal roadside dams may attract tens of thousands of birds, particularly estrildids of 14 species, although streams of doves and weavers throng the surrounding bushes waiting for their turn. A special bird that breeds in this area is the Fire-fronted Bishop, which may be found associating with the thirsty flocks.

About 70km from the start of the Magadi Road is the prehistoric site of Olorgesaillie. There are bandas (rooms with bed, mosquito net, furniture, gas lamp and water, but no food) available here but these may already be occupied if you arrive on a weekend. On moonlit nights Donaldson-Smith's Nightjars may be vocal all night; Heuglin's Coursers call all around but can be difficult to locate and, if there is a lot of grassy cover, the Small Buttonquails may hoot from evening until early morning. The staff put out water for the birds and species such as Blue-capped Cordon-bleus, Crimson-rumped Waxbills and White-bellied Canaries can be numerous visitors, whilst Grey-headed Social Weavers nest in the surrounding trees. It is still another 60km to Lake Magadi, but with so much to see along the road there may be no time to reach the lake on a day trip. As it is necessary to return to Nairobi using the same route, it is advisable to allow plenty of time, as there will be many more avian surprises awaiting discovery.

KEY SPECIES

Resident	Abyssinian Scimitarbill, Black-	Little Rock Thrush, Bare-eyed Thrush,
Small Buttonquail, Heuglin's Courser,	throated Barbet, Red-and-yellow	Lynes's Cisticola, Tiny Cisticola,
Donaldson-Smith's Nightjar,	Barbet, d'Arnaud's Barbet, Bush Pipit,	Banded Parisoma, Schalow's

Wheatear, Scaly Chatterer, Bronze Sunbird, Brubru, Pringle's Puffback, Rufous Sparrow, Swahili Sparrow, Grey-headed Social Weaver, Vitelline Masked Weaver, Chestnut Weaver, Fire-fronted Bishop, Blue-capped Cordon-bleu, Crimson-rumped Waxbill, Straw-tailed Whydah, Pin-tailed Whydah, Eastern Paradise Whydah, Southern Grosbeak Canary, White-bellied Canary, Streaky Seedeater

8 Limuru Ponds (30km)

The Limuru Ponds (Manguo Ponds) are two shallow lagoons northwest of Nairobi just off the main road to Naivasha. These ponds are seasonal and can dry out in times of drought. Although there have been some concerns expressed about personal safety at this site, it is worth stopping en route to or from Naivasha as the ponds can be viewed easily from the roadside and support many aquatic species, including the Maccoa Duck, which is difficult to see elsewhere in the Nairobi area. This region is close to the highest point on the highway before it descends into the rift valley; the roadside culti-vated areas and grassland support a number of high altitude grassland species.

Take the main **Uhuru Highway (A104)** northwest from Nairobi to Naivasha. About 2km before reaching Limuru on the right-hand (east) side of the road, you can pull in and scan the ponds.

KEY SPECIES

Resident
Little Grebe, White-breasted Cormorant, Long-tailed Cormorant, Darter, Goliath Heron, Black-headed Heron, Great Egret, Yellow-billed Stork, African Spoonbill, Egyptian Goose, Spur-winged Goose, White-backed Duck, Fulvous Whistling Duck, Yellow-billed Duck, Red-billed Duck,

Hottentot Teal, Southern Pochard, Maccoa Duck, African Marsh Harrier, Augur Buzzard, Red-knobbed Coot, Common Moorhen, Purple Swamphen, African Jacana, Black-winged Stilt, African Snipe, Whiskered Tern, Red-capped Lark, African Stonechat

Nov–Mar
Northern Shoveler, Garganey, Common Ringed Plover, Little Ringed Plover, Ruff, Marsh Sandpiper, Common Sandpiper, Wood Sandpiper, Green Sandpiper, Yellow Wagtail, Grey Wagtail, House Martin

9 Lake Naivasha (80km)

Lake Naivasha is the closest of the Rift Valley lakes to Nairobi. Even though it is still a considerable distance from the city, this shallow freshwater lake is so good that it is worth making the extra effort to get to it. It is possible to see more species here in an afternoon than would be pos-sible in several months in most temperate countries. Apart from the wetland habitats, the open terrain of the surround-ing countryside and confiding nature of the birds combine to make the birding extremely productive. Virtually every-where you look you will see something new and interest-ing: this is easy birding!

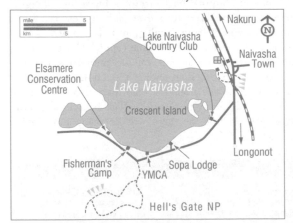

Due to the dense growth of papyrus, it is not possible to walk along the shore of the lake. However, there is a road running southwest from the town of Naivasha which is parallel to the southern shore of the lake although it does not actually run alongside it. There are several turn-offs from this road which run down to the lakeshore, providing access to viewing points and to open water. The riparian vegetation and open wooded areas support a rich community of birds.

Take the main **Uhuru Highway (A104)** northwest from Nairobi to Naivasha. Five kilometres before Naivasha, turn southwest onto the **South Lake Road** at the sign for the **Lake Naivasha Country Club**. The best starting point is from the Lake Naivasha Country Club which has a jetty running out into the lake opposite Crescent Island. This provides an opportunity to scan the lake and to check the surrounding woodland. The open plains north and east of the hotel are also good for birds, particularly larks, finches and plovers. Other access points to the lakeshore can be reached by continuing south on the South Lake Road, turning off for the lake at Sopa Lodge or the YMCA campsite. Entrance fees apply at these places. Fisherman's Camp, about 18km along South Lake Road, is a public area with good birding and local guides available for walks and boat trips. Elsamere Conservation Centre is 5km beyond Fisherman's Camp: it offers excellent cottage accommodation, good food and local guides. The extensive acacia woodland grounds provide superb birding. At the west end of Lake Naivasha, past Elsamere and the Kongoni Game Sanctuary where the tarmac road ends, is a small brackish lake, Lake Oloidien, which is separate from the main lake. There is public access to the shore here, and a host of waterbirds can be seen including many species of Palearctic waders. Check the nearby Leleshawa bush country for the endemic Grey-crested Helmet Shrike.

If you don't have a car, it would still be viable to take a train or matatu to Naivasha, and thence a taxi to Lake Naivasha Country Club. Even to spend a few hours in the vicinity of the club would be well worthwhile.

KEY SPECIES

Resident
Great White Pelican, Pink-backed Pelican, White-breasted Cormorant, Long-tailed Cormorant, Goliath Heron, Purple Heron, Black-headed Heron, Great Egret, Intermediate Egret, Squacco Heron, Yellow-billed Stork, Hadada Ibis, Greater Flamingo, Lesser Flamingo, African Spoonbill, Egyptian Goose, White-backed Duck, Yellow-billed Duck, Red-billed Duck, Hottentot Teal, African Marsh Harrier, African Fish Eagle, Augur Buzzard, Black Crake, Red-knobbed Coot, Common Moorhen, African Jacana, Grey Crowned Crane, Black-winged Stilt, Three-banded Plover, Kittlitz's Plover, Spur-winged Plover, Crowned Lapwing Grey-headed Gull, Whiskered Tern, Ring-necked Dove, Dideric Cuckoo, Fischer's Lovebird, Grey Hornbill, Pied Kingfisher, Malachite Kingfisher, Giant Kingfisher, White-fronted Bee-eater, Rufous-naped Lark, Banded Martin, Rock Martin, White-browed Robin-Chat, Northern Anteater Chat, Black-lored Babbler, White-eyed Slaty Flycatcher, Lesser Swamp Warbler, Green-backed Camaroptera, Rattling Cisticola, Black Cuckoo-shrike, Grey-backed Fiscal, Brubru, Grey-crested Helmet Shrike (rare), Red-winged Starling, Superb Starling, Wattled Starling, Bronze Sunbird, Scarlet-chested Sunbird, Lesser Masked Weaver, Red-headed Weaver, Common Waxbill, Red-billed Firefinch, Purple Grenadier, Brimstone Canary, Yellow-rumped Seedeater

Nov–Mar
Western Marsh Harrier, Lesser Kestrel, Common Ringed Plover, Little Ringed Plover, Little Stint, Ruff, Marsh Sandpiper, Common Sandpiper, Wood Sandpiper, Green Sandpiper, Yellow Wagtail, House Martin, Great Reed Warbler

10 Hell's Gate National Park (90km)

On the eastern flank of the Great Rift Valley, where the eastern highlands fall precipitously to the valley floor, there are a series of geological formations and considerable geothermal activity located within Hell's Gate National Park, which is also an excellent birding spot. The main feature of this park is the Njorowa Gorge which has been cut through the escarpment, and the consequent cliffs are both dramatic and also important nesting sites for many raptors and other cliff-nesting species. Unlike many national parks, you may hike through the park on foot. In combination with Lake Naivasha a visit to Hell's Gate makes a great weekend excursion.

The best way to get to Hell's Gate is via Naivasha. Take the main **Uhuru Highway (A104)** northwest from Nairobi to Naivasha. Five kilometres before Naivasha, turn southwest onto the **South Lake Road** at the sign for the **Lake Naivasha Country Club** and stay on the South Lake Road for about 10km to get to Hell's Gate National Park

KEY SPECIES

Resident
White-backed Vulture, Rüppell's Vulture, Verreaux's Eagle, Tawny Eagle, Lanner Falcon, Peregrine Falcon, Temminck's Courser, Speckled Pigeon, Namaqua Dove, White-browed Coucal, Red-fronted Barbet, White-fronted Bee-eater, Little Bee-eater, Mottled Swift, Nyanza Swift, Grassland Pipit, Rock Martin, Banded Martin, Northern Anteater Chat, Schalow's Wheatear, Winding Cisticola, Rattling Cisticola, Lynes's Cisticola, Pectoral-patch Cisticola, Black-backed Puffback, White-necked Raven, Yellow Bishop, Zebra Waxbill, Common Waxbill, Golden-breasted Bunting, Cinnamon-breasted Bunting

11 Mt Longonot National Park (80km)

Mt Longonot is a (supposedly) dormant volcano, over 2,700m in height, which overlooks Lake Naivasha and the Great Rift Valley. It lies about 90km northwest of Nairobi within a national park. For the fit and energetic, the hike up to the rim of the crater affords fantastic views of the surrounding countryside, and for a birder the climb also enables you to ascend through several vegetation zones. The tropical forest at the foot of the mountain thins out as you ascend to shrub vegetation and thickets of stunted acacia.

Take the main **Uhuru Highway (A104)** northwest from Nairobi to **Naivasha** but branch off onto the old Naivasha highway after Limuru. About 15km after your descent into the Rift Valley turn west at Longonot: just before the railway crossing. The turnoff for the park entrance is about 2km on the left (south).

KEY SPECIES

Resident
Secretarybird, White-backed Vulture, Tawny Eagle, African Harrier-Hawk, Augur Buzzard, White-bellied Bustard, Speckled Pigeon, Namaqua Dove, White-browed Coucal, Red-fronted Barbet, White-fronted Bee-eater, White-throated Bee-eater, Little Bee-eater, Nyanza Swift, Grassland Pipit, Rock Martin, Banded Martin, Northern Anteater Chat, Schalow's Wheatear, Arrow-marked Babbler, Rattling Cisticola, Pectoral-patch Cisticola, Zitting Cisticola, Black-backed Puffback, White-necked Raven, Parrot-billed Sparrow, Common Waxbill, Cinnamon-breasted Bunting

12 Kinangop Plateau (c. 70km)

The Kinangop Plateau lies at the southwest edge of the Aberdare Mountains and is an area of high altitude grassland surrounded by montane forest. This habitat is important for a number of specialised bird species which are endangered or occur very locally in East Africa, but the grassland areas are under considerable threat from agricultural intensification. The overall number of bird species in this habitat is quite low, and visibility can often be poor as a result of the cold, misty weather. However, the site is not far from the main road to Naivasha, and you may pass through here if you are visiting the Kieni Forest. In addition, the area is also good for wintering and passage Palearctic raptors.

Take the main **Uhuru Highway (A104)** northwest from Nairobi to **Naivasha** and just north of **Kijabe** turn east on the flyover onto the **C66** towards **Thika**. This road cuts through the southern flank of the Kinangop Plateau. Any suitable grassland area found while travelling east from the flyover for the next 15km should be checked for pipits and larks, especially Sharpe's Longclaw, which is endemic to the central highlands of Kenya.

KEY SPECIES

Resident		
Mountain Buzzard, Augur Buzzard, Grey Crowned Crane, Harlequin Quail, Black-winged Lapwing, African Snipe, Rufous-naped Lark, Red-capped Lark, Plain Martin, Lesser Striped Swallow, Sharpe's Longclaw, Grassland Pipit,	Long-billed Pipit, African Stonechat, Northern Anteater Chat, Capped Wheatear, Aberdare Cisticola, Wing-snapping Cisticola, Hunter's Cisticola, Tinkling Cisticola, Greater Blue-eared Glossy Starling, Amethyst Sunbird, Eastern Double-collared Sunbird,	Long-tailed Widowbird, Jackson's Widowbird, Brimstone Canary **Nov–Mar** Black Stork, Pallid Harrier, Montagu's Harrier, Eurasian Hobby, Red-throated Pipit

13 Kieni Forest (c. 85km via Naivasha road; c. 60km via Thika)

The Aberdare Mountains to the north of Nairobi form the eastern flank of the Great Rift Valley, and the humid climate at the higher elevations promotes the growth of dense montane forest. Despite the impact of deforestation and other pressures, this area is rich in wildlife and strong efforts are being made to preserve and manage the forests.

In the southern foothills of the Aberdare Mountains, the Kieni Forest forms the southern edge of the more extensive Kikuyu Escarpment Forest and is accessible to birders because the Thika–Kamae Road (C66) runs right through it. This forest supports a rich avifauna, including a number of much sought-after montane forest species such as Olive Ibises and Abbott's Starlings, but the birding can be slow and early mornings are recommended.

Take the main **Uhuru Highway (A104)** northwest from Nairobi to **Naivasha** and just north of **Kijabe** turn east on the flyover onto the **C66**. This road passes through the grasslands of the **Kinangop Plateau** (see **12**) and, after about 20km, passes through some prime forest habitat. A number of trails lead off the road into the interior of the forest, and the best way to explore the area is by stopping and checking at suitable trailheads along this road. You do not need to penetrate far into the forest, and it is inadvisable to leave your car unattended. Security can be a problem here at times and it is best to visit in a group, with a local guide. Continuing on this road will take you into Thika (see overleaf).

KEY SPECIES

Resident		
Olive Ibis, Mountain Buzzard, Ayres's Hawk-Eagle, Rufous-chested Sparrowhawk, African Green Pigeon, Emerald-spotted Wood Dove, Bar-tailed Trogon, Silvery-cheeked Hornbill, Hartlaub's Turaco, Cinnamon-chested Bee-eater, White-	headed Wood Hoopoe, Moustached Tinkerbird, Fine-banded Woodpecker, Mountain Wagtail, Black Saw-wing, Abyssinian Ground Thrush, White-starred Robin, Black-collared Apalis, White-eyed Slaty Flycatcher, White-browed Crombec, Montane White-eye, Black-fronted Bush Shrike, Grey	Cuckoo-shrike, Waller's Starling, Abbott's Starling, African Black-headed Oriole, Montane Oriole, Bronze Sunbird, Golden-winged Sunbird, Streaky Seedeater, Thick-billed Seedeater

14 Thika (40km)

The town of Thika, immortalised in Elspeth Huxley's classic memoir *The Flame Trees of Thika*, is a 45-minute drive from Nairobi. It offers some great birding including several species that are hard to find elsewhere in the central highlands. Here you should visit the Blue Posts Hotel on the outskirts of the town, the grounds of which provide excellent, safe birding. You can wander in the delightful garden as well as dine or have a

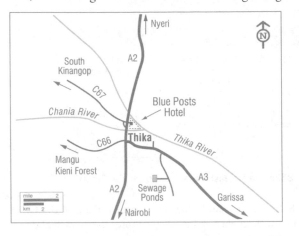

drink on the verandah of the hotel. The grounds are situated at the confluence of the Thika and Chania rivers, and several small but picturesque waterfalls are popular with visitors. A small trail does a circuit alongside the two rivers, and you should look for key species such as Brown-hooded Kingfishers and Grey-olive Greenbuls here. Check the quieter stretches of the river for African Black Ducks and occasional African Finfoots.

Southeast of the town, Thika's Settling Ponds are an excellent site for birds, especially wildfowl and waders but also open country species around the perimeter: they are not easy to find without local knowledge and, theoretically, you are required to seek permission to visit in advance from Thika Municipal Council.

Take the **A2 Highway** northeast from Nairobi to **Thika**. On reaching Thika, a slip road off the dual carriageway takes you directly to the **Blue Posts Hotel**, which is on the right-hand side of the road. Park inside the hotel compound, which is guarded. The circular river trail descends either side of the garden down a steep path but is well worth the effort. To get to the settling ponds, leave Thika eastwards on the A3, through the industrial area. Towards the end of the built-up area, turn right on a dirt track heading south, signposted to the sewage treatment works. After about 3km another track on the right leads directly to the ponds.

From Thika, you can go on to **Kieni Forest** or the **Kinangop Plateau** (both on the C66 – see **12** and **13**): these sites are slightly closer to Nairobi from this direction, but are not quite so easy to find.

KEY SPECIES

Resident
Hadada Ibis, African Black Duck, Yellow-billed Duck, Red-billed Teal, Hottentot Teal, Southern Pochard, Black Kite, African Goshawk, African Finfoot, Spur-winged Plover, Three-banded Plover, African Green Pigeon, Tambourine Dove, Red-eyed Dove, Purple-crested Turaco, Little Swift, Speckled Mousebird, Grey-headed Kingfisher, Brown-hooded Kingfisher, Giant Kingfisher, Cinnamon-chested Bee-eater, Trumpeter Hornbill, Red-fronted Barbet, Spot-flanked Barbet, White-headed Barbet, Lesser Honeyguide, Eastern Honeybird, African Pied Wagtail, Mountain Wagtail, Plain Martin, Wire-tailed Swallow, Lesser Striped Swallow, Black Saw-wing, Zanzibar Sombre Greenbul, Grey-olive Greenbul, Brown-backed Scrub Robin, White-eyed Slaty Flycatcher, African Yellow Warbler, Winding Cisticola, Tawny-flanked Prinia, Grey-backed Camaroptera, Grey-capped Warbler, Abyssinian White-eye, African Penduline Tit, African Paradise Flycatcher, Black-backed Puffback, Rüppell's Long-tailed Starling, Violet-backed Starling, Collared Sunbird, Scarlet-chested Sunbird, Variable Sunbird, Bronze Sunbird, Rufous

Sparrow, White-browed Sparrow-Weaver, Baglafecht Weaver, Spectacled Weaver, African Firefinch, Red-cheeked Cordon-bleu, Bronze

Mannikin, Black-and-white Mannikin, African Citril

Nov–Mar
Northern Shoveler, Garganey, Little

Ringed Plover, Ruff, Marsh Sandpiper, Common Sandpiper, Wood Sandpiper, Green Sandpiper, Yellow Wagtail

Useful contacts and websites

Nature Kenya
www.naturekenya.org
Nature Kenya, P.O.Box 44486, GPO 00100, Nairobi
Tel: +254 20 3749957.
BirdLife Partner in Kenya and publishers of *Kenya Birds*.

African Bird Club
www.africanbirdclub.org
African Bird Club, c/o BirdLife International, Wellbrook Court, Girton Road, Cambridge CB3 0NA, UK.

Langata Giraffe Sanctuary
www.giraffecentre.org
Information about the giraffe sanctuary including nature reserve and trails.

Kenya Wildlife Service
www.kws.org
Informative site with details about all national parks in Kenya.

Ngong Forest Sanctuary
www.ngongforest.org
Information about reserve including site guide, checklist and details of birdwalks.

Kenya Birds
www.kenyabirds.org.uk
Excellent birding website with checklist, site guide and general information.

Birdlife in Kenya
www.safariweb.com
Informative website by Kenya Wildlife Service and Safari Tour Operators.

Kenya Association of Tour Operators
www.katokenya.org

National Museums of Kenya
www.museums.or.ke/gbirds.html

Kenyabirdsnet – email group
groups.yahoo.com/group/kenyabirdsnet/

Kenya Meteorological Department
www.meteo.go.ke

East African Rarities Committee
www.naturekenya.org/Birdcommittee.htm
c/o Department of Ornithology,
NMK, P.O. Box 40658, Nairobi, Kenya
e-mail: kbirds@africaonline.co.ke

ABC Recorder for Kenya: Colin Jackson,
A Rocha Kenya, Box 383, Watamu, 80202, Kenya.

Books and publications

Important Bird Areas of Kenya
Leon Bennun & Peter Njoroge (1999) Nature Kenya
ISBN 9966992111

Annotated Checklist of the Birds of Nairobi
Bill Harvey (1997) Nature Kenya

A Bird Atlas of Kenya
Adrian Lewis & Derek Pomeroy (1988)
A.A. Balkema, ISBN 9061917166

Where to Watch Birds in Africa
Nigel Wheatley (1995) Christopher Helm
ISBN 0713640138

Field Guide to the Birds of East Africa
Terry Stevenson & John Fanshawe (2005)
Christopher Helm, ISBN 0713673478

Birds of Kenya & Northern Tanzania
Dale A. Zimmerman, Don A. Turner & David J. Pearson (1996) Christopher Helm
ISBN 0713639687

Field Guide to the Birds of Kenya and Northern Tanzania
D. A. Zimmerman, D. A. Turner & D. J. Pearson (2005) Christopher Helm, ISBN 0713675500

Illustrated Checklist: Birds of Eastern Africa
Ber van Perlo (1995) HarperCollins
ISBN 0002199378

Grey Hypocolius is a much sought-after bird in Bahrain in winter

Resident
Socotra Cormorant, Black-crowned Night Heron, Western Reef Egret, Chukar (I), Grey Francolin (I), Common Moorhen, Black-winged Stilt, Laughing Dove, Eurasian Collared Dove, Namaqua Dove, Crested Lark, Graceful Prinia, Common Myna (I), White-cheeked Bulbul (I), Red-vented Bulbul (I), House Crow, Red Avadavat (I), House Sparrow

Winter
Great Cormorant, Little Egret, Cattle Egret, Grey Heron, Mallard, Eurasian Teal, Black Kite, Common Kestrel, Common Coot, Eurasian Oystercatcher, Black-headed Gull, Yellow-legged Gull, Steppe Gull, Pallid Swift, Eurasian Skylark, White Wagtail, Common Stonechat, Song Thrush, Lesser Whitethroat, Blackcap, Common Chiffchaff, Eurasian Starling

Passage
Eurasian Turtle Dove, European Bee-eater, Eurasian Hoopoe, Common Swift, Barn Swallow, House Martin, Sand Martin, Common Whitethroat, Willow Warbler, Spotted Flycatcher

Bahrain

Although a country in its own right, not a city, the main island of Bahrain is less than 50km long and nowhere is more than an hour's drive from Manama, the capital, so the whole island is worth including in the city guide. Most of the birding sites are in the northern half of Bahrain Island; the southern half is less likely to be visited by foreign tourists, although access is no longer restricted. Much of Bahrain is barren and rocky desert, with no rivers or permanent waterways. Water is supplied by underground aquifers and the resultant irrigation has permitted some agriculture and, of course, the growth of urbanisation. As with any desert environment, the irrigated plots on the island are highly attractive to migrants, which tend to concentrate in these areas. There are a number of very important mudflats along the northern and eastern coasts which hold important concentrations of migrating and wintering waders although, sad to say, most of the original mangroves and coastal saltmarshes have been reclaimed for development.

A small number of species, quite a high proportion of which are introduced, are resident on the main island. The offshore islands are breeding grounds for large numbers of seabirds, particularly terns and the highly localised Socotra Cormorant.

The greatest species diversity is experienced during migration periods. Located as it is in the Persian Gulf, Bahrain is directly in the path of passage migrants from the Eastern Palearctic which winter in Africa and the Arabian Peninsula. At this time, many vegetated areas such as parks, gardens and bushes alongside irrigation ditches provide cover for a variety of migrant passerines, while the coastal mudflats are thronged with over 30 wader species, several of which are extremely rare in the Western Palearctic.

GETTING AROUND

There are, basically, three options for getting around Bahrain: rented car, taxi or the limited bus service. The general areas of Manama and Maharraq Island are quite well served by bus, and a combination of bus and taxi will probably be sufficient if you wish to explore these areas. In general, a taxi is probably the best option for any point-to-point travel, even if you wish to visit the sites on the east coast or in the interior of the island. After birding, provided you can make your way back to a main road, you will easily hail a passing taxi for your return journey.

If you want to visit several sites in the course of a day, then renting a car is the best option, particularly if you are carrying a telescope, camera, tripod and bottles of water. At some sites, you can simply roll

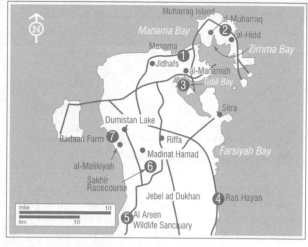

down the car window and, using the car as a mobile hide, approach the birds that way. If you are driving on private property, such as Badaan Farm, always respect restricted areas and don't block any of the main access tracks.

If you have a telescope, you will make good use of it as much of the birding activity will involve scanning flocks of gulls or waders. However, you need to balance the advantage of a telescope against the effort involved in carrying around all of your birding gear in the field. For most of the year, the heat and humidity will sap the energy of even the most committed birder and, if you don't have a car, you may be left cursing the weight of your combined telescope and tripod as you stagger the last kilometre back to your bus-stop or taxi. Luckily, in Bahrain, you are never too far from a shop or café where you can rest and get a cold drink.

Locations

1 Manama and environs

Many of the parks and gardens in the city are good sites for birds. The growth in vegetation provided by constant irrigation provides shelter and feeding opportunities for many migrants and for a small number of resident species. The strip of hotels along the north shore provides several good spots, particularly the Sheraton Hotel, and you can check offshore for gulls and terns. Going further west along the Corniche, will bring you to Bahrain Fort, approximately 4km west of Manama. The shoreline in front of the fort has mudflats and some adjacent scrubby areas which attract migrants and wintering passerines. Several pairs of Barn Owls breed in the holes in the walls of the fort.

Also worth mentioning is the well-known Grey Hypocolius roost in Maqabah, about 10km west of Manama. Take the **Budaiya Road** west from Manama for about 10km to the suburbs of **Saar**. Turn south off Budaiya Road onto **Saar Road** and take the third turn left (east) to get to an area of scattered date palms and thornbushes: one of the few uncultivated plots of land in this area. The birds begin arriving from early evening onwards during the winter months.

KEY SPECIES

Resident
Little Bittern, Kentish Plover, Barn Owl, Desert Lark, Hoopoe Lark, Southern Grey Shrike

Winter
Grey Plover, Lesser Sand Plover, Common Ringed Plover, Little Stint,

Dunlin, Sanderling, Curlew Sandpiper, Common Redshank, Common Greenshank, Eurasian Curlew, Whimbrel, Slender-billed Gull, Lesser Crested Tern, Sandwich Tern, Richard's Pipit, Red-throated Pipit, Water Pipit, Grey Hypocolius, Isabelline Wheatear, Desert Wheatear,

Desert Warbler, Isabelline Shrike

Passage
Tawny Pipit, Yellow Wagtail, Common Redstart, Whinchat, Northern Wheatear, Eastern Olivaceous Warbler, Orphean Warbler, Ménétries's Warbler, Ortolan Bunting

2 Muharraq Island (8km)

The airport is on Al Muharraq Island which lies just northeast of Manama and is accessed by two bridges from the city centre. Although the island is dominated by the airport and its associated buildings, there are several really good sites there within a few minutes drive of the terminal. The airport lagoon, which is favoured by ducks and herons, is just on the north side of the main access road to the airport, and the adjacent vegetation should be checked for passerines. Arad Bay, on the south side of the airport access road, has exposed mudflats at low tide: it can be scanned from the road which runs along the southeast shore of the bay, and from the causeway between Muharraq and Halat an Na'im which lies just offshore. Also at the southeast corner of Arad Bay are the Arad Fort Gardens which are a migrant trap particularly in spring.

North of the airport perimeter lies an area of fodder farms which have good habitat for migrant and wintering passerines. You can then continue south on the Asry Causeway Road for about 8km, stopping at likely spots to scan the mudflats either side of the causeway, and at Asry you can check the lagoon, at the southern end of the road.

KEY SPECIES

Resident
Little Bittern, Osprey, Kentish Plover, Desert Lark, Hoopoe Lark, Southern Grey Shrike

Summer
Saunders's Tern, White-cheeked Tern, White-winged Black Tern, Rufous-tailed Scrub Robin

Winter
Gadwall, Garganey, Northern Pintail, Grey Plover, Pacific Golden Plover, Lesser Sand Plover, Common Ringed Plover, Temminck's Stint, Little Stint, Dunlin, Sanderling, Broad-billed Sandpiper, Curlew Sandpiper, Common Redshank, Spotted Redshank, Common Greenshank, Eurasian Curlew, Whimbrel, Slender-

billed Gull, Lesser Crested Tern, Sandwich Tern, Caspian Tern, Isabelline Wheatear, Desert Wheatear, Desert Warbler, Isabelline Shrike

Passage
Greater Sand Plover, Terek Sandpiper, Bar-tailed Godwit, Ruddy Turnstone, Gull-billed Tern, Common Tern, Whiskered Tern

3 Tubli Bay (7km)

It is one of the more gratifying aspects of urban birdlife to observe the tenacity with which waders will continue to frequent a mudflat no matter how degraded the shoreline becomes through industrial and commercial development: provided sufficient area of mudflat is retained and some undisturbed islands or man-made structures exist as high-tide roosts. Tubli Bay is on the south side of Manama and although much of the area has been reclaimed and built over, there are a number of undisturbed corners of the bay which support a superb variety of some of the Palearctic's most sought after migrant wader species. A bridge linking Sitra Island with Bahrain Island bisects the bay and provides access to Nabi Saleh Island. The inner bay contains a number of viewpoints where you can watch roosting waders, while the outer bay is good for observing feeding gulls, terns and cormorants.

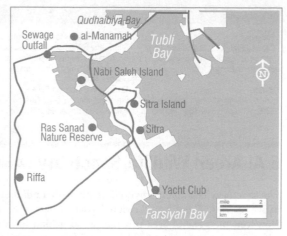

The sewage outfall in the northwest corner of the inner bay is one of the best spots to watch waders, egrets and gulls at close quarters. The Ras Sanad Nature Reserve, at the southwest corner of the inner bay, contains a remnant of the once extensive mangrove swamps, as well as saltmarsh which acts as a wader roost. Try to coincide with an incoming tide and wait for the birds to assemble on the saltmarsh. Other good vantage points are the mudflats between Nabi Saleh Island and the Sitra Bridge and the southwest corner of the outer bay. Continue south to the Yacht Club at the southern tip of Sitra Island, where you can see the resident flamingo flock.

KEY SPECIES

Resident
Little Bittern, Greater Flamingo, Kentish Plover

Summer
Collared Pratincole, Lesser Crested Tern, Saunders's Tern, White-cheeked Tern, White-winged Black Tern

Winter
Glossy Ibis, Eurasian Spoonbill, Grey Plover, Pacific Golden Plover, Lesser Sand Plover, Common Ringed Plover, Temminck's Stint, Little Stint, Dunlin, Sanderling, Broad-billed Sandpiper, Curlew Sandpiper, Ruff, Marsh Sandpiper, Common Redshank, Spotted Redshank, Common Greenshank, Common Snipe, Eurasian Curlew, Whimbrel, Pallas's Gull, Slender-billed Gull, Sandwich Tern, Caspian Tern, Common Kingfisher

Passage
Greater Sand Plover, Terek Sandpiper, Bar-tailed Godwit, Ruddy Turnstone, Red-necked Phalarope, Gull-billed Tern, Common Tern, Whiskered Tern

4 Ras Hayan (22km)

This is a promontory just south of the village of Askar on Bahrain's east coast. There is a fish farm at Ras Hayan, and several points along this stretch of coast are good for observing roosting gulls, terns and waders. The winter gull roost, in particular, contains a good number of 'Herring' Gulls, affording a good opportunity to study identification and perhaps pick out one of the less familiar closely related species or races. You may enter the fish farm at Ras Hayan by seeking permission from the guard at the entrance and then cross the main channel to the promontory. The stretch of coastline on the south shore of the fish farm is most productive.

Take the east coast road south from Manama past Askar as far as **Ras Hayan**. Then, returning north, you can visit the coast road north as far as the village of Askar. Just north of Askar is the headland at Ras Abu Jarjur, a good vantage point for seawatching, and the bay north of Ras Abu Jarjur holds good numbers of gulls, terns and waders in autumn and winter.

KEY SPECIES

Summer
Collared Pratincole, Saunders's Tern, White-cheeked Tern, White-winged Black Tern, Bridled Tern

Winter
Greater Flamingo, Grey Plover, Pacific Golden Plover, Lesser Sand Plover,

Common Ringed Plover, Temminck's Stint, Little Stint, Dunlin, Sanderling, Broad-billed Sandpiper, Curlew Sandpiper, Common Redshank, Spotted Redshank, Common Greenshank, Eurasian Curlew, Whimbrel, Slender-billed Gull, Pallas's Gull, Lesser Black-backed

Gull, Heuglin's Gull, Lesser Crested Tern, Sandwich Tern, Caspian Tern

Passage
Greater Sand Plover, Terek Sandpiper, Bar-tailed Godwit, Ruddy Turnstone, Gull-billed Tern, Common Tern, Whiskered Tern

5 Al Areen Wildlife Sanctuary (30km)

The sanctuary is actually a wildlife park with a number of exotic mammals and some native species. There are restricted parts where visitors on foot are not permitted. However, the general area provides good habitat for a number of desert bird species while the planted areas provide cover for passerines. Despite the fact that visitors get marshalled along on a guided tour bus, there are good birding opportunities along the footpaths between the enclosures and near the administrative buildings.

Take the **Sheik Isa Highway** west from **Manama** and exit south to **Al Zallaq** on the west coast road. Go south from Al-Zallaq towards **Al Markh** for about 2km to the sanctuary entrance on the left-hand side, which is well signposted from Al-Zallaq.

KEY SPECIES

Resident
Cream-coloured Courser, Black-crowned Sparrow-Lark, Lesser Short-toed Lark, Desert Lark, Hoopoe Lark, Southern Grey Shrike

Summer
Rufous-tailed Scrub Robin, Eastern Olivaceous Warbler

Winter
Richard's Pipit, Meadow Pipit, Red-throated Pipit, Water Pipit, Rock Thrush, Pied Wheatear, Isabelline Wheatear, Desert Wheatear, Desert Warbler, Red-breasted Flycatcher, Isabelline Shrike, Masked Shrike

Passage
Lesser Kestrel, Short-toed Lark, Tawny Pipit, Yellow Wagtail, Common Redstart, Whinchat, Northern Wheatear, Orphean Warbler, Ménétries's Warbler, Woodchat Shrike, Ortolan Bunting

6 Sakhir Racecourse (18km)

The racecourse is an important birding site because its small lake is one of the few areas of standing fresh water on the island. As such it is a refuge for a small number of breeding waterfowl and also for migrating ducks. The year-round irrigation promotes the growth of vegetation, which can hold good numbers of migrating passerines.

Take the **Sheik Isa Highway** west from **Manama** and exit south to **Isa Town**. Continue past Isa to **Awali** and turn west at Awali towards **Zellaq**. The racecourse is about 4km west of Awali.

KEY SPECIES

Resident
Little Grebe, Kentish Plover, Hoopoe Lark, Southern Grey Shrike

Summer
Collared Pratincole, Saunders's Tern, Rufous-tailed Scrub Robin, Eastern Olivaceous Warbler

Winter
Black-necked Grebe, Great Crested Grebe, Gadwall, Eurasian Wigeon, Garganey, Northern Pintail, Northern Shoveler, Tufted Duck, Common Pochard, Isabelline Wheatear, Desert Wheatear, Mourning Wheatear, Isabelline Shrike

Passage
Common Quail, Green Sandpiper, Wood Sandpiper, Gull-billed Tern, Common Tern, Whiskered Tern, Blue-cheeked Bee-eater, White-throated Robin, Common Redstart, Red-backed Shrike, Lesser Grey Shrike, Woodchat Shrike

7 Badaan Farm (16km)

Badaan Farm is on the west coast of Bahrain Island, surrounded by a number of interesting areas which contain a variety of habitats. Although the farm itself is private property, birders are normally allowed to visit, after seeking permission at the main entrance. The cultivation of fodder crops, particularly alfalfa, provides a rich food supply for resident

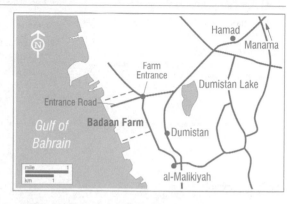

and migrant larks, buntings and other seedeaters, while the vegetated areas provide food and cover for insectivorous species. Potentially any Palearctic passerine could turn up here during migration. In addition, the irrigation channels and run-off ponds attract plovers, crakes and other aquatic species. There is a small island just offshore, viewable from the farm-track, which is good for roosting gulls and herons.

East of the farm, and just off the main road back to Manama, is Nakhl Lawzi (Dumistan) Lake. The water level fluctuates considerably and, when low, exposes substantial expanses of mud particularly at the southern end. When this happens, good numbers of waders can be found here, including White-tailed and Caspian Plovers in spring, if you are lucky. Unfortunately part of the lake has been filled-in for house construction recently so there is some doubt over the site's future.

Take the **Sheik Isa Highway** west from **Manama** and exit south to **Hamad**. Take the road southwest from Hamad towards Dumistan. The Badaan Farm area is just on the northern outskirts of Dumistan.

KEY SPECIES

Resident
Little Bittern, Kentish Plover, Desert Lark, Hoopoe Lark, Southern Grey Shrike

Summer
Collared Pratincole, Rufous-tailed Scrub Robin

Winter
Grey Plover, Pacific Golden Plover,

Lesser Sand Plover, Common Ringed Plover, Temminck's Stint, Little Stint, Dunlin, Sanderling, Broad-billed Sandpiper, Curlew Sandpiper, Common Redshank, Spotted Redshank, Common Greenshank, Eurasian Curlew, Slender-billed Gull, Oriental Skylark, Richard's Pipit, Meadow Pipit, Red-throated Pipit, Water Pipit, Isabelline Wheatear,

Desert Wheatear, Desert Warbler, Isabelline Shrike

Passage
Little Crake, Spotted Crake, Caspian Plover, White-tailed Plover, Greater Sand Plover, Pintail Snipe, Short-toed Lark, Tawny Pipit, Yellow Wagtail, Common Redstart, Whinchat, Northern Wheatear, Ortolan Bunting

Useful contacts and websites

BirdLife Bahrain
www.birdlifemed.org/Contries/barhin/barhin.html
Dr Saeed A. Mohamed,
PO Box 40266, Bahrain. Tel. (+97) 2 17640055.
Email: sam53@batelco.com.bh
BirdLife Representative in Bahrain.

Ornithological Society of the Middle East
www.osme.org
Ornithological Society of the Middle East, c/o The Lodge, Sandy, Bedfordshire SG19 2DL, UK.

Bahrain Bird Report
www.hawar-islands.com

Books and publications

Birds in Bahrain Erik Hirschfeld (1995)
Hobby Publications, ISBN 1872839037

Birds of Bahrain M. Hill and T. Nightingale (1993)
Immel Publishing, London

The Breeding Birds of Hawar – Results of the
1998 Survey. Howard King (1999) Ministry of
Housing Bahrain, ISBN 9990111006

Birds of Southern Arabia Dave Robinson (1992)
Motivate Publishing, ISBN 1873544375

Field Guide to the Birds of the Middle East
R.F. Porter, S. Christensen, P. Schiermacker-Hansen
(2004) Christopher Helm, ISBN 0713670169

Where to Watch Birds in Asia Nigel Wheatley
(1996) Christopher Helm, ISBN 071364303X

• •

TYPICAL SPECIES

Resident
Little Cormorant, Little Egret, Cattle
Egret, Black-crowned Night Heron,
Spotted Dove, Zebra Dove, Red Collared
Dove, Asian Palm Swift, House Swift,
Blue-tailed Bee-eater, Indian Roller,
Coppersmith Barbet, Asian Koel, Greater
Coucal, Common Iora, Common
Tailorbird, Oriental Magpie-Robin, Pied
Fantail, Plain-throated Sunbird, Olive-
backed Sunbird, Streak-eared Bulbul,
Yellow-vented Bulbul, Common Myna,
White-vented Myna, Asian Pied Starling,
Large-billed Crow, Eurasian Tree
Sparrow, Scaly-breasted Munia

Oct–Apr
Chinese Pond Heron, Grey Heron,
Black-capped Kingfisher, Red-rumped
Swallow, Barn Swallow, Yellow Wagtail,
Black Drongo, Black-naped Oriole, Asian
Brown Flycatcher, Yellow-browed
Warbler

*Oriental Magpie-Robin is common in parks and gardens in
Bangkok*

Bangkok

Bangkok is well known to the many birders who visit Thailand as the gateway to
the interior of this bird-rich country. This extremely noisy and congested city
seems hardly likely to induce any birder to spend time there or to use it as a base
to visit sites close by. The city is quite a long distance from any of Thailand's birding
hotspots and a birder's instinct will be to get clear of it as quickly as possible. Certainly,
if you have a few days to spare, it would be worth making a trip to the forests at Khao
Yai, about 200km northeast of the city, or to Kaeng Krachan, about 230km to the
southwest. However, if you have reason to remain in Bangkok for a day or two, there is
some good birding to be had on the coasts either side of the city.

Bangkok is located a little upstream from the mouth of the Chao Praya River and is
surrounded by the Central Plain of Thailand. All of the lowland forest, which was rich

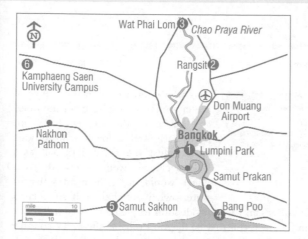

in birdlife, has been cleared from this area for agricultural purposes, and the cultivated areas are relatively poor for birds. Some remnants of mangrove forest still occur along stretches of the coast adjacent to the city and the mudflats, shrimp ponds and coastal marshes in the vicinity of Bangkok remain one of the most important areas in the country for wintering wildfowl. For this reason, a few days spent in the Bangkok area would complement a trip to the forests of the north and northeast, and even if you were also planning to visit the wader hotspots on Peninsular Thailand, there is always a chance to pick up one or two of the rarer species closer to the city.

The best time for wader watching is from November to March, corresponding to the boreal winter, when wildfowl populations are at their highest. This is also when Palearctic migrants such as pipits and buntings can be found in the dried-out rice paddyfields and warblers occur in any rank vegetation.

GETTING AROUND

Given the flat nature of the countryside and the network of canals (klongs) which cut through the city and its outskirts, the provision of cycle lanes would make a delightful way of touring the city and accessing the paddyfields, fish ponds and saltpans on the coast. However, the relentless volume of traffic, noise and exhaust fumes make the prospect of cycling around Bangkok a risky venture at any time of day or night, although the city residents take to the road using all forms of propulsion, with apparent equanimity. There is a very impressive overhead train system (Skytrain) for getting around the city centre but, apart from the parks, it does not serve any of the birding sites outside the city. The metro system, likewise, is limited in scope but future expansion may bring this service to locations beyond the city.

If you don't have a car, the bus services provide a practical way to get to the outskirts, and local taxis and tuk-tuks are both flexible and abundant. There are three main bus stations in the city: Northern and Northeastern (Mochit Mai), Southern and Eastern (Ekkamai), each serving destinations as their names suggest: north, south and east of the city. The northern and eastern terminals are also on the Skytrain route.

If you are going to rent a car, you would be best advised to take public transport out to the airport and collect it there rather than try to navigate your way out of the city from a central pick-up point. It is much easier to orientate yourself onto the highway system from the airport rather than tackle the traffic and street system in Bangkok.

A good map of Bangkok will need to be complemented by a map including the neighbouring provinces of Samut Sakhon and Samut Prakhan, as several of the sites involve travel into these provinces.

Locations

1 Lumpini Park (3km)

Sala Daeng Skytrain Station

KEY SPECIES

Resident	Oct–Apr
Alexandrine Parakeet (I), Coppersmith Barbet, Small Minivet, Black-collared Starling, Vinous-breasted Starling, Scarlet-backed Flowerpecker	Arctic Warbler, Red-throated Flycatcher, Brown Shrike

This well-known park in the centre of the city offers a respite from the noise and bustle of the streets. It is not particularly well-endowed with wooded areas whilst the small artificial lakes lack secluded well vegetated areas which would make them attractive to waterfowl. Nevertheless, it supports a surprising range of species and offers an easy introduction to the country's birdlife for the first-time visitor to Thailand. In fact, this may be the most reliable site in the country for Black-collared Starlings! The traffic noise at any time of day makes it difficult to pick up many of the birds by call, the possible exception being Coppersmith Barbets with their loud, distinctive 'tonking'. Unless you visit very early in the morning, avoid the park at weekends as the crowds are just overwhelming.

The park is located in the southeast of the city centre on Rama IV Road near many of the tourist hotels.

As an alternative, the Queen Sirikit Park/Railway Park near the weekend market is a much larger expanse and also a bit better for birds. The Small Minivet is easy to find there and pretty much all the Lumpini species can be seen apart from Alexandrine Parakeet (which is a feral escapee, anyway).

2 Rangsit (14km)

Bus Nos. 29/34/59

Most of the wetland habitats in the vicinity of Bangkok are saltwater, but Rangsit is one of the closest freshwater habitats to the city and supports a somewhat different range of birds. This is an area of irrigation ponds and lagoons adjacent to rice-paddies. However, with the expansion of the city, much of the cultivated land is becoming built-over and the lagoons themselves increasingly polluted and degraded. Nevertheless, it still retains a good selection of species and because it is close to the airport, it is well-worth a short diversion on your way to or from the car-rental depot. It is also more or less on the way to Khao Yai, so makes a convenient stop-off.

Take **Highway 1** north from Bangkok and continue past the **airport** for about 8km. Exit east onto **Route 305** at the **Nakhon Nayok** exit, which is also signposted for Rangsit. Stay on this road for less than 1km and turn left into Soi 13. Stay on Soi 13 for under 1km until it makes a right-angle bend to the left, and continue straight along a dirt track to reach the lagoons. Several bus routes connect Rangsit with central Bangkok, and Bus No. 59 connects with the airport.

KEY SPECIES

Resident		
Yellow Bittern, Striated Heron, Black-shouldered Kite, White-breasted Waterhen, Ruddy-breasted Crake,	Common Moorhen, Purple Swamphen, Bronze-winged Jacana, Pheasant-tailed Jacana, Red-wattled Lapwing, Black-winged Stilt, Oriental	Pratincole, Green Bee-eater, Plaintive Cuckoo, Zitting Cisticola, Plain Prinia, Dark-necked Tailorbird, Chestnut Munia, Red Avadavat

Oct–Apr		
Lesser Whistling Duck, Eastern Marsh Harrier, Baillon's Crake, Wood Sandpiper, Common Sandpiper,	Common Kingfisher, Siberian Rubythroat, Siberian Stonechat, Black-browed Reed Warbler, Oriental Reed Warbler, Dusky Warbler,	Lanceolated Warbler, Pallas's Grasshopper Warbler, Yellow-breasted Bunting

3 Wat Phai Lom (40km)

Rather surprisingly, birds suffer considerable persecution in Thailand and consequently many colonial-nesting birds can be found breeding in the protected areas of Buddhist temples. The temple at Wat Phai Lom is well known for its breeding colony of Asian Openbills. Its riverside location on the Chao Praya makes it a good spot for other herons and egrets, while the surrounding area provides a mix of habitats including rice paddyfields, scattered trees and patches of bamboo.

Take **Highway 1** north and about 10km past the **airport** turn west onto **Route 306** towards **Pathum Thani**. After about 3km and before reaching Pathum Thani turn north for about 5km to reach Wat Phai Lom. Alternatively, you could take one of the Chao Praya River cruises which depart from Maharat Pier in Central Bangkok, and stop at Wat Phai Lom on their way to Ayuthaya.

KEY SPECIES

Resident		Oct–Apr
Little Grebe, Little Cormorant, Striated Heron, Cinnamon Bittern, Yellow Bittern, Asian Openbill, Cattle Egret, Black-shouldered Kite, Brahminy Kite, White-breasted Waterhen, Ruddy-breasted Crake, Slaty-breasted Rail, Bronze-winged	Jacana, Red-wattled Lapwing, Black-winged Stilt, Greater Painted-snipe, Pied Kingfisher, Green Bee-eater, Asian Emerald Cuckoo, Plaintive Cuckoo, Small Minivet, Zitting Cisticola, Plain Prinia, Dark-necked Tailorbird, Chestnut Munia, Plain-backed Sparrow	Lesser Whistling Duck, Black Kite, Common Sandpiper, Forest Wagtail, Siberian Rubythroat, Siberian Stonechat, Siberian Blue Robin, Red-throated Flycatcher, Yellow-breasted Bunting

4 Bang Poo Mangroves and Pier (34km)

Bang Poo is a coastal town just east of Bangkok, where some stretches of mangrove are still intact and the offshore zone is an important area for wintering gulls. There are disused shrimp ponds which are reverting to mangrove cover and at low tide there are expansive mudflats for loafing gulls, terns and waders. Large gatherings of gulls are quite unusual in the tropics and the offshore gull flock occasionally turns up rarities.

Take **Highway 3** (also known as **Old Sukhumvit Road**) east from Bangkok for 37km and you arrive at the seaside resort of Bang Poo. Despite the fact that this is a military holiday centre, the public may enter the grounds in order to access the ocean pier. You can check the gulls from the tip of the pier, where there is a well-known restaurant. At the base of the pier there is a footpath going east and west through the mangroves and you can scan the mudflats and shoreline through

KEY SPECIES

Resident
Indian Cormorant, Striated Heron, Great Egret, Black-shouldered Kite, Brahminy Kite, White-breasted Waterhen, Common Moorhen, Red-wattled Lapwing, Black-winged Stilt, White-throated Kingfisher, Collared Kingfisher, Plaintive Cuckoo, Golden-bellied Gerygone, Dark-necked Tailorbird

Oct– Apr
Eastern Marsh Harrier, Common Kingfisher, Lesser Sand Plover, Greater Sand Plover, Grey Plover, Pacific Golden Plover, Spotted Redshank, Common Redshank, Common Greenshank, Marsh Sandpiper, Terek Sandpiper, Whimbrel, Black-tailed Godwit, Wood Sandpiper, Common Sandpiper, Ruddy Turnstone, Brown-headed Gull, Black-headed Gull, Heuglin's Gull, Gull-billed Tern, Common Tern, Caspian Tern, Little Tern, Whiskered Tern, White-winged Black Tern

breaks in the vegetation. Also check the shallow ponds and canals which traverse the cultivated plots on the north side of Highway 3 for roosting waders and other aquatic birds. There are regular bus services along Highway 3 between Bangkok (Eastern Bus Terminal) and Chon Buri further east.

5 Samut Sakhon
(48km)

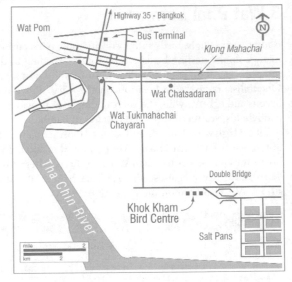

The saltpans and shrimp ponds at Samut Sakhon (also known as Khok Kham) are an excellent site to view the thousands of waders that winter and migrate along Thailand's coastline. This has become one of the most reliable places in Southeast Asia to see Spoon-billed Sandpipers, and several other rare and difficult-to-see species can be found here, including both Asian Dowitchers and Nordmann's Greenshanks.

However, it is not an easy site to work. The saltpans and mudflats are extensive and without a telescope it would be difficult to find the concentrations of birds, let alone identify them. Fortunately, help is available in the form of local restaurant owner Mr Tii whose café, The Khok Kham Bird Centre, acts as an informal bird observatory, where you can check the logbook to find out what's around and where the birds are feeding or roosting. Mr Tii is an accomplished birder and can show you around the saltpans. If you don't have a telescope, he can provide one, so it is worth engaging his services for a few hours. Try to schedule your visit for when the tide is in, as the birds will then be roosting on the saltpans, but bear in mind that as the Gulf of Thailand is a semi-enclosed sea area, the tidal movements are not predictably twice a day: so check the tide tables.

Take **Highway 35** southwest from Bangkok and after about 38km turn south to Samut Sakhon, which is signposted. Drive into the town and turn left at the large monument, take the third turn right and cross over a railway bridge. Continue for about 4km until you come to a T-junction. Bear left at the T-junction and continue for another 4km or so. Mr Tii's café is on the right-hand side of this road. If you pass Mr Tii's, continue for a short distance and then bear right over a small bridge and take the next turn on the right. This will bring you into the saltpan area.

If you don't have a car, you can take the bus from the Southern Bus Terminal to Samut Sakhon and from there get a taxi to Mr Tii's café, which is well known locally.

KEY SPECIES

Resident
Indian Cormorant, Striated Heron, Great Egret, Intermediate Egret, Javan Pond Heron, Pacific Reef Heron, Brahminy Kite, Black-shouldered Kite, White-breasted

Waterhen, Ruddy-breasted Crake, Slaty-breasted Rail, Red-wattled Lapwing, Black-winged Stilt, Little Tern, Collared Kingfisher, German's Swiftlet, Plaintive Cuckoo, Lesser Coucal, Paddyfield Pipit, Zitting

Cisticola, Plain Prinia, Plain-backed Sparrow

Winter
Eastern Marsh Harrier, Lesser Sand Plover, Greater Sand Plover, Grey

Plover, Pacific Golden Plover, Kentish Plover, Little Ringed Plover, Red-necked Stint, Long-toed Stint, Temminck's Stint, Curlew Sandpiper, Broad-billed Sandpiper, Spoon-billed Sandpiper, Spotted Redshank, Common Redshank, Common

Greenshank, Nordmann's Greenshank, Marsh Sandpiper, Terek Sandpiper, Whimbrel, Black-tailed Godwit, Wood Sandpiper, Common Sandpiper, Ruddy Turnstone, Brown-headed Gull, Gull-billed Tern, Common Tern, Caspian Tern,

Whiskered Tern, White-winged Black Tern, Black-capped Kingfisher, Common Kingfisher

Passage
Asian Dowitcher, Great Knot, Red Knot, Oriental Pratincole

6 Kamphaeng Saen University Campus (80km)

There is no real natural woodland around Bangkok and the nearest approximation is to be found on the Kasetsart University Campus at Kamphaeng Saen. The campus park-land resembles gallery forest and contains native lowland vegetation which supports a selection of forest birds. In addition, there is an artificial lake which provides fresh-water habitats, and the Agriculture Faculty has several experimental cultivation plots which attract wintering Palearctic passerines. The university has a thriving Bird Club which means that the campus is well watched and several interesting observations are made each year.

Take **Highway 4** west from Bangkok for about 50km to **Nakhon Pathom**. Turn north on **Route 321** towards Kamphaeng Saen. The campus is on the left-hand side of the road just past Kamphaeng Saen town centre, which is about 30km from Nakhon Pathom.

KEY SPECIES

Resident
Yellow Bittern, Striated Heron, White-breasted Waterhen, Bronze-winged Jacana, Pheasant-tailed Jacana, Red-wattled Lapwing, Green Bee-eater, Fulvous-breasted Woodpecker, Asian Emerald Cuckoo, Eurasian Hoopoe, Plaintive Cuckoo, Small Minivet,

Sooty-headed Bulbul, Zitting Cisticola, Plain Prinia, Long-tailed Shrike, Scarlet-backed Flowerpecker, Chestnut Munia

Oct–Apr
Lesser Whistling Duck, Wood Sandpiper, Common Sandpiper,

Blue-throated Bee-eater, Richard's Pipit, Siberian Rubythroat, Siberian Stonechat, Dusky Warbler, Red-throated Flycatcher, Brown Shrike, Yellow-breasted Bunting

Useful contacts and websites

Bird Conservation Society of Thailand (BCST)
www.bcst.or.th Email: bcst@bcst.or.th
43 Soi Chok Chai Ruam Mitr 29, Vipawadi Rangsit Rd., Din Daeng, Bangkok 10320.
Tel. (+66) 2 6914816/2 6915976.
Principal Thai bird conservation organisation and BirdLife Partner in Thailand.

Thai Water Birds (Thai and English)
www.thaiwaterbirds.com
Birding site with information about Thai wildfowl, includes photos, reports and habitat information.

Birdwatching in Thailand
www.tatnews.org/emagazine/1948.asp
Tourism Authority of Thailand site with useful information about birding in Thailand.

Birding Thailand
http://bwinthailand.homestead.com/bw.html
Nature Trails bird tour company website, packed with useful information about birding in Thailand.

Oriental Bird Club
www.orientalbirdclub.org
P.O.Box 324, Bedford, MK42 0WG, UK.

Tidal Schedule for Bangkok
http://easytide.ukho.gov.uk/EasyTide/EasyTide/ShowPrediction.aspx

Bangkok Transit System – BTS
(English and Thai) www.bts.co.th
Bangkok Skytrain service information.

Bangkok Metro Company (Thai)
www.bangkokmetro.co.th/
Bangkok Subway service information.

Bangkok Mass Transit Authority (English and Thai) www.bmta.co.th
Bus services in Bangkok area.

State Railway of Thailand (English and Thai)
www.railway.co.th
Train services from Bangkok to interior of Thailand.

Books and publications

A Field Guide to the Birds of South-East Asia
Ben F. King *et al.* (1975) Collins, ISBN 0002192063

A Guide to the Birds of Thailand
Boonsong Lekagul & Philip D Round (1991)
Saha Karn Bhaet, ISBN 9748567362

A Field Guide to the Birds of South-East Asia
Craig Robson (2000) New Holland, London
ISBN 1843301180

Field Guide to the Birds of Thailand
Craig Robson (2002) New Holland
ISBN 1843300583

Resident Forest Birds in Thailand
Philip D. Round (1988) ICBP, ISBN 0946888132

Field Checklist of Thai Birds Philip D. Round
(2000) Bird Conservation Society of Thailand

A Photographic Guide to Birds of Thailand
Michael Webster & Chew Yen Fook (1997)
New Holland, ISBN 1853685941

Where to Watch Birds in Asia Nigel Wheatley
(1996) Christopher Helm, ISBN 071364303X

● ●

TYPICAL SPECIES

Resident
Little Grebe, Mallard, Spot-billed Duck, Eurasian Sparrowhawk, Common Kestrel, Oriental Turtle Dove, Eurasian Collared Dove, Spotted Dove, Great Spotted Woodpecker, Sky Lark, Marsh Tit, Coal Tit, Great Tit, Eurasian Nuthatch, Long-tailed Tit, Winter Wren, White-cheeked Starling, Large-billed Crow, Black-billed Magpie, Azure-winged Magpie, Eurasian Jay, Eurasian Tree Sparrow, Grey-capped Greenfinch

Summer
Cattle Egret, Black-crowned Night Heron, Grey Heron, Common Swift, Barn Swallow, Red-rumped Swallow, White Wagtail, Grey Wagtail, Daurian Redstart, Black Drongo

Winter
Northern Pintail, Tufted Duck, Common Pochard, Common Goldeneye, Goosander, Common Buzzard

Beijing

Beijing is on the northern edge of a flat plain occupied by most of the eastern provinces of China. This whole area is very intensively cultivated and much polluted so that birdlife in the areas surrounding the city is heavily depleted. The city extends over a very large area that is for the most part industrial buildings and high density housing. The naturalist needs to look very hard to find the niches occupied by nature in such a hostile urban environment.

Unusually, for an inland conurbation of this size, there is no significant river or other waterway running through the city. However, a number of public parks and historic monuments contain sufficient woodland and some waterways to support a limited but interesting range of resident and migrant species. Nonetheless, the lack of a large wetland site within easy access of the city limits the overall species diversity. For those willing to travel outside the city, but still within the Beijing Municipality, the mountains of the Mongolian Plateau to the north and west offer the highest species diversity, although the overall density of birds is low. Spring and early summer are the best times to visit the mountains, although a small number of endemic species which may be seen there make a trip worthwhile at any time of year.

Despite its inland location there is, in fact, quite a noticeable visible migration through the city's parks

Azure-winged Magpie has a relict distribution, with populations in east Asia and Iberia

and gardens, but for those who may have more than a day to spend birding during spring or autumn, then a trip to Beidaihe (three hours by express, but up to six hours by slow train!) is well worth considering, as even a limited time spent there will be far superior to a stay in Beijing. Beidaihe is one of the migration hotspots in the Northern Hemisphere and needs no introduction to committed birders. It is well described in many publications and trip reports.

GETTING AROUND

The subway system is limited but useful as it serves some of the sites mentioned in this guide. Taxis are cheap and the sites mentioned are well known, and so provided you have the written Chinese address of your destination, the driver will be able to find it. Buses are probably only useful if you are going from terminus to terminus and you will definitely need to have

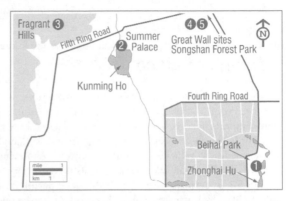

your destination written in Chinese. Your hotel should be able to provide you with location names written in Chinese. Most suburban bus services operate from railway or subway stations, so a trip to the main birding sites in the northwest of the city will involve both modes of transport.

Despite its massive size, complexity and lack of English signage, driving in Beijing is not as daunting a task for the visitor as some other Asian cities. The city is largely laid out as a grid with five concentric ring roads (and a sixth outer ring road under construction), making it generally straightforward to find your way to the outskirts. Just be prepared for the large distances to be traversed and the featureless aspect of many of the buildings, making it difficult to navigate by any identifiable landmarks.

Beijing is vast and certain locations such as the Summer Palace or the Fragrant Hills will be 'off the map' on any large-scale map, so you will probably need to purchase both a large-scale map for navigating the central locations and a small-scale map of Beijing Municipality to locate the more distant sites.

All public areas in Beijing are very crowded, even early in the morning when crowds of people are out practising their Tai-chi exercises or even open-air ballroom dancing! Also walking off the footpath onto the grass or through shrubbery may be frowned upon so be prepared to be ticked off if caught doing so! Most public parks and gardens in Beijing open around 06.30 and close at dusk.

For the most part, the climate in Beijing is very dry, with rain mainly in the summer months, so no special clothing is necessary except comfortable walking shoes.

Locations

1 Beihai Park (City Centre)

Beihai Park is just west of the Forbidden City, which could be considered the city centre. It contains a large lake, with a wooded islet (Jade Island) which is reached by a bridge. The southern half of the lake is popularly used as a paddle-boat pond and is very

KEY SPECIES

Resident
Vinous-throated Parrotbill

Summer
Eastern Crowned Warbler,
Yellow-rumped Flycatcher,
Yellow-billed Grosbeak

Winter
Dusky Thrush, Siberian

Accentor

Passage
Eurasian Hoopoe, Olive-backed Pipit, Red-flanked
Bluetail, Red-throated
Flycatcher, Dusky Warbler,
Pallas's Warbler, Yellow-browed Warbler

disturbed but the northern half is quieter and better for birds. Despite the fact that the park is hemmed in on all sides by a huge bustling city, at certain times it can host a surprising variety of birds. Climb the steps to the White Dagoba Buddhist Shrine on Jade Island and this affords you an excellent view at treetop level.

2 Summer Palace (15km)   Bus No. 375

The Summer Palace is a lavish visitor attraction, with extensive monuments and gardens. From a birder's perspective, the best locations are Kunming Lake near the 'new' Summer Palace, and the wooded slopes of Longevity Hill on the north side of the lake. In the absence of any other significant wetland habitat in the Beijing area, the lake provides a valuable birding resource. Wintering wildfowl, including Baer's Pochards, occur on the lake, for as long as it remains ice-free, and the surrounding park is good for passerines. Just beyond the park perimeter are some cultivated plots and paddyfields which are also worth checking for herons and waders on migration.

The Summer Palace is in the northwest of the city sandwiched between the **Fourth** and **Fifth Ring Roads**. Several bus routes serve the park or you can take a taxi from Xizhimen subway station.

KEY SPECIES

Resident
Eurasian Sparrowhawk, Grey-headed
Woodpecker, Grey-capped
Woodpecker, Vinous-throated
Parrotbill, Chinese Penduline Tit,
Blue Magpie, Meadow Bunting

Summer
Yellow Bittern, Purple Heron, Chinese
Pond Heron, Garganey, Ruddy-

breasted Crake, Common Kingfisher,
Oriental Reed Warbler, Black-browed
Reed Warbler, Eastern Crowned
Warbler, Yellow-rumped Flycatcher,
Yellow-billed Grosbeak

Winter
Ruddy Shelduck, Falcated Duck,
Baer's Pochard, Common Goldeneye,
Smew, Goosander, Common

Buzzard, Siberian Accentor

Passage
Oriental Honey-buzzard, Eurasian
Hoopoe, Olive-backed Pipit, Red-flanked Bluetail, Dusky Warbler,
Pallas's Warbler, Yellow-browed
Warbler

3 Fragrant Hills (20km)   Bus No. 318/333

This park (called Xiang Shan in Chinese), a former hunting estate for the emperors, is about 4km west of the Summer Palace and is probably the closest stretch of open countryside to Beijing. Consequently, this is doubtless the best birding location accessible from Central Beijing, and a variety of woodland birds, including a few Chinese endemics, can be found here. The habitat is mainly mixed woodland with some ornamental lakes. By taking the chairlift to the top of the slopes, you give yourself the benefit of height as you descend through the wooded slopes. (Otherwise, you end up with a crick in your neck from squinting upwards all the time at silhouettes in the tree canopy!). The hills rise to over 500m, making them one of the highest points in this incredibly flat city, and the upper slopes are clad with pinewoods, with more open scrubby areas at the summit.

The Fragrant Hills Park is on the northwest side of the city on **XiangShan Lu** which can be accessed from the **Fifth Ring Road**. If using public transport, the best way to get there is to take the bus (No. 333) from the Summer Palace or bus (No. 318) from Pingguoyuan subway station. Alternatively, all taxi drivers will be familiar with the location of XiangShan.

KEY SPECIES

Resident
Grey-headed Woodpecker, Père David's Laughingthrush, White-browed Chinese Warbler, Vinous-throated Parrotbill, Blue Magpie, Daurian Jackdaw, Meadow Bunting

Summer
Common Cuckoo, Siberian Stonechat, Eastern Crowned Warbler,

Yellow-rumped Flycatcher, Black Drongo, Yellow-billed Grosbeak

Winter
Common Buzzard, Siberian Accentor

Passage
Oriental Honey-buzzard, Eastern Marsh Harrier, Eurasian Hobby, Amur Falcon, Eurasian Hoopoe, Olive-

backed Pipit, Red-flanked Bluetail, Scaly Thrush, Red-throated Flycatcher, Dusky Warbler, Radde's Warbler, Pallas's Warbler, Yellow-browed Warbler, Little Bunting

Further away

4 **Great Wall** (70km)

Badaling railway station
Bus No. 916/919

Most visitors to Beijing will probably take a trip to the Great Wall, despite its distance from the city. Whether by organised tour or travelling independently, a visit to the Great Wall provides an opportunity to look for birds on the wooded slopes leading up to the wall. There are several access points from Beijing, the best known of which is Badaling. This site has the advantage of being very well served by public transport and tourist buses, so it is easy to get to and it also has a chairlift up to the wall. However it is, of course, very crowded, even on the trails up to the wall. If you visit Badaling, take the west route (rather than the east) up to the wall, as it is steeper, but therefore less-frequented by people. Another well-known access point is at Mutianyu which also has a chairlift. You could ride it up to the Wall and then hike back down through the wood-land. You should also check the wooded areas around the tourist restaurant. Although it

is a little more distant from Beijing than Badaling, the mixed woodland cover at Mutianyu is more extensive and the crowds are smaller.

If you have your own trans-port, you could also con-sider visiting Huanghuacheng, which is further east of Badaling, but much less tour-isty, and with a little more variety in terms of habitat as there is a lake (Jintang Lake) adjacent to the wall.

To get to **Badaling**, take the main road north out of Beijing (**Deshengmenwei**) for approximately 70km. To

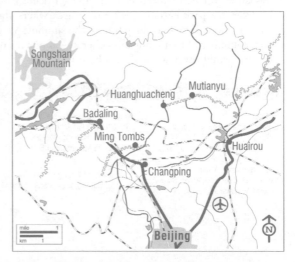

get to **Huanghuacheng**, take **Deshengmenwei** as far north as **Changping**, and fork right towards the **Ming Tombs**. Continue northeast to **Jiuduhe** and from there, turn north to Huanghuacheng, which is about 10km from Jiuduhe.

To get to **Mutianyu**, take the **Airport Expressway** to **Huairou** and then turn north in Huairou to Mutianyu, passing the Huairou Reservoir on the way. Mutianyu is the most distant of the three sites, being about 70km from Beijing.

There are several tourist bus services, as well as a train service, from Beijing North railway station (Xishimen) to Badaling, and both tourist and regular bus services (Bus No. 916) to Mutianyu, but no direct service to Huanghuacheng.

KEY SPECIES

Resident
Eurasian Sparrowhawk, Common Pheasant, Grey-headed Woodpecker, Grey-capped Woodpecker, White-backed Woodpecker, Père David's Laughingthrush, White-browed Chinese Warbler, Yellow-bellied Tit, Vinous-throated Parrotbill, Godlewski's Bunting, Meadow Bunting, Yellow-throated Bunting, Rock Bunting

Summer
Black Kite, Forest Wagtail, Eastern Crowned Warbler, Blyth's Leaf Warbler, Chestnut-flanked White-eye, Yellow-rumped Flycatcher, Black-naped Oriole, Hair-crested Drongo, Yellow-billed Grosbeak

Winter
Common Buzzard, Upland Buzzard, Siberian Accentor, Brambling, Pallas's Rosefinch

Passage
Red-throated Flycatcher

5 Songshan Forest Park (90km)

Songshan Mountain is one of the highest peaks in the Beijing area and is a protected forest park. Although it is rather distant from the city, it is worth considering for a visit as it offers the possibility of seeing some of the species more typical of the montane forest in this part of China. It is possible to drive to a height of 700m, and from there you could hike to the summit at just over 2,000m. The best time to visit is in early summer and this would make an ideal two-day trip if you have a weekend to spare, particularly if you want to be on the mountainside at dawn.

Take the **Badaling Expressway** north from Beijing and continue past Badaling to **Yanqing**. Then go north on the **Zhangjiakou Highway** for about 25km. The Songshan Forest Park entrance is on the right-hand (north) side of the road. You can drive from the park entrance to a resort area where there are hotels and restaurants and from where there is a hiking trail to the summit. As this is also a popular hot-spring resort, a number of tour companies operate excursions to this site, so a weekend trip is still quite feasible even if you do not have a car.

KEY SPECIES

Resident
Eurasian Sparrowhawk, Common Pheasant, Koklass Pheasant, Hill Pigeon, Grey-headed Woodpecker, Chinese Thrush, Père David's Laughingthrush, White-browed Chinese Warbler, Songar Tit, Snowy-browed Nuthatch, Eurasian Nutcracker, Blue Magpie, Red-billed Chough, Daurian Jackdaw, Godlewski's Bunting, Meadow Bunting, Yellow-throated Bunting

Summer
Eurasian Hobby, Large Hawk-Cuckoo, Oriental Cuckoo, Crag Martin, Long-tailed Minivet, Siberian Stonechat, Asian Stubtail, Eastern Crowned Warbler, Blyth's Leaf Warbler, Yellow-browed Warbler, Narcissus Flycatcher, Yellow-rumped Flycatcher, Hair-crested Drongo

Useful contacts and websites

Conserving China's Biodiversity
(English and Chinese)
www.chinabiodiversity.com
General conservation website with information
about birds and sites in China.

China Birding (English)
www.chinatibettravel.net/cnbirds
Bird tours website with information about birding in
China.

Oriental Bird Club
www.orientalbirdclub.org
P.O.Box 324, Bedford, MK42 0WG, UK

Beijing Public Transport Holdings
(English and Chinese) www.bjbus.com
Bus services in Beijing City and Municipality.

Beijing Subway (English and Chinese)
www.bjsubway.com

Books and publications

A Field Guide to the Birds of China
John MacKinnon and Karen Phillipps (2000)
Oxford University Press, ISBN 0198549407

A Field Guide to the Birds of South-East Asia
Craig Robson (2000) New Holland
ISBN 1853683132

**A Photographic Guide to Birds of China
Including Hong Kong** John MacKinnon and
Nigel Hicks (1996) New Holland, ISBN 1859749690

Where to Watch Birds in Asia
Nigel Wheatley (1996) Christopher Helm
ISBN 071364303X

● ●

Saunders's Tern flying past the unique Burj Al Arab Hotel

TYPICAL SPECIES

Resident
Little Grebe, Western Reef Heron, Grey Francolin, Black-winged Stilt, Kentish Plover, Slender-billed Gull, Laughing Dove, Green Bee-eater, Indian Roller, Eurasian Hoopoe, Crested Lark, White-cheeked Bulbul, Red-vented Bulbul, Graceful Prinia, Arabian Babbler, Purple Sunbird, Southern Grey Shrike, Common Myna, House Crow, Indian Silverbill

Winter
Great Cormorant, Cattle Egret, Little Egret, Great Egret, Grey Heron, Mallard, Eurasian Teal, Western Marsh Harrier, Eurasian Oystercatcher, Common Redshank, Common Sandpiper, Pallas's Gull, Yellow-legged Gull, Pallid Swift, Sky Lark, White Wagtail, Siberian Stonechat, Bluethroat, Black Redstart, Isabelline Wheatear, Song Thrush, Isabelline Shrike

Passage
Purple Heron, Garganey, Eurasian Turtle Dove, European Bee-eater, Blue-cheeked Bee-eater, Sand Martin, Barn Swallow, Common Nightingale, Common Redstart, Isabelline Wheatear, Pied Wheatear, Common Whitethroat, Common Chiffchaff, Willow Warbler, Spotted Flycatcher, Isabelline Shrike, Woodchat Shrike

Dubai

The United Arab Emirates (UAE) is one of the most popular birding destinations in the Middle East, and Dubai is a prime position for visiting many of the country's top sites. The coastline of the Arabian Gulf is an important wintering and migration staging area for Palearctic waders, with islands supporting internationally important seabird colonies. Many passerines which are very rare in the Western Palearctic winter in significant numbers in the UAE, and its strategic location at the southeast corner of the Arabian Peninsula makes it a migration hotspot which routinely produces rarities.

Summer is generally quiet and, of course, the intense heat and humidity discourage field activity, even for the most enthusiastic birder. There is a rather small range of resident species, a high proportion of which are naturalised exotics, but there is usually a good variety of waders present and certain breeding seabirds, including Socotra Cormorants and White-cheeked Terns, which would enliven an early morning or evening visit to the coast.

The rapid development of the Gulf States has resulted in large areas of the coastal strip becoming built-up, with hotels and high-rise buildings, as well as residential and commercial development. Plenty of open areas remain, however, while irrigation and greening has created habitats that attract many passage migrants and wintering species. Hotel gardens are always a good place to start and regularly turn up surprises, rarities even.

Dubai's location on the Gulf coast of the UAE places it within easy reach of some of the best coastal sites and within a short drive of some productive and scenic desert and mountain sites. Additionally, a number of man-made wetlands, parks and golf courses in or close to the city centre have emerged in recent years as excellent birding spots, and have become an essential stop for every visiting bird-tour group. Winter trips are invariably productive, but migration periods offer the best possibility in terms of species diversity. The range of species encountered differs somewhat between spring and autumn and many birders may want, or need, to make two trips to the UAE in order to 'clean up' completely. A serious birding trip would ideally last eight to ten days and involve visiting sites further inland, around Al Ain in particular, and on the Gulf of Oman coast. Nevertheless, the short-term visitor restricted to the area around Dubai will find more than enough birding opportunities, at any time of year.

GETTING AROUND

If it were not for the extremely hot and humid conditions which Dubai experiences for most of the year, it would be quite possible to see many species simply by relying on public transport. There is a bus service to a number of sites, but it is advisable to take taxis (all metered) around the city area. The prospect of walking from bus stop to birding site or of waiting for a bus in the oppressive heat makes an air-conditioned car a much more attractive proposition. Also, as several of the sites are open desert or irrigated fields, a car makes a very useful mobile hide with which to approach feeding birds. Although getting stuck off-road is easily done, a 2WD car is perfectly adequate for sites around Dubai. Unfortunately, despite all the new road construction, or perhaps because of it, this can be a very congested city and traffic jams are frequent. Always allow extra time for

a round trip. A popular and cost-effective option to consider is to contact one of the resident expatriate birders, all of whom will provide information for free, or guide for a reasonable fee (and who will usually pick up and drop you off at your hotel).

Several of the sites involve travelling into or through neighbouring emirates, but there are no restrictions and signposting is mostly good so you shouldn't get lost, at least not for long. In any case, English is widely spoken so you can always get directions.

Many of the sites are on private property, so please always ask for permission to enter. Visiting birders are a well-known phenomenon in Dubai and access is rarely, if ever, refused. Make sure that you do not enter restricted areas (or use your camera nearby) or block any tracks or entrances. Take care when parking off the roadside as this is where you are likely to get bogged down. Undoubtedly the best time for birding is early morning and late afternoon: the midday temperatures and accompanying haze make birding somewhat arduous, and most passerines, at any rate, become inactive then. Coastal sites require timing your visit to coincide with a rising or high tide: tide times are given in all daily English language newspapers, usually provided free outside your hotel door in the morning. The largely open terrain means that a telescope is more or less essential if you are to stand a chance of enjoying the full variety of birdlife that this superb birding destination can offer, so don't forget to pack it. Also, bear in mind that Dubai Airport has the largest Duty Free Shop in the world, so you might even pick up a bargain!

Locations

1 Al Safa Park (6km)   Bus Nos. 7/12/98

This city park is one of the best migrant traps on the Arabian Peninsula. If you are visiting Dubai during spring or autumn, then an early morning visit is likely to yield excellent results. The presence of a boating pond provides some wetland habitat to attract aquatic species such as herons, ducks and terns. Many common resident species can be also seen here.

Take the main highway, **Sheikh Zayed Road**, southwest from Dubai towards Abu Dhabi. The park is signposted off to the right about 5km from the Dubai Trade Centre, and opens from 08.00. There is a nominal charge.

KEY SPECIES	
Resident Shikra, Red-wattled Lapwing, Alexandrine Parakeet, Rose-ringed Parakeet, Asian Pied Starling **Winter** Black-crowned Night Heron, Common Pochard, Tufted Duck, Common Kingfisher	**Passage** Gull-billed Tern, Tree Pipit, Yellow Wagtail, Grey Wagtail, Rufous-tailed Scrub Robin, Marsh Warbler, Eastern Olivaceous Warbler, Ménétries's Warbler, Semi-collared Flycatcher, Red-breasted Flycatcher, Golden Oriole and rarities

2 Khor Dubai (4km)   Bus Nos. 16/61

Khor Dubai is a tidal creek providing a habitat for large numbers of waders, gulls, terns and herons right in the heart of the city. The site is a nature reserve and is now surrounded by a perimeter fence. Mangroves have been successfully planted, but some management of these is now being undertaken to keep areas open for waders. There are three purpose-built hides, currently open until 16.00 daily except Fridays (the weekend). Leica telescopes and binoculars are available to use free of charge in each of these. It is best to visit at high tide, as waders are distant at other times.

Take the **Hatta Road** east from Dubai, and follow signs for the **Ras al Khor Wildlife Sanctuary**, frustratingly located just across the carriageway so you are compelled to go some way before getting back on the right side of the road.

The wooded, but private, area of Zabeel, immediately to the west of the Khor, attracts raptors in winter. Scanning the skies from mid-morning onwards can be rewarding, with Booted Eagles, Black Kites and Oriental Honey Buzzards all regularly sighted, sometimes all at once. The Khor itself connects to the sea through the centre of the city and is flanked on its west side by the Creekside Park, a leafy park which supports some commoner species, often has falls of migrants, and provides a vantage point to check the creek narrows for gulls and terns.

KEY SPECIES

Resident Greater Flamingo, Osprey, Gull-billed Tern, Caspian Tern **Winter** Eurasian Spoonbill, Common Shelduck, Northern Pintail, Northern Shoveler, Black Kite, Oriental Honey	Buzzard, Greater Spotted Eagle, Booted Eagle, Pied Avocet, Grey Plover, Pacific Golden Plover, Lesser Sand Plover, Common Ringed Plover, Little Stint, Dunlin, Curlew Sandpiper, Terek Sandpiper, Bar-tailed Godwit, Black-tailed Godwit, Eurasian Curlew,	Caspian Gull, Caspian Tern, Sandwich Tern **Passage** Greater Sand Plover, Broad-billed Sandpiper, Saunders's Tern (and also many of those species listed here as winter visitors)

3 Mushrif National Park (15km)   Bus No. 11M

In many areas of the UAE desert the natural vegetation has been depleted by a progressive lowering of the water-table and by overgrazing. Mushrif Park is a fenced reserve area of native Ghaf tree *Prosopis cineraria* parkland which supports a typical community of native birds. Apart from the resident and wintering passerines, the main birding interest in this site is that it is a reliable year-round location to see Pallid Scops Owls, usually near the administration buildings at dusk. The park is just 12km southeast of Dubai Airport and makes a convenient stop-off if returning a rental car, or when returning from a day's birding at the sites along the Hatta Road.

Take the **Airport Road** (which runs along the west side of the airport and Terminal 1) south from Dubai towards **Al Khawaneej**. The park is sign-posted off to the right-hand side of the road, and there is a small, easily missed, sign at the turning into the park itself.

KEY SPECIES

Resident Pallid Scops Owl, Arabian Babbler **Summer** Rufous-tailed Scrub Robin, Chestnut-shouldered Petronia **Winter** Oriental Honey-buzzard,	Eurasian Sparrowhawk, Desert Lesser Whitethroat **Passage** European Roller, Eastern Olivaceous Warbler, Ménétries's Warbler, Desert Lesser Whitethroat, Semi-collared Flycatcher, Masked Shrike, Golden Oriole

4 Dubai Pivot Fields (20km)   Bus No. 16

The area of the Al Awir sewage treatment plant combines several adjacent sites and over the course of a day spent here, at any time of year, you are likely to run into fellow-birders. The Pivot Fields have become renowned as one of the UAE's premier sites, attracting large numbers of wintering passerines, waders, herons and raptors, and providing an unsurpassed opportunity to find several tricky species. During migration, there is a constant turnover of migrating birds passing through these fields, and they are then well worth a daily visit. It would be hard to find anywhere else in the world where Richard's and Blyth's Pipits, Tawny and Long-billed Pipits, and Water and Buff-bellied Pipits can all be seen regularly in a single day, along with several races of Yellow Wagtail. Obviously,

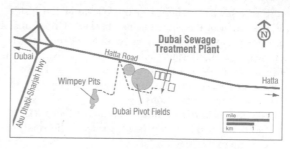

the task of identification is easier in spring, but in autumn and winter there is the challenge of having to separate, to name but a few, Oriental Skylarks from Sky Larks, Northern Wheatears from Isabelline Wheatears, juvenile Pallid Harriers from Montagu's Harriers, and usually thoroughly confusing, and often skulking, larks, pipits and buntings in profusion. It is advisable to take a telescope. The Pivot Fields are private property but normally a polite request to the staff at the entrance will enable immediate entry.

Adjacent to the Pivot Fields are the Al Warsan lakes, formerly Wimpey Pits, which are flooded sand quarries filled with treated wastewater from the sewage plant. These attract grebes, herons, terns, waders and hirundines. Needless to say, the abundant insect life in this whole area means that any bushes or areas of vegetation should be checked thoroughly for chats, warblers and shrikes.

Take the main road southeast to **Hatta** and about 2.5km after passing the cloverleaf interchange with the main Abu Dhabi/Sharjah highway, turn right at the sign for the Sewage Treatment Facility. After turning right, make an almost immediate left-hand turn and this road, running alongside the fields, leads to the entrance about 500m further on, on the corner on the left-hand side. If, instead of making an immediate left turn, you continue straight on for a kilometre, you are brought level with the Warsan lakes on the right. A sealed road takes you across and close to the lakes. They were formerly much larger, and there is a danger that they will be infilled or landscaped and thus lose much of their attraction.

KEY SPECIES

Resident
Red-wattled Lapwing, Bank Myna

Summer
Caspian Reed Warbler

Winter
Black-necked Grebe, Glossy Ibis, Eurasian Wigeon, Northern Pintail, Ferruginous Pochard, Pallid Harrier, Common Ringed Plover, Little Ringed

Plover, White-tailed Plover, Eurasian Curlew, Common Snipe, Pintail Snipe, Chestnut-bellied Sandgrouse, Egyptian Nightjar, Common Kingfisher, Oriental Skylark, Richard's Pipit, Tawny Pipit, Red-throated Pipit, Blyth's Pipit, Long-billed Pipit, Water Pipit, Buff-bellied Pipit, Isabelline Wheatear, Desert Wheatear

Passage
White Stork, Garganey, Montagu's Harrier, Pallid Harrier, Pacific Golden Plover, Little Stint, Temminck's Stint, Curlew Sandpiper, Ruff, Marsh Sandpiper, Whiskered Tern, Yellow Wagtail, Citrine Wagtail, Northern Wheatear, Ménétries's Warbler

5 Emirates Golf Course (25km)

The combination of irrigated fairways, small lakes and dense shrubbery make the Emirates Golf Course one of the best birding sites in the UAE. Many migrant passerines, including rare vagrants, are seen here in spring and autumn and a rich variety of wintering birds include some eagerly sought-after Palearctic species which are difficult to find elsewhere. There is a track around the inside of the perimeter fence which allows much of the course to be checked. Unfortunately, but understandably, birders are not permitted to roam the course more freely. You will need to check-in at the entrance gatehouse but, unless a major tournament is in progress (usually in early February), birders are readily granted access. There is a birders' logbook at the gatehouse, so always consult it to see what's around and, of course, reciprocate by leaving a record of your observations.

Take the main **Abu Dhabi** road out of Dubai (**E11**). After about 25km you will see the golf course on the left (south) side of the road, opposite the **Hard Rock Café**, where you exit as signposted.

KEY SPECIES

Resident
Shikra, Red-wattled Lapwing

Summer
Cream-coloured Courser, Little Ringed Plover, Whimbrel

Winter
Black-necked Grebe, Striated Heron,

Eurasian Sparrowhawk, Eurasian Curlew, Common Snipe, Pintail Snipe, Common Kingfisher, Water Pipit

Passage
Garganey, Caspian Plover, Pacific Golden Plover, Little Stint, Temminck's Stint, Ruff, Whimbrel,

Whiskered Tern, Tree Pipit, Yellow Wagtail, Rufous-tailed Scrub Robin, Marsh Warbler (spring), Eastern Olivaceous Warbler, Ménétries's Warbler, Semi-collared Flycatcher (spring)

Further Away

6 **Khor al Beidah** (40km)

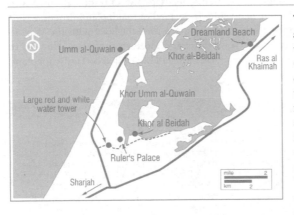

The coastal mudflats at Khor al Beidah are one of the best wader haunts in the Arabian Gulf. The Great Knot, for example, occurs no closer to the Western Palearctic as a regular winter visitor. It is also a completely reliable site for Crab Plovers. This is a very extensive site, which may at first appear rather daunting. There are many inaccessible creeks and small islands which are hidden by mangroves and would require a boat to reach them. Luckily, there is an access track between the dunes which brings you to an open area with an unhindered view of a large part of the mudflats, including the best, but spread out, high-tide roost. Sinaiya Island, beyond, supports a large colony of Socotra Cormorants. The sand dunes hold wheatears, larks and pipits. To obtain the best views, it is best to visit at or around high tide.

Take the main road from Dubai towards **Sharjah/Ajman**, following signs for **Umm al-Quwain** and **Ras al Khaimah**. As you drive past Ajman you will notice Khor Ajman on your left just after some traffic lights and speed humps: check for Crab Plovers from the road. Continue for 11km until you reach the junction for Umm al-Quwain, to the left, and Falaj al Mualla to the right (south). Turn left at this junction towards Umm al-Quwain. After 3km turn right down to a roundabout with a large red and white water tower. Turn right at this roundabout onto a sandy track which runs alongside the perimeter wall of the Ruler of Umm al Quwain's palace, but which eventually leads to the beach and southern shoreline of the Khor. (A more direct route on a blacktop road from this same roundabout will soon be ready.) Other more direct routes across the saltflats from the main Sharjah to Ras al Khaimah road, which you will have just left, may also be passable with care. If you stay on the road down into Umm al-Quwain, you can continue north to the tip of the city itself where the breakwater and beach are particularly good for Socotra Cormorants, gulls, terns and waders.

An alternative site which is good for both waders and gulls is Dreamland Beach, further east along the Ras al-Khaimah highway. Return to the main Sharjah to Ras al-Khaimah highway and continue east for about 15km. Turn off the highway, using the U-turn, onto a track down to the beach just beyond the Dreamland Aqua Park. This area can be excellent for waders.

KEY SPECIES

Resident
Socotra Cormorant, Black-crowned Sparrow-Lark, Hoopoe Lark

Summer
Cream-coloured Courser, White-cheeked Tern, Saunders' Tern, Lesser Short-toed Lark

Winter & passage
Greater Spotted Eagle, Crab Plover, Grey Plover, Greater Sand Plover, Lesser Sand Plover, Common Ringed Plover, Great Knot, Little Stint, Dunlin, Broad-billed Sandpiper, Curlew Sandpiper, Bar-tailed Godwit, Eurasian Curlew, Whimbrel, Terek Sandpiper, Ruddy Turnstone, Yellow-legged Gull, Heuglin's Gull, Pallas's Gull, Gull-billed Tern, Tawny Pipit, Desert Wheatear, Desert Warbler

7 Qarn Nazwa (50km)

This is a famous stakeout for Pharaoh Eagle Owls and for other rocky desert species. Take the main road east to **Hatta**, crossing the **Al Habab** roundabout after about 45km. Qarn Nazwa, 8km further on, is a rocky outcrop about 80m high to the south of the road, surrounded by attractive red dunes. Turn off the main road onto a sealed road running along the west side of the Qarn; park and walk. The Acacia trees and slopes around the back of the hill, at its northeastern corner, regularly hold Variable Wheatears and Plain Leaf Warblers. The owls are mostly seen towards dusk halfway down on the west side, often, very usefully, perching and calling from the ridge crest. Although the hill is now fenced off, and you are unable to ramble over it, most species can still be found with ease.

KEY SPECIES

Resident
Pharaoh Eagle Owl, Little Owl, Rock Martin, Brown-necked Raven

Winter
Long-billed Pipit, Yellow-vented Bulbul, Blue Rock Thrush, Rufous-tailed Wheatear, Variable

Wheatear, Orphean Warbler, Desert Warbler, Desert Lesser Whitethroat, Plain Leaf Warbler, House Bunting

Passage
Rufous-tailed Rock Thrush, Upcher's Warbler, Pale Rockfinch

8 Ghantoot (58km)

The plantation at **Ghantoot** is hallowed ground, though not of turf but an irrigated stand of native trees, being known as the most reliable site in the UAE for the Grey Hypocolius. Flocks are regularly present in spring, small numbers sometimes being present from early winter onwards.

To get to Ghantoot, take the **E11** towards Abu Dhabi. Take the exit, which is signposted for Ghantoot (or **Ghantut**) and **Al Jazira Resort**, 58km from the Dubai Trade Centre. Turn left at the next roundabout for Al Jazira Resort, and then right at the following roundabout: signposted 'Bungalows'. The plantation is immediately on your right. It is best to drive under the arch ahead and round the back of the 'Hill', between it and a wall, parking at the back of the plantation itself. Walk in to the right. Grey Hypocolius are usually found within two hundred metres of the hill. Late in the day is generally best. Although birds may still be at roost early in the day here, they do not usually remain to feed. It is possible to phone or pick up taxis to and from the resort 15 minutes' walk away.

The Ghantoot Polo Club, off the first roundabout, is good for Chestnut-bellied Sandgrouse and Cream-coloured Coursers, although access is generally not permitted. Luckily, the Coursers can easily be 'scoped' from the top of the man-made hill immediately by the plantation, while the sandgrouse often fly over calling noisily.

KEY SPECIES

Resident
Chestnut-bellied Sandgrouse, Rock Martin, Green Bee-eater

Winter
Cream-coloured Courser, Orphean Warbler, Desert Lesser Whitethroat, Plain Leaf Warbler

Passage
Grey Hypocolius, Rufous-tailed Scrub Robin, Ménétries's Warbler

Useful contacts and websites

BirdLife Middle East UAE
www.birdlifemed.org/Contries/uae/uae.html
BirdLife Partner for Middle East Region with section on United Arab Emirates.

Arabian Wildlife www.arabianwildlife.com
General site with information about birds

UAE Interact – Twitcher's Guide
www.uaeinteract.com/nature/bird/index.asp
Local birding website with checklist, information about habitats and recent reports.

Ornithological Society of the Middle East
www.osme.org
Ornithological Society of the Middle East, c/o The Lodge, Sandy, Bedfordshire SG19 2DL, UK.

Emirates Bird Records Committee
c/o P.O. Box 45553, Abu Dhabi, UAE
The EBRC maintains the national checklist and offers assistance and guiding service for visiting birders. Simon Aspinall (EBRC Chairman), email: hud-hud10@emirates.net.ae

Tommy Pedersen's website
www.tommypedersen.com
Tommy Pedersen (EBRC recorder), email: tommy777@emirates.net.ae
Personal website with photo gallery, news and updated birding site maps and other information.

Dubai Municipality Public Transport
www.vgn.dm.gov.ae/DMEGOV/dm-mp-transportation
Bus Services in Dubai area.

Books and publications

The Breeding Birds of the United Arab Emirates Simon Aspinall (1996) Hobby Publications ISBN 1872839045

The Birds of the United Arab Emirates Colin Richardson (1990) Hobby Publications ISBN 1872839002

The Shell Birdwatching Guide to the United Arab Emirates Colin Richardson and Simon Aspinall (1998) Hobby Publications ISBN 1872839053

Birds of Southern Arabia
Dave Robinson (1992) Motivate Publishing ISBN 1873544375

Field Guide to the Birds of the Middle East
R.F. Porter, S. Christensen, P. Schiermacker-Hansen (2004) Christopher Helm, ISBN 0713670169

Where to Watch Birds in Asia
Nigel Wheatley (1996) Christopher Helm ISBN 071364303X

Important Bird Areas in the Middle East
M I Evans (1994) BirdLife International ISBN 0946888280

Hong Kong

Although it has now been overtaken by many larger cities in terms of the number of its inhabitants, Hong Kong still constitutes one of the most densely populated cities in the world – a consequence of its history and geography. This makes very little room for birds and their habitats yet, surprisingly, Hong Kong offers some excellent birding and is a popular destination for several bird tours. The reason, of course, is the phenomenal spring wader migration through the area which offers some of the most sought-after wader species in the world. What is interesting is that most of the wetland habitats are actually man-made as a result of fish-farming

Black Kite is a familiar sight in most Asian cities

and agricultural activities, and that conservation efforts are now in place to maintain them as such in the face of threats of drainage and infill for urban construction. Even the forested areas in the interior of the New Territories are largely regenerated secondary growth, following years of deforestation.

Its location just south of the Tropic of Cancer means that within the relatively small area of the Hong Kong Special Autonomous Region (SAR) it is possible to see an interesting blend of tropical and temperate zone species, particularly during migration periods. By contrast, the summer (June to August) can be rather poor as the small number of resident species is augmented by just a few summer visitors. Situated on the eastern flank of the vast Pearl River Delta, the city is strategically located on an important migration flyway and provides a vital staging point for waders, wildfowl and many passerines.

TYPICAL SPECIES

Resident
Little Grebe, Chinese Pond Heron, Black-crowned Night Heron, Pacific Reef Heron, Little Egret, Great Egret, Black Kite, Common Moorhen, Spotted Dove, House Swift, Red-whiskered Bulbul, Chinese Bulbul, Oriental Magpie-Robin, Masked Laughingthrush, Common Tailorbird, Great Tit, Japanese White-eye, Crested Myna, Black-collared Starling, Eurasian Magpie, Collared Crow, Large-billed Crow, Tree Sparrow, Scaly-breasted Munia

Winter
Great Cormorant, Grey Heron, Eurasian Wigeon, Eurasian Teal, Eastern Marsh Harrier, Common Coot, Common Redshank, Common Greenshank, Black-headed Gull, Yellow-legged Gull, Eurasian Wryneck, Yellow Wagtail, White Wagtail, Grey Wagtail, Common Stonechat, Bluethroat, Dusky Warbler, Yellow-browed Warbler, Pallas's Warbler

Passage
Garganey, Japanese Sparrowhawk, Common Kestrel, Oriental Pratincole, Greater Sand Plover, Curlew Sandpiper, Red-necked Stint, Gull-billed Tern, Indian Cuckoo, Barn Swallow, Eyebrowed Thrush, Black-browed Reed Warbler, Oriental Reed Warbler, Japanese Paradise Flycatcher

GETTING AROUND

Your options for getting around Hong Kong depend very much on whether you are staying on Hong Kong Island or on the mainland at Kowloon. If on the island, you are better off relying on public transport which is very comprehensive and efficient and provides the fastest, most convenient way of commuting to the birding sites on the mainland. A car would be a real liability on the congested streets of Hong Kong. The subway (MTR) provides easy access between Hong Kong Island and several points on

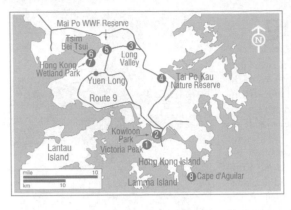

the mainland, and it interconnects with the commuter rail system (KCR) which fans out from Kowloon, serving many of the sites in the New Territories. In addition, there are several bus companies and a Light Rail System (LRT) serving both Kowloon and the New Territories, which connect with the commuter rail stations. There is also a vast network of minibus services which serve every corner of the SAR, although you will need to get some local assistance to make use of them. You can get an Octopus Smart Card in KCR and MTR stations, and in convenience stores such as 7-11, which is valid for travel on the subway, trains, buses and ferries. Unlike mainland China, most signposting and destination boards are bilingual (both Chinese and English).

If you have a car, you can visit several sites in the New Territories in the course of a day's outing from Kowloon. Route 9 starts and ends at Kowloon and loops around the New Territories, providing the quickest access to many of the birding sites. However once you exit the highway, it would be advisable to have a good road map for navigation. On the other hand, most of the sites are well known, well watched and developed with ecotourism in mind, so trails, signposting, water fountains and information boards are of a high standard and make it all very easy for the visiting birder.

Locations

1 **Victoria Peak** (3km)   *(Bus No. 15/15B/15C)*

Victoria Peak at over 500m is the highest point on Hong Kong Island and one of the best known tourist sites. It tends to get very crowded, especially at weekends, but most tourists tend to congregate around the Peak Tower and souvenir shops. If you break away from the crowds and walk the loop circuit of Lugard Road and Harlech Road, this brings you through some well-wooded areas of the hillside. An alternative path leads from the Peak Lookout Restaurant along Pok Fu Lam Reservoir Road downhill towards the reservoir. Although the birdlife includes nothing exceptional, this site is very conveniently located for anyone staying on Hong Kong Island and, as it is a firmly established part of the tourist trail, it is most likely that you will visit the Peak at some point in your stay.

KEY SPECIES

Resident
Chinese Goshawk, Asian Koel, Greater Coucal, Hwamei, Greater Necklaced Laughingthrush, Black-throated Laughingthrush, Blue Whistling Thrush, Blue Magpie, Fork-tailed Sunbird, White-rumped Munia

Winter
Oriental Turtle Dove, Olive-backed Pipit, Siberian Rubythroat, Red-flanked Bluetail, Daurian Redstart, Japanese Thrush, Pale Thrush, Scaly Thrush, Asian Stubtail, Asian Brown Flycatcher

Passage
Fork-tailed Swift, Black-winged Cuckoo-shrike

The best way to get to Victoria Peak is to take the Peak Tram which departs from a terminal at the west end of Hong Kong Park, which is also a site worth visiting early in the morning. The Peak is also served by First Bus No. 15 and CityBus Nos. 15, 15B and 15C from the Central District.

2 Kowloon Park (6km)

(Bus Nos. 7/12/98)

It must have something to do with skyscrapers, but modest city parks in the midst of huge skyscraper canyons, such as Central Park in New York, seem to be particularly good at attracting transient birds, and Kowloon Park seems to fit this category quite nicely. As well as migrants, it supports a good few resident species and is well worth a visit if you are staying in Kowloon. The park itself is rather ornamental but most of the trees are indigenous species and there is a pond for waterfowl, although nearly all the ducks are not wild. Spring and autumn are, of course, the best periods for birding in the park and The Hong Kong Birdwatching Society hosts regular field trips here at these times. This is a good way to meet local birders, and if you are new to birding in Southeast Asia it provides an excellent opportunity to familiarise yourself with the commoner members of this rich and sometimes bewildering avifauna.

The park entrance is next to the Mosque on Nathan Road, the main north/south thoroughfare of Kowloon Peninsula. The Tsim Sha Shui KCR and MTR station is conveniently located at the southeast corner of the park.

KEY SPECIES

Resident
White-breasted Waterhen, White-breasted Kingfisher, Common Kingfisher, Asian Koel, Greater Coucal, Hwamei, Black-throated Laughingthrush, Blue Magpie, Chestnut-tailed Starling (escape), White-rumped Munia, Fork-tailed Sunbird.

Winter
Oriental Turtle Dove, Olive-backed Pipit, Siberian

Rubythroat, Rufous-tailed Robin, Japanese Thrush, Grey-backed Thrush, Pale Thrush

Passage
Red Collared Dove, Pale-legged Leaf Warbler, Eastern Crowned Leaf Warbler, Arctic Warbler, Yellow-rumped Flycatcher, Narcissus Flycatcher, Mugimaki Flycatcher, Asian Brown Flycatcher

3 Long Valley (30km)

(Bus No. 76K)

Long Valley is a wet and dry agricultural area near the northern border of the SAR, which is cultivated for vege-tables. These fields provide very productive feeding opportunities for passerines and other species, particularly in winter and during migra-tion. The area is very open, with no appreciable cover from which to approach the birds, so the best way to 'do' this site is to walk along the

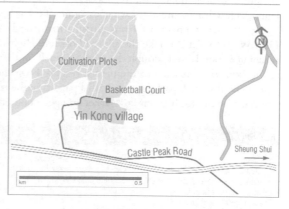

dykes between the paddies and cultivated plots checking the field margins and ditches. A telescope would be very beneficial here.

The fields are just north of the village of **Yin Kong** which is on **Route 9**. Take **Route 1** northeast from **Kowloon** to **Sha Tin**, then **Route 9** to **Fanling**. At Fanling, exit south onto **Castle Peak Road**, which is signposted Kwu Tung. After about 2km turn off Castle Peak Road into the village signposted Yin Kong and park near the public toilets or Basketball Court on the north side of the village, from where you can walk into the fields.

The nearest KCR station is at Sheung Shui about 1km away, from where you can get a taxi, red minibus No. 17 or Bus No. 76K, which operates along Castle Peak Road.

KEY SPECIES

Resident
White-breasted Waterhen, Bonelli's Eagle, Greater Painted-snipe, White-breasted Kingfisher, Common Kingfisher, Plain Prinia, Long-tailed Shrike, White-rumped Munia

Winter
Japanese Quail, Little Ringed Plover, Green Sandpiper, Common Sandpiper, Pintail Snipe, Common Snipe, Oriental Turtle Dove, Red-throated Pipit, Olive-backed Pipit, Richard's Pipit, Siberian Rubythroat, Red-flanked Bluetail, Zitting Cisticola, Black-faced Bunting, Little Bunting

Passage
Schrenck's Bittern, Cinnamon Bittern, Cattle Egret, Black-shouldered Kite, Eurasian Hobby, Oriental Pratincole, Wood Sandpiper, Red Collared Dove, Citrine Wagtail, Red-rumped Swallow, Pallas's Warbler, Great Reed Warbler, Brown Shrike, Grey-headed Bunting, Chestnut Bunting, Yellow-breasted Bunting, White-shouldered Starling

4 Tai Po Kau Nature Reserve (22km)   (Bus Nos. 70/72/73A/74A)

This is a forest reserve in the eastern part of the New Territories. It is a fairly mature secondary growth forest and supports a fine selection of resident species as well as being good for migrants. There are a number of trails through the forest, but those generally best for birding are the Red and Blue Trails. Regeneration has enabled several forest species to become re-established in the Hong Kong area, though some, such as the populations of Blue-winged Minlas and Yellow-cheeked Tits, derive from escapes from captivity. However, others such as the Scarlet Minivets and Chestnut Bulbuls are natural recolonisers. The best strategy is an early morning walk up the hill from Tai Po Road and then along the two lower trails, starting on the Blue and then cutting to the Red along the signposted shortcut. This brings you back around to the small dam, where all trails begin. The Brown and Yellow Trails scale the valley on either side of the stream, but forest quality there is poorer. Check the birders' logbook at the Warden's Hut which will inform you of what's around, if it has been updated recently.

Take **Route 1** northeast from **Kowloon** towards Sha Tin, then **Route 9 (Tolo Highway)** to **Tai Po**. Take the sliproad off Tolo Highway for **Tai Po Market** and, on reaching a roundabout, double-back on yourself on the road marked Tai Po Kau. After about 2km, the entrance to the nature reserve is on the right-hand (west) side of the road, behind a small parking area with a pagoda in the middle. If using public transport you can take the KCR to Tai Po Market and get a taxi or bus to the entrance.

KEY SPECIES

Resident
Crested Serpent Eagle, Chinese Goshawk, Asian Koel, Greater Coucal, Great Barbet, Chestnut Bulbul, Scarlet Minivet, Grey-chinned Minivet, Orange-bellied Leafbird, Greater Necklaced Laughingthrush, Black-throated Laughingthrush, Hwamei, Blue Whistling Thrush, Silver-eared Mesia (I), Red-billed Leiothrix (I), Blue-winged Minla (I), White-bellied Yuhina, Yellow-cheeked Tit (I), Velvet-fronted Nuthatch (I), Fire-breasted Flowerpecker, Scarlet-backed Flowerpecker, Fork-tailed Sunbird, Blue Magpie, White-rumped Munia

Summer
Chestnut-winged Cuckoo, Large Hawk-Cuckoo, Hainan Blue Flycatcher

Winter
Oriental Turtle Dove, Olive-backed Pipit, Red-flanked Bluetail, Daurian Redstart, Rufous-tailed Robin, Eyebrowed Thrush, Grey-backed Thrush, Japanese Thrush, Pale Thrush, Scaly Thrush, Asian Stubtail, Grey-headed Flycatcher, Asian Brown Flycatcher

Passage
Forest Wagtail, Siberian Blue Robin, Black-winged Cuckoo-shrike, Arctic Warbler, Pale-legged Leaf Warbler, Narcissus Flycatcher, Mugimaki Flycatcher, Yellow-rumped Flycatcher

5 **Mai Po WWF Reserve** (25km)

Mai Po is one of the best-known birding sites in the world. Located in the north-west of the New Territories, Mai Po is a collection of shrimp ponds which are periodically flooded and drained producing a rich roosting and feeding area for many wild-fowl, waders and herons. The site is adjacent to inner Deep Bay which has a huge expanse of tidal mudflats, and Mai Po becomes a roosting area for the birds when the tide is in.

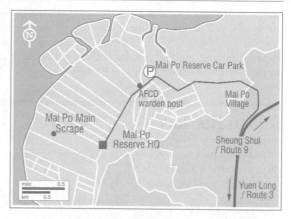

The site is managed by WWF, and access is strictly controlled by government permit. However, the reserve is very well equipped with hides and boardwalks, permitting close observation of roosting waders. During spring and autumn, most birders and photographers visit the hides at high tide and visit the boardwalk hide, located at the edge of Deep Bay in the mangroves, on the falling tide. A tide level of 2.2m entirely covers the mudflats. The spring and autumn wader migrations are phenomenal and include many of the Palearctic's most sought-after species. This is one of the few places in the world where the rare Spoon-billed Sandpiper is seen annually. A week would not be too long to spend visiting the reserve at peak migration time in April and October as the numbers of waders and other species are so large, and the traffic of flocks through the site produces a turnover of birds every day.

There is a perception that getting authorisation to visit Mai Po is very difficult and impractical for the short-term visitor. However, this is not now the case. Guided tours for the public are conducted at weekends and on holidays: however, these produce few opportunities for serious birdwatching and are not recommended. If you wish to obtain a permit for the reserve, you should contact WWF as far in advance as possible, in order to book one of the permits available at the reserve. There is a daily charge for this service and use of the permit requires a deposit. The reserve can be contacted at maipo@wwf.org.hk, or by phone at +852 24716306. Should a permit not be available, it is possible to obtain one from the Agriculture, Fisheries and Conservation Department of the government, but you will need to apply sufficiently far in advance to

KEY SPECIES

Resident
Spot-billed Duck, White-breasted Waterhen, White-breasted Kingfisher, Common Kingfisher, Pied Kingfisher, Greater Coucal, Yellow-bellied Prinia, Plain Prinia, Black-collared Starling, Long-tailed Shrike, White-rumped Munia

Winter
Dalmatian Pelican, Black-faced Spoonbill, Eurasian Spoonbill, Common Shelduck, Northern Shoveler, Mallard, Northern Pintail, Osprey, Pied Avocet, Grey Plover, Pacific Golden Plover, Kentish Plover, Dunlin, Spotted Redshank, Marsh Sandpiper, Eurasian Curlew, Saunders's Gull,

Oriental Turtle Dove, Black-capped Kingfisher, Olive-backed Pipit, Siberian Rubythroat, Red-flanked Bluetail, White-cheeked Starling, Red-billed Starling, Little Bunting, Black-faced Bunting, Chinese Penduline Tit

Passage
Intermediate Egret, Lesser Sand Plover, Great Knot, Long-toed Stint, Broad-billed Sandpiper, Spoon-billed Sandpiper, Sharp-tailed Sandpiper, Whimbrel, Asian Dowitcher, Wood Sandpiper, Terek Sandpiper, Ruddy Turnstone, Caspian Tern, White-winged Black Tern, White-shouldered Starling

allow for processing time and postage of the permit. AFCD can be contacted at mailbox@afcd.gov.hk

Take **Route 3** or **5** northwest from Kowloon, then west and north on **Route 9** and exit west at the **Fairview Park Exit** onto the slip road (**Castle Peak Road**). Go north on Castle Peak Road for about 2km to Mai Po village and turn left (west) at the signpost for Mai Po WWF Reserve. If using public transport, take the KCR train to Sheung Shui or Long Ping and from there get a taxi or minibus to Mai Po. The reserve entrance is about 1.5km from Mai Po village.

6 **Tsim Bei Tsui** (34km)

If you experience difficulty gaining access to Mai Po, then the mudflats and fish ponds at Tsim Bei Tsui offer a reasonable alternative. Tsim Bei Tsui is a village on the southern shore of Deep Bay and is part of the same complex tidal wetland system as Mai Po. It is also less acutely dependent on tidal conditions than Mai Po as there is a smaller expanse of mudflats, and with a telescope you can scan the waders at low or mid-tide. This can be done from the road near the police post at the beginning of the fence, or from the top of the adjacent hill. It is also possible to walk along the border fence road towards and through the fish ponds inland. There is much to see in

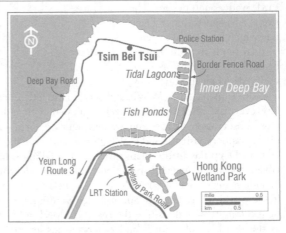

this area, and the fish ponds can contain many birds during migration and winter seasons.

Take **Route 3** or **5** northwest from Kowloon, then west and north on **Route 9** and follow the signs towards Tuen Mun. After about 28km, exit north on Hung Tin Road Interchange towards **Lau Fau Shan**. Continue north from Lau Fau Shan on **Deep Bay Road** along the shore of the Outer Deep Bay to Tsim Bei Tsui. If you are using public transport you could take the KCR to Yuen Long, from where you could get a taxi or green-roofed Minibus No. 35 from Tai Fung St to Sha Kiu or Tsim Bei Tsui. Alight where the bus turns around.

KEY SPECIES

Resident
Purple Heron, Spot-billed Duck, White-breasted Waterhen, White-breasted Kingfisher, Common Kingfisher, Pied Kingfisher, Plain Prinia, Long-tailed Shrike, White-rumped Munia

Winter
Dalmatian Pelican, Black-faced Spoonbill, Intermediate Egret, Common Shelduck, Northern Shoveler, Mallard, Northern Pintail, Osprey, Pied Avocet, Grey-headed Lapwing, Grey Plover, Pacific Golden Plover, Kentish Plover, Dunlin, Spotted Redshank, Marsh Sandpiper,

Eurasian Curlew, Common Sandpiper, Saunders's Gull, Oriental Turtle Dove, Black-capped Kingfisher, Olive-backed Pipit, Siberian Rubythroat, Red-flanked Bluetail, Daurian Redstart, Chinese Penduline Tit, White-cheeked Starling, Little Bunting

Passage
Great Knot, Long-toed Stint, Broad-billed Sandpiper, Sharp-tailed Sandpiper, Whimbrel, Asian Dowitcher, Wood Sandpiper, Terek Sandpiper, Black Drongo, White-shouldered Starling, Yellow-breasted Bunting

7 Hong Kong Wetland Park (28km) *(LRT 705/706/761)*

This is another wetland site which offers an alternative to Mai Po, if you have little time to spare and are unable to get into the former at short notice. At time of writing the park is not yet fully developed and offers limited access to the wetland site, but the Phase 2 expansion will provide a network of nature trails, boardwalks, a bird observation tower and several strategically located hides. The wetland is being reconstructed from abandoned fish ponds on the south shore of Deep Bay and is almost surrounded by the high rises of Tin Shui Wai New Town.

Take **Route 3** or **5** northwest from **Kowloon** to **Yuen Long** and exit north on the **Long Tin Road** interchange towards **Tin Shui Wai**. After about 3km, Long Tin Road becomes **Wetland Park Road** and the entrance to the park is on the right-hand (east) side of the road. The park is easy to get to by public transport as there is a Light Rail (LRT) station on Wetland Road.

KEY SPECIES

Resident	Winter	Passage
White-breasted Waterhen, White-breasted Kingfisher, Common Kingfisher, Pied Kingfisher, Greater Painted-snipe, Yellow-bellied Prinia, Plain Prinia, White-rumped Munia	Northern Shoveler, Mallard, Northern Pintail, Osprey, Pied Avocet, Little Ringed Plover, Spotted Redshank, Marsh Sandpiper, Eurasian Curlew, Common Sandpiper, Oriental Turtle Dove, Black-capped Kingfisher, Olive-backed Pipit, Siberian Rubythroat, Chinese Penduline Tit	Cinnamon Bittern, Broad-billed Sandpiper, Sharp-tailed Sandpiper, Whimbrel, Wood Sandpiper, Terek Sandpiper

Useful contacts and websites

Hong Kong Birdwatching Society (English)
www.hkbws.org.hk
GPO Box 12460, Hong Kong. Tel. (852) 2377 4387.
Local BirdLife Partner.

Hong Kong Outdoors (English)
www.hkoutdoors.com
Martin Williams' personal website with information about birding around Hong Kong

Mai Po Nature Reserve (English)
www.wwf.org.hk/eng/maipo
WWF website for Mai Po Nature Reserve.

Hong Kong Nature Net (English)
www.hknature.net
General natural history website about Hong Kong with information about birds.

Oriental Bird Club www.orientalbirdclub.org
P.O. Box 324, Bedford, MK42 0WG, UK

Hong Kong Bird Records Committee
www.hkbws.org.hk/record.html
c/o Hong Kong Bird Watching Society, G.P.O. Box 12460, Hong Kong. Email: recorder@hkbws.org.hk
Hong Kong Birdline: Tel. (852) 2667 4537.

Tidal Schedule for Hong Kong
www.hko.gov.hk/tide/etide_main.htm

KMB (English and Chinese) www.kmb.hk
Bus Services in Kowloon and the New Territories.

KCRC Services (English and Chinese)
www.kcrc.com
Local Commuter Train and Light Rail services in Kowloon and the New Territories.

MTR Corporation (English)
www.mtrcorp.com
Metro Services between Hong Kong Island, Lantau Island and Kowloon.

Books and publications

Birds of Hong Kong and South China
Clive Viney, Karen Phillipps and Lam Chiu Ying (2005, 8th edition) Information Services Department
ISBN 962020347X

The Avifauna of Hong Kong
G.J. Carey, *et al.* (2001) Hong Kong Bird Watching Society, ISBN 9627508020

A Field Guide to the Birds of China
John MacKinnon and Karen Phillipps (2000)
Oxford University Press, ISBN 0198549407

A Field Guide to the Birds of South-East Asia
Craig Robson (2000) New Holland
ISBN 1853683132

A Photographic Guide to Birds of China Including Hong Kong
John MacKinnon and Nigel Hicks (1996)
New Holland, ISBN 1853687642

A Photographic Guide to Birds of Hong Kong
Hong Kong Birdwatching Society (2005)

Birding Hong Kong: A Site Guide
David Diskin (1995) Privately published.

Where to Watch Birds and Other Wildlife in Hong Kong and Guangdon T. J. Woodward and

G. J. Carey (1996) Privately published.
ISBN 9628508415

Where to Watch Birds in Asia
Nigel Wheatley (1996) Christopher Helm
ISBN 071364303X

● ●

KEY SPECIES

Resident
Little Black Cormorant, Little Egret, Great Egret, Intermediate Egret, Black-crowned Night Heron, Cattle Egret, White-breasted Waterhen, Spotted Dove, Island Collared Dove, Plaintive Cuckoo, Edible-nest Swiftlet, Linchi Swiftlet, House Swift, Collared Kingfisher, Striated Swallow, Pacific Swallow, Common Iora, Pied Triller, Sooty-headed Bulbul, Yellow-vented Bulbul, Common Tailorbird, Golden-bellied Gerygone, Pied Fantail, Oriental Magpie-Robin, Long-tailed Shrike, Ashy Drongo, Bar-winged Prinia, Olive-backed Sunbird, Javan Myna, Black-naped Oriole, Scarlet-headed Flowerpecker, Eurasian Tree Sparrow, Javan Munia, Scaly-breasted Munia

Collared Kingfisher is the most common kingfisher on Java

Jakarta

The Indonesian Archipelago is one of the world's great centres of endemism and, as a country, Indonesia has the most diverse avifauna in Asia. However, its size and the vast number of islands, some of which are themselves rich in endemics, make 'doing' Indonesia a formidable undertaking requiring several months or even a year for the most dedicated birder. Consequently, most birders arriving in Jakarta will wish to get clear of the city as quickly as possible in order to hit the hotspots without any delay.

In truth, the city has few good birding sites within its precincts, and a visiting birder will need to be prepared to travel for a few hours further afield to find any reasonable birding habitat. This large, densely populated urban area is on the northwest coast of Java, the most populous island in the world, where unfortunately human population growth and urban development have vastly depleted both habitats and birdlife.

The city lies on a broad sweeping bay which was formerly lined with mangroves, almost all of which have been cleared away. However, some of the coral islands off the coast of Jakarta still retain their mangroves and intertidal habitats and they can be visited on day trips from the city's waterfront. The low-lying marshy land around Jakarta, drained by the Ciliwung River, was once covered by swamp forests but has long-since been cleared for rice cultivation and subsequently for urban development. There is virtually no lowland forest left on Java, and even national parks that contain some primary forest are under threat from unauthorised logging.

The situation is slightly better in relation to the montane forest which covers the slopes of the volcanoes south of Jakarta, although getting there involves a rather lengthy journey. I have included Gunung Gede-Pangrango National Park, about two hours' drive from the city, as it is, arguably, the best site on Java. There are about 24 endemics on Java and most can be seen here. If time permits, it is also feasible to visit the coastal lowland forests at Carita National Park in western Java (but this site is not included here).

GETTING AROUND

Jakarta is unusual as a coastal city in that many of the best sites involve lengthy journeys inland. Hence you will spend more of your day travelling rather than birding if you are relying on public transport. Having a car makes a big difference to your mobility and places the more distant sites well within a day's trip of the city, and even offers the opportunity of visiting multiple sites. It can also provide some respite from the extreme heat and the heavy downpours of rain, which are typical of Java's

climate. Driving within the city is not advisable, as it is extremely congested and very confusing for a visitor. If possible try to navigate to the Outer Ring Road which encircles the city. This has major interchanges with the other main toll roads: Jalan Prof. Sediyatmo going towards the airport and the coastal sites; Jalan Merak going towards the west coast; and Jalan Jagorawi going south to the mountains.

Without a car you will need to use trains or buses. There are three main bus terminals: Kalideres in the west of the city, serving the west coast; Pulo Gadung in the east, serving central and eastern Java; and the Kampung Rambutan terminal in south Jakarta, serving all points south. These terminals are quite distant from the city centre but at least are easily reached by taxi. Apart from the bus and train services, there is an extensive informal system of taxis, minibuses (bemos), motorcycles and converted lawnmowers, all of which combine to make the whole island of Java accessible and convenient.

Most birders visit Java during the dry season, between April and October, although even during this period you can expect some heavy showers. The wet season can produce very difficult conditions for birding, and even under the forest canopy the trails can be extremely muddy and slippery. You can easily buy some cheap rainwear locally. This is more portable and easily deployable than an umbrella, which can be awkward in the forest, but this needs to be balanced against the noise created when walking.

Make sure you obtain a map which covers western Java (Jawa bagian barat) as well as the city of Jakarta.

Locations

1 Muara Angke Nature Reserve (13km)

The reserve is now a rather small wetland site at the mouth of the Muara Angke River in the western suburbs of Jakarta. Although increasingly encroached upon by urban development, the marsh still supports a reasonable selection of wetland birds and is a reliable site for Sunda Coucals. Some attempts are being made to restore the mangroves and there is a boardwalk (sometimes in a state of disrepair) and a bird observation tower. The coastal zone along the northern perimeter of the reserve is a good lookout point to scan for terns and other seabirds.

Take the main road west out of Jakarta (**Jalan Prof. Sediyatmo**) and exit north at **Pluit Karang** interchange towards the **Pantai Indah Kapuk** residential area. Follow the main road (**Jalan Pluit Karang**) through this area, passing through a large arch and crossing a bridge over the Muara Angke River. Turn right at the first round-about onto **Pantai Indah Utara 2**. The reserve entrance is on the right-hand (north) side of this road, a little way beyond the roundabout. There are several bus and minibus routes from the main bus terminals. **Patas Bus Route P6B** serves the area from **Kampung Rambutan** terminal.

KEY SPECIES

Resident
Little Black Cormorant, Darter, Cinnamon Bittern, Purple Heron, Great Egret, Javan Pond Heron, Striated Heron, Black-crowned Night Heron, Milky Stork, Lesser Whistling Duck, Sunda Teal, Barred Buttonquail, Ruddy-breasted Crake, White-browed Crake, Watercock, White-breasted Waterhen, Common

Moorhen, Purple Swamphen, Javan Plover, Pink-necked Green Pigeon, Red-breasted Parakeet, Sunda Coucal, Savanna Nightjar, Small Blue Kingfisher, Sacred Kingfisher, Blue-tailed Bee-eater, Fulvous-breasted Woodpecker, Sunda Woodpecker, Zitting Cisticola, Clamorous Reed Warbler, Olive-backed Tailorbird, Plain Prinia, Bar-winged Prinia, Pied

Fantail, Mangrove Whistler, Golden-bellied Gerygone, White-breasted Woodswallow, Racket-tailed Treepie, Javan Myna

Oct–Apr
Black Bittern, Wood Sandpiper, Common Sandpiper, Whiskered Tern, White-winged Black Tern, Oriental Reed Warbler

2 **Pulau Rambut** (35km)

The Kepulauan Seribu or Thousand Islands Archipelago off Jakarta Bay, with its undisturbed mangrove and marine habitats, provides an invaluable haven for wildlife. Most of the islands are uninhabited and a number of them are wildlife sanctuaries. The sea area around the islands has been designated as the Kepulauan Seribu Marine National Park.

Pulau Rambut is one of the more important islands for birds as it provides a secure breeding site for herons and other aquatic species. There are a number of trails across the island and an observation tower, where you can observe the heronries at fairly close quarters without disturbing the birds. There is also an impressive roost of Large Flying Foxes in the mangroves. At low tide the sandy shores support small flocks of waders. A permit is required to visit the island, which can be obtained on arrival.

KEY SPECIES

Resident
Little Black Cormorant, Darter, Christmas Island Frigatebird, Lesser Frigatebird, Purple Heron, Great Egret, Intermediate Egret, Pacific Reef Heron, Black-crowned Night Heron, Milky Stork, Black-headed Ibis, Glossy Ibis, White-bellied Sea Eagle, Slaty-breasted Rail, Great Crested Tern, Green Imperial Pigeon, Pied Imperial Pigeon, Asian Koel, Collared Kingfisher,

Oriental Magpie-Robin, Bar-winged Prinia, Golden-bellied Gerygone, Plain-throated Sunbird, Olive-backed Sunbird, Black-naped Oriole, Hair-crested Drongo, White-breasted Woodswallow, Javan Myna

Oct–Apr
Pacific Golden Plover, Common Greenshank, Whimbrel, Eurasian Curlew, Common Sandpiper

You can access the island by taking a ferry to the nearby island of Untung Jawa, and from there you need to charter a boat or take a tourist boat to Pulau Rambut. Ferries for **Untung Jawa** depart from **Tanjung Pasir**, a coastal town about 30km northwest of Jakarta. Take the main road west out of Jakarta (**Jalan Merak**) and, after about 20km, turn **north** on the link road for the **airport** but continue on towards Tanjung Pasir. There are bus and minibus services to Tanjung Pasir from Kalideres bus terminal.

Alternatively, if you are with a group, there is also the possibility of chartering a boat at Marina Jaya Ancol in Jakarta.

There are few other good birding sites close to the city, but if your time or travel is restricted, the **Kemayoran Urban Park** just northeast of the city centre offers some birding possibilities. So does **Ragunan Zoo**, which is somewhat further afield, being about 20km south of the city centre: exit south from the **Outer Ring Road** onto **Jalan Harsono**.

Further Away

3 Bogor Botanical Gardens (60km)

A trip to the Botanical Gardens in Bogor (Kebun Raya Bogor) is a popular day-trip for many Jakartans. This town is south of Jakarta in the foothills of the Parahyangan mountain range which runs from east to west along the central spine of Java. If you want to experience a good selection of Javan forest birds without enduring a trip deep into the interior of the island, then this is a very worthwhile option. There is a good variety of native trees and shrubs, especially flowering species which attract many birds, so always check these when they are in bloom or when fruiting. A lake in the gardens and a river running through them provide some aquatic habitats. The only disadvantage of this site is that it does not open until 08:00 and it gets very crowded at weekends. If you are planning a major birding trip to the hotspots of Java and perhaps the other islands, you are unlikely to see many birds here that cannot be found elsewhere, but otherwise these gardens provide a gentle introduction to the birds of Indonesia for those who are new to birding in Southeast Asia.

The gardens are in the centre of Bogor, near the railway station (northwest corner) and bus terminal (southeast corner). Take the main toll road south from Jakarta (**Jagorawi Highway**). If you are using public transport, Bogor is well served by train from **Gambir Station**, the main station in Jakarta city centre, or by bus from **Kampung Rambutan** terminal. If travelling by bus, catch one that goes to **Baranangsiang** bus terminal in Bogor: the gardens are only a few minutes walk from here.

KEY SPECIES

Resident
Black-crowned Night Heron, Pink-necked Pigeon, Grey-cheeked Pigeon, Black-naped Fruit-Dove, Blue-crowned Hanging Parrot, Plaintive Cuckoo, Linchi Swiftlet, Blue-eared Kingfisher, Coppersmith Barbet, Fulvous-breasted Woodpecker, Sooty-headed Bulbul, Black-winged Flycatcher-Shrike, Bar-winged Prinia, Olive-backed Tailorbird, Hill Blue Flycatcher, Little Spiderhunter, Plain-throated Sunbird, Plain Flowerpecker, Scarlet-headed Flowerpecker, Oriental White-eye, Black-naped Oriole, Large-billed Crow, Java Sparrow

Oct–Apr
Oriental Cuckoo, Grey Wagtail, Siberian Thrush, Arctic Warbler, Asian Brown Flycatcher, Yellow-rumped Flycatcher, Brown Shrike

4 Puncak Pass (90km)

Puncak Pass lies along the main road to the southeast of Bogor, shortly before the turn-off to Cibodas. This area has only recently begun to become popular with birders, mainly due to its easy access, notably the **Telaga Warna Nature Reserve**. Although most of the species seen here can also be found in Gunung-Gede National Park, it may be worth visiting Puncak from Jakarta or Bogor if time is limited. September to October is the best time to watch thousands of raptors migrating south, often at close range.

From Bogor, take **Highway 2** towards **Bandung**. After about 30km the road crosses Puncak Pass. **Cibulao** village is surrounded by tea plantations and forest patches. If using public transport, take a bus or minibus from Jakarta or Bogor and ask to be dropped off at **Telaga Warna**, near Rindu Alam restaurant. This is a good site for some Javan endemics, such as Pygmy Tit and Javan Tesia. A local guide is needed to go deeper into the forest, but the lake and its surroundings are worth a look. The Raptor Conservation Society may be able to arrange trips to look for Javan Hawk-Eagles, Changeable Hawk-Eagles and Black Eagles. Contact them by email or at their offices near the gate of Gunung Gede NP.

KEY SPECIES

Resident
Black Eagle, Crested Serpent Eagle, Javan Hawk-Eagle, Changeable Hawk-Eagle, Spotted Kestrel, Plaintive Cuckoo, Chestnut-breasted Malkoha, Linchi Swiftlet, Collared Kingfisher, Javan Kingfisher, Blue-eared Kingfisher, Flame-fronted Barbet, Brown-throated Barbet, Ashy Bulbul, Orange-spotted Bulbul, Black-winged Flycatcher-Shrike,

Lesser Shortwing, Sunda Robin, Sunda Forktail, Sunda Whistling Thrush, Javan Tesia, Olive-backed Tailorbird, Mountain Warbler, Sunda Warbler, Indigo Flycatcher, Little Pied Flycatcher, Horsfield's Babbler, Pygmy Wren-Babbler, White-bibbed Babbler, Crescent-chested Babbler, Javan Fulvetta, Pygmy Tit, Great Tit, Blue Nuthatch, White-flanked Sunbird, Little Spiderhunter, Blood-

breasted Flowerpecker, Scarlet-headed Flowerpecker, Oriental White-eye, Ashy Drongo, Javan Munia

Oct–Apr
Oriental Honey-buzzard, Chinese Goshawk, Japanese Sparrowhawk, Oriental Cuckoo, Grey Wagtail, Siberian Thrush, Arctic Warbler, Asian Brown Flycatcher, Yellow-rumped Flycatcher, Brown Shrike

5 Cibodas Botanical Gardens (100km)

Although these gardens are quite distant from the city, they are easily visited in the course of a day trip. Moreover, they are by the entrance to Gunung Gede National Park, a major birding hotspot, so you might consider visiting this site for a day or weekend, particularly if you have a car at your disposal.

Situated at an altitude of almost 1,400m, these gardens (Kebun Raya Cibodas) enjoy a much cooler climate than the lowlands around Jakarta and have been planted with many temperate zone species. A good selection of montane birds can be seen here, including a number of Javan

endemics, although it would be better to work the Gunung Gede trails to see a greater variety. However, some species such as Pygmy Tits and White-flanked Sunbirds are much easier to see in the gardens.

There is a good network of paths inside the gardens with the forest edge being the best area. The gardens are adjacent to the Cibodas Golf Course, and the pathway alongside the golf course towards the Gunung Gede mountain trail also has good forest-edge habitat. The golf course itself provides a good vantage to watch for raptors over the montane forest, and this is perhaps the easiest way to see Javan Hawk-Eagles.

There is a modest fee for entry, but the gardens open at 06:00 so you have a good opportunity to visit the site early in the morning when bird activity is at its highest and before the crowds have built up. Beware that the gardens are usually packed at week-ends! A visit to the gardens is easily combined with the Gunung Gede trails (see 6).

Take the main toll road south from Jakarta (**Jagorawi Highway**) to **Bogor**. At Bogor turn east onto the road towards **Bandung** (**Highway 2**). After about 30km, after crossing the Puncak Pass and just before reaching **Cipanas**, you will reach the village of **Desa Cimacan**; turn south here for about 5km to get to Cibodas. When you get to Cibodas, check in at **Freddy's Homestay** for information or if you plan to stay overnight. This is a guesthouse and restaurant, less than a kilometre north of the entrance, which maintains a birders' log of recent sightings at Cibodas and Gunung Gede. There are several other places to stay here and local guides can be hired to find you the birds.

If travelling by public transport, take a direct bus to Bandung via Puncak, and ask to be dropped off at Perempatan Cibodas (Desa Cimacan). From here, take an 'angkot' (small, yellow minibus) directly to Kebun Raya Cibodas.

KEY SPECIES

Resident
Black Eagle, Crested Serpent Eagle, Javan Hawk-Eagle, Spotted Kestrel, Plaintive Cuckoo, Chestnut-breasted Malkoha, Yellow-throated Hanging Parrot, Linchi Swiftlet, Collared Kingfisher, Flame-fronted Barbet, Brown-throated Barbet, Ashy Bulbul, Orange-spotted Bulbul, Black-winged Flycatcher-Shrike, Lesser Shortwing, Sunda Robin, Sunda Forktail, Sunda Whistling Thrush, Olive-backed Tailorbird, Mountain Warbler, Sunda Warbler, Indigo Flycatcher, Little Pied Flycatcher, Horsfield's Babbler, Pygmy Wren-Babbler, White-bibbed Babbler, Crescent-chested Babbler, Javan Fulvetta, Pygmy Tit, Great Tit, Blue Nuthatch, White-flanked Sunbird, Little Spiderhunter, Blood-breasted Flowerpecker, Scarlet-headed Flowerpecker, Oriental White-eye, Ashy Drongo, Javan Munia

Oct–Apr
Grey Wagtail, Siberian Thrush, Eastern Crowned Leaf Warbler, Asian Brown Flycatcher, Mugimaki Flycatcher

---

6 Gunung Gede-Pangrango National Park (100km)

Although this national park is rather distant from the city, it is one of the most important birding sites on the island of Java and in Indonesia as a whole. Gede and Pangrango are two volcanoes to the south of Jakarta surrounded by montane primary forest. This area supports almost all of the Javan endemic species and is one of the last outposts for many of them. It can be visited in the course of a (longish) day's birding from Jakarta, but ideally should be visited over two days (or more) to give you the opportunity to be on-site at dawn for some of the mountain's specialities.

Gede is almost 3,000m high and the hiking trail from Cibodas Botanical Gardens traverses several vegetation zones from submontane forest to the high-altitude meadows at the rim of the craters. The hiking trail is over 11km long with numbered posts along the way, and would require a full day to undertake this tough climb in both directions. However, most of the mountain's specialities can be seen on the lower parts of the trail, making a climb to the top seem less attractive. Many birders cover this trail in two days by camping out near the summit, and this gives time to explore some of the side trails and to spend time in the best birding habitats. If you are restricted to just a day, then the best sites are the Blue Lake which is less than an hour's hike along the main trail from Cibodas; the Cibereum Waterfalls which is also about an hour from Cibodas and a good evening stakeout for Salvadori's Nightjars and Waterfall Swifts; and the Hot Springs area which is about two hours from Cibodas and at an altitude of over 2,000m. The only species that are usually seen above the Hot Springs are Rufous Woodcocks, Wedge-tailed Pigeons, Dark-backed Imperial Pigeons, Horsfield's Thrushes, Island Thrushes and Volcano Swiftlets: these last can be seen above the rim of Gunung Gede crater itself.

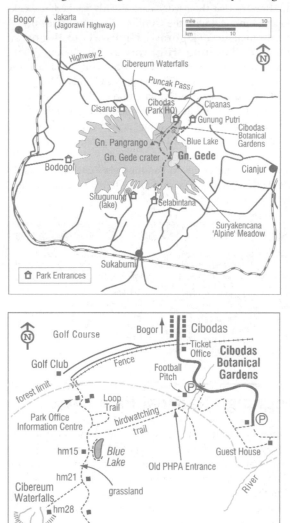

On the way back down the mountain, if you have time you could take the 'Birdwatching Trail' which branches off to the right (east) below the Blue Lake and leads into the Cibodas Botanical Gardens. This is a much quieter trail and less-frequented by visitors. However, it is much more overgrown and very steep in places, making for difficult birding, but when the park is swarming with day-trippers it does provide some relief from the crowds.

Needless to say, it is best to visit the park on weekdays, as the trails can be jam-packed with noisy visitors at weekends. Despite its tropical location, the higher elevations of Gunung Gede can be very cold in the early morning, so bring some extra layers of clothing and even gloves if you have them!

The directions are the same as for Cibodas Botanical Gardens, as this is the main entrance to the park and you will need to check in here in order to obtain a visitor's permit. Also check the logbook at Freddy's Homestay for recent sightings and, if you wish, you can hire a bird guide here.

KEY SPECIES

Resident
Black Eagle, Crested Serpent Eagle, Changeable Hawk-Eagle, Javan Hawk-Eagle, Spotted Kestrel, Chestnut-bellied Partridge, Red Junglefowl, Rufous Woodcock, Pink-necked Pigeon, Wedge-tailed Pigeon, Dark-backed Imperial Pigeon, Barred Cuckoo-Dove, Little Cuckoo-Dove, Oriental Cuckoo, Rusty-breasted Cuckoo, Javan Scops Owl, Collared Scops Owl, Salvadori's Nightjar, Waterfall Swift, Volcano Swiftlet, Blue-tailed Trogon, Flame-fronted Barbet, Brown-throated Barbet, Crimson-winged Woodpecker, Orange-backed Woodpecker, Banded Broadbill, Lesser Cuckoo-shrike, Sunda Cuckoo-shrike, Orange-spotted Bulbul, Sunda Bulbul, Sunda Minivet, Lesser Shortwing, White-browed Shortwing, Sunda Robin, Sunda Forktail, White-crowned Forktail, Javan Cochoa, Sunda Whistling Thrush, Sunda Thrush, Horsfield's Thrush, Island Thrush, Sunda Warbler, Mountain Warbler, Mountain Tailorbird, Javan Tesia, Sunda Bush Warbler, Indigo Flycatcher, Snowy-browed Flycatcher, Little Pied Flycatcher, Grey-headed Canary-Flycatcher, Rufous-tailed Fantail, Horsfield's Babbler, Chestnut-backed Scimitar Babbler, Eye-browed Wren-Babbler, Pygmy Wren-Babbler, White-bibbed Babbler, Crescent-chested Babbler, Rufous-fronted Laughingthrush, White-browed Shrike-Babbler, Chestnut-fronted Shrike-Babbler, Javan Fulvetta, Spotted Crocias, Blue Nuthatch, White-flanked Sunbird, Little Spiderhunter, Grey-breasted Spiderhunter, Blood-breasted Flowerpecker, Mountain White-eye, Javan Grey-throated White-eye, Ashy Drongo, Lesser Racket-tailed Drongo, Tawny-breasted Parrotfinch, Pin-tailed Parrotfinch, Mountain Serin

---

7 Pulau Dua (90km)

Pulau Dua and the adjacent Pulau Satu are two small islands lying just offshore in Banten Bay. They are separated from the mainland by sandflats and mangroves but you can walk across to the islands from the shore. There are a number of important breeding colonies of herons on the islands, and the sandflats support significant numbers of wintering and migrating waders. Pulau Dua is also the best site in Java for the endemic Javan White-eye. Although uninhabited there are trails through areas of former cultivation and there is an observation tower near the heronry.

Take the toll road west from Jakarta towards **Merak** and exit north at **Serang** to **Karangatun**. At Karangatun you can walk across to the island at low tide but you can take a boat trip from the harbour if you are unsure or unaware of the tide schedule. If travelling by public transport, you can take a bus from Kalideres Terminal to Serang, and from there take a bemo to Karangatun. There is also a train service to Serang from Tanah Abang Station, on the west side of Jakarta city centre, but it is rather infrequent.

KEY SPECIES

Resident
Little Black Cormorant, Darter, Cinnamon Bittern, Purple Heron, Great-billed Heron, Javan Pond Heron, Pacific Reef Heron, Striated Heron, Milky Stork, Glossy Ibis, Black-headed Ibis, Sunda Teal, Brahminy Kite, Barred Buttonquail, Slaty-breasted Rail, Zebra Dove, Island Collared Dove, Small Blue Kingfisher, Sacred Kingfisher, Blue-tailed Bee-eater, Zitting Cisticola, Grey-headed Canary-Flycatcher, Javan White-eye, White-breasted Woodswallow, Large-billed Crow, Asian Pied Starling, Scaly-breasted Munia

Oct–Apr
Osprey, Pacific Golden Plover, Lesser Sand Plover, Red-necked Stint, Sanderling, Common Greenshank, Whimbrel, Wood Sandpiper, Common Sandpiper, Whiskered Tern, White-winged Black Tern

---

Useful contacts and websites

Indonesian Ornithologists' Union (IdOU)
www.pili.or.id/kukila
Informal ornithological group and publishers of the journal *Kukila*.

Rare Bird Records
Kukila, P.O. Box 146, Bogor 16001, Indonesia.
Email: kukila@pili.or.id

BirdLife International – Indonesia Programme
www.birdlife.org/worldwide/national/indonesia
PO Box 310/Boo, Bogor 16003, Indonesia.
Email: birdlife@burung.org

Birding in Indonesia
www.indo.com/birding/
Rob Olivier's personal website with section dealing with birding on Java.

Oriental Bird Club www.orientalbirdclub.org
P.O.Box 324, Bedford, MK42 0WG, UK.

Raptor Conservation Society (RCS)
Email: raptorcs@hotmail.com or
u_suparman@yahoo.com

Tidal Schedule for Jakarta Bay
www.easytide.ukho.gov.uk/EasyTide/EasyTide/
ShowPrediction.aspx

Books and publications

Birding Indonesia
Paul Jepson (1997) Periplus Editions
ISBN 1569521336

Field Guide to the Birds of Java and Bali
John MacKinnon (1990) Gadjah Mada University
Press, ISBN 9794200921

A Field Guide to the Birds of Borneo,
Sumatra, Java and Bali
John MacKinnon and Karen Phillipps (1993)
Oxford University Press, ISBN 0198540353

A Field Guide to the Birds of South-East Asia
Craig Robson (2000) New Holland
ISBN 1853683132

A Photographic Guide to the Birds of
Indonesia Morten Strange (2002) Christopher
Helm, ISBN 0713664045

Where to Watch Birds in Asia
Nigel Wheatley (1996) Christopher Helm
ISBN 071364303X

Greater Racket-tailed Drongo looks as if it has
two smaller birds chasing it when it flies

TYPICAL SPECIES

Resident
Grey Heron, Spotted Dove, Zebra Dove, Asian Palm Swift, House Swift, Common Flameback, Brown Barbet, Gold-whiskered Barbet, White-breasted Kingfisher, Blue-throated Bee-eater, Asian Koel, Greater Coucal, Pacific Swallow, Common Iora, Common Tailorbird, Oriental Magpie-Robin, Plain-throated Sunbird, Yellow-vented Bulbul, Common Myna, Javan Myna, Asian Glossy Starling, Black-naped Oriole, House Crow, Large-billed Crow, Eurasian Tree Sparrow, Scaly-breasted Munia

Winter
Little Egret, Common Kingfisher, Blue-tailed Bee-eater, Fork-tailed Swift, Barn Swallow, Richard's Pipit, Yellow Wagtail, Brown Shrike, Tiger Shrike

Kuala Lumpur

Kuala Lumpur is in the state of Selangor on the western plain of the Malay Peninsula about equidistant (40km) from the coast and from the Main Range, the chain of highlands which form the central spine of the peninsula. The lowlands of the Malay Peninsula were once dominated by forest, but the drive in from the airport to KL will reveal that much of this has been cleared to make way for tin mining and rubber and oil-palm plantations. However, wherever plantations or tin mines have been abandoned, secondary-growth forest quickly re-establishes itself and can support a good variety of lowland forest edge species. By contrast, the highlands have been better preserved and there are some good areas of montane forest within an easy drive of KL.

The coast directly adjacent to KL at Port Klang has been developed largely for industry and port activities, but the northern part of coastal Selangor contains some excellent mudflats and mangrove swamps which are important internationally as migration staging points and wintering sites for thousands of waders.

Most birders visit Malaysia in February or March. The resident birds are at their most vocal then, the winter migrants are still present as well as being augmented by passage migrants, and the weather during this inter-monsoonal period is drier. On the other hand, the activity of many of the forest birds is dictated by the flowering and fruiting seasons of a variety of different trees in the forest, which vary throughout the year, so birding in these forests can be productive at any time.

Apart from the central city parks and green spaces, the good sites require you to travel some distance outside the city, but the travel time is well worth it because of the quality of the birding.

GETTING AROUND

Getting to the best sites in the KL area is most easily accomplished by car. Even if you only have a day or two to spare, you should try to obtain the use of a car since you will otherwise waste a lot of time travelling by public transport. Once clear of the city the driving is relatively easy and directions are well signposted. A car also gives you the opportunity to travel in

the heat of the day so that you may visit more than one site.

There is a good commuter rail system (KTM, LRT and Monorail) within KL. It also serves the surrounding towns but unfortunately none of the birding sites apart from the central city parks and open spaces. The bus service can fill in all of the gaps, but services are rather slow and infrequent and attempting to get to any of the distant sites involves changing buses and makes for generally slow progress. There are several bus stations in KL, the biggest of which, and the one which serves the birding destinations, is the Pudu Raya Terminal on Jalan Pudu about 100m from Chinatown.

Local taxis are flexible in taking tourists to out-of-town locations but could be expensive unless you have the option of sharing with a companion. Within the city precincts, there are several city bus services, but without a detailed knowledge of the city, these are of limited use to the visitor and you are better off using taxis.

If you are planning a trip to any of the more distant birding sites, dress appropriately for heat and biting insects and bring your rain gear with you at all times. Leeches will be a problem on wet trails in the mountainous areas.

Locations

1 Lake Gardens (City Centre) *(KL Sentral Railway Station)*

This central city park offers a good opportunity for the birder who is new to Southeast Asia to become familiar with all the common species before embarking on a more intensive birding trip. There are no birds here that cannot be seen at any of the good birding sites, but if time does not allow you to venture outside the city limits, then you can, at least, encounter the typical species here. There is some good secondary growth woodland and a lake, as well as more open parkland. There is even a bird park within the gardens, featuring many Malaysian birds, so if you are not planning to visit the interior of Malaysia, you can see what birds you are missing! The Lake Gardens are about 500m north of the KL Sentral Railway Station.

KEY SPECIES

Resident		Oct–Apr
Cinnamon Bittern, Striated Heron, White-breasted Waterhen, Crested Goshawk, Large-tailed Nightjar, Rufous Woodpecker, Pied Triller,	Olive-winged Bulbul, Little Spiderhunter, Greater Racket-tailed Drongo, Scarlet-backed Flowerpecker, Red-eyed Bulbul	Black Baza, Oriental Honey-buzzard, Japanese Sparrowhawk, Arctic Warbler, Asian Brown Flycatcher, Purple-backed Starling

2 Bukit Nanas Forest Park (City Centre) *(Bukit Nanas Metrorail Station)*

Bukit Nanas is a small hill just north of the city centre where the KL Tower is located. The hillsides are forested and the area is preserved as a forest park. Although small and surrounded by the city, the forest supports some lowland forest species. There are two forest trails at the northern end of the park but, although there are a few streams, it lacks a good-sized pond to attract aquatic species. The park opens from 07.00 and is well worth visiting if you are staying at any of the nearby hotels.

KEY SPECIES

Resident	Oct–Apr
Crested Goshawk, Large-tailed Nightjar, Pied Triller, Pied Fantail, Olive-winged Bulbul, Little Spiderhunter, Greater Racket-tailed Drongo, Scarlet-backed Flowerpecker	Japanese Sparrrowhawk, Asian Brown Flycatcher, Purple-backed Starling,

3 Rimba Ilmu Botanic Gardens (6km) *(Universiti LRT Station)*

The Botanic Gardens are located on the University of Malaya campus on the southwest side of the city. Many indigenous trees and shrubs have been planted in the gardens which attract a good selection of forest birds. Within the campus there is a lake and other small watercourses, which are worth checking for kingfishers and small herons.

Take the main road southwest from KL (**Highway 2**) towards the coast and about 1km past the **Sprint Highway Interchange** turn north on **Jalan Universiti**. The entrance to the university campus is about 600m on the right-hand side of the road. The gardens are on the northeast side of the campus. The Universiti LRT Station is about 2km southwest of the campus and on the far side of the highway, but there is a shuttle bus service to the campus.

KEY SPECIES

Resident		**Oct–Apr**
Cinnamon Bittern, Striated Heron, White-breasted Waterhen, Crested Goshawk, Large-tailed Nightjar, Blue-crowned Hanging Parrot, Coppersmith Barbet, Rufous Woodpecker, Grey-rumped Treeswift,	Plaintive Cuckoo, Rusty-breasted Cuckoo, House Swift, Pied Triller, Dark-necked Tailorbird, Olive-winged Bulbul, Little Spiderhunter, Greater Racket-tailed Drongo, Scarlet-backed Flowerpecker	Black Baza, Oriental Honey-buzzard, Japanese Sparrowhawk, Common Sandpiper, Arctic Warbler, Asian Brown Flycatcher, Purple-backed Starling

4 Templer Park (25km)   (Bus No. 66)

Templer Park is a former mining site in the foothills of the Main Range which has been turned into a public recreational area. Its main physical feature is the series of waterfalls which cut through the hills. The popular areas of the park are the swimming pool, campgrounds and golf course, as well as the waterfalls, but it is surrounded by lowland forest reserves and several trails lead from the park into the forest. While the public areas get very crowded, you can quickly leave the crowds behind by following one of the forest trails from the first car park through areas of bamboo thickets and dense forest.

Take the main highway north (**Highway 1**) towards **Ipoh**. The park entrance is located on the right-hand side of the highway, about 6km past the turnoff for **Batu Caves** (a well-known tourist site). Make sure that you head for this, the second entrance, not the first, which is about 3km further back towards Batu Caves. The bus for Kuala Kubu Baharu (Len Omnibus No. 66), from Pudu Raya passes right by the entrance.

KEY SPECIES

Resident		Flycatcher, Grey-headed Canary-Flycatcher, Ruby-cheeked Sunbird, Purple-naped Sunbird, Scarlet-backed Flowerpecker, Orange-bellied Flowerpecker
Crested Goshawk, Large-tailed Nightjar, Emerald Dove, Blue-crowned Hanging Parrot, Blue-eared Barbet, Crimson-winged Woodpecker, Grey-and-buff Woodpecker, Whiskered Treeswift, Grey-rumped Treeswift, White-bellied Swiftlet, Drongo Cuckoo, Chestnut-breasted	Malkoha, Scarlet Minivet, Blue-winged Leafbird, Cream-vented Bulbul, Olive-winged Bulbul, Grey-cheeked Bulbul, Striped Tit-Babbler, Dark-necked Tailorbird, White-rumped Shama, Chestnut-naped Forktail, White-bellied Yuhina, Golden-bellied Gerygone, Black-naped Monarch, Asian Paradise	**Winter** Forest Wagtail, Siberian Blue Robin, Eastern Crowned Leaf Warbler, Ferruginous Flycatcher

5 Old Gombak Road (25km)

One of the tourist sites to the north of KL, the Batu Caves mark the start of the River Gombak Valley. If you are visiting the caves, then it would be worth driving further along the 'old' Gombak Road, which passes through some good lowland and submontane forest. Birding is good all along the main road and the short trails running along the many streams. One of the best areas is around the University of Malaya Field Studies Centre at Km 30.

Take the main highway north (**Highway 1**) towards **Ipoh**, and after about 20km turn right (east) onto the **Middle Ring Road II** (**MRR2**) towards Batu Caves. Pass the entrance for Batu Caves until you come to a T-junction which is the old Gombak Road. Turn left, heading north, and the Field Studies Centre is about 10km past the T-junction. The Batu Caves are well served by buses from KL, but none serve the full length of the old Gombak Road and you would be better off getting a taxi from the caves.

KEY SPECIES

Resident
Crested Serpent Eagle, Emerald Dove, Blue-crowned Hanging Parrot, Rhinoceros Hornbill, Gold-whiskered Barbet, Blue-eared Barbet, Crimson-winged Woodpecker, Buff-necked Woodpecker, Grey-and-buff Woodpecker, Whiskered Treeswift, Grey-rumped Treeswift, White-bellied Swiftlet, Chestnut-breasted Malkoha, Red-billed Malkoha, Scarlet Minivet, Blue-winged Leafbird, Cream-vented

Bulbul, Olive-winged Bulbul, Red-eyed Bulbul, Spectacled Bulbul, Striped Tit-Babbler, Hairy-backed Babbler, Grey-headed Babbler, Dark-necked Tailorbird, Rufescent Prinia, White-rumped Shama, Chestnut-naped Forktail, Spotted Fantail, Greater Racket-tailed Drongo, White-bellied Yuhina, Sultan Tit, Velvet-fronted Nuthatch, Black-naped Monarch, Asian Paradise Flycatcher, Ruby-cheeked Sunbird, Purple-naped

Sunbird, Little Spiderhunter, Long-billed Spiderhunter, Yellow-breasted Flowerpecker, Crimson-breasted Flowerpecker, Orange-bellied Flowerpecker.

Winter
Forest Wagtail, Grey Wagtail, Siberian Blue Robin, Eastern Crowned Leaf Warbler, Asian Brown Flycatcher, Ferruginous Flycatcher, Yellow-rumped Flycatcher

6 **Forest Research Institute** (16km) (Kepong Railway Station)

The Forest Research Institute of Malaysia (FRIM) is an area of parkland northwest of the city centre which was once a mining area but has been replanted with indigenous forest. It has good secondary forest and is contiguous with Bukit Lagong Forest Park, creating a rich habitat for many forest-edge species. There are several watercourses within the FRIM, providing habitats for aquatic as well as forest species. There are several nature trails and also an upper canopy walkway which connects with the trails system in Bukit Lagong Forest. The canopy walkway is open daily from 09.30 to 14.30 except on Mondays and Fridays. It is not, however, the best place for birding as only a maximum of four people are allowed on each bridge and platform at all times, and you are expected to move on and not hog the walkway.

Take **Highway 1** north and after about 5km exit west onto the **Middle Ring Road II** (**MRR2**) towards Kepong. After about 5km, turn north at the traffic light intersection in Kepong, just before the flyover to the **LDP Expressway**. The institute entrance is on the right-hand side.

KEY SPECIES

Resident
Striated Heron, Crested Goshawk, Crested Serpent Eagle, White-breasted Waterhen, Large-tailed Nightjar, Blue-eared Kingfisher, Blue-crowned Hanging Parrot, Rufous Woodpecker, Grey-rumped Treeswift,

Plaintive Cuckoo, Pied Triller, Dark-necked Tailorbird, Red-eyed Bulbul, Olive-winged Bulbul, Little Spiderhunter, Greater Racket-tailed Drongo, Scarlet-backed Flowerpecker, Brown Barbet

Oct–Apr
Black Baza, Oriental Honey-buzzard, Japanese Sparrowhawk, Common Sandpiper, Arctic Warbler, Asian Brown Flycatcher, Purple-backed Starling

Further Away

7 Kuala Selangor Nature Park (70km) *(Bus 141)*

Kuala Selangor is the best coastal location within easy access of KL. The mouth of the Selangor River broadens into a muddy estuary at this point and the south bank has been preserved as a nature park. There are extensive mangrove forests, mudflats, and dense scrub vegetation. There is a shallow lake in the centre of the park and several drainage canals. Part of the mangrove swamp has been drained and the drier conditions have permitted a

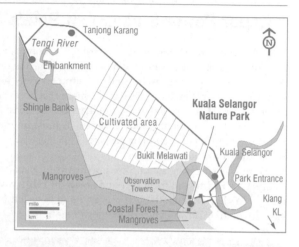

flourishing growth of coastal scrub forest which has, in fact, increased species diversity. The park has a concrete boardwalk trail into the mangroves. Additionally, there are tower hides overlooking the lake where you can also scan the treetops. There are accommodation huts within the park, and if you want to experience a night in the jungle, complete with animal howls in the night but with the comforts of a town just five minutes away, then this is worth trying, especially since you will then be on-site at dawn, the most productive time of the day.

Outside the park, the wooded hillside of **Bukit Melawati** to the north of the park entrance has good habitat for woodland species, and White-bellied Sea-eagles nest on the radio mast.

Take **Highway 2** west from KL to **Klang** and from there go north on **Highway 5** to **Kuala Selangor**, which is about 40km from Klang. The entrance to the park is on the left-hand side of the road into town, before you come to the bridge over the river.

If you are unable to access the mudflats at Kuala Selangor, then the estuary and mudflats at **Tanjong Karang** offer a good alternative. This estuary is about 15km north of Kuala Selangor, so it is a short drive if you have a car. Drive north from Kuala Selangor to the town of Tamjong Karang, crossing the bridge over the **River Tengi** as you enter the town. Turn left after crossing the bridge and drive for about 2km until you reach the sea embankment.

KEY SPECIES

Resident
Lesser Adjutant, Striated Heron, Purple Heron, Black-crowned Night Heron, Great Egret, Intermediate Egret, White-bellied Sea Eagle, Brahminy Kite, Black-shouldered Kite, White-breasted Waterhen, Slaty-breasted Rail, Red Junglefowl, Stork-billed Kingfisher, Collared Kingfisher, Mangrove Pitta, Lesser Coucal, Chestnut-breasted Malkoha, Laced Woodpecker, Pied Triller, Golden-bellied Gerygone, Pied Fantail, Ashy Tailorbird, Rufous-tailed Tailorbird, Mangrove Whistler, Zitting Cisticola, Yellow-bellied Prinia, Mangrove Blue Flycatcher, Ruby-cheeked

Sunbird
Winter
Watercock, Greater Sand Plover, Grey Plover, Pacific Golden Plover, Little Ringed Plover, Red-necked Stint, Curlew Sandpiper, Common Redshank, Common Greenshank, Marsh Sandpiper, Terek Sandpiper, Eurasian Curlew, Whimbrel, Black-tailed Godwit, Wood Sandpiper, Common Sandpiper, Pintail Snipe, Common Snipe, Little Tern, White-winged Black Tern, Black-capped Kingfisher, Common Kingfisher, Asian Brown Flycatcher, Oriental Reed Warbler, Arctic Warbler

At high tide the waders roost on isolated shingle banks close to the beach. Try to find a concealed spot on the embankment behind the beach before the tide comes in all the way and wait for the birds to congregate.

8 Kapar Power Station Ash Ponds (54km)

This site has come to the notice of birders in comparatively recent times. It is now recognised as one of the most important wader sites on the west coast of Malaysia. Access, at present, is restricted and the site may only be visited with special permission. However, it is included here in the hope that access may be more readily available in the future. There are extensive mudflats on the coast at Kapar, and at high tide the waders roost in the holding ponds. This site is most important during the boreal winter and during spring and autumn passage. There are also year-round residents such as herons and other aquatic species.

The power station is about 8km northwest of Kapar town. Take **Highway 2** west from KL to **Klang** and from there go north on **Highway 5** for about 14km to Kapar. Drive a further 4km north on Highway 5 and turn left into **Jalan Tok Muda**. About 4km down the road you will see the power station.

KEY SPECIES

Resident
Yellow Bittern, Striated Heron, Purple Heron, Great Egret, Intermediate Egret, White-bellied Sea Eagle, Brahminy Kite, White-breasted Waterhen, Slaty-breasted Rail, Stork-billed Kingfisher, Collared Kingfisher, Lesser Coucal, Dollarbird, Ashy

Tailorbird, Zitting Cisticola, Yellow-bellied Prinia, Ruby-cheeked Sunbird

Winter
Greater Sand Plover, Grey Plover, Pacific Golden Plover, Little Ringed Plover, Red-necked Stint, Curlew Sandpiper, Common Redshank,

Common Greenshank, Marsh Sandpiper, Terek Sandpiper, Eurasian Curlew, Whimbrel, Black-tailed Godwit, Wood Sandpiper, Common Sandpiper, Pintail Snipe, Common Snipe, Little Tern, White-winged Black Tern, Black-capped Kingfisher, Oriental Reed Warbler

9 Bukit Fraser Nature Reserve (100km)

Although this site is rather distant from KL, it is worth making the extra effort to get here as it is one of the most outstanding birding sites in Southeast Asia. If you have a car you can get to Bukit Fraser in under two hours early in the morning and see a lot in the course of a day.

Fraser's Hill (Bukit) is a hill resort located at about 1,500m. The surrounding montane forest is a rich ecosystem supporting a wide diversity of resident and also wintering bird species. There are eight trails which traverse the resort hillsides and vary in length between 1km (Rompin Trail) and 6km (Pine Tree Trail). Most birders regard the Bishop's Trail (2km) as the best for birds. Stretches of these trails can be very steep and also slippery after rain, and if you lack adequate footwear you may prefer simply to walk the tarmac roads which fan out from the resort. This strategy has the added benefit of avoiding the leech-infested forest tracks.

Apart from the trails at the hilltop, the road up to Fraser's Hill traverses the transition zone between lowland and montane forest and supports a different community of birds to that at the higher elevations. The area around the Gap Resthouse (800m elevation) is renowned for the quality of the birding. The road up from the Gap Resthouse to Fraser's Hill (about 8km) ascends the hill and affords excellent views of the slopes at tree-top level, enabling you to see some of the upper canopy species. If you arrive early in the morning or have stayed overnight at the Gap Resthouse, start at the Gap and work your way up to Fraser's Hill. As you ascend, montane species make their appearance above 1,100m. Birds are active throughout the day on the hilltop due to the cooler climate, unlike at the Gap, where activity generally dies down during the midday heat.

Take **Highway 1** north from KL and turn northeast at **Kuala Kubu Baharu** onto **Highway 55** for about 30km until you reach the Gap. There are two one-way roads towards Bukit Fraser from the Gap: the old road which is for downhill traffic and the 'new' road for uphill traffic. Unfortunately the new road has often been closed for repairs and it's best to check at the Gap Resthouse for the current situation. There is a twice daily bus service from Kuala Kubu Baharu to Bukit Fraser but the schedule would not permit much time on site.

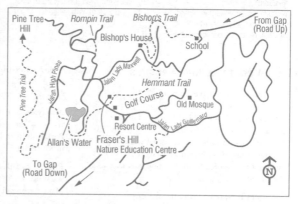

KEY SPECIES

Resident
Crested Serpent Eagle, Blyth's Hawk-Eagle, Orange-breasted Trogon, Red-headed Trogon, Blue-crowned Hanging Parrot, Green-billed Malkoha, Red-bearded Bee-eater, Blue-eared Barbet, Fire-tufted Barbet, Rufous Piculet, Maroon Woodpecker, Crimson-winged Woodpecker, Greater Yellownape, Lesser Yellownape, Wreathed Hornbill, Rhinoceros Hornbill, Helmeted Hornbill, White-bellied Swiftlet, Grey-rumped Treeswift, Silver-breasted Broadbill, Streaked Spiderhunter, Bar-winged

Flycatcher-shrike, Javan Cuckoo-shrike, Grey-chinned Minivet, Scarlet Minivet, Blue-winged Leafbird, Orange-bellied Leafbird, Lesser Racket-tailed Drongo, Greater Racket-tailed Drongo, Bronzed Drongo, Asian Fairy Bluebird, Verditer Flycatcher, Hill Blue Flycatcher, Asian Paradise Flycatcher, Stripe-throated Bulbul, Ochraceous Bulbul, Black-headed Bulbul, Black-crested Bulbul, Sultan Tit, Striped Tit-Babbler, White-bellied Yuhina, Chestnut-capped Laughingthrush, White-hooded Babbler, Silver-eared Mesia, Blue-

winged Minla, Mountain Fulvetta, Black-and-crimson Oriole, Slaty-backed Forktail, Dark-necked Tailorbird, Green Magpie, Buff-vented Flowerpecker, Yellow-breasted Flowerpecker

Winter
White-throated Needletail, Ashy Minivet, Siberian Thrush, Arctic Warbler, Mugimaki Flycatcher, Eastern Crowned Leaf Warbler, Asian Brown Flycatcher, Dark-sided Flycatcher, Ferruginous Flycatcher

---

10 Tanjung Tuan (120km)

If you happen to be in Kuala Lumpur during the months of February to April or September to October, you could consider visiting Tanjung Tuan on the Negeri Sembilan coast, south of KL. Thousands of migrating raptors cross the narrow Straits of Malacca from Sumatra to Malaysia in spring and vice-versa in autumn. One of the best places to observe this passage is at Tanjung Tuan. This site is watched regularly by members of the Malaysian Nature Society during the migration period. Large numbers of Oriental Honey-buzzards, Black Bazas, Grey-faced Buzzards, Chinese Goshawks and Japanese Sparrowhawks pass this point during spring and autumn migration. Apart from the migrating raptors, Tanjung Tuan Forest Reserve is also an important area of remnant coastal forest which supports a good selection of species, so is worth a visit at any time of year.

Take **Highway 2** west to **Klang** and turn south on **Highway 5** to **Port Dickson**. Take the coast road south from Port Dickson to Tanjung Tuan. Follow the coast road out to the lighthouse. The entrance to the forest reserve is on the left-hand side. The lighthouse area is the best vantage point for the raptor migration.

KEY SPECIES

Resident
White-breasted Waterhen, White-bellied Sea Eagle, Large-tailed Nightjar, Black-capped Kingfisher,

Asian Palm Swift, Dollarbird, Black-headed Bulbul, White-headed Munia

Passage
Grey-faced Buzzard, Oriental Honey-buzzard, Black Baza, Japanese Sparrowhawk, Chinese Goshawk

Useful contacts and websites

The Malaysian Nature Society
www.mns.org.my
JKR 641, Jalan Kelantan, 50480 Kuala Lumpur, Malaysia. Tel. 603 2287 9422. Fax. 603 2287 8773. Malaysian environmental organisation website with information about birding in Malaysia. Their shop (about 500m from KL Sentral Station) stocks a good selection of birding gear.

Fraser's Hill Nature Education Centre
www.wwfmalaysia.org/fhnec

Kuala Selangor Nature Park
www.naturepark.freeservers.com
Information about habitat, birds and other wildlife.

Oriental Bird Club
www.orientalbirdclub.org
P.O.Box 324, Bedford, MK42 0WG, UK

Broadbill – Birdwatcher's Site
www.members.at.infoseek.co.jp/bluebonnet/malaysia-e.htm
Japanese website (in English) with visitor information about birding in Malaysia.

Malaysian Birds & Birding Sites
www.malaysianbirds.myphotos.cc
KT Khoon's personal website packed with photos and information about birding sites in Malaysia.

Kaisoon Online
www.members.fortunecity.com/kaisoon

Kingfisher Tours SDN BHD
Email: kingfishertours@hotmail.com
Contact: Mano Tharmalingam/Dennis Yong, Tel. 603 2142 1454. Fax. 603 2142 9827.
Suite 1107, 11th Floor, Bangunan Yayasan Selangor, Jalan Bukit Bintang, 55100 Kuala Lumpur.
Bird tour company which provides short organised guided bird tours.

Tidal Schedule for coast of Selangor
(Pelabuhan Kelang)
www.kjc.gov.my/htdocs2/data/IDM26004.html

Malaysia Central
www.mycen.com.my/search/transport.html
Portal site with links to all major transport operators in KL region.

Pudu Raya Bus Terminal
www.udaholdings.com.my/puduraya/asal.html
Main bus terminal serving birding sites.

Books and publications

A Birdwatchers Guide to Malaysia
John Bransbury (1993) Waymark
ISBN 0646145592

A Photographic Guide to Birds of Peninsular Malaysia and Singapore
Geoffrey Davison and Chew Yen Fook (1995)
New Holland, ISBN 1853685135

A Field Guide to the Birds of West Malaysia and Singapore Allen Jeyarajasingam & Alan Pearson (1999) Oxford University Press
ISBN 0198549628

A Field Guide to the Birds of South-East Asia
Craig Robson (2000) New Holland
ISBN 1843301180

A Field Guide to the Birds of South-East Asia
Ben F. King, Edward C. Dickinson, Martin W. Woodcock (1975) Collins, ISBN 0002192063

Where to Watch Birds in Asia
Nigel Wheatley (1996) Christopher Helm
ISBN 071364303X

Zebra Doves are common open-country birds around Manila

Manila

Most birders with an interest in the avifauna of Southeast Asia will eventually visit Manila, as the Philippine Archipelago is one of the most outstanding areas of endemism in the world and Manila is its main gateway. Manila is on the west coast of Luzon, the largest of the Philippine Islands and also home to a number of species unique to the island. Unfortunately, most of these endemics are highly dependent on the unique forest ecosystems of the island and due to extreme deforestation are now restricted to isolated and ever-decreasing forest remnants. A Manila-based birder has a limited, though certainly feasible, opportunity to see some of these unique birds, by undertaking a day trip outside the city. Some of the commoner endemics can be seen within the city parks and open spaces. In addition, the Philippines are a favoured wintering location for a number of Palearctic species and are strategically located for waders migrating from Siberia to Australia. The shores of Manila Bay provide habitats for these birds but they are fast disappearing as more land is reclaimed for urban development.

Most birders visit the Philippines between February and April, when the weather is driest and many Palearctic species can also be seen wintering or on spring passage. Few, of course, spend much time in Manila as there is a considerable distance to be covered if you wish to visit the hotspots in the north of Luzon or indeed the other islands of the archipelago. However, even birders on a short visit to Manila will find much to interest them in the course of a day trip outside the city at any time of year. Within one or two hours' drive of the city you can reach the wooded foothills of the Sierra Madre Mountains in the east, the lowland marshes of the Pampanga River valley to the north, and the forested volcanic peaks (extinct and non-extinct) to the south and west.

GETTING AROUND

Like all major cities of Southeast Asia, Manila has undergone enormous expansion in recent years and to a large extent its transport infrastructure lags behind the pace of development. This presents significant challenges to the visiting birder. On the one

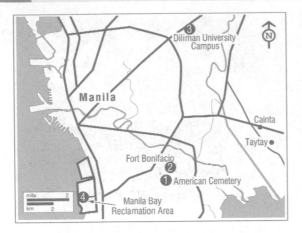

hand, large volumes of traffic and restricted entry to the city centre on certain days of the week make driving within the Metro Manila area inadvisable. On the other hand, some of the out of town sites require a car for access if you are on a tight schedule. If you have a car, avoid using it in the city, as the Light Rail Transit systems (LRT and MRT) can get you from the outskirts to many of the key points within the centre, and the stations are well served by feeder buses, taxis and the ubiquitous Filipino jeepneys. There are future plans to extend the LRT system and the commuter rail lines to serve the coastal areas south of the city and this will make a convenient way to link the city and airport with the coastal sites along the southern shore of Manila Bay.

Depending on where you are staying, you should be able to navigate to the Epifanio de Los Santos Avenue (EDSA). This is the key orbital route circling the city beginning near the coast at Pasay and arcing through the outer city to Quezon City. It also interchanges with the North and South Luzon Expressways and the Manila/Cavite Coastal Road.

If you don't have a car, you will have to rely on the public bus services to get to the more distant locations. The buses are frequent and cover most of the areas very comprehensively but because there is no central bus station in the city from which they operate, the visitor requires some local assistance to locate the correct terminal. Each bus company, of which there are many, operates from a different terminal from the others. The main bus terminals are the EDSA Terminal in Pasay to the south of the city, from which many buses serving the south and west of Luzon operate; the Avenida Terminal in the north central part of the city, for buses serving northern and central Luzon; and the Cubao Terminal in Quezon City, serving points north and east. Many of the bus terminals are served by LRT, and local taxis and jeepneys provide a feeder service at the destination, so it is quite possible to visit all the birding sites by public transport alone, although travelling time will take a lot longer.

Finally, it is worth a visitor making contact with the Wild Bird Club of the Philippines (see p. 123) as they often have outings within the Metro Manila area and may also be able to offer assistance, such as getting earlier entry into the American Cemetery.

Locations

1 American Cemetery (13km) (J28 Jeepney Terminal)

As there are rather few green open spaces in Manila, perhaps it is because of the relative peace and tranquillity in the grounds of the cemetery that so many birds can be found here. There are some small wooded areas with mature trees but very little bushy, unkempt undergrowth of the type favoured as cover by birds. Moreover, the lawns are very neat and manicured and, one would think, unlikely to offer much food or cover for foraging birds but, amazingly, this is one of the best birding sites in Manila. Perhaps it is the short grass that makes it easy to see one of the special birds at this site – the Barred Rail. This and several endemic species can be seen here with ease, as well as numerous migrants.

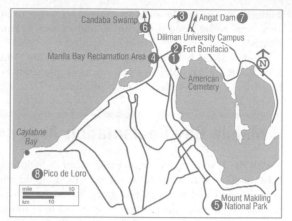

The cemetery is on **McKinley Road** in the southeast of the city and not too far from the airport. It is easy to reach by jeepney or taxi. The cemetery does not open until 09.00 and you will need to produce a passport or some form of identification as well as sign-in.

KEY SPECIES

Resident	Oct–Apr	
Barred Rail, White-eared Brown Dove, Pied Bushchat, Lowland White-eye	Chinese Goshawk, Japanese Sparrowhawk, Blue Rock Thrush,	Chestnut-cheeked Starling

2 Fort Bonifacio (13km)   (J11/J16 jeepney)

Fort Bonifacio is near the American Cemetery and it houses another cemetery – The National Heroes Cemetery (Libingan ng mga Bayani) which is also good for birds.

Parts of this cemetery have long grass and scrub thickets which support a different range of birds such as Barred Buttonquail. There are also some wet marshy areas and irrigation ponds. A visit to both cemeteries is worthwhile as they are close together and complement each other in habitat and birds.

The cemetery is on **McKinley Road** and is conveniently served by both jeepney and taxi.

KEY SPECIES

Resident
Yellow Bittern, Barred Buttonquail, White-eared Brown Dove, Philippine Coucal, Golden-bellied Gerygone, Pied Bushchat, Lowland White-eye

Oct–Apr
Blue Rock Thrush

3 Diliman University Campus (13km)   (Quezon MRT Station Line 3)

The University of the Philippines campus at Diliman is worth a visit if you are staying in Quezon City and your time is limited. The campus contains open areas of parkland and woodland which approximate to forest edge habitat, and it supports a good selection of lowland forest birds. This site also offers an ideal opportunity to familiarise yourself with the typical birds of the Philippines before undertaking a more intensive visit to one of the hotspots in the interior of Luzon. The campus is just east of **Elliptical Road**, a well-known

KEY SPECIES

Resident
Yellow Bittern, Barred Buttonquail, White-breasted Waterhen, White-eared Brown Dove, Philippine Coucal, Pied Bushchat

landmark in central Quezon City and is easy to get to by jeepney. The most productive areas tend to be along the southern and eastern perimeter where the best habitat is located.

Another site worth checking in the area is the Ninoy Aquino Park, just west of Elliptical Road. This park in some ways complements Diliman Campus as it contains a freshwater lagoon which attracts aquatic species.

4 Manila Bay Reclamation Area (10km)   *(EDSA LRT Station Line 1)*

Manila Bay is an important wader migration and wintering area. It consists of extensive tidal mudflats with some relict mangroves along the shoreline south of Manila. Large areas of mudflat have been reclaimed to create fish ponds and salt pans and, more recently, sections of the bay have been filled-in as more land is required for construction. Over time, of course, this habitat will probably be lost, but for the moment while the land is under reclamation, the large open areas of infill are attractive to birds as feeding and roosting areas.

The best and most accessible area at present is **Pasay City** just south of Manila city centre and not too far from the airport. Take the **EDSA** south to where it intersects with **Macapagal Boulevard**. Continue west on the EDSA extension and take the first turn right (north) onto **PEA Road One**. Follow the road all the way to the **Seaside Boulevard** which runs alongside the **Libertad Channel** and the shoreline of the bay. This extensive area is part construction site and part reclaimed land. There are open areas of grassland, gravel workings and large earthworks. All these are worth checking for grassland birds, roosting waders and herons.

KEY SPECIES

Resident
Yellow Bittern, Cinnamon Bittern, Striated Heron, Purple Heron, Philippine Duck, White-browed Crake, Barred Rail, Plain Bush-hen, Oriental Pratincole, Australasian Grass Owl, Lesser Coucal, Zitting Cisticola, Golden-headed Cisticola

Winter
Lesser Sand Plover, Pacific Golden Plover, Kentish Plover, Red-necked Stint, Long-toed Stint, Common Redshank, Common Greenshank, Marsh Sandpiper, Terek Sandpiper, Wood Sandpiper, Common Sandpiper, Pintail Snipe, Common Snipe, Siberian Rubythroat, Clamorous Reed Warbler, Oriental Reed Warbler

5 Mt Makiling National Park (56km)

This is one of the closest locations to Manila where substantial tracts of primary forest still remain. It is at the rear of the University of the Philippines campus and, in addition to the national park, the Botanical Gardens and the grounds of the College of Forestry are also good for birds. The mountain reaches an altitude of about 1,100m and there is an 8km hiking trail from the campus to the summit. If you want to, you can drive about halfway along this trail and while this may save time in getting to the summit, walking the lower elevations can turn up some good birds. In addition to the mountain trail, if you have time check out the Centre for Philippine Raptors in the Botanic Gardens, which has one of the few captive Great Philippine Eagles in the world: probably the closest you will come to this magnificent species.

Take **Highway 1 (South Luzon Expressway)** south to **Los Baños**. At the main crossroads in the centre of the town turn right and drive for about 5km towards the university campus (known locally as The Forestry). The trail starts at the southern perimeter of the campus. There are bus services to Los Baños operating from the EDSA Bus Terminal.

KEY SPECIES

Resident
Crested Serpent Eagle, Philippine Falconet, Spotted Buttonquail, Barred Rail, Plain Bush-hen, White-eared Brown Dove, Black-chinned Fruit Dove, Emerald Dove, Luzon Bleeding-heart, Guaiabero, Philippine Hawk-Cuckoo, Drongo Cuckoo, Red-crested Malkoha, Scale-feathered Malkoha, Purple Needletail, Glossy Swiftlet, Pygmy Swiftlet, Coppersmith Barbet,

Indigo-banded Kingfisher, Spotted Kingfisher, Luzon Hornbill, Striated Swallow, Yellow-wattled Bulbul, Philippine Bulbul, Ashy Thrush, Golden-headed Cisticola, Grey-backed Tailorbird, Tawny Grassbird, White-browed Shama, Blue-headed Fantail, Black-naped Monarch, Yellow-bellied Whistler, White-breasted Woodswallow, Elegant Tit, Striped-sided Rhabdornis, Plain-throated

Sunbird, Purple-throated Sunbird, Flaming Sunbird, Thick-billed Flowerpecker, Red-keeled Flowerpecker, Pygmy Flowerpecker, Yellowish White-eye, Balicassiao, Large-billed Crow, White-bellied Munia

Winter
Ashy Minivet, Eye-browed Thrush, Blue Rock Thrush

Further Away

6 **Candaba Swamp** (70km)

The Candaba Swamp is basically the flood plain of the River Pampanga which floods seasonally. Through the creation of impoundments for the flood waters, a network of fish ponds and rice-paddies has been created. Although these are variable in extent and quality from year to year they comprise nevertheless one of the best freshwater sites within a day's birding from Manila. The areas of permanent water promote the growth of reedbeds and other aquatic vegetation which provide cover for breeding and wintering aquatic species, while the seasonally flooded rice-paddies provide rich feeding for ducks and herons.

Take the **North Luzon Expressway** north to the **Santa Rita Exit**, where you exit east towards **Candaba**. Drive through **Plaridel** and after 10km, turn left under a flyover (**Jollibee** on right) onto **Baliuag-Candaba Road**. Drive 10km to **Bahay Pare**. It would be worthwhile when there to ask how to get to the Mayor of Candaba's house. That will take you to the ponds with the ducks.

There are bus services to Candaba from Manila operating from the Avenida Bus Terminal on Rizal Avenue (near the D. José LRT Station).

KEY SPECIES

Resident
Little Grebe, Yellow Bittern, Black Bittern, Striated Heron, Cinnamon Bittern, Purple Heron, Cattle Egret, Philippine Duck, Barred Rail, Purple Swamphen, White-browed Crake, White-breasted Waterhen, Pheasant-tailed Jacana, Oriental Pratincole,

Greater Painted-snipe, Island Collared Dove, Blue-tailed Bee-eater, Pied Bushchat, Zitting Cisticola, Clamorous Reed Warbler

Winter
Eurasian Teal, Garganey, Northern Shoveler, Northern Pintail, Eurasian Wigeon, Common Pochard, Tufted

Duck, Eastern Marsh Harrier, Black-winged Stilt, Pacific Golden Plover, Long-toed Stint, Common Greenshank, Marsh Sandpiper, Wood Sandpiper, Common Sandpiper, Pintail Snipe, Common Snipe, Oriental Reed Warbler, Streaked Reed Warbler

7 **Angat Dam** (60km)

One of the main reservoirs which supply Manila is at the southern end of the Sierra Madre Mountains which run along the eastern flank of Luzon island. Strangely enough, the reservoir itself is not particularly interesting for birds: it is the surrounding lowland forest, protected as part of the watershed, that is most interesting ornithologically.

However, like many parts of the Philippines, the forest has been over-exploited and large tracts have been degraded and replaced by bamboo scrub. Two main trails are worth checking within the forested area: the Ridge Trail and the Reservoir Trail.

Take the **EDSA** expressway to **Quezon City** and exit northeast towards **Igay** which is about 25km north of Quezon City. Continue north to **Bigti** and turn east to **Hilltop** village at the entrance to the **Angat Watershed Area**. There is a barrier at the entrance, but birders are not normally refused entry. The main road into the area leads to the Ridge Trail, which is approximately 10km long, the first 3km of which are quite poor in terms of habitat and birds. After the first km you can turn right onto the Reservoir Trail which leads down to the reservoir through some good quality forest.

Bear in mind that there have been concerns expressed recently about personal safety in this area and visitors are advised to consult local information sources before travelling here. In general, it is better to obtain permission from the Watershed Department of the National Power Corporation before visiting (www.napocor.gov.ph). People have been refused entry.

KEY SPECIES

Resident
Philippine Serpent Eagle, Philippine Falconet, Red Junglefowl, White-eared Brown Dove, Black-chinned Fruit Dove, Emerald Dove, Luzon Bleeding-heart, Guaiabero, Philippine Hawk-Cuckoo, Drongo Cuckoo, Red-crested Malkoha, Scale-feathered Malkoha, Rufous Coucal, Purple Needletail, Glossy Swiftlet, Pygmy Swiftlet, Coppersmith Barbet, Philippine Kingfisher, Indigo-banded Kingfisher, Spotted Kingfisher, Sooty

Woodpecker, Philippine Trogon, Luzon Hornbill, Striated Swallow, Yellow-wattled Bulbul, Philippine Bulbul, Ashy Thrush, Golden-headed Cisticola, Grey-backed Tailorbird, Tawny Grassbird, White-browed Shama, Blue-headed Fantail, Black-naped Monarch, Yellow-bellied Whistler, White-breasted Woodswallow, Elegant Tit, White-fronted Tit, Striped-sided Rhabdornis, Plain-throated Sunbird, Purple-throated Sunbird, Flaming Sunbird,

Thick-billed Flowerpecker, Red-keeled Flowerpecker, Pygmy Flowerpecker, Yellowish White-eye, Balicassiao, White-lored Oriole, Large-billed Crow, Coleto

Winter
Ashy Minivet, Olive-backed Pipit, Eye-browed Thrush, Blue Rock Thrush

Passage
Grey-faced Buzzard

8 Pico de Loro (86km)

Although rather distant from the city, this site is worth a visit, particularly if you are staying on the south side of Manila or your visit takes you to the tourist resort of Caylabne Bay. The Pico de Loro is a well-known rock formation along a ridge of low forested hills to the southwest of Manila. It is set within the Mataas na Gulod National Park and offers some excellent forest birding. The Pico de Loro is accessed by way of a hiking trail which starts from the Department of the Environment and Natural Resources Station (DENRS) at Puerto Azul. The hike to the summit takes you through some good forest habitat.

Take the main coastal highway south from Manila towards **Ternate**. At Ternate take the road towards **Caylabne Bay Resort** which passes through **Puerto Azul**. At Puerto Azul, the Pico de Loro Trail commences at the DENR Station which is designated by a magnet sign (the hills exhibit some magnetic phenomena). If you continue west on the road from Puerto Azul towards Caylabne Bay Resort, the road passes through some good forest habitat and at Caylabne Bay Resort, there is a freshwater lagoon which is good for the endemic Philippine Duck. This site is less easy to visit if you are relying on public transport: although there is a bus service from Manila to Ternate, this is infrequent and not reliable.

KEY SPECIES

Resident
Philippine Duck, Crested Serpent Eagle, Brahminy Kite, Philippine Falconet, Red Junglefowl, White-eared Brown Dove, Yellow-breasted Fruit Dove, Guaiabero, Philippine Hawk-Cuckoo, Red-crested Malkoha, Scale-feathered Malkoha, Lesser Coucal, Purple Needletail, Glossy Swiftlet, Pygmy Swiftlet, Whiskered Treeswift, Coppersmith Barbet, White-throated Kingfisher, Indigo-banded Kingfisher, White-bellied Woodpecker, Greater Flameback, Luzon Hornbill, Yellow-wattled Bulbul, Philippine Bulbul, Ashy Thrush, Golden-headed Cisticola, Grey-backed Tailorbird, White-browed Shama,, Blue-headed Fantail, Black-naped Monarch, Yellow-bellied Whistler, White-breasted Woodswallow, Philippine Fairy Bluebird, Elegant Tit, Striped-sided Rhabdornis, Blackish Cuckoo-shrike, Orange-bellied Flowerpecker, Red-keeled Flowerpecker, Pygmy Flowerpecker, Yellowish White-eye, Balicassiao, Asian Glossy Starling, Large-billed Crow, Coleto

Oct–Apr
Chinese Goshawk, Ashy Minivet, Olive-backed Pipit, Blue Rock Thrush

Useful contacts and websites

Wild Bird Club of the Philippines
www.birdwatch.ph
Birding organisation which is active in the Manila area.

Rare Bird Records
www.birdwatch.ph/articles/rarebirdsdoc.html
Email bird_records_PH@yahoogroups.com

Haribon Foundation
www.haribon.org.ph
National conservation organisation and BirdLife partner in the Philippines.
Suites 401-404 Fil-Garcia Tower, Kalayaan Avenue, Diliman QC. Tel. 436 4363, 434 8237, 436 2756.

Oriental Bird Club
www.orientalbirdclub.org
P.O.Box 324, Bedford, MK42 0WG, UK

Tidal Schedule for Manila Bay
www.easytide.ukho.gov.uk/EasyTide/EasyTide/ShowPrediction.aspx

National Power Corporation
www.napocor.gov.ph
Controls access to Angat Dam area.

Light Rail Transit Authority (LRTA)
www.lrta.gov.ph
Light Rail transit system in Metro Manila area.

Metrostar Mass Rail Transit (MRT)
www.dotcmrt3.gov.ph
Elevated railway which operates along EDSA ring road.

Transit Desk
www.dotc.gov.ph/actioncenter/transitdesk.htm
Portal site with information about main bus and rail services operating out of Manila.

Books and publications

A Guide to the Birds of the Philippines
Robert S. Kennedy *et al.* (2000) Oxford University Press, ISBN 0198546688

Photographic Guide to the Birds of the Philippines
Tim Fisher & Nigel Hicks (2000) New Holland ISBN 1859745105

The Birds of The Philippines
E. C. Dickinson, R. S. Kennedy and K. C. Parkes (1991) British Ornithologists' Union ISBN 0907446124

A Field Guide to the Birds of South-east Asia
Craig Robson (2000) New Holland ISBN 1843301180

A Field Guide to the Birds of South-East Asia
Ben F. King, Edward C. Dickinson, Martin W. Woodcock (1975) Collins, ISBN 0002192063

Where to Watch Birds in Asia
Nigel Wheatley (1996) Christopher Helm ISBN 071364303X

Mumbai (Bombay)

Alexandrine Parakeet is a conspicuous and noisy resident of the parks in Mumbai

This vast city extends across seven islands which have been connected together and infilled since the 19th century to form one large island: Salsette Island. Nevertheless, there are still some pockets of the original mangrove coastline and mudflats which have escaped development, thus far. The city is well off the track of many bird tour groups as it is quite distant from the birding hotspots of northern India and the southwest. However, many birders visiting Goa will probably pass through Mumbai en route and will find some good birding opportunities if they have any reason to linger in the city. There are still many birds to be seen wherever there are parks or large gardens, despite the density of the human population, and this is reflected throughout India, perhaps as a result of the great tolerance that people have for birds, and animals in general.

Apart from the coastal sites, the city is about a two-hour drive from the foothills of the Western Ghats, a mountain range which extends right along the western flank of India and is rich in birdlife, particularly where the forest cover has been preserved. In addition to the resident birds which can be seen near the city, the coastal mudflats and Mumbai Harbour are visited in the boreal winter by large numbers of waders, gulls and terns, and the wooded areas then support Palearctic passerines. The winter months are the best time for birding in Mumbai, not just because the variety of birds is greater, but because the weather is a little cooler and much drier. The monsoon season from June to September produces extremely wet weather but, even then, there are always gaps in the downpours which permit some birding.

GETTING AROUND

Many sites require considerable travelling time because of the sheer size of the city. Moreover, because most tourists and other visitors will stay in the city centre at the southern end of Salsette Island, also known as South Mumbai, a significant distance has to be covered to reach the best birding sites which are located near the north end of the island. If you intend to confine your birding activity to Salsette Island, then the public transport system is perfectly adequate, and the numerous taxis and auto-rickshaws will easily fill in any gaps, although the

TYPICAL SPECIES

Resident
Little Cormorant, Little Egret, Cattle Egret, Black Kite, Brahminy Kite, Spotted Dove, White-throated Kingfisher, Alexandrine Parakeet, Rose-ringed Parakeet, Green Bee-eater, Asian Koel, Greater Coucal, House Swift, Rufous Woodpecker, Brown-headed Barbet, Coppersmith Barbet, Dusky Crag Martin, Red-rumped Swallow, Black Drongo, Greater Racket-tailed Drongo, Jungle Babbler, Oriental Magpie-Robin, Red-vented Bulbul, Red-whiskered Bulbul, Common Tailorbird, Oriental White-eye, House Crow, Large-billed Crow, Common Myna, Asian Pied Starling, Eurasian Golden Oriole, Black-hooded Oriole, Purple Sunbird, House Sparrow, Chestnut-shouldered Petronia, Scaly-breasted Munia

Oct–Apr
Lesser Flamingo, Black-shouldered Kite, Black-tailed Godwit, Brown-headed Gull, Black-headed Gull, Grey Wagtail, Yellow Wagtail, Striated Swallow, Barn Swallow, Ashy Drongo, Red-breasted Flycatcher, Long-tailed Shrike, Rosy Starling,

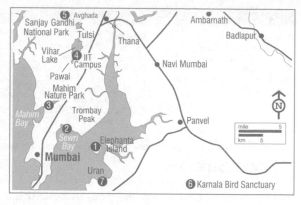

auto-rickshaws are not allowed to operate in South Mumbai. There are three suburban train lines, although the trains are so crowded that they present a formidable obstacle for a birder who is fully laden with telescope, tripod and other birding gear. If travelling lightly, however, the trains are manageable during off-peak times. The Western Line operates from Churchgate along the west side of Salsette Island; the Central Line operates from the main train station (CST) and serves the central and eastern side of Salsette Island; and the Harbour Line, which also operates from the CST, serves New Mumbai (Navi Mumbai) on the mainland. There is also an extensive bus service which, in addition to serving the the city centre, also operates a number of east/west cross-island routes. Many tourists also use the ferries to cross Mumbai Harbour to Elephanta Island in the bay and to destinations on the mainland. Most of the ferry services operate from the quayside at the Gateway to India, one of the most well-known tourist sites in the city.

If you have a car and want to visit sites on the mainland, your biggest challenge will be negotiating the traffic congestion in South Mumbai. The key routes out of the city centre are the Western Express Highway and the Eastern Express Highway, located on either side of Salsette Island. By taking the Eastern Express Highway, and branching off east at Chembur, you can take the Thana Creek Bridge over to the mainland and the Pune Highway. If you are travelling to the mainland, you will probably need to obtain two maps, one which covers Salsette Island and another which covers New Mumbai and the adjacent areas of Maharashtra State.

Locations

1 Elephanta Island (10km)

Elephanta Island at the mouth of Mumbai Harbour is not particularly interesting, ornithologically, but it is so near the city centre, and the boat trip out to the island affords such a good opportunity to see the birds of the bay, that it is worth undertaking the trip. As it is also one of the prime tourist attractions in Mumbai, the likelihood is that you will probably visit the island anyway.

The main attractions are the Hindu Cave Temples and other shrines which attract many pilgrims and tourists, and the island can get very crowded at certain times. The first boat departs from the Gateway to India at 09.00 and this will get you onto the island before the crowds build up. There may be some restrictions on the boat trips during the monsoon season. The island itself is well wooded and while the vast bulk of visitors follow the path round to the Cave Temples, there are plenty of opportunities to branch off onto less frequented side tracks. If the tide is out when you land on the island, check the mudflats on either side of the small harbour for waders and herons. This is quite a good site for viewing raptors during winter, including White-bellied Sea Eagles.

Resident	Oct–Apr	
Indian Cormorant, Striated Heron, Indian Pond Heron, Small Minivet, Ashy Prinia, Black-naped Monarch, Asian Paradise Flycatcher, Tickell's Blue Flycatcher	Grey Heron, Western Reef Egret, White-bellied Sea Eagle, Osprey, Eurasian Oystercatcher, Little Ringed Plover, Lesser Sand Plover, Little Stint, Curlew Sandpiper, Common Redshank, Common Greenshank, Eurasian Curlew, Black-tailed Godwit,	Bar-tailed Godwit, Terek Sandpiper, Wood Sandpiper, Common Sandpiper, Ruddy Turnstone, Pallas's Gull, Heuglin's Gull, Slender-billed Gull, Gull-billed Tern, Whiskered Tern, Eurasian Hoopoe, White Wagtail, Forest Wagtail

2 Sewri Bay (7km)   *(Sewri Railway Station/Harbour Line)*

Mumbai has an extensive coastline, both to seaward and on the east side of Salsette Island. There are several tidal creeks which once supported extensive mangrove forests but have long since been cleared for urban development. One area on the eastern shoreline, Sewri Bay, has escaped development and efforts are being made to restore the mangroves there. At present there seems to be a perfect balance in terms of the area of mangroves to mudflat, so that waders and other birds have adequate roosting and feeding areas. This area is at its best from October to April when migrating and wintering birds are present in greatest numbers, but its proximity and convenience to the city centre makes this site worth a visit at any time, and it is especially famous for the huge gatherings of flamingos.

Take the **Shahid Bhagat Singh Road** north along the east coast and the southern extremity of the bay is accessible on the right-hand side of the road after about 5km. If using public transport, take a train to Sewri and go east from the train station to the harbour area, where you may watch from the jetty.

Resident		
Striated Heron, Indian Pond Heron, Great Egret, Intermediate Egret	Black-winged Stilt, Pied Avocet, Pacific Golden Plover, Grey Plover, Little Ringed Plover, Lesser Sand Plover, Greater Sand Plover, Little Stint, Curlew Sandpiper, Sanderling, Whimbrel, Eurasian Curlew, Black-tailed Godwit, Bar-tailed Godwit,	Common Redshank, Marsh Sandpiper, Common Greenshank, Terek Sandpiper, Wood Sandpiper, Common Sandpiper, Pallas's Gull, Heuglin's Gull, Slender-billed Gull, Gull-billed Tern, Whiskered Tern, White Wagtail
Oct–Apr		
Purple Heron, Grey Heron, Western Reef Egret, Greater Flamingo, Osprey,		

3 Mahim Nature Park (15km)   *(Sion Railway Station/Harbour Line)*

This is a restored woodland park on the south shore of Mahim Creek, located on what was once landfill. The park comprises a rich mix of woodland, scrub and estuarine habitats and consequently supports a diverse range of species. No doubt, as the woodland matures and becomes more extensive, it will support a wider range of species. As it is quite close to the city centre, it offers some good birding but it is only open from 09.00, limiting the amount of early-morning birding that you can do. At the shoreline along Mahim Creek, there is good growth of mangroves and open areas of mudflat which, though extremely polluted, nonetheless draw a few waders.

The park is located in the central part of Salsette Island and is easily reached by taxi or public transport. The entrance is on the north side of the Bandra – **Sion Link**

Road. Several bus routes operate along this road and connect with Sion Railway Station and Bandra Railway Station which are nearby.

KEY SPECIES

Resident	Oct– Apr	
Indian Pond Heron, White-breasted Waterhen, Red-wattled Lapwing, Asian Palm Swift, White-throated Fantail, Purple-rumped Sunbird	Grey Heron, Little Ringed Plover, Little Stint, Common Redshank, Common Greenshank, Wood Sandpiper, Common Sandpiper, Gull-	billed Tern, Whiskered Tern, Blyth's Reed Warbler, Greenish Warbler

4 Indian Institute of Technology Campus (25km)
(Kanjur Road Railway Station/Central Line)

The campus of the IIT in north Mumbai is an excellent site to familiarise yourself with the birds of the Indian subcontinent. All of the common birds of lowland India, and some less common ones, find habitats here. The campus is contiguous with the Sanjay Gandhi National Park (beware of stray Leopards!), and it also has a large lake (Powai Lake) adjoining its western limits, adding to its habitat diversity. The best areas are in the northwest sector beyond the accommodation blocks, and the lakeside parkland near the Devi Temple. As with all tropical mixed woodland, always seek out any blooming or fruiting trees and shrubs as they inevitably attract birds.

The campus can be reached by taking the **Eastern Express Highway** as far north as **Kanjur Marg**, about 22km north of the city centre, and turning west towards **Powai**. The campus is well signposted on the right-hand (north) side of the road (**Shankaracharya Road**). If travelling by public transport, there is a suburban train service to Kanjur Road (Kanjur Marg), and you can take a taxi from there.

KEY SPECIES

Resident	
Indian Pond Heron, Intermediate Egret, White-breasted Waterhen, Purple Swamphen, Red-wattled Lapwing, Black-headed Cuckoo-shrike, Ashy Prinia, White-throated Fantail,	Thick-billed Flowerpecker, Black-headed Munia
	Oct–Apr
	Western Marsh Harrier, Common Coot, Whiskered Tern, Blyth's Reed Warbler, Booted Warbler, Greenish Warbler

5 Sanjay Gandhi National Park (40km)
(Borivili Railway Station/Western Line)

Sanjay Gandhi (Borivili) National Park is an extensive mixed forest park in the northern part of Salsette Island and an excellent birding spot which is very accessible from the city centre. A wide range of birds and other wildlife thrive in this park despite the fact that it is practically surrounded by urban development. There are a couple of large lakes which support good numbers of aquatic birds and contribute to the superb species diversity. The park consists of a collection of low hills, crossed by hiking trails, and the archaeological site at Kanheri Caves, a well-known tourist attraction. The Bassein tidal creek cuts through the northern end of the park and is forested with mangroves on either bank. Several of the more interesting trails require a permit. This can be obtained quite easily at the Conservator of Forests Office at the park entrance or by phoning ahead (tel. (91) 22-8860362/(91) 22-8860389).

The park lies between the two main routes north out of Mumbai, the **Western Express Highway** and the **Eastern Express Highway**. Access is from either expressway, but the main entrance is off the Western Expressway at Borivili. This is the most popular access point for tourists, as it leads to the main recreational areas and to the Kanheri Caves. The birder should follow the entrance road south (permit required)

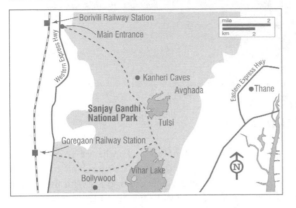

leading from near the caves to the two lakes, **Lake Vihar** and **Lake Tulsi**. The park is well served by public transport, but without a car or a cooperative taxi-driver you won't be able to cover too much distance. There are railway stations at Borivili and at several other stops along the western flank of the park. There is a bus service from Borivili Station to the Kanheri Caves, and it would be a fairly easy hike from the caves to Lakes Vihar and Tulsi, and from there to the southern park entrance at Goregaon (close to the Bollywood Film Studios). You could take a taxi from the park entrance to the station at Goregaon to get the train back to the city.

KEY SPECIES

Resident
Indian Pond Heron, Lesser Whistling Duck, Spot-billed Duck, Indian Peafowl, Grey Junglefowl, Jungle Bush Quail, White-breasted Waterhen, Red-wattled Lapwing, Plum-headed Parakeet, Asian Palm Swift, Indian Grey Hornbill, Black-rumped Flameback, Black-headed Cuckoo-shrike, Common Woodshrike, Orange-headed Thrush, Malabar Whistling Thrush, White-rumped Shama, Rufous Treepie, Small Minivet, Ashy Prinia, White-throated Fantail, Black-naped Monarch, Tickell's Blue Flycatcher, Gold-fronted Leafbird, Puff-throated Babbler, Indian Scimitar Babbler, Pale-billed Flowerpecker, Thick-billed Flowerpecker, Black-headed Munia

Oct–Apr
Asian Openbill, Osprey, Common Coot, Green Sandpiper, Forest Wagtail, Spangled Drongo, Blue-capped Rock Thrush, Blyth's Reed Warbler, Greenish Warbler, Asian Paradise Flycatcher

6 **Karnala Bird Sanctuary** (65km)   *(Panvel Railway Station/Harbour Line)*

The fort at Karnala is a well-known historic site at the top of a 370m hill. The surrounding forest has been preserved as a bird sanctuary and supports a good variety of resident forest birds as well as Palearctic migrants. This is the closest point to Mumbai where you can see some of the species more typical of the Western Ghats mountain range. There are several hiking trails through the forest, and the ascent up the hill provides opportunities to look out over the forest canopy and see some of the upper-canopy species, such as hornbills, at eye-level. The fort itself is occupied by a breeding pair of Peregrine Falcons, of the local subspecies (*F. p. peregrinator*) also known as the Shaheen Falcon. Despite the rather small area of forest, it has the advantage of being very manageable in a day trip from Mumbai, and it is also very conveniently located if you are travelling the road between Mumbai and Pune.

Take the **Thana Creek Bridge** east towards '**New**' **Mumbai** on the east bank of Thana Creek and continue east on the **Pune Highway**. After about 50km, and about

3km past **Panvel**, turn right (south) towards Karnala on the **Goa Highway (NH-17)**. The fort and bird sanctuary are signposted about 12km south of the turn-off. There is a bus and train service from Mumbai to Panvel, from where you can take a taxi to Karnala.

KEY SPECIES

Resident
Crested Serpent Eagle, Shikra, Peregrine Falcon, Grey Junglefowl, Yellow-footed Pigeon, Pompadour Green Pigeon, Plum-headed Parakeet, Vernal Hanging Parrot, Indian Grey Hornbill, Rufous Woodpecker, Yellow-crowned Woodpecker, Brown-capped Woodpecker, Black-rumped Flameback, Brown-headed Barbet, Large Cuckoo-shrike, Rufous Treepie,

Scarlet Minivet, Small Minivet, White-rumped Shama, Malabar Whistling Thrush, Orange-headed Thrush, White-throated Fantail, Black-naped Monarch, Brown-cheeked Fulvetta, White-browed Bulbul, Indian Scimitar Babbler, Gold-fronted Leafbird, Bronzed Drongo, Ashy Drongo, Greater Racket-tailed Drongo, Common Woodshrike, Purple-rumped Sunbird, Crimson Sunbird,

Crimson-backed Sunbird, Pale-billed Flowerpecker

Oct–Apr
White-eyed Buzzard, Blyth's Reed Warbler, Blue-capped Rock Thrush, Western Crowned Warbler, Greenish Warbler, Verditer Flycatcher, Spangled Drongo, Common Rosefinch

7 Uran (77km)

One of the best coastal sites in Maharashtra State lies just across Mumbai Harbour. The coastline from Uran to the Jawaharlal Nehru Port Trust (JNPT) area is an expanse of mudflats and salt-marsh which is good for gulls, terns, waders and herons. In addition to the saltwater habitats, there are a number of freshwater ponds and marshy areas, particularly the lagoon opposite the Police Station at JNPT, which support a number of aquatic species. Also, the open areas of cultivation and scrub along the road between JNPT and Uran are good for a range of open-country species such as raptors and passerines. The area is being progressively reclaimed for urban and commercial developments but there are still large open areas of bird-rich habitat.

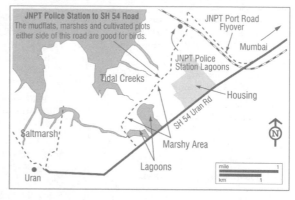

Take the **Thana Creek Bridge** east towards '**New**' **Mumbai** on the east bank of Thana Creek, and turn south on **SH 42A (aka Palm Beach Road)** for about 20km to the intersection with **SH 54**. This is the **Uran Road**. Continue west along this road for about another 8km until you reach the flyover which is the main access road into JNPT Port. Turn right (north) at the flyover and after 1km take the third turn left. Continue for a few hundred metres and cross the railway line. The JNPT Police Station is on the other side of the railway track and the JNPT lagoon is opposite the police station. There is a gravel track on the right-hand side after you cross the railway which provides access to the lagoon.

If you go west from the Police Station for about 2km, the road passes through a mosaic of mudflats, marshes, cultivation plots and shallow ponds. Flocks of birds can be found anywhere along this road where feeding and roosting opportunities exist. The road eventually curves south to join the SH 54 Uran road. At Uran, the road eventually runs alongside the shoreline, where you can scan the tidal creeks and saltmarshes.

KEY SPECIES

Resident
Little Grebe, Indian Cormorant, Cinnamon Bittern, Indian Pond Heron, Great Egret, Intermediate Egret, Lesser Whistling Duck, White-breasted Waterhen, Purple Swamphen, Pheasant-tailed Jacana, Bronze-winged Jacana, Common Kingfisher, Ashy-crowned Sparrow-Lark, Oriental Skylark, Paddyfield Pipit, Ashy Prinia, Plain Prinia, Zitting Cisticola, Baya Weaver, Black-breasted Weaver

Oct–Apr
Purple Heron, Grey Heron, Western Reef Egret, Black-headed Ibis, Glossy Ibis, Eurasian Spoonbill, Eurasian Teal, Garganey, Northern Pintail, Northern Shoveler, Common Gadwall, Spot-billed Duck, Booted Eagle, Western Marsh Harrier, Osprey, Common Coot, Black-winged Stilt, Pied Avocet, Pacific Golden Plover, Grey Plover, Little Ringed Plover, Lesser Sand Plover, Greater Sand Plover, Little Stint, Curlew Sandpiper,

Ruff, Whimbrel, Eurasian Curlew, Black-tailed Godwit, Bar-tailed Godwit, Common Redshank, Marsh Sandpiper, Common Greenshank, Common Snipe, Green Sandpiper, Wood Sandpiper, Common Sandpiper, Pallas's Gull, Heuglin's Gull, Slender-billed Gull, Gull-billed Tern, Caspian Tern, Whiskered Tern, Greater Short-toed Lark, White Wagtail, Citrine Wagtail, Common Stonechat, Pied Bushchat, Bluethroat

Useful contacts and websites

Bombay Natural History Society
www.bnhs.org
Hornbill House, Dr Sálim Ali Chowk, Shaheed Bhagat Singh Road, Mumbai 400 023
Environmental NGO and BirdLife Partner in India.

Birds of Bombay
www.groups.yahoo.com/group/birdsofbombay
Mailing list for birders interested in birds of Mumbai.

Oriental Bird Club
www.orientalbirdclub.org
P.O.Box 324, Bedford, MK42 0WG, UK

Borivili National Park
www.borivlinationalpark.com

Tidal Schedule for Mumbai Harbour
www.mumbaiport.gov.in

BEST Undertaking
www.bestundertaking.com/transport
Bus service operator throughout Mumbai.

Western Railways
www.westernrailwayindia.com
Suburban rail service in Western Mumbai.

Central Railway
www.centralrailwayonline.com
Suburban rail service in central and eastern Mumbai, as well as sites on the mainland.

Books and publications

A Birdwatchers' Guide to India
Krys Kazmierczak & Raj Singh (2001) 2nd edition.
Oxford University Press, ISBN 1871104084

Birds of Mumbai
Sunjoy Monga (2003) India Book House
ISBN 8175083913

The Common Birds of Bombay EHA (1999)
ISBN 8176220353

Helm Identification Guides: Birds of the Indian Subcontinent
Richard Grimmett, Carol Inskipp, Tim Inskipp (1998) Christopher Helm, ISBN 0713640049

Pocket Guide to the Birds of the Indian Subcontinent R. Grimmett, C. Inskipp, T. Inskipp (2001) Christopher Helm, ISBN 0713663049

A Field Guide to Birds of the Indian Subcontinent
Krys Kazmierczak and Ber van Perlo (2000)
Pica Press, ISBN 1873403798

Birds of South Asia: The Ripley Guide
P. C. Rasmussen & J. C. Anderton (2005)
Smithsonian Institution & Lynx Edicions
ISBN 8487334660

A Photographic Guide to the Birds of India
Bikram Grewal, Bill Harvey and Otto P. Fister (2002)
Christopher Helm, ISBN 0713664037

The Book of Indian Birds (12th edition)
Salim Ali (1996) Oxford University Press
ISBN 0195637313

Where to Watch Birds in Asia
Nigel Wheatley (1996) Christopher Helm
ISBN 071364303X

New Delhi

The capital of India lies beside the south-flowing Yamuna River in the Gangetic Plain of northern India. It is within a few hours reach of several well-established birding sites such as Naini Tal in the foothills of the Himalaya, and Bharatpur, Ranthambhore and Corbett National Parks, but those detailed here are within two hours' drive and so are readily accomplished in a day's visit.

Although it contains over 14 million people, the Delhi area has a remarkable number of interesting sites within the city itself and nearby in Haryana State. Much of Haryana is now intensively cultivated, but even the fields can be productive for birds, particularly if they are flooded for rice or have just been harvested. In addition, there are still remnant wetlands along the Yamuna and the larger canals, and there are important wetland sanctuaries at Bhindawas, Sultanpur and Badkhal Lake. The state has a bird list of 560 species, of which 465 have been seen within two hours of Delhi and 451 in the city itself: this gives it the second biggest capital city list in the world, after Nairobi.

Within the city there are a surprising number of green open spaces. Some are well-used parks but others closely approximate to the natural vegetation that once occupied the rocky 'Ridge' that runs through the centre of the city. In addition, many roads are lined with a variety of well-established trees, and the more affluent housing colonies have mature, if small, gardens. Apart from resident species Delhi has a number of interesting summer visitors which move north out of the peninsula to breed before and during the summer monsoon. More striking is the number of winter visitors and passage migrants, for the city is well positioned on one of the major Asian flyways.

The spectacular Eurasian Hoopoe is frequently encountered in Delhi

TYPICAL SPECIES

Resident
Greater Flamingo, Little Egret, Cattle Egret, Great Egret, Black Ibis, Painted Stork, Black Kite, Indian Peafowl, Black-winged Stilt, Red-wattled Lapwing, Yellow-wattled Lapwing, River Lapwing, Yellow-footed Pigeon, Alexandrine Parakeet, Rose-ringed Parakeet, Blossom-headed Parakeet, House Swift, White-throated Kingfisher, Eurasian Hoopoe, Brown-headed Barbet, Coppersmith Barbet, Dusky Crag Martin, Brown Rock Chat, Oriental Magpie-Robin, Indian Robin, Striated Grassbird, Common Tailorbird, Graceful Prinia, Rufous-fronted Prinia, Long-tailed Shrike, Rufous Treepie, House Crow, Bank Myna, Common Myna, Brahminy Starling, Asian Pied Starling, Red-whiskered Bulbul, Common Babbler, Striated Babbler, Large Grey Babbler, Jungle Babbler, Purple Sunbird, Black-breasted Weaver, Streaked Weaver

Summer
Yellow Bittern, Cinnamon Bittern, Black Bittern, Comb Duck, White-eyed Buzzard, Watercock, Pheasant-tailed Jacana, Bronze-winged Jacana, Greater Painted-snipe, Blue-cheeked Bee-eater, Blue-tailed Bee-eater, Eurasian Golden Oriole, Rosy Starling

Winter
Bar-headed Goose, Ruddy Shelduck, Mallard, Eurasian Teal, Red-crested Pochard, Ferruginous Pochard, Greater Spotted Eagle, Imperial Eagle, Bonelli's Eagle, White-tailed Lapwing, Bimaculated Lark, Greater Short-toed Lark, Olive-backed Pipit, Rosy Pipit, Long-tailed Minivet, Bluethroat, Lesser Whitethroat, Brooks's Leaf Warbler, Hume's Warbler, Red-throated Flycatcher, Red-breasted Flycatcher, Rufous-tailed Shrike

Delhi has a continental climate with winter temperatures dropping into single figures in January. It may rain occasionally then, but much more restricting is the thick fog that delays airline flights by hours and, additionally, can severely limit birdwatching if you are unlucky. Unfortunately, the worst days are not predictable, and the agricultural areas of the neighbouring state of Haryana are usually the worst affected. From late March to June it can be very hot, with temperatures peaking well above 40°C. The monsoon, if it is going to arrive, breaks in late June or early July and although it is humid, the temperatures are lower then. It does not rain every day during the monsoon and usually for only a few hours when it does. The monsoon begins to peter out erratically in August but the weather remains hot and humid well into October when there is a spell of pleasant weather until winter sets in, in December.

GETTING AROUND

There is now a new Metro service in New Delhi but it is limited to two lines and does not serve any of the outlying birding sites. However, future expansion of this service may provide access to some of the areas west of the city. In general, mass public transport is best avoided but it is easy to get a three-wheeler auto, popularly known as a 'scooter', for city sites and to arrange a taxi for half-day trips or longer. Special

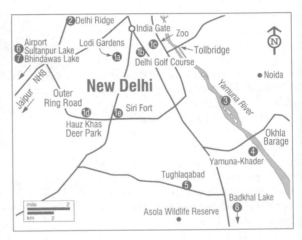

rates can be negotiated and your hotel will help you do this. Generally speaking hiring transport is very reasonable compared with most of the rest of the world, and you will have the advantage of a Hindi-speaking (and one hopes English-speaking) driver with you. It is not advisable to drive yourself unless you are used to driving in India. The roads are clogged with a myriad of conveyances and assorted animals, and driving discipline, let alone etiquette, is non-existent. If you want to go it alone, the trains in India offer a perfectly adequate service to tourists and several sites are accessible within a day's train ride of New Delhi. Most trains operate from the New Delhi Railway Station although a few still operate out of the Old Delhi Station, so check before travelling. For birding in Delhi itself, use the latest edition of the Eicher City Map to find your way around and, if possible, a map of the neighbouring state of Haryana. If you are in Delhi on a Sunday (and sometimes on a Saturday as well) check out the Northern India Bird Network email group by the previous Friday, and you may find that this very active group is meeting somewhere. The Yamuna River is one of the most regular spots, and the group welcomes birding visitors to the capital to join it.

Locations

1 Delhi City Parks (City Centre)

Delhi is full of parks of all sizes and they are easily located on most maps. Some in the centre, such as the **Lodi Gardens (1a)**, are well manicured but contain many old indigenous trees. Pay particular attention to Peepul, Banyan and Neem trees for Yellow-footed Pigeons, three species of parakeet and both Coppersmith and Brown-headed Barbets. These parks, as well as large gardens, hotels and institutional grounds, are usually rich in birds, including the national bird, the ubiquitous Indian Peafowl. Similarly, almost all the environs of the ancient monuments, the **Delhi Golf Course (1b)** – if you can gain access – and the riverside ghats are also good places to find birds. **Delhi Zoo (1c)** has a large breeding waterbird colony (particularly impressive during the monsoon), breeding Eurasian Thick-knees and wintering Olive-backed Pipits. All such places are good to wander in to see the commoner north Indian species, often at close quarters. Most older buildings have Dusky Crag Martins, Brown Rock Chats, nesting parakeets and Rock Doves: Wallcreepers have wintered on some of them. During spring passage almost anything can turn up but look out especially for the large flocks of Rosy Starlings and Common Rosefinches on fruiting mulberry trees and Sulphur-bellied and Western Crowned Leaf Warblers. A few White-rumped Vultures still roost and nest near **Claridges Hotel**. The large **Deer Park at Hauz Khas (1d)** and the nearby **Siri Fort Park (1e)** are both historically significant and ornithologically interesting because there is still fairly natural vegetative cover. In winter several species of warbler and birds of prey can be found. To find these parks take **Aurobindo Marg** south from the **Inner Ring Road** and they are both located close to the **Outer Ring Road**. The zoo (closed on Fridays) is between the **Mathura Road** and the Ring Road south of **Purana Quila**. A glance at a map of Delhi will soon show you other areas with potential.

KEY SPECIES

Resident
Indian Pond Heron, Spot-billed Duck, Oriental Honey Buzzard, Black Kite, White-rumped Vulture, Shikra, Eurasian Thick-knee, Red-wattled Lapwing, Yellow-footed Pigeon, House Swift, Alexandrine Parakeet, White-throated Kingfisher, Black-rumped Flameback, Brown-headed Barbet, Coppersmith Barbet, Dusky Crag Martin, White-browed Wagtail, Small Minivet, Oriental Magpie-Robin, Indian Robin, Brown Rock Chat, Common Tailorbird, Long-tailed Shrike, Rufous Treepie, Bank Myna, Common Myna, Brahminy Starling, Asian Pied Starling, Red-whiskered Bulbul, Common Babbler, Large Grey Babbler, Jungle Babbler, Purple Sunbird

Winter
Peregrine Falcon, Olive-backed Pipit, Bluethroat, Blue Rock Thrush, Lesser Whitethroat, Hume's Warbler, Sulphur-bellied Warbler, Red-throated Flycatcher, Red-breasted Flycatcher, Common Rosefinch, Wallcreeper (rare)

Summer
Green Bee-eater, Common Hawk-Cuckoo, Asian Koel, Eurasian Golden Oriole

Passage
Rosy Starling, Western Crowned Leaf Warbler

2 Delhi Ridge (5km)

This famous area, studied intensively by Tony Gaston 30 years ago, is now rather fragmented and degraded in parts. But what remains is in theory protected and still holds some interesting birds. It is a combination of thick thorn scrub, rocks and patches of woodland. Most of Delhi's land birds can be found here with effort, and it can be excellent for overhead raptors. The Central Ridge is the last known site in Delhi for the very local and apparently rapidly declining Marshall's Iora, but it has not been seen or heard there for several years.

The easiest way to access this area is by the first road on your left off **Willingdon Crescent** (renamed **Mother Theresa Crescent**), soon after the **Sardar Partel Road** junction, by the **Gandhi Murti**. Drive (or better walk) straight through to the **Polo Clubhouse**. You can park there and, after checking the field, take one of the paths into the jungle. The best trail, which leads to mature ridge woodland, is right at the end of the Polo Ground and usually deeply littered with stable straw. Watch out for polo horses exercising at speed! The uphill tracks take you to the Buddha Jayanti Smarak Park which is also worth a visit and another way in (from Vandemataram Marg) but the fence has few gaps in it. Another option, in the southern part of the Ridge, is the Sanjay Van Park which lies south of IIT and the Qutab Institutional Area. Take **Tara Crescent Road** off the **New Mehrauli Road** and just past the new **DFIDI** (old **USAID**) office you will see a gate. Excellent footpaths take you through a range of habitats including small lakes along the dammed stream. Seventy species in two hours are easily possible here in winter.

KEY SPECIES

Resident
Oriental Honey-buzzard, White-rumped Vulture, Jungle Bush Quail, Indian Peafowl, Eurasian Thick-knee, Yellow-wattled Lapwing, Yellow-crowned Woodpecker, Small Minivet, Indian Robin, Grey-breasted Prinia, White-eared Bulbul, Large Grey Babbler, Chestnut-shouldered Petronia

Summer
Common Hawk-Cuckoo, Asian Koel, Eurasian Golden Oriole

Winter
Booted Eagle, Tawny Pipit, Long-billed Pipit, Long-tailed Minivet, Bluethroat, Red-throated Flycatcher, Red-breasted Flycatcher, Lesser Whitethroat, Hume's Whitethroat,

Eastern Orphean Warbler, Brooks's Leaf Warbler, Common Rosefinch

Passage
Asian Paradise Flycatcher, Verditer Flycatcher, Blyth's Reed Warbler, Booted Warbler, Sykes's Warbler, Greenish Warbler, Western Crowned Leaf Warbler, Rosy Starling, White-capped Bunting

3 Yamuna River (10km)

Undoubtedly, both banks of the Yamuna and its associated wetlands are the most productive places in the city, with a bird list of over 330 species. During peak periods, seeing over 120 species in a day is not difficult. Given its position on a major flyway, there are regular records of rarities and almost anything can turn up, particularly in winter and on passage. The river is justly famous for duck and goose flocks with up to 20,000 birds of 20 species, and there is a roosting winter gull flock of up to 10,000 birds of at least four species. In addition, a wide range of wetland species including most of India's herons and egrets, resident Greater Flamingos and many migrant wader species can be seen. The winter roosts of hirundines, wagtails, mynas and starlings probably exceed a 'lakh' (100,000) and are a wonderful spectacle at dusk, especially if they are being harried by falcons or Marsh Harriers. In summer the range of species is no less impressive and the fly-pasts of breeding bitterns of three species are often spectacular. In the late 1990s the enigmatic Bristled Grassbird bred just after the monsoon rains, and was recorded again in 2005.

The best area to visit is between the new flyover and the barrage. From **Okhla village** take the **river road** from the great **Sand Mound** (marked on Eicher maps) and view the river and marshes from the several small tree-lined bunds that strike out eastwards. This area is really only good in winter and best in the evening. At the eastern end of the barrage you will notice a narrow curved spit that strikes out northwards. There is a footpath on it and from the end it is often possible to have excellent views of the duck and flamingo flocks if you use the bushes as a hide.

A little further on towards Noida, a narrow road branches north following the river.

If the water is reasonably high and not too clogged with water hyacinth, this can give excellent views of many species, especially in the morning. The end of this narrow road coincides with the start of the new flyover and a raised bund heads westwards to the Temple, exactly opposite the Okhla Sand Mound. This is probably the most productive walk of all for variety, with some very extensive reed-beds to look over. It is here

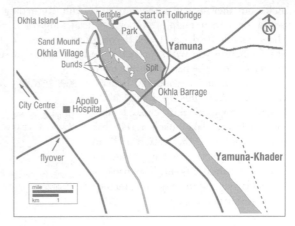

that the largest winter roosting flocks can be observed, including a Marsh Harrier roost of up to 40 birds.

The river is orientated north/south so it is important to take account of the rising and setting sun. Thus, observations from the east bank can be difficult in the evening (and vice versa). From mid-December to mid-February, because of the thick fog in Delhi, it may be better to wait until mid-morning if you want to use a telescope on the duck flocks, for instance. Also remember that a flood barrage at Okhla controls the flow and therefore water depth can vary considerably from day to day, obviously affecting the numbers, variety and distribution of the birds.

KEY SPECIES

Resident
Greater Flamingo, Purple Heron, Red-necked Falcon, Black Francolin, Brown Crake, Purple Swamphen, White-tailed Stonechat, Striated Grassbird, Yellow-bellied Prinia, Striated Babbler, Red Avadavat, Scaly-breasted Munia, Black-headed Munia, Black-breasted Weaver, Streaked Weaver

Summer
Yellow Bittern, Cinnamon Bittern,

Black Bittern, Cotton Pygmy-goose, Comb Duck, Watercock, Greater Painted-snipe, Pheasant-tailed Jacana, Bronze-winged Jacana, Oriental Pratincole, Small Pratincole, River Tern, Black-bellied Tern, Pied Cuckoo, Grey-bellied Cuckoo, Blue-cheeked Bee-eater

Winter
Black-necked Grebe, Greylag Goose, Bar-headed Goose, Ruddy Shelduck, Western Marsh Harrier, Pied Avocet,

River Lapwing, White-tailed Lapwing, Pallas's Gull, Caspian Gull, Rosy Pipit, Tree Pipit, Citrine Wagtail, Bluethroat, White-crowned Penduline Tit, European Starling, Common Rosefinch, Crested Bunting

Passage
Eurasian Hobby, Pale Sand Martin, Paddyfield Warbler, Blyth's Reed Warbler

4 Yamuna-Khader Area (13km)

This area is the southern extension of the west bank of the Yamuna, south of the barrage. The best way to enter is by the narrow tree-lined road that turns south from the main barrage (Noida) road just before the barrage on the Delhi side. Follow this for about 2km, where you meet a T-junction with a metalled road. Turning left you will quickly see a track off to the left again. This is drivable and takes you almost to the Yamuna bank. You have to walk the last 100m. This area can be very good for waders if water levels are not too high but this is unpredictable as it depends on the opening of the barrage: if the river floods it will be very muddy! Much of the area is now vegetable fields but there is still some tamarisk scrub. Returning to the road you can continue south for 3km and explore other left-branching tracks to the river as well as scanning the vegetable fields and trees. If you go west on one of the new roads just south from the T-junction you will soon reach the high bank that surrounds the fly-ash pit of the power station. The bank is visible from the junction and beyond the houses. You can climb this bank and walk round a huge hyacinth-choked lake. Although the 'lake' itself rarely has much of interest, the walk round can be very productive and gives you an excellent vantage point. The bank is particularly good for migrants, including occasional wheatears.

KEY SPECIES

Resident
Darter, Black Ibis, Egyptian Vulture, Brown Crake, River Lapwing, Sand Lark, Graceful Prinia, Yellow-eyed Babbler

Summer
Cotton Pygmy-goose, Small Pratincole, Bronze-winged Jacana

Winter
Long-legged Buzzard, Steppe Eagle, Lesser Sand Plover, Dunlin, Curlew Sandpiper, Ruff, Common Redshank, Black-tailed Godwit, Water Pipit, Rosy Pipit, Booted Warbler, Rufous-tailed Shrike

5 Tughlaqabad-Asola (10km)

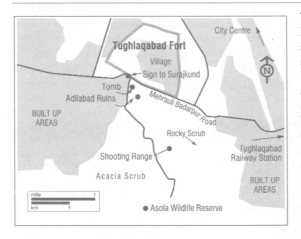

This historic area of south Delhi is rich in dry country species. Much of it is thorn scrub and trees on undulating rocky hills and almost anywhere is worth a look although it can be hard work. Tughlaqabad is on the west/east **Mehrauli-Badarpur Road** and well marked on all maps. The most accessible area is the waste field and surrounding scrub between the Adilabad Fort (worth entering) and Ghiyauddin Tughlaq's Tomb. The main fort is no longer very productive as squatters have encroached upon much of the area. There is a high fee to enter the fort itself. The road to **Surajkund** just west of the tomb takes you to the entrance to **Asola Wildlife Reserve**, about 2km south on your right. There is no entrance charge and the best thing to do is park inside the gate and walk round on the numerous paths. Unfortunately it does not open until 08:30 although there is a resident watch-person who will probably let you in earlier with the right encouragement. It closes at 18:00 and, surprisingly, on Sundays. There are many, quite approachable, Nilgai and Jackal here and there is a Blackbuck breeding scheme.

KEY SPECIES

Resident
Black-shouldered Kite, Jungle Bush Quail, Eurasian Thick-knee, Yellow-wattled Lapwing, Painted Sandgrouse, Sirkeer Malkoha, Indian Bushlark, Dusky Crag Martin, Brown Rock Chat,

Rufous-fronted Prinia, Bay-backed Shrike, Chestnut-shouldered Petronia

Winter
Common Kestrel, Long-billed Pipit, Blue Rock Thrush, Variable Wheatear, Desert Wheatear, Isabelline Wheatear,

Lesser Whitethroat, Hume's Whitethroat, Eastern Orphean Warbler, Sulphur-bellied Warbler, Yellow-eyed Babbler, Common Babbler, Southern Grey Shrike

Further Away

6 Sultanpur Lake National Park (45km)

Sultanpur Jheel is south-west of Delhi beyond Gurgaon on the Faruknagar Road. The park covers a total area of over 200 hectares including the Flats, which are outside the park boundaries. It has a bird list of over 330 species and was famous in the past for its large flocks of wintering wildfowl, cranes and pelicans. Unfortunately, farming and housing development, together with a series of poor monsoons, inter-

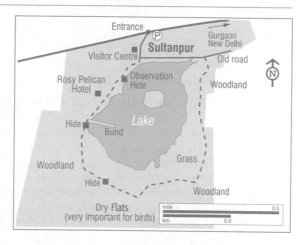

rupted the natural flow into the jheel, and there was a period when it was largely dry. Since 2001 there have been attempts to recover the situation by pumping water in from the nearest canal. At the car park there is an entrance booth and soon after a brick walkway of about 2km goes all the way round. If you leave the walkway to the south and cross a broken chain link fence you come into a dry, grassy area known as the Flats. Although the national park is rather small it offers some of the best birding near Delhi with rarities regularly being found. It has a significant waterbird colony on the islands, with up to 150 pairs of Painted Storks in winter. The jheel and its surrounding grassland are the main focus of attention but the woodlands and the southern Flats can be very productive. The Flats is the only site in the area for Indian Coursers, which are present from December to June.

Leave Delhi by **NH8** (the **Jaipur Road**) and continue through the new **DLF** developments, bypassing Gurgaon. At the **Sonha/Alwar** roundabout turn right and when in the middle of **old Gurgaon Town** turn left and continue straight until you reach open country. Nine kilometres short of Sultanpur you cross the **Basai** rail crossing. Follow the road straight taking a left fork twice. It usually takes an hour to drive to Sultanpur from south Delhi but the road can be very busy at peak times. Accommodation and meals are available at the adjacent Rosy Pelican Hotel. En route, and particularly after the Basai crossing, look out for Eurasian Thick-knees, Yellow-wattled Lapwings and Black Ibises in the fields and Brown Rock Chats on the buildings. Basai itself, briefly the gem of Haryana birding, dried up in late 2003.

KEY SPECIES

Resident
Darter, Purple Heron, Black-crowned Night Heron, Eurasian Spoonbill, Painted Stork, Black-necked Stork, Oriental Honey-buzzard, Black Francolin, Grey Francolin, Barred

Buttonquail, Sarus Crane, Purple Swamphen, Chestnut-bellied Sandgrouse, Collared Scops Owl, Dusky Eagle-Owl, Indian Bushlark, Small Minivet, Common Woodshrike, Large Grey Babbler, Sind Sparrow,

Red Avadavat, Indian Silverbill

Summer
Comb Duck, Greater Painted-snipe, Indian Courser, Red Collared Dove, Green Bee-eater, Blue-cheeked Bee-eater, Blue-tailed Bee-eater, Pied

Cuckoo, Common Hawk-Cuckoo, Grey-bellied Cuckoo, Ashy-crowned Sparrow-Lark

Winter
Greylag Goose, Bar-headed Goose, Gadwall, Garganey, Northern Shoveler, Eurasian Wigeon, Ferruginous Pochard, Western Marsh Harrier, Pallid Harrier, Eurasian

Sparrowhawk, Greater Spotted Eagle, Imperial Eagle, Bonelli's Eagle, Booted Eagle, Peregrine Falcon, Common Crane, Water Rail, Baillon's Crake, Indian Courser, Ruff, Common Redshank, Short-eared Owl, Eurasian Wryneck, Bimaculated Lark, Greater Short-toed Lark, Tawny Pipit, Long-billed Pipit, Tree Pipit, Olive-backed Pipit, Rosy Pipit, Long-tailed Minivet,

Dark-throated Thrush, Siberian Stonechat, Variable Wheatear, Desert Wheatear, Isabelline Wheatear, Moustached Warbler, Lesser Whitethroat, Hume's Whitethroat, Eastern Orphean Warbler, Brooks's Leaf Warbler, Red-throated Flycatcher, Red-breasted Flycatcher, Thick-billed Flowerpecker, Rufous-tailed Shrike, Spanish Sparrow

7 Bhindawas Lake Bird Sanctuary (80km)

The sanctuary, to the west of Delhi and about 40km west of Sultanpur, is an artificial creation to control flooding and has two escape channels into the canal system. In 2004 two new cross bunds were made restricting the wetland considerably and creating a public park near the heronry. It is not yet known what effect this will have. Unfortunately, poor monsoons and the increasing demands for irrigation water do leave it dry for long periods so it is best to check the delhibird.org website under recent sightings to see what conditions are like before travelling. Attempts are made periodically to pump water in.

The bird list at Bhindawas is over 270 species. The main attraction is the large mixed flocks of wildfowl and the large number of wading birds. There has been a very large heronry on the babul-covered 'island' which is active in the summer. The heronry also has the largest known breeding colony of Sind Sparrows in Haryana (again in summer). The bird spectacle, when water levels are sufficient, can keep you engrossed for hours, from one of the two watch towers or from the bunds. Otherwise it is necessary to drive slowly round stopping at vantage points to scan the wetland. Unfortunately, most of the planted trees are eucalyptus, and passerines are both less numerous and less varied than at Sultanpur. An added attraction is the large herd of Nilgai.

It is probably best to approach the site via **Jhajjar** and **Chuchakwas** for the first time. Carry on from **Sultanpur** to turn right, in **Faruknagar**, onto the Jhajjar road. At Jhajjar turn left for Chuchakwas. At the main temple here turn left down a dirt road for 7km until you reach the 11km, circular, earth bund, which is usually drivable. There is accommodation at the Forest Rest House but it needs to be booked through the Forest Department. The journey time from Delhi is just over two hours depending on the traffic.

Along some of the roads to Bhindawas there are interesting, small, dry scrub areas that usually repay a visit.

KEY SPECIES

Resident
Darter, Little Cormorant, Indian Cormorant, Great Cormorant, Glossy Ibis, Black-headed Ibis, Black Ibis, Purple Heron, Black-crowned Night Heron, Indian Pond Heron, Eurasian Spoonbill, Woolly-necked Stork, Black-necked Stork, Oriental Honey-buzzard, Red-necked Falcon, Sarus Crane, Purple Swamphen, Eurasian Thick-knee, Common Kingfisher, Pied Kingfisher, White-browed Fantail, Common Woodshrike, Large Grey Babbler

Summer
Lesser Whistling Duck, Cotton Pygmy-goose, Comb Duck, Spot-billed Duck, Greater Painted-snipe, Bronze-winged Jacana, Pheasant-tailed Jacana, Pied Cuckoo, Green Bee-eater, Blue-cheeked Bee-eater, Blue-tailed Bee-eater, Eurasian Golden Oriole, Sind Sparrow

Winter
Great White Pelican, Greylag Goose, Bar-headed Goose, Ruddy Shelduck, Gadwall, Garganey, Northern Shoveler, Eurasian Wigeon, Ferruginous Pochard, Osprey, Western Marsh Harrier, Hen Harrier, Greater Spotted Eagle, Imperial

Eagle, Bonelli's Eagle, Booted Eagle, Peregrine Falcon, Common Crane, Common Coot, Pied Avocet, Northern Lapwing, White-tailed Lapwing, Ruff,	Common Redshank, Olive-backed Pipit, Rosy Pipit, Citrine Wagtail, Lesser Whitethroat, Hume's Warbler, Red-throated Flycatcher, Red-	breasted Flycatcher, European Starling

8 Badkhal Lake (75km)

(Faridabad Railway Station)

This attractive reservoir is close to Faridabad, about an hour's drive south of South Delhi past Asola and Surajkund. The Haryana border is reached soon after Asola and just before Surajkund. After that the road runs through the rocky thornbush of the Southern Ridge and many new housing developments. The lake is well signposted as it is a tourist attraction with boating and a hotel, the Grey Falcon. Thus it is best avoided on winter weekends.

Park in the car park, and climb onto the dam wall. This is an excellent vantage point of the lake although the pedal boats are worth hiring for a close approach to the wild-fowl. The large trees on the bund and the old fish ponds and reedbeds to the left are worth looking at carefully. At the southern end is a productive area of marshy pasture while the cliffs to the west are worth scanning. The more adventurous can explore the rocky areas around and beyond the Grey Falcon Hotel, another good vantage point, which is to the north of the dam. From there a footpath leads down to an ancient but active temple in a shady valley. The rocky areas have similar bird species to Asola but are less disturbed and more extensive. All in all the area is good for a number of species that are scarce or difficult to find elsewhere near Delhi.

KEY SPECIES

Resident	Winter	Summer
Asian Openbill, Cotton Pygmy-goose, Bronze-winged Jacana, Pheasant-tailed Jacana, Spotted Dove, Painted Sandgrouse, Common Kingfisher, Pied Kingfisher	Great Crested Grebe, Red-crested Pochard, Ferruginous Pochard, Common Snipe, Grey-headed Canary-Flycatcher, Rufous-tailed Wheatear, Citrine Wagtail, Yellow Wagtail, Rosy Pipit	Green Bee-eater, Blue-tailed Bee-eater, Blue-cheeked Bee-eater, Greater Painted-snipe, Asian Paradise Flycatcher

Useful contacts and websites

Northern India Bird Network (delhibird)
www.delhibird.net
Birding group website which is packed with information about birding in the New Delhi area and includes site information, checklists, recent sightings and also supports the delhibird email group: http://groups.yahoo.com/group/delhibird

Waterbirds of India
www.kolkatabirds.com/waterbird.htm
Birding website for northeast India with information about Sultanpur and Yamuna River.

Oriental Bird Club
www.orientalbirdclub.org
P.O.Box 324, Bedford, MK42 0WG, UK.

Delhi Metro
www.delhimetrorail.com
Metro service in New Delhi and suburbs.

Indian Railways
www.indianrail.gov.in/
Train services to sites outside New Delhi

Books and publications

A Photographic Guide to the Birds of India
Bikram Grewal, Bill Harvey & Otto Pfister (2002)
Christopher Helm, ISBN 0713664037

Pocket Guide to the Birds of the Indian Subcontinent Richard Grimmett, Carol Inskipp & Tim Inskipp (2002) Christopher Helm
ISBN 0713663049

Field Guide to the Birds of Northern India
Richard Grimmett & Tim Inskipp (2003)
Christopher Helm, ISBN 0713651679

A Birdwatchers' Guide to India
Krys Kazmierczak & Raj Singh (2001) 2nd edition
Oxford University Press, ISBN 1871104084

The Atlas of the Birds of Delhi and Haryana
Bill Harvey, Bikram Grewal & Nikhil Devasar (2006)
Rupa & Co. ISBN 8129109549

Birds of Delhi
Ranjit Lal (2003) Oxford University Press
ISBN 0195672194

A Field Guide to the Birds of the Indian
Subcontinent Krys Kazmierczak & Ber van Perlo
(2000) Pica Press, ISBN 1873403798

Where to Watch Birds in Asia
Nigel Wheatley (1996) Christopher Helm
ISBN 071364303X

Seoul

Seoul is located on the Hangang River approximately 25km from the coast. However, the course of the river takes a further 50km to reach the sea and is therefore not tidal. The city itself is built over several wooded hills and the creation of public parks to incorporate these hills has created a varied and accessible amount of habitat for birds. Like all Asian cities, Seoul is a bustling metropolis which will immediately induce in a visiting birder from the West a burning desire to get out of the city as far and as fast as possible. However, by exploring the riparian and forest habitats of the city parks and riverside walk, the true richness of Seoul's avifauna becomes apparent. Winter or summer, there are a number of locations within easy reach of the city centre which can provide some excellent birding. The climate is temperate, but summers can be hot and wet and winters have occasional severely cold spells, with sub-zero temperatures. Spring and autumn are the most pleasant seasons.

The Korean Peninsula is a compact and convenient location to see a wide variety of Eastern Palearctic species. Moreover, its geographical position on the flyways north to the Siberian Arctic provides a superb opportunity

White-tailed Eagle is a winter possibility on the Hangang River

KEY SPECIES

Resident
Striated Heron, Black-crowned Night Heron, Little Egret, Great Egret, Grey Heron, Mallard, Spot-billed Duck, Common Kestrel, Common Pheasant, Common Coot, Black-tailed Gull, Oriental Turtle Dove, Sky Lark, Black-backed Wagtail, Brown-eared Bulbul, Daurian Redstart, Great Tit, Coal Tit, Marsh Tit, Japanese White-eye, White-cheeked Starling, Bull-headed Shrike, Large-billed Crow, Carrion Crow, Black-billed Magpie, Azure-winged Magpie, Eurasian Tree Sparrow,

Eurasian Siskin, Yellow-throated Bunting

Summer
Fork-tailed Swift, Barn Swallow, White Wagtail, Grey Wagtail, Black-naped Oriole

Winter
Little Grebe, Great Cormorant, Eurasian Wigeon, Eurasian Teal, Northern Pintail, Tufted Duck, Common Pochard, Goosander, Black-headed Gull, Vega Gull, Buff-bellied Pipit, Dusky Thrush, Goldcrest, Daurian Jackdaw, Rook, Brambling, Rustic Bunting

to witness migration. In recent years, it has emerged as a truly world class location to observe wader migration, and it is hoped that increased international recognition of this fact will stem the reclamation of tidal habitats for development purposes. Although it is an inland city, Seoul is located in close proximity to some superb coastal habitat. Most travellers are likely to arrive by air and the international airport is located right on Yeongjeong Island, where huge concentrations of waders can be seen. The business traveller can easily visit two city locations in a day using the excellent subway system and can use any extra time to visit Incheon.

GETTING AROUND

The best way to travel within the city limits is to use the subway system. The service is frequent, fast and economical and all destination signs are in English. Although Seoul has a well-defined Downtown area there are additional commercial, shopping, residential and hotel districts right across the city which means that rush-hour traffic heads in many different directions. Consequently, if you are travelling during peak periods you can expect to encounter a crowded subway whatever direction you are going in.

Taxis are flexible although a little expensive and, given the language barrier, can be difficult to use for birding purposes, unless you have a specific destination in mind and have it written in Korean. Your hotel will be able to assist you with this. Likewise bus services are not so practical as destination signs are usually only in Korean and they are susceptible to traffic conges-

tion. Even so, for some of the coastal sites, you will need to get a provincial bus, most of which depart from the Seoul Express Bus Terminal on the south bank of the Hangang, although some services operate from Sincheon Terminal on the north side of the river: both are served by the subway. All train services out of Seoul operate from Seoul Station which is located centrally and is also served by subway.

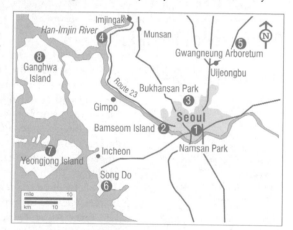

If you have a car and plan to do your own driving, be advised that Seoul and its hinterland are densely populated with many expressways and interchanges, which can be confusing for a visitor as not all signs carry an English translation. The sites to the north of Seoul are best accessed by getting onto the Outer Circular Expressway (Route 100) and turning north at the relevant interchange. This is easier than trying to navigate through the centre of the city. The coastal sites, on the other hand, can be reached by getting onto the Olympic Expressway, which runs parallel to the south bank of the Hangang. However, this option will set your agenda for the day, as there is no bridge over the river downstream from Seoul, so you won't be able to visit an inland site without returning to the Circular Expressway.

You will need both a good city map of Seoul for the city sites and a map of Gyeonggi-do Province for the coastal and interior sites and, of course, subway maps, which can be picked up at your hotel desk or at any airport information kiosk.

Locations

1 Namsan Park (City Centre)

(Dongguk Subway Station)

This is basically a smaller version of Bukhansan Park but right within the city centre and less arduous a climb. All of the species at Bukhansan can be seen here although in smaller numbers and, as the park is much smaller, it is hard to find areas that are undisturbed. It can be accessed from the Dongguk University Subway Station (Line 3). Take the road that leads all the way up to the Seoul Tower, and you will find several trails that lead off this road which will be less frequented by joggers. There is also a quiet trail around the perimeter of the university sports ground.

KEY SPECIES

Resident
Chinese Goshawk, Japanese Sparrowhawk, Japanese Pygmy Woodpecker, Grey-headed Woodpecker, Great Spotted Woodpecker, Varied Tit, Long-tailed Tit, Winter Wren, Eurasian Nuthatch, Vinous-throated Parrotbill, Eurasian Jay, Grey-capped Greenfinch

Summer
Common Cuckoo, Dollarbird, Red-rumped Swallow, Scaly Thrush, Eastern Crowned Warbler, Blue-and-white Flycatcher, Yellow-rumped Flycatcher, Chinese Grosbeak

2 Bamseom Island (City Centre)

(Yeoinaru Subway Station)

KEY SPECIES

Resident
Chinese Goshawk, Common Kingfisher, Vinous-throated Parrotbill

Passage
Common Sandpiper, Common Redshank, Great Reed Warbler

Winter
Great Crested Grebe, Common Goldeneye, Smew, Common Buzzard, Yellow-legged Gull, Slaty-backed Gull, Chinese Penduline Tit

This is an island in the Han River which is inaccessible and has been left undisturbed, but can be viewed from the Seogangdaegyo Bridge. It is an area of rich riparian habitat that is probably at its best in winter, when the surrounding waterway attracts large numbers of ducks. It also provides a perch for raptors including, on occasion, a White-tailed Eagle. The nearest subway station is Yeoinaru on Line 5. During the winter (Dec–Feb) there is a ferry service to the island which enables viewers to watch the birds at close quarters. This ferry departs several times a day from Yeouido Park on the south bank of the river. On winter weekends local birders set their telescopes up at a watchpoint at Hangang Riverside Park on the south bank of the river so that they can watch the birds in midstream and on the island.

3 Bukhansan Park (5km)

(Gyeongbokgung Subway Station)

This is a forested hill 342m high, just north of the city centre, with quite extensive mixed woodlands on its slopes. Although it is very popular with visitors, it is possible to find quiet secluded parts in which to watch birds. In general, the mixed woodland at the lower elevation is better for birds than the coniferous forest higher up the slopes. If you are not planning to visit any forest areas in the interior of the country during your stay in Seoul, then a visit to Bukhansan will be well worth while.

Go north from the Gyeongbokgung Subway Station (Line 3) to enter the park and then follow the trails up the slope. Plan to keep birding in a northerly direction as there are a number of plush hotels further up the slopes where you could get a taxi to bring you back to the subway if all the walking and climbing is too exhausting.

KEY SPECIES

Resident
Chinese Goshawk, Japanese Sparrowhawk, Japanese Pygmy Woodpecker, Grey-headed Woodpecker, Great Spotted Woodpecker, Varied Tit, Long-tailed Tit, Winter Wren, Eurasian Nuthatch, Vinous-throated Parrotbill, Eurasian Jay, Grey-capped Greenfinch

Summer
Common Cuckoo, Dollarbird, Red-rumped Swallow, Scaly Thrush, Eastern Crowned Warbler, Blue-and-white Flycatcher, Yellow-rumped Flycatcher, Chinese Grosbeak

Winter
Common Buzzard, Siberian Accentor, Hawfinch, Black-faced Bunting

4 Han-Imjin River (50km)

This is an extensive wetland site at the confluence of the Han and Imjin rivers, which in winter attracts large flocks of geese and cranes as well as many other birds. It is really only practical to visit if you have a car, as much of the birding is done by cruising Route 23 along the east bank of the Han River, northwest out of Seoul. However, the Imjin River forms part of the border with North Korea and it is not always permitted to stop at likely points on the road from which to scan the river, so some patience is needed. The surrounding flat countryside, particularly harvested rice-fields, are important for wintering geese and cranes.

Take **Route 23** northwest from Seoul: the 30km stretch of this road between the **Outer Circular Expressway** and **Imjingak** passes through a variety of habitats and has the Han and Imjin rivers on the east side of the road over much of its distance. However, due to the military presence, exercise some caution when pulling over to scan for birds. At Imjingak, there is a visitor centre with a flat roof which is a viewpoint for tourists over the Demilitarised Zone (DMZ) and North Korea, but it is equally good as a vantage point for scanning the surrounding flat countryside for geese and cranes. If you don't have a car, you could get a tour bus to Imjingak, and explore the general area just south of the DMZ.

KEY SPECIES

Resident
Mandarin, Long-billed Plover, Common Kingfisher, Vinous-throated Parrotbill, Meadow Bunting

Passage
Eurasian Spoonbill, Osprey, Common Sandpiper, Common Redshank, Spotted Redshank, Eurasian Hoopoe, Great Reed Warbler, Japanese Bush Warbler, Chestnut Bunting, Black-faced Bunting

Winter
Great Crested Grebe, Swan Goose, Bean Goose, Greater White-fronted Goose, Ruddy Shelduck, Common Goldeneye, Smew, Red-breasted Merganser, Scaly-sided Merganser, Cinereous Vulture, White-tailed Eagle, Steller's Sea Eagle, White-naped Crane, Red-crowned Crane, Yellow-legged Gull, Chinese Penduline Tit, Siberian Accentor

5 Gwangneung Arboretum (25km)   (Uijeongbu Subway Station – Line 1)

Although it contains many exotic trees and shrubs, there are substantial stands of indigenous forest in the arboretum, and it is one of the best locations near Seoul to see the full range of the forest birds which are typical of the Korean Peninsula. Additionally, its location along the banks of a river and the presence of two small lakes mean that there are also habitats for aquatic and riparian species. Unfortunately the arboretum is closed at weekends, so this is a site only worth considering during weekdays, and even then it does not open until 09.00. There are several trails within the arboretum and it is also worth checking the riverbank near the car park as it is a reliable site for Solitary Snipe.

Take the **Dongbu Expressway** north from Seoul, which becomes **Highway 3**, as far north as **Uijeongbu**. Turn east on **Route 43** towards **Pocheon** and turn south after about 5km towards **Gwangneung**. The arboretum entrance is on the right-hand side of the road. There is a bus service (No. 21) to the arboretum from Uijeongbu bus terminal, or simply take a taxi from the subway station.

KEY SPECIES

Resident
Chinese Goshawk, Japanese Sparrowhawk, Black Woodpecker, Japanese Pygmy Woodpecker, Grey-headed Woodpecker, Great Spotted Woodpecker, White-backed Woodpecker, Varied Tit, Long-tailed Tit, Winter Wren, Brown Dipper, Eurasian Treecreeper, Eurasian

Nuthatch, Vinous-throated Parrotbill, Eurasian Jay

Summer
Common Cuckoo, Oriental Cuckoo, Ruddy Kingfisher, Dollarbird, Red-rumped Swallow, Siberian Blue Robin, Scaly Thrush, Pale Thrush, Grey-backed Thrush, Eastern

Crowned Warbler, Blue-and-white Flycatcher, Yellow-rumped Flycatcher, Chinese Grosbeak

Winter
Common Buzzard, Upland Buzzard, Solitary Snipe, Green Sandpiper, Siberian Accentor, Hawfinch, Pallas's Rosefinch

6 Song Do (40km)   *(Dongchun Subway Station) (Bus Nos. 8/16)*

Song Do is an area of mudflats surrounded by reclaimed land just southwest of Incheon. This site holds numerous waders and is also excellent in winter for gulls when both Saunders's and Relict Gulls can be seen. Unfortunately, the future of this site is uncertain as there is ongoing construction work to reclaim this vast area for development. Because of its location it can be possible to get better views of the birds at this site than at the more extensive mudflats at other sites, as access routes to the land reclamation works create roads which are adjacent to the mudflats and to freshwater lagoons which act as roosting sites for waders.

Take **Route 77** south from **Incheon** to **Song Do Resort**, and turn west to access the reclaimed zone. The area is served by subway and also a bus service from the airport.

KEY SPECIES

Winter
Ruddy Shelduck, Northern Shoveler, Red-breasted Merganser, Eastern Marsh Harrier, Peregrine Falcon, Eurasian Oystercatcher, Dunlin, Eurasian Curlew, Ruddy Turnstone, Herring Gull, Saunders's Gull, Relict Gull, Slaty-backed Gull

Passage
Chinese Egret, Lesser Sand Plover, Grey Plover, Great Knot, Red-necked Stint, Sharp-tailed Sandpiper, Grey-tailed Tattler, Common Redshank, Spotted Redshank, Common Greenshank, Marsh Sandpiper, Terek Sandpiper, Black-tailed Godwit, Bar-tailed Godwit, Whimbrel, Far Eastern Curlew

7 Yeongjong Island (60km)   *(Bus Nos. 600/601/602)*

Seoul's international airport is built on Yeongjong Island, and although much of the island and surrounding mudflats have been reclaimed, there are still substantial mudflats where huge numbers of birds occur on passage and in winter, and all within a five-minute taxi ride of the airport. As with any mudflat habitat, it is best to to coincide with a high tide, otherwise the birds will be scattered over a huge area and will be very distant. A telescope is recommended, in all conditions. The tidal mudflats to the south hold the largest concentration of waders and here it is possible to see Nordmann's Greenshank, as the site has a very good track record for this rare and elusive species. This area is also a regular site for Chinese Egrets, Black-faced Spoonbills, Saunders's Gulls and Red-crowned Cranes, all of which are endemic to the region and extremely rare on a global basis.

Yeongjong Island can be accessed from Seoul by the **Seoul Incheon Airport Expressway bridge**, or by ferry from Wolmido Island. There is a perimeter road around Yeongjeong Island, and the best way to see the birds is to take a taxi from the airport terminal either to circle the island, if you are on a tight schedule, or to arrange a pickup point later. You could also consider renting a bicycle at the Wolmido ferry terminal. There are woodland areas in the northeast of the island around Mt Baekunsan which are good for passerines. In fact, if you are staying at one of the airport hotels you could see most of the birds of the Seoul area without leaving the island.

KEY SPECIES

Resident
Common Kestrel, Kentish Plover, Common Kingfisher, Winter Wren, Meadow Bunting

Summer
Intermediate Egret, Cattle Egret, Little Ringed Plover, Little Tern, Black-capped Kingfisher

Winter
Ruddy Shelduck, Falcated Duck,

Northern Shoveler, Red-breasted Merganser, Upland Buzzard, Eastern Marsh Harrier, Peregrine Falcon, Eurasian Oystercatcher, Dunlin, Eurasian Curlew, Ruddy Turnstone, Saunders's Gull, Relict Gull, Common Gull, Herring Gull, Slaty-backed Gull, Siberian Accentor, Black-faced Bunting, Chestnut Bunting, Pallas's Reed Bunting

Passage
Chinese Egret, Black-faced Spoonbill, Red-crowned Crane, Lesser Sand Plover, Grey Plover, Great Knot, Red-necked Stint, Sharp-tailed Sandpiper, Grey-tailed Tattler, Common Redshank, Spotted Redshank, Common Greenshank, Nordmann's Greenshank, Terek Sandpiper, Black-tailed Godwit, Bar-tailed Godwit, Whimbrel, Far Eastern Curlew

8 Ganghwa Island (55km)

No account of birding in Korea would be complete without a mention of Ganghwa Island. This island northwest of Seoul and almost on the border with North Korea is one of the premier sites in Korea, if not the world, for waders and other aquatic birds. It is considered *the* site for Black-faced Spoonbills and Chinese Egrets, which breed there, and is also a superb site in winter and at migration time for waders, cranes and wildfowl. The interior of the island is also a rich habitat of wooded hillsides, cultivated areas and rice-fields and supports a varied selection of passerines including thrushes, buntings, and woodland species.

The island is a popular weekend destination for visitors from Seoul and is becoming increasingly developed, but logistically it is somewhat more difficult for a birder without a car. There is a direct bus service to Ganghwa from the Shincheon Bus Terminal

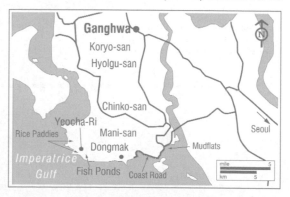

(served by subway Line 2) but on arrival at Ganghwa you will need to take a taxi or another bus to get to the south of the island. The main sites for waders are the mudflats to the east of Dongmak, which is a beach resort on the south coast. With this site, like all of the vast mudflats on the west coast of Korea, it is vital to visit on a rising tide in order to view the birds at a reasonable distance. Further west from Dongmak are the rice-paddies and fish ponds of Yeocha-Ri.

KEY SPECIES

Resident
Common Kestrel, Kentish Plover, Common Kingfisher, Zitting Cisticola, Winter Wren, Meadow Bunting,

Summer
Intermediate Egret, Chinese Egret, Cattle Egret, Black-faced Spoonbill, Common Moorhen, Little Ringed Plover, Common Sandpiper, Little Tern, Black-capped Kingfisher, Japanese Bush Warbler, Great Reed Warbler

Winter
Greater White-fronted Goose, Bean Goose, Ruddy Shelduck, Common Gadwall, Northern Shoveler, Red-breasted Merganser, Upland Buzzard, Eastern Marsh Harrier, Peregrine Falcon, Red-crowned Crane, Eurasian Oystercatcher, Common Ringed Plover, Dunlin, Sanderling, Eurasian Curlew, Ruddy Turnstone, Common Gull, Herring Gull, Slaty-backed Gull, Saunders's Gull, Siberian Accentor, Black-faced Bunting, Japanese Reed Bunting, Pallas's Reed Bunting

Passage
Garganey, Lesser Sand Plover, Grey Plover, Great Knot, Red-necked Stint, Sharp-tailed Sandpiper, Grey-tailed Tattler, Curlew Sandpiper, Common Redshank, Spotted Redshank, Common Greenshank, Marsh Sandpiper, Terek Sandpiper, Black-tailed Godwit, Bar-tailed Godwit, Whimbrel, Far Eastern Curlew, Grey-backed Thrush, Tristram's Bunting

Useful contacts and websites

Birding Korea
www.wbkenglish.com

Tour2Korea.com
www.english.tour2korea.com/03Sightseeing/Theme Tours/visiting.asp
Tourist website with section on birdwatching in Korea.

Korean Wild Birds Society (Korean)
www.kwbs.or.kr

Birds of Korea (English and Korean)
www.home.megapass.co.kr/~skua

Bird's Whisperer's Home (English and Korean)
www.aves.birdinkorea.net

Oriental Bird Club www.orientalbirdclub.org
P.O.Box 324, Bedford, MK42 0WG, UK

Korean National Arboretum
www.koreaplants.go.kr

Seoul Metropolitan Rapid Transit Corp.
www.smrt.co.kr/english_smrt/index.jsp
Information about subway service in Seoul area.

Korail (Korean and English)
www.korail.go.kr
Train services out of Seoul.

Tidal Schedule for Incheon
www.easytide.ukho.gov.uk/EasyTide/EasyTide/ShowPrediction.aspx

Books and publications

A Field Guide to the Birds of Korea
Woo-Shin Lee, Tae-Hoe Koo & Jin-Young Park (2000) LG Evergreen Foundation
ISBN 8995141506

A Field Guide to the Birds of South-East Asia
Craig Robson (2000) New Holland
ISBN 1853683132

Where to Watch Birds in Asia
Nigel Wheatley (1996) Christopher Helm
ISBN 071364303X

A Black-crowned Night Heron flies over the modern Shanghai skyline

Shanghai

Shanghai is situated just south of the mouth of the Yangtze River, which feeds into the East China Sea half way between Beijing and Hong Kong. One of the largest and most heavily populated cities in the world, Shanghai is a modern, exciting and vibrant place. One thing it is not, however, is a birdwatching Mecca! At first sight the city appears to be a mass of concrete and glass with no suitable habitats for birds or any other wildlife for that matter, and there are only about 20 regular birdwatchers here out of a population of over 15 million. Despite this, Shanghai's situation on the east coast of China means that interesting birds can be found if you know where to look, particularly during migration times.

The city is criss-crossed with many tributaries from the Huangpu River, where you can see wagtails, herons and kingfishers. There are many small and medium-sized parks and gardens but these are of limited scope for the keen birdwatcher. There are plans to create fifteen 'Wildlife Reserves' around the city over the next ten years but it remains to be seen whether these will appear.

There are really only four reasonable birdwatching areas within the city limits: Shanghai Zoo, the Botanic Gardens, the Gongqing Forest Park and Song Qing Ling's Mausoleum. In order to see more natural habitats, with a greater abundance of bird life, it is necessary to travel to outlying areas such as Chongming Island, Taihu Lake or Bin Hai. If a visitor has the time to travel and can stay overnight for one or two days there are several interesting nature reserves to the west of Hangzhou city which is 140km south of Shanghai, and much further south and west is Yellow Mountain. About 400km (five hours by bus) north of Shanghai is the fantastic Yancheng Nature Reserve, where many east China specialities such as Reed Parrotbills, Oriental Storks, Relict Gulls and Black-faced Spoonbills can be seen at the right time of year. If you do decide to stay overnight away from Shanghai don't forget your passport: you'll need it to check in to a hotel.

GETTING AROUND

The biggest difficulty facing the foreign visitor in getting around Shanghai is language. The vast majority of people do not speak English so don't expect to be able to stop and ask anyone for directions. Fortunately most of the roads are signposted in English as well as Chinese, so first of all buy an English language map showing the outlying areas as well as the city centre. If you can hire a personal Chinese/

English interpreter through your hotel concierge, so much the better, and to avoid any misunderstandings due to language difficulties it is always best to utilise your hotel staff for any telephone arrangements with local travel agents, bird guides and drivers.

The underground system is modern and cheap but covers a very limited area, although more lines are under construction. Public transport buses are also cheap but are best avoided unless you know your specific route number and destination name in Chinese. Tourist buses to some of the birding sites depart from the Shanghai Stadium Bus Terminal in the south of the city and served by Metro Line 1. Taxis are more expensive but still cheap by Western standards, and so are the best means of transport within the main parts of the city. Very few taxi drivers speak English so either use an interpreter or use your map to show the driver where you are and where you want to go.

Car hire is possible from Pudong International Airport: however you will not be allowed out of the city limits, which makes this option much less attractive. Also a combination of congested roads, highway rules that make Alice's Wonderland look normal, and reckless, unpredictable local drivers make driving here a dangerous and far from relaxing experience! Hiring a car with a driver does remain an option, though, and this can easily be arranged through your hotel or other travel agencies.

All the parks and gardens are well laid out with paved footpaths, so comfortable walking shoes are quite adequate for these locations and a good pair of binoculars is the only optical aid you will need. The winter is quite cold but rarely drops much below freezing. Summer is hot and can be very humid, so carry a bottle of drinking water and wear a hat. Spring and autumn tend to offer the most pleasant climate for birding, the clearest skies and the greatest diversity of birds. The best time to visit most places is first thing in the morning: not only are there more birds about but you also avoid the crowds. Unless you are of a particularly gregarious nature try to avoid weekends and especially public holidays. The Shanghaiese are apparently allergic to rain and consequently the most undisturbed Shanghai birding can be achieved on wet days.

Locations

---

1 **Song Qing Ling's Mausoleum** (10km)

This small park in the west of the city is worth visiting in autumn and winter as it contains a high density of fruiting trees which can attract large numbers of thrushes, as well as a variety of *Phylloscopus* warblers and flycatchers. One or two Peregrines roost on the 'UFO' tower just to the west and hunt over the park in the late afternoon. After or during rain, spring visits can be rewarding, too. The modest entrance fee seems to put off many local people from using the park and as a result it is surprisingly quiet considering its location. One drawback is that access for non-residents is between 08.30 and 17.00 only (last entry strictly 16.30).

To get there, take a taxi to the junction of **Hong Qiao Road/Song Yuan Road** and alight 100m along Song Yuan Road. The park is on the opposite side if the road. A visit here can be conveniently combined with a trip to the nearby zoo.

KEY SPECIES

Winter
Peregrine Falcon, Pale Thrush, Scaly Thrush, Red-flanked Bluetail

Passage
Daurian Redstart, Japanese Thrush, Yellow-browed Warbler, Eastern Crowned Leaf Warbler

2 Shanghai Zoo (13km) (Dong Wu Yuan)

The zoological gardens cover approximately 1.5km^2 and were formerly known as Western Suburb Park, a cultural and entertainment park. As zoos go this is quite a pleasant place: most of the animals seem to be well looked after, although there are still too many iron bars and concrete floors. The zoo boasts many mature trees, gardens, green spaces and a large lake, creating a variety of habitats for wild birds. There seem to be more Azure-winged Magpies and Yellow-billed Grosbeaks here than anywhere else in Shanghai.

The zoo is not too far from the city centre on **Hong Qiao Road** and all the taxi drivers know how to get there, so it is best to travel by taxi. The journey time is approximately 30 minutes. There is a modest entrance fee and opening times are 08.30 to 17.30.

KEY SPECIES

Resident
Mandarin Duck, Brownish-flanked Bush Warbler, Azure-winged Magpie, Yellow-billed Grosbeak

Summer
Little Egret, Cattle Egret, Chinese Pond Heron

Winter
Little Grebe, Eurasian Teal, Garganey, Common Pochard, Spot-billed Duck, Red-flanked Bluetail, Pale Thrush

Passage
Thrushes, *Phylloscopus* warblers, Flycatchers

3 Shanghai Botanic Gardens (12km) (Zhi Wu Yuan)

A well-established garden but kept very 'tidy'. However, there are still neglected corners worth seeking out and as with other sites it is best to visit first thing in the morning. There are two good birding spots on the south side of the river: the weedy area of golden rod and willow in the north-west, which is good for Eurasian Woodcock; and the

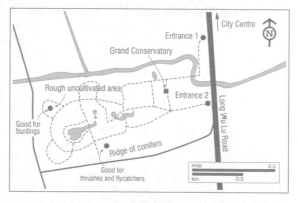

City Centre
N
Entrance 1
Grand Conservatory
Rough uncultivated area
Entrance 2
Long Wu Lu Road
Good for buntings
Ridge of conifers
Good for thrushes and flycatchers
mile 0.5
km 0.5

nearby 'valley' beneath the ridge of conifers, which is a regular spot for Pale and Scaly Thrushes. If you are just birding, you need only pay the cheaper entrance fee. There is a higher fee for access to the Tropical Houses and the Bonsai Garden.

Located on **Long Wu Road**, south of the city centre, you can best get there by taxi. You can also take the subway (Metro Line 3) to Shilong Road Station but, if you do, you are then faced with a 25-minute walk to the garden entrance. Entry is officially from 07.00 but it is possible to enter earlier if you can

KEY SPECIES

Resident
Woodcock, Japanese White-eye

Winter
Eurasian Sparrowhawk, Common Coot, Water Rail, Daurian Redstart, Pale Thrush, Scaly Thrush

Passage
Common Sandpiper, Terek Sandpiper, Asian Stubtail

encourage the guards/ticket vendors! Last entry is at 17.00 but the gates do not close until dusk and the last hour or so of the day is typically free of people.

4 Century Park (8km) (Shi Ji Jong Yuan) *(Century Park Metro Station/Line 2)*

Century Park is the largest of all the parks in Shanghai, covering an area of over 140ha in the east of the city. You will need a full day to see it all. A wide variety of habitats have been created here including deciduous and coniferous woodland, lakes, ponds and streams, open grassland and gardens. There is a 'Bird Island' in the south covering around 2ha and surrounded by water. The park is also intersected by the Zhangjia River, which further increases the potential for a greater number of bird species.

To get there from the city, take **Metro Line 2** to **Century Park Station** (travel time is about 25 minutes from the city centre). There is a small entrance fee. To get there from Pudong Airport, take the high-speed train (maglev) to the end of the line (Longyang Road), change to Metro Line 2 and then go two stops to Century Park. The Magnetic Levitation train is the fastest train in the world and takes only eight minutes to travel the 28km from the airport to Longyang Road Station.

KEY SPECIES

Summer
Asian House Martin

Winter
Spot-billed Duck, Eurasian Teal, Common Coot, Common Moorhen, Water Rail, Pale Thrush, Scaly Thrush, Rustic Bunting, Yellow-throated Bunting, Chestnut-breasted Bunting, Yellow-breasted Bunting

5 Gongqing Forest Park (12km) (Sen Lin Gong Yuan) *(Bus No. 8)*

Situated on the west bank of the Huangpu River in the northeast of the city, this is considered by some to be the best of the sites within the city limits. It is much less of a manicured garden and has a more 'natural' feel about it. If you avoid following the crowds and the paths (the former generally follow the latter) you can easily find surprisingly quiet and undisturbed spots from which to watch birds.

You can get there by taxi but it is just as quick, easy and cheaper to go by bus. From tour bus Terminal 5 at Shanghai Stadium, take **bus number 8**. Its final destination is Gongqing Forest Park, located on **Jungong Road**. Buses leave at regular intervals throughout the day starting at 07.45.

KEY SPECIES

Winter
Daurian Redstart, Pale Thrush, Scaly Thrush, Rustic Bunting, Yellow-throated Bunting, Chestnut-breasted Bunting, Yellow-breasted Bunting

Passage
Thrushes, *Phylloscopus* warblers, Flycatchers

Further Away

6 Chongming Island – Dongtan Bird Nature Reserve (60km)

(Bus No. 5)

This is by far the best birdwatching area near the city, and so if you can only make one birding trip during your stay in Shanghai, this is the one to take! There are no entry fees nor permits required. The island is at the mouth of the Yangtze River and Dongtan Bird Reserve is at the very tip of the eastern end, facing the East China Sea. It is possible to visit the reserve from Shanghai in a single day but an overnight stay makes for a more relaxed time.

As silt is washed down the Yangtze River it is deposited at the end of Chongming Island at an extraordinary rate causing the island to grow many metres each year. It is this silt that has created the tidal saltmarshes of Dongtan Bird Reserve and one of the wintering grounds of the rare Hooded Crane. As the island has increased in size the Chinese authorities have reclaimed much of this saltmarsh by building large earth dykes creating areas of fish ponds and reedbeds on the inland side of the dykes. Although not providing as good an environment for many of the wetland and wading birds as the marshland, it is nonetheless an excellent place for birding and provides an alternative place to go if the tide is covering the saltmarsh.

To get there take bus number 5 from tour bus Terminal 5 at **Shanghai Stadium**. Buses leave every ten minutes. Travel to the port terminal at the end of the line (Bao Yang Matou). Purchase a one-way ferry ticket to Bao Zhen, the port on Chongming Island, from the ticket office. There are three types of ferry: fast, slow and very slow (the slowest takes about an hour and a half). On the slower ferries your ticket shows your seat number, on the fast ferry it's every man for himself!

From Bao Zhen port take a private minibus (these are cheaper than the taxis) to Dong Tan wetland viewing point (Shi Di Guan Guang Zhan). From here you can walk along the path on top of the dykes or, weather, tide and footwear permitting, walk out onto the marsh itself. To enter the reedbed area it is better to stop about 1km before the wetland viewing point and walk along one of the many paths criss-crossing the fish ponds.

The distance from the port is approximately 50km, the travel time is one and a half hours, and the cost about 60 RMB one way. Ask the minibus driver to pick you up later (do not pay him for the outward journey until he returns). The last ferry leaves at 19.30.

Note: a journey to Chongming Dongtan should not be undertaken lightly, especially for the travelling birder without Mandarin or a Mandarin-speaking guide, as the travel arrangements are convoluted and the travelling time far greater than the distance would suggest (up to four hours one way). There are a few restaurants on Chongming Island but no facilities of any kind close to the nature reserve.

KEY SPECIES

Resident
Sooty Tern, Zitting Cisticola, Reed Parrotbill, Common Moorhen

Summer
Yellow Bittern, Cinnamon Bittern, Little Egret, Cattle Egret, Chinese

Pond Heron, Purple Heron, Black-capped Kingfisher, Oriental Reed Warbler

Winter
Black Stork, Eurasian Spoonbill, Tundra Swan, Swan Goose, Bean

Goose, Lesser White-fronted Goose, Greylag Goose, Spot-billed Duck, Eurasian Teal, Northern Shoveler, Falcated Duck, Gadwall, Eurasian Wigeon, Garganey, Common Pochard, Goosander, Common Crane,

Hooded Crane, Eurasian Curlew, Whimbrel, Red-throated Pipit, White-throated Rock Thrush, Blue Rock Thrush, Japanese Thrush	**Passage** Great Egret, Black-faced Spoonbill, Great Knot, Red-necked Stint, Sharp-tailed Sandpiper, Grey-tailed Tattler, Spotted Redshank, Common	Greenshank, Common Sandpiper, Terek Sandpiper, Wood Sandpiper, Marsh Sandpiper, Whimbrel, Bar-tailed Godwit, Black-tailed Godwit

7 Bin Hai Golf Course and coastline (60km)

The Pudong coastline is heavily threatened by development but still attracts a variety of interesting species, particularly waders and reedbed specialists. Bin Hai Golf Course offers fairly easy access to a significant (but decreasing) area of reedbed, which still held Reed Parrotbills in spring 2004, plus an area of mudflats and some woodland and rough areas for migrant passerines. It is not too far from Pudong Airport and well worth a detour if you are travelling to or from the airport.

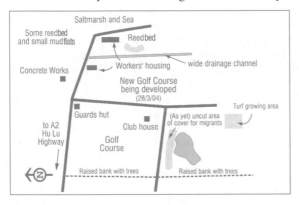

The easiest way to get there is by taxi, though it is rather expensive. Ask your hotel to write out the name and directions in Mandarin characters. If your driver is not confident from the start, get out and try another! In case of loss of confidence half way through the journey (not uncommon!) ask him/her to head to the airport and seek further directions from the information desk there. The golf course receptionists are happy to arrange taxis for birders returning to the city centre. It is worth buying a drink in the restaurant to maintain goodwill and mentioning that you are birding in order to raise awareness of the ornithological importance of the site.

KEY SPECIES

| **Resident**
Little Egret, Intermediate Egret, Plain Prinia, Zitting Cisticola, Reed Parrotbill

Summer
Oriental Reed Warbler, Black-browed Reed Warbler | **Winter**
Great Bittern, Great Egret, Eurasian Wigeon, Eurasian Teal, Hen Harrier, Common Gull, Heuglin's Gull, Vega Gull, Buff-bellied Pipit, Chinese Penduline Tit, Pallas's Reed Bunting, Reed Bunting | **Passage**
Osprey, Garganey, Lesser Sand Plover, Pacific Golden Plover, Broad-billed Sandpiper, Marsh Sandpiper, Terek Sandpiper, Sharp-tailed Sandpiper, Little Curlew, Richard's Pipit, Siberian Rubythroat, Bluethroat, Yellow-browed Bunting |

8 Taihu Lake Scenic Area (120km) (YuanTouZhu)

To the west of Shanghai is a vast area of lakes dominated by Taihu Lake, which is over 60km across and covers an area of more than 2,000km². To tour the surrounding area would take several weeks but to get a taste of it in a short period of time you can take a train from **Shanghai to Wuxi City** and from there take a bus or taxi to **Yuan TouZhu**. There is an entrance fee to Yuan TouZhu, which includes travel on the 'road train' and a boat trip to the 'Three Island Mountains'. By all means go on the road train to get further into the park but don't get on the boat, it's a complete waste of time! As with other locations in China, try to avoid going the same way as everybody else. The best birding is to be found by climbing up through the forested hills facing the lake.

You travel from Shanghai Railway Station to Wuxi City. Rail tickets are purchased from outside the main station and various other outlets, but it is best to ask your hotel concierge to buy return tickets for you. Travel time is a little over one hour. If you have a Chinese speaker with you a taxi is fine but otherwise the bus is actually easier. The taxi rank is situated below ground directly opposite the main railway station entrance. The bus station is also right outside the railway station. Take bus number 1: they leave at regular intervals throughout the day and terminate at Yuan TouZhu scenic area.

KEY SPECIES

Resident
Common Pheasant, Great Spotted Woodpecker, Brownish-flanked Bush Warbler, Blue Magpie, Black-billed Magpie, Long-tailed Tit, Black-throated Tit

Summer
Little Egret, Cattle Egret, Pheasant-tailed Jacana, Large Hawk-Cuckoo, Asian Paradise Flycatcher

Winter
Swan Goose, Bean Goose, Lesser White-fronted Goose, Greylag Goose. Spot-billed Duck, Eurasian Teal, Northern Shoveler, Falcated Duck, Gadwall, Eurasian Wigeon, Garganey, Common Pochard, Goosander, Rustic Bunting, Yellow-throated Bunting, White-capped Bunting, Yellow-breasted Bunting

Useful contacts and websites

Conserving China's Biodiversity (English and Chinese) www.chinabiodiversity.com
General conservation website with information about birds and sites in China.

China Birding (English)
www.chinatibettravel.net/cnbirds
Bird tours website with information about birding in China.

Oriental Bird Club
www.orientalbirdclub.org
P.O.Box 324, Bedford, MK42 0WG, UK

Ms Gan Xiao
Tel. (+86) 21 55072140
Mobile: 13501651984
Email: 632023086@fudan.edu.cn
Mr Zhang Lin Mobile: 13764195979
Email: zhanglinas@hotmail.com
Local bird guides.

Shanghai Metro Map
www.smartshanghai.com/en/travel/metromap.php

Books and publications

A Field Guide to the Birds of China
John MacKinnon and Karen Phillipps (2000)
Oxford University Press
ISBN 0198549407

A Photographic Guide To Birds of China
John MacKinnon and Nigel Hicks (1996)
New Holland
ISBN 1859749690

A Field Guide to the Birds of South-East Asia
Craig Robson (2000) New Holland
ISBN 1853683132

Where to Watch Birds in Asia
Nigel Wheatley (1996) Christopher Helm
ISBN 071364303X

• •

Singapore

It can sometimes be hard to believe that this highly developed 21st century city, with its skyscrapers, crowded expressways and sprawling industrial estates, is actually a tropical island in every sense. If you look carefully there are pockets of rainforests, mangroves and tidal creeks still preserved amidst all the urbanisation. Of course, many species have been rendered extinct on the island by habitat loss, but Singapore still retains a good representation of Southeast Asia's avifauna as well as some interesting species more typical of the Australasian region, together with many Palearctic

White-vented Myna is a familiar resident at the Botanic Gardens

migrants during the boreal winter and during migration periods. The abandonment of agriculture and quarrying on the island has allowed the regeneration of secondary-growth woodland, which can be quite rapid in a tropical climate, and this offers hope that new sites can be created to compensate for lost habitat. The climate is hot and very humid year-round, with average daytime temperatures ranging from 26°–30°C. Short but heavy showers occur all year but especially between November and January.

The period between September and November is a peak time for birding, as the autumn migration is underway with many Palearctic waders, terns and passerines passing through. Northbound migration between February and April is equally good, especially as this coincides with the start of the breeding season for resident species. The island is strategically located at the end of the Malay Peninsula, which acts as a migration flyway. Many waders winter in Singapore and several Siberian passerines which would be difficult to see elsewhere, or inhabit remote breeding grounds, can be seen quite easily in Singapore's urban parks.

Singapore Airport provides a temporary stop-off for many birders on their way to the birding hotspots of Southeast Asia, but if you have even half a day to spare, it is worth investigating the island's birdlife. In fact nowhere on the main island of Singapore is more than an hour away from the central district and a lot of ground can be covered in the course of a weekend. If you are staying longer, then you don't have to limit yourself to Singapore Island as it is relatively close to Malaysia and some of the Indonesian islands, and the frequent ferry services make these bird-rich areas very accessible.

TYPICAL SPECIES

Resident
Grey Heron, Black-crowned Night Heron, Brahminy Kite, Common Moorhen, Spotted Dove, Zebra Dove, Red Collared Dove, Asian Palm Swift, House Swift, Sunda Woodpecker, Banded Woodpecker, Collared Kingfisher, Asian Koel, Red-breasted Parakeet, Pacific Swallow, Common Iora, Common Tailorbird, Plain-throated Sunbird, Olive-backed Sunbird, Red-whiskered Bulbul, Yellow-vented Bulbul, Abbott's Babbler, Common Myna, Hill Myna, White-vented Myna, Asian Glossy Starling, Black-naped Oriole, House Crow, Large-billed Crow, Eurasian Tree Sparrow, Scaly-breasted Munia

Sep–Apr
Little Egret, Cattle Egret, Great Egret, Common Kingfisher, Barn Swallow, Yellow Wagtail

GETTING AROUND

On a small compact island like Singapore Island, you really don't need a car unless you want a mobile shelter from the heat and the sun, although most open-air amenities in Singapore are never too far from drinking water and a refreshment stall. If you are driving, equip yourself with a good street map. The expressways which criss-cross the island can whisk you very quickly to the general area that you want to get to, but once you have exited the expressway, navigating the dense road system can be quite a challenge. The expressways, of course, can be very busy during peak periods and, if you are planning to drive over the Johor Causeway to Malaysia, this can be very congested at weekends. Electronic Road Pricing (ERP) is in place on several of the expressways and on the approach roads to the central business district.

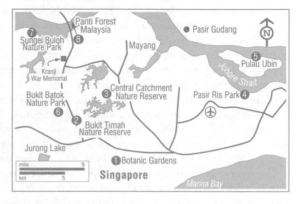

To avoid all the hassle of driving, there is a very good public transport system provided by a combined train (MRT) and light rail (LRT) system. Where the rail system does not reach, there are frequent bus services and many of the rail stations are served by feeder bus routes. The public transport system is shared by two companies, SBS and SMRT, which fan out from the central business district on the south coast of the main island. SBS mainly serves the south and east of Singapore and SMRT mainly serves the north and west, but there is considerable overlap. All the bus route numbers above 600 are SMRT services, and those under 300 are generally SBS services, but again there is considerable overlap so always check which operator serves your destination. However, both services are integrated to the extent that one ticket (Transitlink) is valid with both operators and there is a network of bus interchange terminals throughout the island.

Locations

1 **Botanic Gardens** (3km)

The Botanic Gardens are on the outskirts of the city and contain many indigenous trees and shrubs that are a magnet for breeding and wintering birds. The three lakes in the grounds provide some aquatic habitats, and good birds have been recorded here. It also has a good reputation as a reliable site for Spotted Wood Owls, although some local knowledge or assistance would be required to find a roosting bird. If you only have a few hours to

KEY SPECIES

Resident
Yellow Bittern, Striated Heron, Lesser Whistling Duck, White-breasted Waterhen, Pink-necked Pigeon, Dollarbird, Long-tailed Parakeet, Blue-crowned Hanging Parrot, Stork-billed Kingfisher, White-throated Kingfisher, Spotted Wood Owl, Fork-tailed Swift, Grey-rumped Treeswift, Edible-nest Swiftlet, Indian Cuckoo, Greater Coucal, Pied Triller, Oriental Magpie-Robin, Dark-necked Tailorbird, Scarlet-backed Flowerpecker, Javan Munia

Sep–Apr
Oriental Honey-buzzard, Common Sandpiper, Blue-tailed Bee-eater, Forest Wagtail, Ashy Minivet, Siberian Blue Robin, Asian Brown Flycatcher, Purple-backed Starling

spare in Singapore, then this is the best location to find many of the common birds of Southeast Asia. It opens early in the morning, from 05.00, so you may visit at times when it is uncrowded and the birds are at their most active. The gardens are on **Cluny Road**, just north of the city centre, and are easy to get to by taxi or bus.

2 Bukit Timah Nature Reserve (8km)

(SBS Bus 170) (SMRT Bus 67/171/184/182)

Just as Singapore is an island between Malaysia and Sumatra, Bukit Timah is an island in the midst of an urban jungle. This hill is the highest point on Singapore Island and is capped by tropical rainforest, the last remnant of the forest that once covered the island. This area was preserved for the purposes of water catchment and today provides a rich habitat for rainforest species. Unfortunately the Bukit Timah Expressway (BKE) cuts right through the forested area, separating Bukit Timah Reserve from the adjacent Central Catchment Reserve, thus reducing the extent of the forest and partly fragmenting and isolating the bird populations. Nevertheless, this forest supports a wide variety of species, some of which would be difficult to find in the more remote and extensive forests of Malaysia or Indonesia, yet are regularly seen in this city centre nature reserve. Four trails fan out from the visitor centre and all should be visited if time permits as they are comparatively short and wind their way to the summit

The reserve is on the east side of **Upper Bukit Timah Road**. Going west on the **Pan Island Expressway (PIE)**, exit north (Exit 26A) onto Upper Bukit Timah Road and then take the next exit east at the roundabout (**Ewart Circus**) onto **Hindhede Drive**, which leads to the park entrance.

KEY SPECIES

Resident
Changeable Hawk-Eagle, Emerald Dove, Pink-necked Pigeon, Laced Woodpecker, Common Flameback, Red-crowned Barbet, Lineated Barbet, Coppersmith Barbet, Dollarbird, Violet Cuckoo, Chestnut-bellied Malkoha, Blue-rumped Parrot, Blue-crowned Hanging Parrot, Lesser Green Leafbird, Blue-winged Leafbird, Greater Racket-tailed Drongo, Asian Fairy Bluebird, Straw-headed Bulbul, Black-crested Bulbul, Olive-winged Bulbul, Striped Tit-Babbler, Little Spiderhunter, Purple-throated Sunbird, Crimson Sunbird, Oriental Magpie-Robin, Dark-necked Tailorbird, Scarlet-backed Flowerpecker, Orange-bellied Flowerpecker

Sep–Apr
Japanese Sparrowhawk, Blue-tailed Bee-eater, Fork-tailed Swift, Hooded Pitta, Forest Wagtail, Ashy Bulbul, Siberian Blue Robin, Arctic Warbler, Asian Brown Flycatcher, Blue-and-white Flycatcher, Ferruginous Flycatcher, Tiger Shrike, Purple-backed Starling

Passage
Oriental Honey-buzzard

3 Central Catchment Nature Reserve (10km)

(SMRT Bus 167/855/980)

The hills in the centre of Singapore Island act as a rainfall catchment area, and there are three sizeable reservoirs surrounded by some good areas of secondary-growth forest. The reserve area is over 100 hectares and there are several kilometres of nature trails, one of which (**Petaling Trail**) features a treetop suspension bridge (open daily 09.00–17.00 except Mondays), which places you at eye-level with the upper canopy species. Naturally, it is best to visit early in the morning when the birds are most active and there are no noisy crowds. The park opens as early as 05.30. There are several other very good spots for birds, including the shores of the McRitchie Reservoir and the area of woodland to the east of Singapore Island Country Club.

There are several access points to the reserve. Take the **Pan Island Expressway (PIE)** northwest from the city centre to the interchange at **Adam Road** and exit north on Adam Road which runs into **Lornie Road**. The entrance to the reserve is about 2km further along Lornie Road on the left-hand (north) side. From the car park, there is a trail system going east which loops around the McRitchie Reservoir and leads to the treetop suspension bridge: a round-trip of about 10km. The trails at the west end of the car park follow the perimeter of the Country Club Golf Course and after about 6km lead to the Sime Track, which is one of the more productive birding areas. If driving, you

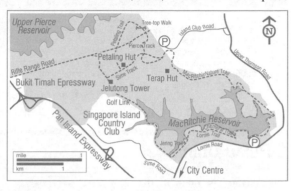

can get to the Petaling Trail and **Upper Pierce Reservoir** quicker by staying on Lornie Road for a further kilometre and then turning north on **Upper Thomson Road**. Continue north for about three kilometres and turn east on **Island Club Road**. There is a car park at the end of this road and an entrance to the Lower Pierce Reservoir.

KEY SPECIES

Resident		Sep–Apr
Oriental Honey-buzzard, Changeable Hawk-Eagle, White-bellied Sea Eagle, Emerald Dove, Pink-necked Pigeon, Rufous Woodpecker, Laced Woodpecker, Common Flameback, Red-crowned Barbet, Coppersmith Barbet, Dollarbird, White-throated Kingfisher, Blue-throated Bee-eater, Large-tailed Nightjar, Violet Cuckoo, Chestnut-bellied Malkoha, Blue-rumped Parrot, Blue-crowned Hanging Parrot, Lesser Green	Leafbird, Blue-winged Leafbird, Ashy Drongo, Greater Racket-tailed Drongo, Asian Fairy Bluebird, White-rumped Shama, Oriental Magpie-Robin, Straw-headed Bulbul, Black-crested Bulbul, Olive-winged Bulbul, Cream-vented Bulbul, Short-tailed Babbler, Striped Tit-Babbler, Little Spiderhunter, Dark-necked Tailorbird, Tiger Shrike, Purple-throated Sunbird, Crimson Sunbird, Scarlet-backed Flowerpecker, Orange-bellied Flowerpecker	Japanese Sparrowhawk, Blue-tailed Bee-eater, Black-capped Kingfisher, Indian Cuckoo, Fork-tailed Swift, Forest Wagtail, Ashy Bulbul, Siberian Blue Robin, Arctic Warbler, Eastern Crowned Warbler, Asian Paradise Flycatcher, Asian Brown Flycatcher, Ferruginous Flycatcher, Blue-and-white Flycatcher, Purple-backed Starling

4 Pasir Ris Park (34km)

(Pasir Ris MRT Station)

This is an area of preserved and restored mangroves lying between two river outlets on the northeast coast of Singapore Island. There are no mudflats to speak of but a narrow beach along the shoreline allows you to scan for seabirds and, frequently, White-bellied Sea Eagles. This park is quite near Singapore Airport and could be easily visited, by taxi, in the course of a four-hour stopover. The park is well supplied with wooden boardwalks through the mangroves and an observation tower. The best mangrove and birding area is along the west bank of the Sungei Tampines River, where the observation tower is located.

From the **Tampines Expressway** (**TPE**) in the northeast sector of Singapore Island, exit north at **Pasir Ris Interchange**. Ignore the first park on the right-hand side (**Pasir Ris Town Park**), which is a largely ornamental amenity park, and continue north to Pair Ris Park. The entrance is on Pasir Ris Green opposite the **Pasir Ris Bus Interchange**.

5 **Pulau Ubin** (10km)   *(SBS Bus 2/19/29 to Changi Village)*

Pulau Ubin is one of the inhabited islands off the north coast of Singapore Island. This is a popular tourist destination and 'bumboats' convey tourists from Changi Beach to the main village at Ubin. You can rent a bicycle here and this is the ideal way to get around the island, as it is over 1,000ha in extent. If you don't feel up to cycling, you can avail yourself of the island taxi

service. It is much less developed than the rest of Singapore and supports a number of species which are absent from the main island. In addition, the boat trip to the island can be good for wintering gulls and terns. There is a wide variety of habitats on the island including secondary-growth forest, mangroves, cultivated areas, tidal mudflats, fish ponds and flooded quarry pits. A day on the island will yield a good selection of species typical of these habitats, without wasting a lot of time commuting from one good spot to the other, as can be the case on the main island.

The island ferries depart from **Changi Village Jetty** (near the bus terminal) which is just north of the airport, and take about ten minutes to get to the main jetty at **Ubin**. From Ubin, take the main road (Jalan Ubin) north towards the campsite at Noordin Village. This is about 2km long and leads through a variety of habitats including forest, mangroves, ponds and cultivation. An alternative route is to turn east off Jalan Ubin after about 1km and take the unsurfaced track towards Chek Jawa, the easternmost tip of Pulau Ubin. There are extensive tidal mudflats across this area which are worth checking, but they are dependent on tidal conditions and access is controlled. Check at the National Parks Information Office near the jetty at Ubin for the latest information.

If you are on a stopover at the airport and don't have sufficient time to visit Pulau Ubin, bear in mind that Changi Beach, just east of the jetty, is also a good birding spot for waders and terns, while the gardens of the Meridien Hotel in Changi Village are good for woodland species.

KEY SPECIES

Resident
Yellow Bittern, Purple Heron, Oriental Honey-buzzard, Changeable Hawk-Eagle, White-bellied Sea Eagle, Red Junglefowl, Emerald Dove, Red Collared Dove, Pink-necked Pigeon, Oriental Pied Hornbill, Laced Woodpecker, Common Flameback, Dollarbird, White-throated Kingfisher, Blue-throated Bee-eater, Large-tailed Nightjar, Greater Coucal, Lesser Coucal, Golden-bellied Gerygone, White-rumped Shama, Oriental Magpie-Robin, Olive-winged Bulbul, Short-tailed Babbler, Striped Tit-Babbler, Mangrove Whistler, Dark-necked Tailorbird, Tiger Shrike,

Purple-throated Sunbird, Crimson Sunbird, Scarlet-backed Flowerpecker, Orange-bellied Flowerpecker

Sep–Apr
Black Bittern, Schrenck's Bittern, Great Egret, Intermediate Egret, Greater Sand Plover, Grey Plover, Pacific Golden Plover, Little Ringed Plover, Red-necked Stint, Long-toed Stint, Broad-billed Sandpiper, Curlew Sandpiper, Ruff, Spotted Redshank, Marsh Sandpiper, Terek Sandpiper, Eurasian Curlew, Whimbrel, Black-tailed Godwit, Bar-tailed Godwit, Wood Sandpiper, Common

Sandpiper, Pintail Snipe, Common Snipe, Ruddy Turnstone, Black-headed Gull, Great Crested Tern, Lesser Crested Tern, Black-naped Tern, Little Tern, White-winged Black Tern, Black-capped Kingfisher, Blue-tailed Bee-eater, Fork-tailed Swift, Indian Cuckoo, Forest Wagtail, Grey Wagtail, Ashy Bulbul, Siberian Blue Robin, Arctic Warbler, Eastern Crowned Warbler, Asian Paradise Flycatcher, Asian Brown Flycatcher, Purple-backed Starling

Passage
Oriental Honey-buzzard

6 Bukit Batok Nature Park (8km)

(SBS Bus 61/66/157) (SMRT Bus 178/852)

There are numerous public parks in Singapore, most of which are rather too small or too neat to support a wide variety of species, but the Bukit Batok Nature Park has some good regenerated forest areas, as well as a flooded quarry pond which, although rather deep, provides some habitat for aquatic species. It has many surfaced pathways through the park and if your time in Singapore is limited, and you are not equipped to undertake a more onerous forest trek, then this park will introduce you to a good cross-section of Singapore's birds. As with all urban parks, an early morning visit, preferably midweek, is best.

The park is on the west side of **Upper Bukit Timah Road**. Going west on the **Pan Island Expressway** (**PIE**), exit north (Exit 26A) onto Upper Bukit Timah Road and then take the second turn west onto **Old Jurong Road** which leads to **Bukit Batok East Avenue**. The park entrance is on the right-hand side of this road.

KEY SPECIES

Resident
Yellow Bittern, Striated Heron, Lesser Whistling Duck, White-breasted Waterhen, Pink-necked Green Pigeon, Dollarbird, Long-tailed Parakeet, Blue-crowned Hanging Parrot, Lineated Barbet, White-throated Kingfisher, Grey-rumped Treeswift, Edible-nest Swiftlet, Drongo Cuckoo, Pied Triller, Straw-headed

Bulbul, White-crested Laughingthrush, Dark-necked Tailorbird, Crimson Sunbird, Scarlet-backed Flowerpecker, Javan Munia

Sep–Apr
Blue-tailed Bee-eater, Forest Wagtail, Grey Wagtail, Ashy Minivet, Siberian Blue Robin, Asian Brown Flycatcher, Yellow-rumped Flycatcher

7 Sungei Buloh Nature Park (27km)

(SMRT Bus 925)

Sungei Buloh is a tidal creek on the north coast of Singapore Island with a fairly large area of intact mangroves and tidal mudflat. This is the best wetland site in Singapore and well worth a visit, particularly during the boreal winter, when large numbers of waders can be seen. There is an excellent system of nature trails, boardwalks and several well-sited bird hides, as well as an observation tower. The habitats include both saltwater and freshwater. The abundant wildlife includes mud-skippers and numerous very large Water Monitor lizards.

There are three distinct zones within the park and all three should be visited if time permits. The closest to the car park and entrance includes both a mangrove boardwalk and a loop trail (Route 1) which provides access to two observation platforms from where you can scan the open mudflats. The second zone lies in the central part of the park and includes several lagoons and shrimp ponds surrounded by mangroves. There is a trail (Route 2) around the perimeter of the lagoons, with several bird hides and a tower hide overlooking the main tidal channel. An area of freshwater pools at the west end of the reserve can be accessed by a third trail (Route 3).

Take the **Bukit Timah Expressway (BKE)** north towards the **Johor Causeway** and exit west at the last interchange before the Causeway (**Woodlands Avenue**) towards **Kranji**. At Kranji, turn north on Kranji Road and cross the **Kranji Reservoir Dam**. On the west side of the dam turn north on **Neo Tiew Crescent**. The park entrance is on the right-hand side of this road. There is a bus service (SMRT No. 925) from Kranji MRT Station.

KEY SPECIES

Resident
Yellow Bittern, Striated Heron, Purple Heron, White-breasted Waterhen, White-browed Crake, Slaty-breasted Rail, White-bellied Sea Eagle, White-throated Kingfisher, Stork-billed Kingfisher, Greater Coucal, Large-tailed Nightjar, Banded Woodpecker, Laced Woodpecker, Common Flameback, Paddyfield Pipit, Golden-bellied Gerygone, Pied Fantail, Ashy Tailorbird, Oriental White-eye, Zitting Cisticola, Yellow-bellied Prinia, Tiger Shrike, Javan Munia

Sep–Apr
Black Bittern, Great Egret, Intermediate Egret, Watercock, Osprey, Black Baza, Black-winged Stilt, Greater Sand Plover, Lesser Sand Plover, Pacific Golden Plover, Little Ringed Plover, Red-necked Stint, Long-toed Stint, Broad-billed Sandpiper, Curlew Sandpiper, Ruff, Spotted Redshank, Marsh Sandpiper, Terek Sandpiper, Eurasian Curlew, Whimbrel, Black-tailed Godwit, Bar-tailed Godwit, Wood Sandpiper, Common Sandpiper, Pintail Snipe,

Common Snipe, Ruddy Turnstone, Little Tern, White-winged Black Tern, Black-capped Kingfisher, Asian Brown Flycatcher, Oriental Reed Warbler, Arctic Warbler, Brown Shrike

Passage
Great Knot, Oriental Pratincole, Gull-billed Tern

Useful contacts and websites

Singapore Bird Group
www.nss.org.sg/wildbirdsingapore
c/o Nature Society (Singapore), The Sunflower, 510 Geylang Road, #02 – 05, Singapore 389466.
Tel. (+65) 741 2036. Fax.(+65) 741 0871. Local BirdLife Partner, publisher of the annual journal *Iora*.

Singapore Botanic Gardens
www.sbg.org.sg

National Parks Board, Singapore
www.nparks.gov.sg

Sungei Buloh Wetlands Reserve
www.naturia.per.sg/buloh

Wild Singapore www.wildsingapore.com
General natural history website about Singapore with information about birds.

Oriental Bird Club
www.orientalbirdclub.org
P.O.Box 324, Bedford, MK42 0WG, UK

Singapore Bird Group Records Committee
www.nss.org.sg/wildbirdsingapore
Nature Society (Singapore), c/o Blk 870 Woodlands St 81, #06-304, Singapore 730870.

Tidal Schedule for Singapore
www.getforme.com/environment_watertidetable. htm

SBS Transit www.sbstransit.com.sg
Bus, Metro and Light Rail Services (South and East Singapore).

SMRT Corporation www.smrt.com.sg
Bus, Metro and Light Rail Services (North and West Singapore).

Books and publications

Birds: An Illustrated Field Guide to the Birds of Singapore
Lim Kim Seng & Dana Gardner (1997) Sun Tree Publishing, ISBN 9813066008

The Birds of Singapore
C. Briffett & S. Supari (1994) Oxford University Press, ISBN 0195886062

A Field Guide to the Birds of West Malaysia and Singapore Allen Jeyarajasingam and Alan Pearson (1999) Oxford University Press ISBN 0198549628

Birds of Singapore
C. Hails & F. Jarvis (1987) Times Editions, Singapore, ISBN 9971400995

A Guide to the Common Birds of Singapore
C. Briffett (1986) Singapore Science Centre, Singapore

Pocket Checklist of the Birds of the Republic of Singapore Lim Kim Seng (1999) Nature Society (Singapore) ISBN 981041918X

A Field Guide to the Birds of South-East Asia
Craig Robson (2000) New Holland ISBN 1853683132

A Field Guide to the Birds of South-East Asia
Ben F. King, Edward C. Dickinson, Martin W. Woodcock (1975) Collins, ISBN 0002192063

Where to Watch Birds in Asia
Nigel Wheatley (1996) Christopher Helm ISBN 071364303X

• •

Taipei

Taiwan has become a popular destination with birding tours and independent birders alike due to the interest in the endemic species of the island. As a result, more and more birders pass through Taipei. The city is at the northern tip of the island and rather distant from the Central Mountain Ranges where most of the endemics are found. However there is plenty to interest birders in the vicinity of the city itself.

Taipei is dominated by the Danshui river system, being surrounded on three sides by the Danshui and its tributaries, the Xindian and the Jilong. The original flood plain has long since been reclaimed and built over, but there are still some excellent wetland sites by the river banks which have been preserved as flood buffer zones and which hold significant numbers of waterfowl. The foothills of the Central Mountain Ranges are within an hour's drive of the city and, in winter, they support some of the mid-elevation forest species, including a number of endemics. They provide the visiting birder with a good introduction to the forest birds of Taiwan without having to travel too far into the interior.

Most of the woodland species are resident and can be seen at any time of the year, although in winter there is greater species diversity as the higher altitude species descend to the lower elevations. However, for more general birding, the spring and autumn migration produces the most species as the wetlands accommodate large numbers of migrating waders and wildfowl. Winter is also very good, not just for the wintering ducks, but also for a number of Eastern Palearctic passerines which winter on the island, including some of the most sought-after vagrants to

The tiny Japanese White-eye is a common resident in Taipei

Western Europe and North America. Additionally, a small number of summer migrants breed on the island, although they are difficult to see close to Taipei.

TYPICAL SPECIES

Resident
Black-crowned Night Heron, Little Egret, Cattle Egret, Oriental Honey-buzzard, Black Kite, Crested Serpent Eagle, Crested Goshawk, White-breasted Waterhen, Common Moorhen, Common Sandpiper, Black-browed Barbet, Common Kingfisher, House Swift, Oriental Turtle Dove, Spotted Dove, Red Collared Dove, White-bellied Green Pigeon, White Wagtail, Barn Swallow, Pacific Swallow, Black-naped Monarch, Blue Rock Thrush, White-vented Myna (I), Common Myna (I), Light-vented Bulbul, Black Bulbul, Japanese White-eye, Streak-breasted Scimitar-

Babbler, Rufous-capped Babbler, Grey Treepie, Black-billed Magpie (I), Black Drongo, Eurasian Tree Sparrow, White-rumped Munia, Scaly-breasted Munia

Summer
Oriental Cuckoo

Winter
Grey Heron, Great Egret, Intermediate Egret, Striated Heron, Osprey, Common Kestrel, Pacific Golden Plover, Little Ringed Plover, Kentish Plover, Lesser Sand Plover, Marsh Sandpiper, Common Greenshank, Wood Sandpiper, Red-necked Stint, Black-backed Wagtail, Yellow Wagtail,

Grey Wagtail, Olive-backed Pipit, Siberian Rubythroat, Red-flanked Bluetail, Daurian Redstart, Siberian Stonechat, Scaly Thrush, Pale Thrush, Brown-headed Thrush, White-cheeked Starling, Manchurian Bush Warbler, Oriental Reed Warbler, Yellow-browed Warbler, Arctic Warbler, Brown Shrike, Black-faced Bunting

Passage
Chinese Goshawk, Grey-faced Buzzard, Red-necked Phalarope, Whiskered Tern, White-winged Black Tern, Striated Swallow

GETTING AROUND

Getting to any location in Taiwan can be a challenge for the non-Chinese speaker, as place names and directions are not always indicated in English. Moreover, many English place names can differ between maps and signposts and are dependent on which system of romanisation has been used to derive the English spelling. So a little bit of guesswork is sometimes required to ascertain a place name.

The lowland sites around Taipei are easy to get to by public transport as the Metro (MRT) will quickly get you from any city centre location to the outskirts. The bus system is much more comprehensive and serves many towns in the surrounding mountains but destinations are indicated in Chinese, so it is essential that you obtain an English bus guidebook and familiarise yourself with the route number. Most of the provincial buses depart from the bus terminals near the central Taipei Main Station. If

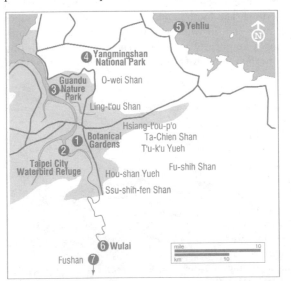

you are short on time, you could also consider booking a taxi for a day or half-day. This is more expensive than public transport but offers complete flexibility, particularly if you want to be on-site in the mountains by dawn. Even just as a means of travelling between sites, taxis are direct and much more convenient. As with all East Asian countries, always carry your hotel's or office's business card which you can show to a taxi driver in order to get back.

If you have a car, then you have much more versatility,

particularly for visiting the mountain sites, where distances are longer and where there would otherwise be much uphill hiking involved. Driving in the city has all the stresses that you would expect for a large congested urban area. The key through-route is Zhongshan Road which runs north/south through the centre and connects at each end with the expressway system. Make sure you obtain a map which includes all of Taipei County in addition to the Taipei metropolitan area.

Locations

1 Taipei Botanical Gardens (City Centre)

(Chiang Kai-Shek Memorial Hall MRT Station)
(Bus Nos. 1, 12, 204, 242)

KEY SPECIES

Resident
Malayan Night Heron, Asian Glossy Starling (I)

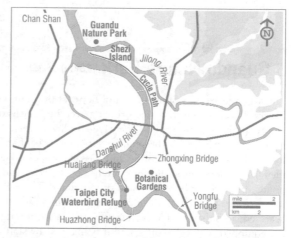

The Botanical Gardens are just south of the city centre near the National Museum of History and are very conveniently located for anyone staying in the centre. Apart from providing an oasis of calm in the midst of the city, the gardens allow anyone unfamiliar with the birds of East Asia to encounter the commoner species in less stressful conditions than birding in tropical forests. For this reason, many visiting bird tours pay a visit to the gardens at the start of their trip. However, they have also earned a reputation as the best place in East Asia to see the very shy Malayan Night Heron, a small number of which are resident here and can be seen much more readily than in their more typical habitat of dense rainforest. The gardens are at the **intersection of Nanhai Road and Heping West Road**, just two blocks southwest of Chiang Kai-shek Memorial Hall, a well-known city landmark.

2 Taipei City Waterbird Refuge (5km) *(Lungshan Temple MRT Station)*

As Taipei is at the confluence of two sizeable rivers, the Danshui and the Xindian, there is an extensive area of open water which is important for wintering wildfowl. A linear park and cycle path run for almost 20km along the city side of the river banks. This affords an opportunity to scan the river and sandbars for wildfowl and other aquatic birds. The stretch from the Zhongxing Bridge on the Danshui to the Yongfu Bridge on the Xindian has been designated as a winter wildfowl refuge. The best area is around the Huajiang Bridge.

Huajiang Bridge is in the southwest of the city, not far from the popular tourist sites at Snake Alley and Lungshan Temple. From Lungshan Temple MRT Station, go west on

Heping West Road towards **Huajiang Bridge**. If numbers of wildfowl are low at Huajiang Bridge, you could also check Huazhong Bridge, which is about 2km upstream from Huajiang Bridge. If you have a bicycle (which can be rented at several outlets in the city) you might consider cycling along the cycle path which runs along the Danshui river bank as far north as Shezi 'Island', at the confluence of the Danshui and Jilong Rivers, a distance of about 15km. This low-lying area has greater expanses of mudflats and is good for waders.

KEY SPECIES

Resident	**Winter**	Falcon, Dunlin, Black-headed Gull,
Spot-billed Duck, Oriental Skylark,	Gadwall, Eurasian Wigeon, Mallard,	Naumann's Thrush
Long-tailed Shrike, Black-collared	Northern Shoveler, Northern Pintail,	**Passage**
Starling (I), Crested Myna, Zitting	Garganey, Baikal Teal, Eurasian Teal,	Little Curlew
Cisticola, Yellow-bellied Prinia, Plain	Common Pochard, Tufted Duck,	
Prinia	Eastern Marsh Harrier, Peregrine	

---

3 Guandu Nature Park (28km) *(Guandu MRT Station) (Fuxinggang MRT Station)*

This wetland area is just northwest of Taipei at the confluence of the Danshui and Jilong Rivers. It was formerly a rice-growing area but cultivation was abandoned when the nature park was set up and the rice-paddies reverted to freshwater marshes. The Danshui River is tidal at this point and its muddy banks are exposed at low tide, providing feeding areas for wintering and passage waders. There is an extensive growth of mangroves along the shallower stretches of the river and, in addition to the freshwater ponds, there is a substantial stand of woodland. Overall, the resultant mosaic of habitats provides a very rich diversity of species in a rather small area which is very easily accessed from the city. The park is very well laid out with nature trails and hides, which are equipped with telescopes, and it is manned by staff and volunteers from the Wild Bird Society of Taipei. If your visit to Taipei is brief, then Guandu provides a very quick and convenient way to do some birding close to the metropolis. The best times to visit are during winter and in migration periods, but its ease of access makes it worth a look at any time. It is also popular with local birders and so provides a valuable opportunity to make contact and get updated on the latest news and advice.

Take **Zhongshan Road** north and cross over the **Zhongshan Bridge**. Continue north to **Dadu Road** and turn left (west). Continue west on Dadu Road, and turn left on Zhixing Road just before the turn-off for **Guandu Temple**, a well-known tourist spot. The car park and the park entrance are on the left-hand side of the road. Unfortunately, the park does not open until 09.00. There is a small entrance fee. If travelling by Metro, the MRT station is about 10 or 15 minutes' walk north of the park entrance. The main birding areas at Guandu are the Nature Center; the Embankment, including the southwest and southeast Birdwatching Areas; and along both sides of the Guizikengxi Construction Road which is accessed either by following the Embankment or from the Fuxinggang MRT Station.

KEY SPECIES

Resident	breasted Crake, Lesser Coucal,	Oriole, Black-collared Starling (I),
Little Grebe, Yellow Bittern, Sacred	Greater Painted-snipe, Black-winged	Crested Myna, Plain Martin, Zitting
Ibis (I), Spot-billed Duck, Ruddy-	Stilt, Long-tailed Shrike, Black-naped	Cisticola, Golden-headed Cisticola,

Yellow-bellied Prinia, Plain Prinia, Vinous-throated Parrotbill, Black-headed Munia (I), White-headed Munia (I)

Winter
Great Bittern, Chinese Pond Heron, Falcated Duck, Eurasian Wigeon,

Mallard, Northern Shoveler, Northern Pintail, Garganey, Baikal Teal, Eurasian Teal, Tufted Duck, Eastern Marsh Harrier, Peregrine Falcon, Water Rail, Common Snipe, Green Sandpiper, Long-billed Dowitcher, Dunlin, Black-tailed Gull, Black-headed Gull, Black-capped

Kingfisher, Bluethroat

Passage
Black-faced Spoonbill, Oriental Stork, Black-tailed Godwit, Eurasian Curlew, Spotted Redshank, Common Redshank, Sharp-tailed Sandpiper

---

4 Yangmingshan National Park (22km)   *(Bus Nos. 230/260)*

Yangmingshan is a cluster of low mountains just north of Taipei, the slopes of which are partially clad with the deciduous forest typical of the lower elevations. Although not as rich, ornithologically, as the higher mountain ranges of the interior, nevertheless Yangminshan supports a sizeable population of the endemic Taiwan Magpie, as well as a good range of other forest species. This national park extends over a large area (11,000ha), so visitors on foot will only be able to cover a small portion. Bus services from Taipei operate on the main road through the park and can bring you directly to some of the better areas. The Tatun Natural Park is a subsection of the national park which occupies the western sector. There are two routes in this area which should be your priority: the Butterfly Corridor and the Bird Watching Trail. In addition, the area around the Park Headquarters and Yangming Park can also be good, particularly early in the morning.

Take **Zhongshan Road** north, cross over the **Zhongshan Bridge** and turn onto **Yangde Boulevard** for about 15km climbing a rather winding road to the park entrance. Continue north through the park on **Highway 2A** for about 4km to the turn-off for **Highway 101A**. Turn left (west) onto Highway 101A and continue in this direction for about 2km to the trailhead for the Butterfly Corridor on the left-hand side of the road. Continue through the Butterfly Corridor to the start of the Bird Watching Trail. The combined trail is a round-trip of about 6km. If using public transport, you can take the bus as far as the turn-off for Highway 101A. Also accessible by bus is Lengshuikeng, where there are trails which are equally good for birding.

KEY SPECIES

Resident
Besra, Chinese Bamboo Partridge, Oriental Skylark, Taiwan Whistling Thrush, Brownish-flanked Bush Warbler, Hwamei, Spot-breasted Scimitar-Babbler, Dusky Fulvetta, Grey-cheeked Fulvetta, White-bellied Yuhina, Taiwan Magpie, Vinous-throated Parrotbill

5 Yehliu (30km)

The Yehliu promontory juts out from the north coast of Taiwan into the East China Sea and has become one of the best migrant watchpoints in Taiwan, having produced many rarities over the years. The Yehliu Scenic Area is a well-known tourist site because of the unusual rock formations found there. However, for birders it is the scrubby areas that are the most interesting as many migrants can be found here in the right weather. For-

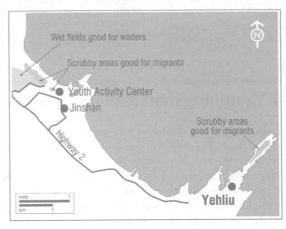

tunately, the kind of drizzly, overcast conditions suitable for a fall of passerine migrants are not unusual on the northeast coast of Taiwan. Obviously, spring and autumn are the best times to observe migration, but in winter it is worth scanning the coast for waders and herons. Summer can also be good for seabirds, if seawatching conditions are right. If you are unlucky enough to coincide with a typhoon in late summer, this site offers the possibility of some productive seawatching during the storm and perhaps some wind-blown passerine migrants in its aftermath.

Take **Highway 2A** through the **Yangmingshan NP** towards **Jinshan** on the coast. Turn south on **Highway 2** towards **Keelung** and Yehliu is located about 10km south on this road. There are provincial bus services which depart from the Taipei Main Station and serve the coastal towns between Jinshan and Keelung, all of which stop at Yehliu. This is a popular tourist site and many guided tours from Taipei visit the area. Jinshan itself is worth a look. This coastal town also attracts migrants: check the scrubby area around the Youth Activity Center and the surrounding fields for waders.

KEY SPECIES

Resident
Streaked Shearwater, Brown Booby, Pacific Reef Egret, Taiwan Whistling Thrush, Hwamei, Striated Prinia

Summer
Lesser Frigatebird, Great Crested Tern, Black-naped Tern, Little Tern

Winter
Peregrine Falcon, Great Cormorant

Passage
Chinese Egret, Grey-tailed Tattler, Ruddy Turnstone, Herring Gull, Common Tern, Eurasian Hoopoe, Fork-tailed Swift, Japanese Paradise Flycatcher, Japanese Thrush, Grey-streaked Flycatcher, Asian Brown Flycatcher, Ferruginous Flycatcher, Narcissus Flycatcher, Mugimaki Flycatcher, Blue-and-white Flycatcher, Japanese Robin, Asian Stubtail,

Lanceolated Warbler, Dusky Warbler, Pallas's Warbler, Pale-legged Leaf-Warbler, Eastern Crowned-Warbler, Brambling, Eurasian Siskin, Tristram's Bunting, Little Bunting, Yellow-browed Bunting, Rustic Bunting, Yellow-throated Bunting, Yellow-breasted Bunting, Chestnut Bunting, Yellow Bunting

6 Wulai (29km)

Wulai is a tourist resort which is famous for its waterfall. It is south of Taipei in the foothills of the Central Mountain Ranges. The hill forests in this area support the full range of low and mid-elevation woodland species including several of Taiwan's endemics.

From the city centre, take **Roosevelt Road** south onto the Taipei-Ilan Highway (**Highway 9**). Turn right (south) at **Xindian** onto the Xindian-Wulai Road (**Highway 9A**), and continue through the town of Wulai for about 1km to the waterfall area, which is well signposted. There is a car park near the waterfall from where you can walk along the road by the side of the river (Route 107) towards Fushan. The area around the waterfall is very touristy and can be crowded at weekends. Continue south for a further kilometre and you will come to a suspension bridge over the Nanshih Stream. Cross over the bridge to a forested track, which continues into the valley. This track leads to Nei-tung Forest Recreation Area, where there is a network of hiking trails.

If you stay on Route 107, you can drive as far south as **Fushan**, approximately 17km further on. This road passes through good forest habitat, so it is worth stopping off at promising birding sites.

If you are relying on public transport, there is a bus service from the Taipei Main Station to Wulai. The bus station in Wulai is about 1km from the waterfall. If you arrive by bus, there is also a birding trail which is signposted and commences at the car park beside the bus station. This runs along the Tonghou Stream, a tributary of the Nanshih, through a cemetery, which can be quite productive.

KEY SPECIES

Resident
Besra, Black Eagle, Chinese Bamboo Partridge, Grey-capped Woodpecker, Grey-chinned Minivet, Bronzed Drongo, Brown Dipper, Taiwan Whistling Thrush, Plumbeous Water Redstart, Little Forktail, Varied Tit, Spot-breasted Scimitar Babbler,

Dusky Fulvetta, Grey-cheeked Fulvetta, White-bellied Yuhina, Taiwan Magpie, Large-billed Crow, Maroon Oriole, Plain Flowerpecker

Summer
Silver-backed Needletail

Winter
Collared Owlet, Asian House Martin, Vivid Niltava, White-tailed Robin, Green-backed Tit, Yellow Tit, Rusty Laughingthrush, Steere's Liocichla, White-eared Sibia, Taiwan Yuhina, Eurasian Jay, Fire-breasted Flowerpecker

Further Away

7 Fushan (47km)

Some of the best lowland forest in the Taipei area is in the vicinity of Fushan village, 17km south of Wulai and bordering several important nature reserves and wildlife refuges. Before you get to Fushan, you will need to obtain a permit from the police checkpoint on the road into the village: this just requires you to present your passport and a small fee.

The Fushan-Baling Trail is a hiking trail from Fushan village into Takuanshan Forest Preserve and is about 17km long, but of course, there is no need to walk all the way. This is a good site for Taiwan Partridges but, as with all shy gamebirds, it is often the first birders of the day to walk the trail who will get to see the birds, at least at weekends when there are many hikers. To locate this trail, look for the suspension bridge that crosses the stream at Fushan village.

An alternative is the Fushan-Hapen Trail which leads through some good habitat from Fushan village to the Fushan Botanical Garden, another excellent site for hill-forest birds. The trail is found by crossing the concrete bridge in Fushan village and following the paved road to a dead-end, where a series of concrete steps mark the start. While in Fushan village, you should also check the fish farm as it is one of the most regular sites in Taiwan for Tawny Fish Owls. Enquire from one of the staff at the fish farm about the presence of the birds.

KEY SPECIES

Resident
Besra, Black Eagle, Taiwan Partridge, Chinese Bamboo Partridge, Swinhoe's Pheasant, Grey-capped Woodpecker, Tawny Fish Owl, Grey-chinned Minivet, Bronzed Drongo, Brown Dipper, Taiwan Whistling Thrush, Plumbeous Water Redstart, Varied Tit, Rusty Laughingthrush,

Spot-breasted Scimitar Babbler, Dusky Fulvetta, Grey-cheeked Fulvetta, White-bellied Yuhina, Taiwan Magpie, Large-billed Crow, Maroon Oriole, Plain Flowerpecker

Summer
Silver-backed Needletail

Winter
Collared Owlet, Asian House Martin, Snowy-browed Flycatcher, Vivid Niltava, White-tailed Robin, Green-backed Tit, Yellow Tit, Black-throated Tit, Rufous-faced Warbler, Steere's Liocichla, White-eared Sibia, Taiwan Yuhina, Eurasian Jay, Fire-breasted Flowerpecker

Useful contacts and websites

Wild Bird Federation Taiwan www.bird.org.tw
Wild Bird Federation Taiwan, 1F #3 Lane 36, Chinglung St., Taipei, Taiwan, R.O.C.
National bird conservation organisation and local BirdLife Partner.

Wild Bird Society of Taipei www.wbst.org.tw
1F #3 Lane 160, Fu-Shing S. Rd. Sec. 2, Taipei.
Local birding organisation which sells field guides and bird recordings, as well as organising local guides.

Natural Kingdom Inc.
www.natural-kingdom.com.tw
#8 Lane 380, Fu-Shing N. Rd., Taipei.
Birding supplies and bird tour operator. Website contains useful information about birding in Taiwan.

Birding in Taiwan
www.geocities.com/RainForest/9003

Feng Huang Ku Bird Park
www.taiwanbird.fhk.gov.tw/home.htm
General birding website, with information on Taiwanese birds, sound recordings and video clips.

Taiwan's Ecological Conservation
www.gio.gov.tw/info/ecology/English

Yangmingshan National Park
www.ymsnp.gov.tw

Guandu Nature Park www.gd-park.org.tw

Oriental Bird Club
www.orientalbirdclub.org
P.O. Box 324, Bedford, MK42 0WG, UK

Tidal Schedule for Taiwanese Coast
www.marine.cwb.gov.tw/CWBMMC/tidefocE.html

Metro Taipei www.english.trtc.com.tw
Taipei Rapid Transit Corporation. Information about Metro services (MRT) in Taipei area.

Books and publications

A Field Guide to the Birds of Taiwan
Chia-hsiong Wang et al. (1991) Taiwan Wild Bird Information Centre & Wild Bird Society of Japan
ISBN 9579578001

A Field Guide to the Birds of Taiwan
James Wan-Fu Chang (1985) Cheng & Tsui
ISBN 0917056434

A Field Guide to the Birds of China
John MacKinnon and Karen Phillipps (2000)
Oxford University Press, ISBN 0198549407

A Field Guide to the Birds of South-East Asia
Craig Robson (2000) New Holland
ISBN 1853683132

A Field Guide to the Birds of South-East Asia
Ben F. King, Edward C. Dickinson, Martin W. Woodcock (1975) Collins, ISBN 0002192063

A Photographic Guide to Birds of China Including Hong Kong
John MacKinnon and Nigel Hicks (1996)
New Holland, ISBN 1853687642

Where to Watch Birds in Asia
Nigel Wheatley (1996) Christopher Helm
ISBN 071364303X

Tokyo

Tokyo is located at the head of Tokyo Bay and stretches out across the Kanto Plain. For the visitor it represents a vast sprawling conurbation and an unlikely habitat for birds. However, first appearances can be deceptive. The Western visitor will immediately be struck by the absence of House Sparrows and abundance of Tree Sparrows, and by the incongruous sight of Large-billed Crows patrolling amongst the skyscrapers. City parks provide oases

Large-billed Crow is a familiar urban resident in Tokyo

of habitat for a variety of species and are surprisingly calm and removed from the bustle which lies just outside the gates. Lakes and ponds in the parks can provide ice-free refuges for wildfowl during the winter, and although Tokyo Bay has been largely reclaimed to provide development land for the port, there are still some remnant mudflats which provide feeding for waders.

Japan has a unique avifauna which combines many species of the East Asian landmass with a number of island endemics, but the scattered nature of the archipelago together with the distance from north to south means that any comprehensive birding trip requires a significant amount of travel. There are a number of good birding locations within a day's journey of Tokyo, but a business or leisure visitor is less likely to have enough time available. Nevertheless, within easy reach of the city, there are several sites which support Japanese endemics as well as other species whose ranges are localised to this part of Asia. Whether you are visiting from the eastern, western or southern hemispheres, you will find much to interest you in the selection of Japanese avifauna which can be seen in and around Tokyo.

TYPICAL SPECIES

Resident
Little Grebe, Great Cormorant, Black-crowned Night Heron, Little Egret, Great Egret, Grey Heron, Spot-billed Duck, Black Kite, Common Moorhen, Oriental Turtle Dove, House Swift, Black-backed Wagtail, Brown-eared Bulbul, Great Tit, Japanese White-eye,

White-cheeked Starling, Large-billed Crow, Carrion Crow, Eurasian Tree Sparrow, Grey-capped Greenfinch

Summer
Barn Swallow, Asian House Martin

Winter
Eurasian Wigeon, Eurasian Teal,

Mallard, Northern Pintail, Tufted Duck, Common Pochard, Common Kestrel, Common Coot, Common Sandpiper, Black-headed Gull, Black-tailed Gull, Grey Wagtail, Pale Thrush, Brown-headed Thrush, Dusky Thrush, Hawfinch

GETTING AROUND

Tokyo has an excellent public transport system which means that most locations are conveniently near a railway or subway station, although trains are very crowded at times. Fares are very reasonable and information is usually displayed in English as well as Japanese, making navigation comprehensible, although the rail map does require careful study to familiarise yourself with its complexity of overhead trains, subways, interchanges and different operators. *The Tokyo Transit Book*, published by *The Japan Times*, is well worth obtaining. Although equally comprehensive, the bus system does not often display information in English and, moreover, is susceptible to getting caught

up in the city traffic, so it is preferable to use the trains and subways. Although business and leisure travellers will most probably be staying at a hotel in one of Tokyo's central districts, bear in mind that this is such a vast city that traffic flows in many different directions at all hours and it is therefore best to try to avoid travelling at peak periods.

The taxi service, unfortunately, is not suitable for birders, as it is very expensive and, apart from the main city parks, if you do not know the exact address in Japanese or how to get there, the taxi driver will have difficulty finding it.

Unless you have some considerable experience of driving in Tokyo, it is not recommended. Apart from the slow average speed you are likely to achieve, and the density of the streets and expressways which traverse the city, signposting is generally in Japanese and trying to drive and navigate using a bilingual map is probably not worth the stress. Since the public transport system is so comprehensive, a car is less of a requirement than in almost any other city in the world. Try to obtain a bilingual map that covers both Tokyo and the neighbouring prefectures of Chiba, Ibaraki and Kanagawa.

Birding in Tokyo is not really weather dependent, but the tide level does make a big difference when birding at any of the Tokyo Bay locations. Comfortable walking shoes are adequate for all of the following locations.

Locations

1 **Meiji Shrine** (West Central Tokyo)   (Harajuku Station)

This popular tourist spot is mentioned in all tourist guides. The park (Yoyogi Park) surrounding the shrine contains many woodland species. Get the subway or train to Harajuku Station and walk into the park following the main driveway to the shrine. There are several paths off the main drive which lead to quiet corners of the woods, enabling you to listen and watch for the birds. This is a very popular destination for tour parties but most tourists restrict themselves to the shrine itself, so you do not encounter many people in the wooded driveways. However some official traffic is permitted, so be careful of passing cars. The dense foliage can make for difficult birding, particularly as you try to pinpoint the source of a strange call. Beware of the squirrels, which can make a variety of bird-like sounds! There are two water bodies in the park: a pond adjacent to the north entrance and another in the Iris Garden. Both are worth a visit, especially in winter when they hold Mandarin Ducks. There is a small entrance charge into the Iris Garden.

KEY SPECIES

Resident
Japanese Pygmy Woodpecker, Common Kingfisher, Varied Tit, Azure-winged Magpie

Winter
Mandarin Duck, Daurian Redstart, Red-flanked Bluetail, Japanese Bush Warbler, Rustic Bunting, Black-faced Bunting

2 Ueno Park (North Central Tokyo) *(Ueno Station)*

This is a large park in the north-central part of Tokyo, adjacent to the city zoo. Most importantly, it contains a large area of open water, where rare ducks tend to appear in winter. The quieter areas of the park contain the usual woodland birds, but the lakes provide the best opportunity to see a number of waterfowl, particularly in winter. There is a Great Cormorant colony within the zoo which is visible from outside. Ueno Station is adjacent to the park entrance. The best areas are towards the southwestern end. If visiting the park, it is also worth checking the Imperial Palace Gardens, about 2km south of Ueno and a couple of stops along the loop line (Yamanote Line). The palace itself is normally closed to the public, but it is surrounded by a system of moats and gardens which are accessible. In winter, the moat provides open water for waterfowl including ducks and herons, while the gardens contain Azure-winged Magpies.

KEY SPECIES

Resident
Mute Swan, Japanese Pygmy Woodpecker, Common Kingfisher, Varied Tit, Azure-winged Magpie

Winter
Mandarin Duck, Northern Shoveler, Smew, Vega Gull

3 Oi Bird Park (10km) *(Ryutsu Senta Monorail Station)*

This is one of the best sites in Tokyo and if you have a day to spare, this is probably the priority. It is an area of the eastern shore of Tokyo Bay near Haneda Airport which, although surrounded by reclaimed land, has been left intact with areas of marsh and mudflat. It is run by the Wild Bird Society of Japan and members are on hand at weekends to assist with information. A variety of freshwater species can be seen in the ponds, and waders occur on the mudflats. The area is very well vegetated so, in addition to the waterfowl, walking the trails in the Bird Park offers opportunities of seeing a number of passerine species, particularly those, such as finches and buntings, which might not be found in the wooded city parks. The park is well laid out with trails and hides.

Take the **Monorail** from **Hamamatsu-cho Station** to **Ryutsu Centre** and as you leave the station, turn right and cross the bridge over the canal. From the bridge, you have a chance to see Black-tailed Gulls and, in summer, Little Terns. Continue on for two blocks, following the rather obscure signs, and you come to the Bird Park entrance on the right. There is a small entrance fee.

KEY SPECIES

Resident
Common Coot, Black-winged Stilt, Kentish Plover, Common Sandpiper, Black-tailed Gull, Common Kingfisher, Japanese Skylark, Zitting Cisticola, Bull-headed Shrike

Summer
Little Ringed Plover, Little Tern, House Swift, Great Reed Warbler

Winter
Gadwall, Northern Shoveler, Greater Scaup, Dunlin, Common Gull, Vega Gull, Slaty-backed Gull, Buff-bellied Pipit, Daurian Redstart, Rustic Bunting, Black-faced Bunting

Passage
Lesser Sand Plover, Red-necked Stint, Grey-tailed Tattler, Marsh Sandpiper, Common Greenshank, Ruddy Turnstone, Terek Sandpiper, Wood Sandpiper, Black-tailed Godwit, Bar-tailed Godwit, Eurasian Curlew, Whimbrel, Common Snipe, Red-throated Pipit

Northeast Tokyo Bay

If you have more than half a day available, you can do several good Tokyo Bay locations in that time, provided you synchronise your visit with an incoming tide. Kasairinkai Park, Funabashi Keihinkoen (Funabashi Seaside Park), and Yatsu-higata Tideland Park are all accessible along the **Keiyo Line** from **Tokyo Station**. Low tide times for

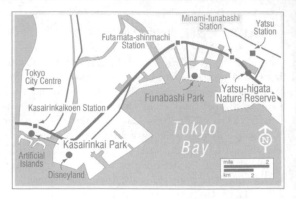

Funabashi and Kasairinkai are the same as for Tokyo Bay (birding at low tide at Funabashi is more productive than at Kasairinkai). Yatsu's low tide is delayed by about two hours, due to its distance from the open waters of the bay and the narrow inlet channels, so you should schedule the three sites accordingly.

4 Kasairinkai Park (10km)   *(Kasairinkaikoen Station)*

A public park on the shoreline of Tokyo Bay. In addition to providing access to the bay in order to scan for wintering ducks and grebes, there are two freshwater ponds which attract wintering ducks and herons. There are hides set up between the two ponds, providing some cover. Although the park tends to be heavily used by the public, the best birding areas are comparatively quiet. There are two offshore islands of reclaimed land, one of which is accessible by footbridge. The outer island is used as a roost site by waders, and they can be 'scoped from the shoreline.

This site is very easy to get to by public transport. **Kasairinkaikoen Station** is at the entrance to Kasairinkai Park: simply walk straight up the broad causeway to the bay. If you happen to be visiting Disneyland which is adjacent, then this site might provide some respite!

5 Yatsu-higata Tideland Park (20km)   *(Minami-funabashi Station)*
(Yatsu Station)

Seventeen minutes and five stations further along the Keiyo Line is Minami-funabashi Station, only a five-minute walk from Yatsu-higata Tideland Park: turn right at the turnstile and right again at the sidewalk. Use the pedestrian walkway which crosses under the expressway, then continue in the same direction along the expressway away from the station, and you'll see the mudflats on your left. This is an area of mudflats connected to the sea by two tidal channels. It has been developed with public walkways and hides, which enable you to scan the mudflats, and it provides an excellent facility for viewing waders and herons. There is a footpath right around the perimeter of the lagoon, a distance of about 3.5km. A telescope is probably necessary to appreciate the full potential of the location, but there are some public telescopes set up in the visitor centre. Spring, autumn and winter are when birds are present in highest numbers, although like any wetland it has potential at all times of the year. It is important to time your visit with a rising tide, which concentrates the birds in the lagoon. At high tide, however, there is little roosting space available and the birds vacate the lagoon.

If you are driving, take **Highway 357** northeast from Tokyo towards **Chiba**. At **Minami-funabashi**, turn right and continue for about 1km to the park entrance. An alternative and easier route by train is to take the **Keisei No.1 Line** from Ueno Station to **Yatsu Station**, from where the mudflats are a short walk: follow the sign-posted directions in English. There is also a train service from Tokyo Station to Narita Airport which stops at Yatsu, making this a convenient stop-off en route to the airport.

6 Funabashi Park (18km)   *(Futatamata-shinmachi Station)*

Futatamata-shinmachi is between Kasairinkaikoen Station and Minami-funabashi Station on the Keiyo Line, one station back from Minami-funabashi. From this station, Funabashi KK is a 30-minute walk: go towards the orange 'Eneos' gas station, turn right, then right again under the railway line, and go straight until you reach the bay and turn left with the road. Otherwise take a taxi. At the park entrance, you can catch a bus back to the train station. Funabashi KK is known for its wintering flock of Eurasian Oystercatchers and is excellent then for shorebirds and waterfowl. A telescope is a boon for searching the waterfowl rafts for rarities.

KEY SPECIES

Resident
Common Kestrel, Black-winged Stilt, Kentish Plover, Common Kingfisher, Japanese Skylark, Zitting Cisticola, Bull-headed Shrike

Summer
Intermediate Egret, Cattle Egret, Little Ringed Plover, Little Tern, House Swift, Great Reed Warbler

Winter
Great Crested Grebe, Black-necked Grebe, Greater Scaup, Smew, White-winged Scoter, Gadwall, Northern Shoveler, Eastern Marsh Harrier, Peregrine Falcon, Eurasian Oystercatcher, Grey Plover, Dunlin, Common Gull, Common Gull, Vega Gull, Slaty-backed Gull, Saunders's Gull, Buff-bellied Pipit, Daurian Redstart, Rustic Bunting, Reed Bunting, Black-faced Bunting

Passage
Lesser Sand Plover, Grey Plover, Great Knot, Red Knot, Red-necked Stint, Sharp-tailed Sandpiper, Grey-tailed Tattler, Common Sandpiper, Wood Sandpiper, Marsh Sandpiper, Common Greenshank, Ruddy Turnstone, Terek Sandpiper, Black-tailed Godwit, Bar-tailed Godwit, Eurasian Curlew, Far Eastern Curlew, Whimbrel, Common Snipe

7 Tama River – Upstream (30km)   *(Seisekisakuragaoka Station)*

The Tama River forms the southern boundary of the Tokyo City district, and flood control zones along its banks have resulted in some relatively undisturbed riparian habitat. The midstream gravel banks provide habitat for roosting gulls and other water-fowl and are also a reliable site for Long-billed Plovers. This is primarily a winter birding location and, in addition to the waterfowl, it is also good for open-country passerines, such as Japanese Wagtails and buntings. The Tokyo chapter of the Wild Bird Society of Japan holds bird walks along this stretch of the river on the first Sunday of every month.

It is easily accessible by train from central Tokyo on the Keio Line from Shinjuku Station. On arriving at Seisekisakuragaoka, the river bank is a ten-minute walk east from the station. The footpath runs along the river embankment and you can walk for several kilometres either upstream or downstream. A reasonable walk through one of the better stretches would be along the south bank downstream from Seisekisakuragaoka to Noborito Station (Nambu Line), although you will need to cross over to the north bank temporarily in order to pass a tributary inlet.

If you are driving, take the **Chuo Expressway** towards **Hachioji**, and exit south at the first exit after you cross the Tama River.

KEY SPECIES

Resident		Winter
Common Buzzard, Green Pheasant, Common Kestrel, Long-billed Plover, Common Sandpiper, Common Kingfisher, Japanese Skylark, Japanese Wagtail, Zitting Cisticola, Winter Wren, Azure-winged Magpie,	Bull-headed Shrike, Meadow Bunting **Summer** Striated Heron, Intermediate Egret, Little Ringed Plover, Common Sandpiper, Little Tern, Great Reed Warbler	Smew, Gadwall, Northern Shoveler, Buff-bellied Pipit, Daurian Redstart, Black-faced Bunting, Reed Bunting

8 Tama River Estuary (18km) _(Kojimashinden Station)_

The Tama River flows into Tokyo Bay at Kawasaki and, although the area is very built up, there are some remaining mudflats, which are accessible from the south shore of the estuary. This location is quite close to Tokyo's domestic airport at Haneda, so if you have some time to kill, or are collecting or dropping off a hire car, then this is a convenient site to visit. By rail, take the Keikyu (Keihin Kyuko) Line from Shinagawa to Kawasaki, and change trains within the station (the Daishi Line) to Kojimashinden Station (the terminus). The river is about ten minutes' walk north of the station: turn left onto the road beyond the exit and walk straight to the embankment. Walk along the south shore of the estuary towards the mouth of the river.

If you are driving, take the **Expressway No. 1** (**Kaigen-dori**) south towards **Kawasaki**.

KEY SPECIES

Resident		
Black Kite, Common Kestrel, Common Kingfisher, Japanese Skylark, Zitting Cisticola, Bull-headed Shrike **Winter** Great Crested Grebe, Greater Scaup, Smew, Black-winged Stilt, Common	Gull, Slaty-backed Gull, Buff-bellied Pipit, Daurian Redstart, Black-faced Bunting, Reed Bunting, Meadow Bunting **Passage** Lesser Sand Plover, Kentish Plover, Grey Plover, Pacific Golden Plover, Red-necked Stint, Dunlin, Grey-tailed	Tattler, Common Sandpiper, Common Greenshank, Terek Sandpiper, Black-tailed Godwit, Bar-tailed Godwit, Common Snipe, Eurasian Curlew, Far Eastern Curlew, Whimbrel, Little Tern

Further Away

9 Mount Takao (45km) _(Takaosanguchi Station)_

This is a well-known birding site northwest of Tokyo, at an elevation of almost 600m. It presents a good opportunity to see some woodland species, both winter and summer. It is very popular with visitors and best avoided at weekends: early weekday mornings are best. As well as resident woodland species, this site can be good for migrant warblers and flycatchers in spring and early summer.

It can be reached from Shinjuku Station by taking the **Keio Line** to **Takaosanguchi Station**. At this station, you can take a cable car to a point at mid-elevation, where there is a network of several trails fanning out and ascending through mixed woodland to the top of the mountain.

KEY SPECIES

Resident
Common Buzzard, Northern Goshawk, Japanese Green Woodpecker, Japanese Pygmy Woodpecker, Great Spotted Woodpecker, Japanese Wagtail, Grey Wagtail, Varied Tit, Coal Tit, Long-tailed Tit, Japanese Bush Warbler, Azure-winged Magpie, Eurasian Jay, Japanese Grosbeak, Meadow Bunting

Summer
Eastern Crowned Warbler, Asian Stubtail, Scaly Thrush, Blue-and-white Flycatcher, Narcissus Flycatcher

Winter
Daurian Redstart, Red-flanked Bluetail, Goldcrest, Winter Wren, Rustic Bunting, Grey Bunting

10 Mount Fuji area (85km)   *(Gotemba Railway Station)*

Although Mount Fuji is rather distant from Tokyo it is such a well-known area, and included in the itinerary of so many tours, that it is quite likely that you may visit it during your stay in the city. Mount Fuji encompasses a very wide area and there are a number of sites within the locality which have some birding potential. As the mountain is 3,700m high, it includes several vegetational zones which all support distinctive birds. The foothills and lower elevations are mainly deciduous woodland, which gives way at higher altitude to coniferous forest and ultimately to alpine vegetation above the treeline. It is possible to drive to an altitude of about 2,000m and from there to hike to the summit, but the rather small number of birds to be found there would not justify the considerable effort. Also worth checking are the lakes which surround Mount Fuji. The most interesting from a birder's perspective is Lake Yamanaka. This is near the southern approach road to the mountain.

Take the **Tomei Expressway** southwest towards **Nagoya**, but turn north after about 50km at **Gotemba** to **Lake Yamanaka**. The road to Fuji runs alongside the lakeshore at this point and there are several lay-by areas from which to scan the lake. The drive from Lake Yamanaka up to the base camp for mountain climbers (Fifth Station), at about 2,000m, ascends through some of the best forest habitat, and it is worth stopping wherever there is a place to park to check likely-looking habitat. Bear in mind, however, that private vehicles are not permitted up the last several kilometres of the road for several weeks in high summer. If using public transport, you can take a train to Gotemba and a bus from the train station up to the Fifth Station, or take one of the tourist buses which depart from Shinjuku Station in Tokyo.

KEY SPECIES

Resident
Common Buzzard, Northern Goshawk, Japanese Green Woodpecker, Japanese Pygmy Woodpecker, Great Spotted Woodpecker, Japanese Wagtail, Grey Wagtail, Red-flanked Bluetail, Goldcrest, Varied Tit, Coal Tit, Willow Tit, Long-tailed Tit, Japanese Accentor, Winter Wren, Eurasian Nuthatch, Eurasian Nutcracker, Azure-winged Magpie, Eurasian Jay, Japanese Grosbeak

Summer
Fork-tailed Swift, Olive-backed Pipit, Japanese Robin, Siberian Blue Robin, Scaly Thrush, Arctic Warbler, Eastern Crowned Warbler, Asian Stubtail, Japanese Paradise Flycatcher, Blue-and-white Flycatcher, Narcissus Flycatcher

Winter
Daurian Redstart, Brambling, Rustic Bunting, Grey Bunting

11 Ukishima Marsh (40km)   *(Sawara Station)*

The area northeast of Tokyo in Chiba and Ibaraki Prefectures looks rather like the Netherlands: it is flat and low-lying, with irrigation canals and even windmills, but instead of polders growing tulips and vegetables, the fields are flooded for rice culti-

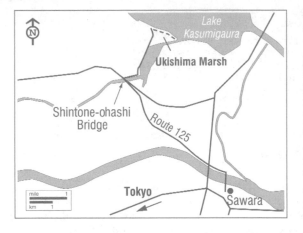

vation. The area is drained by the Tone River and some of the remnant marshes still remain: one such is the reedbed at Ukishima. This marsh is an extension of a much bigger wetland complex along the southern shore of Lake Kasumigaura. Although somewhat distant from Tokyo, it is quite close to Narita Airport, and consequently well worth a visit if you have spare travelling time. The marsh supports one of the richest communities of reedbed species in the Tokyo area, and the areas of open water attract large numbers of wildfowl in winter. The rice-paddies can also be good for waders when lowered water levels coincide with migration. The best area is the bridge at Ukishima, where there is a hide and a neat little speaker system, with bird recordings of the songbirds which inhabit the reedbed. On the west side of this bridge there is a road which leads to Lake Kasumigaura and runs parallel to a reed-fringed channel. This road to the lakeshore offers some of the best habitat.

Take the **Higashi-Kanto Expressway** and exit north at **Sawara**. Take **Route 125** northwest towards **Tsuchiura**. Ukishima Marsh is about 4km on the right-hand (north) side of the road. This site is not so easy to get to by public transport, as it is about 5km from the nearest train station (Sawara). A taxi from the station would be expensive but not budget-breaking; you could then also arrange to be picked up again later in the day.

KEY SPECIES

Resident
Osprey, Common Buzzard, Common Kestrel, Common Sandpiper, Common Kingfisher, Japanese Skylark, Zitting Cisticola, Japanese Marsh Warbler, Bull-headed Shrike, Meadow Bunting

Summer
Yellow Bittern, Intermediate Egret, Greater Painted-snipe, Little Ringed Plover, Common Sandpiper, Great Reed Warbler, Black-browed Reed Warbler, Japanese Reed Bunting

Winter
Smew, Gadwall, Northern Shoveler, Common Goldeneye, Eastern Marsh Harrier, Hen Harrier, Green Sandpiper, Wood Sandpiper, Northern Lapwing, Grey-headed Lapwing, Buff-bellied Pipit, Japanese Wagtail, Daurian Redstart, Black-faced Bunting, Reed Bunting

Passage
Pacific Golden Plover, Red-necked Stint, Common Greenshank, Marsh Sandpiper, Spotted Redshank, Black-tailed Godwit, Common Snipe, Latham's Snipe, Eurasian Curlew, Whimbrel, Little Tern, Sand Martin

Useful contacts and websites

Wild Bird Society of Japan (English and Japanese) www.wing-wbsj.or.jp
WBSJ, Aoyama Flower Bldg., 1-1-4 Shibuya, Shibuya ku, Tokyo 150.
Japanese bird conservation organisation and BirdLife partner for Japan.

Japanese Society for Preservation of Birds (English and Japanese) www.jspb.org
3F Dai-10 Tanaka Building, 54-5, Wada 3 chome, Suginami-ku, Tokyo 166-0012. Tel. 03 5378 5691.

Ornithological Society of Japan (English and Japanese) www.wwwsoc.nii.ac.jp/osj
Bird research organisation and publisher of the *Japanese Journal of Ornithology*.

Broadbill (English and Japanese)
www.members.at.infoseek.co.jp/bluebonnet/
japanesee.htm

Birds of Japan (English)
www.ca.geocities.com/kantorilode

Virtual Birding in Tokyo (English and Japanese)
www.fsinet.or.jp/~bird

Bird Songs in Japan (English and Japanese)
www.midopika.cool.ne.jp

Yatsu-higata Nature Observation Center
(English and Japanese)
www.city.narashino.chiba.jp/~yatsu-tf/english
Information about the Yatsu-higata mudflats including
bird reports and tide tables.

Tokyo Nature Weekly (English and Japanese)
www.pro.tok2.com/~tokyonature/english
General natural history website with information
about birds and organised bird walks within the city.

Oriental Bird Club
www.orientalbirdclub.org
P.O.Box 324, Bedford, MK42 0WG, UK

East Japan Railway Company (English)
www.jreast.co.jp
Train services to outer suburban areas of Tokyo.

japan-guide.com (English)
www.japan-guide.com
Portal site with links to all of the main bus, train and
subway services operating in Tokyo.

Tidal Chart for Tokyo Bay and Yatsu-higata
www.city.narashino.chiba.jp/~yatsu-tf/english/tide-
graph

Books and publications

A Birdwatcher's Guide to Japan
Mark Brazil (1995) Kodansha Europe
ISBN 0870118498

A Field Guide to the Birds of South-East Asia
Craig Robson (2000) New Holland
ISBN 1853683132

The Birds of Japan
Mark A. Brazil (1991) Christopher Helm
ISBN 0713680067

A Field Guide to the Birds of Japan
K. Sanobe (editor) (1982) Wild Bird Society of Japan
ISBN 0870117467 (Out of print)

A Field Guide to the Birds of Korea
Woo-Shin Lee, Tae-Hoe Koo & Jin-Young Park
(2000) LG Evergreen Foundation
ISBN 8995141514
(Contains all Japanese avifauna except endemics,
pelagics and tropical southern-island species. Range
maps include Japan.)

Where to Watch Birds in Asia
Nigel Wheatley (1996) Christopher Helm
ISBN 071364303X

A Birder's Guide to Japan
Jane Washburn Robinson (1987) Ibis Publishing Co
ISBN 0788154605

Auckland

Auckland is the gateway to New Zealand for most visitors and 1.2 million people live in the greater Auckland area – more than a third of the entire country's population. Built on a narrow isthmus between two island-studded harbours, Auckland is renowned for its beauty with 100km of coast-line crammed with stunning

New Zealand Dotterel on the shoreline of Auckland Harbour

beaches, 23 regional parks, 3 marine reserves and a landscape dotted with 48 volcanic cones. With all the water surrounding the region, Auckland's residents boast the largest boat ownership per capita in the world. It is no surprise that Auckland is called 'The City of Sails'. Summer is from December to February and winter from June to August. Temperatures range from 24°C in February to 14°C in July.

The birding opportunities around Auckland are good for those with a couple of days to spare. During the boreal winter, large numbers of waders from Asia visit the muddy bays of the North Island, and the

TYPICAL SPECIES

Resident
Mallard (I), Black Swan (I), Pacific Black Duck, Spotted Dove (I), Welcome Swallow, Dunnock (I), Song Thrush (I), Eurasian Blackbird (I), Grey Gerygone, Grey Fantail, Silver-eye, Tui, Australasian Magpie (I), European Starling (I), Common Myna (I), Chaffinch (I), European Greenfinch (I), European Goldfinch (I), House Sparrow (I)

seabirds of the Southern Ocean are at their breeding stations on headlands and offshore islands. This is the best time to visit the city. The few endemic landbirds are largely restricted to remote areas and offshore islands where the threat from introduced mammalian predators has been eliminated or reduced. The European visitor to Auckland will be struck by the much more numerous introduced European species and the paucity of native species. However, a significant number of endemics can be seen, including several very rare birds at sites close to the city.

GETTING AROUND

A car is necessary for many of the best birding sites, as they are quite distant and if your time is limited you are going to need the speed and flexibility. If you decide to use public transport there are rail and bus options to consider which connect Auckland with the towns nearest to the sites, but the

'final mile' will require you to hike, hitch or cycle. You can buy a Flexipass that will allow you many hours of travel around Auckland and throughout New Zealand. Trains and most buses terminate at Britomart Transport Centre in central Auckland, which is located conveniently beside the terminal for the ferries which serve the offshore islands.

Driving in New Zealand is generally stress-free, by all accounts. There is no orbital motorway around Auckland but all the main highways converge at one point near the city centre so finding your way out of the city or across the city is relatively easy.

Locations

1 Manukau Harbour (12km) (Bus Nos. 332/338/375)

This is an excellent wader viewing area with paths, plantings and hides provided by the water treatment company which manages the nearby Mangere Wastewater Treatment Plant. It is very close to Auckland International Airport, so is handy for those just arriving or leaving. There is a seven-kilometre coastal walkway around the harbour with bird hides at each end.

Take a bus or taxi from the airport to **Mangere** town centre and then go north for about 1km on **Greenwood Road** to the wastewater plant, where there is a car park and access to the coastal walkway.

KEY SPECIES

Summer		Winter
White-fronted Tern, Little Tern, Wrybill, Pacific Golden Plover, South Island Oystercatcher, Double-banded Plover, Red Knot, Sharp-tailed Sandpiper, Terek Sandpiper, Red-	necked Stint, Curlew Sandpiper, Far Eastern Curlew, Marsh Sandpiper, Whimbrel, Hudsonian Godwit, Black-tailed Godwit, Bar-tailed Godwit (eastern race), Ruddy Turnstone	Royal Spoonbill, Wrybill, Pied Oystercatcher, Double-banded Plover, Bar-tailed Godwit, Red Knot

2 Waitakere Regional Park (30km)

The Waitakere Ranges, northwest of Auckland, are extensively covered with temperate rainforest, mostly of secondary growth. This is a large area criss-crossed by hiking trails and with many tourist and camping facilities. One of the best sites to visit is the Ark in the Park in the Cascades area. This is a beautiful area of rainforest close to Auckland with many, very large Kauri trees, where a restoration project is protecting 2,000ha by intensive control of mammalian predators. Reintroducing Stitchbirds here is being considered. There are good signposted walks in the area and the Upper Kauri Walk is recommended.

Go west on the **North Western Motorway** (State Highway 16) and take **Exit 2** onto **Great North Road** towards **Titirangi**. Drive through Titirangi village and at the roundabout, take **Scenic Drive**. The visitor centre is on the left, 5km along Scenic Drive.

KEY SPECIES

Resident
New Zealand Pigeon, New Zealand Kaka, Grey Gerygone, Whitehead, Grey Fantail, Tomtit, New Zealand Robin (I)

3 Muriwai Beach (40km)

You will probably see Australian Gannets from the boat to Tiritiri Matangi (and certainly on a pelagic trip in Hauraki Gulf) but if neither of these is possible then Muriwai Beach gives you the chance to see them breeding. The birds arrive at Oaia Island and Otakimiro Point during August. Breeding starts in early summer with eggs hatching in November and December. From January to March the chicks are plentiful and make great viewing. The birds start to depart in late April through May.

Muriwai Beach is situated 40 minutes northwest of Auckland City by road but is impossible to reach by public transport alone. Leave Auckland City via the **North Western Motorway** (State Highway 16 West). The motorway ends at Westgate Shopping Centre. Turn left at the traffic lights and continue along SH16, through the villages of Kumeu and Huapai to **Waimauku**. At Waimauku turn left into Muriwai Road. Follow the road to its intersection with Motutara Road. Turn right into Motutara Road and follow it past the Muriwai Lodge Store and Wuz's Restaurant, taking the second road on the left (Waitea Road) and then the first road right into **Maori Bay**. Leave your car in the Maori Bay car park and follow the signs to the observation platforms overlooking the Gannets.

KEY SPECIES

Summer
Australian Gannet (plus other seabirds during rough weather)

4 Lake Kereta (70km)

This sand dune lake lies northwest of Auckland on South Head, the southern arm of Kaipara Harbour. Australasian Bitterns and Fernbirds occasionally breed in the dense raupo surrounding the main lake at Kereta. The Shelley Beach road is worth taking to scan the bay for waders and terns. A colony of Royal Spoonbills is often seen in this coastal area too.

Take the **South Kaipara Head** road north out of **Helensville** and turn left into **Wilson's Road** (just after the right-hand turning for Shelley's Beach). The lake is at the end of this gravel road. The best site is a smaller pond on the right just before the main lake, which is good for New Zealand and Australasian Grebes. You can park in the open area near the Department of Conservation sign. Walk through the gate and along the wooded track to view the pond.

If you have time, you can drive a little further north from Lake Kereta to **Papakaunui Spit** and **Waionui Inlet** at the top of South Head on the Kaipara Harbour. This reserve has good viewing of migratory waders in summer and the rare endemic race of the Fairy Tern. This is one of only four breeding sites for Fairy Terns in New Zealand, with a population of about 40 birds.

From Lake Kereta, return to South Head Road and follow it north for another 10km, taking the left turn after Lake Otatoa into the pine forest area. Follow the unsealed road for another 5km to the edge of Waionui Inlet. You can climb the hill to your right for good views of the inlet and the spit.

KEY SPECIES

Lake Kereta		Papakaunui Spit
New Zealand Grebe, Australasian Grebe, Little Black Cormorant, Little Pied Cormorant, Great Cormorant,	Australasian Bittern, Royal Spoonbill, Grey Teal, New Zealand Scaup, Paradise Shelduck, Caspian Tern, Fernbird	Variable Oystercatcher, Fairy Tern, migrant waders in summer

5 Wenderholm Regional Park (48km)   *(Bus No. 895)*

This site is a popular picnic area. It has a safe beach and scenic walks and is a great family destination, so you should try to arrive before the crowds. The park is on a steep ridge overlooking Waiwera. It sits between the mouths of the Puhoi and Waiwera Rivers and comprises a forested headland, a large sandspit and saltmarsh areas. The main attraction is the Banded Rail and there are opportunities for very close views of this species. The best site is the area of mangroves on the left as you enter the car park. High tide is best as the area of dry mangroves is then much reduced. It is also a good site to see the North Island race of the New Zealand Robin: there are nearly 40 on the

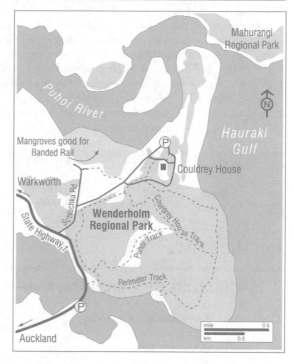

headland, and the best place to see them is on the top ridge track in the bush. The headland is a mainland island site where all mammalian predators are controlled. The robins were first introduced there in 1998 from Tiritiri Matangi. Tui are also plentiful.

From Auckland, head north on the **Northern Motorway** (State Highway 1). Wenderholm Regional Park is just north of **Waiwera**: turn right at the signs. Straka's Pond, a very productive wetland just before Wenderholm inland from Waiwera, which supports a good variety of wetland species, is worth visiting if time permits.

KEY SPECIES

Little Black Cormorant, Banded Rail, Red-breasted Dotterel, New Zealand Pigeon, Eastern Rosella (I), Australasian Pipit, New Zealand Robin, Tui

---

6 Tiritiri Matangi (50km)   *(Bus Nos. 896/897/899)*

Located just 4km from the end of Whangaparaoa Peninsula, Tiritiri Matangi Island is one of New Zealand's most important and exciting conservation projects. Since 1994, this 220-hectare island has been cared for by volunteers who have planted over 280,000 trees. The island is now 60 per cent forested with the remainder being left as grassland for species such as the Takahe. All mammalian predators have been eradicated and a number of species of threatened and endangered endemic birds have been successfully introduced. Nowhere else in New Zealand can you readily walk among so many indigenous species in such significant quantities. Guided walks around the island are recommended and are available at a very reasonable cost, the proceeds contributing to the island's restoration.

Ferry times can change but the vessel usually departs from both downtown Auckland (09.00) and Gulf Harbour (served by Bus Nos. 896/897/899) (09.45) and allows up to 150 visitors per day to visit the island. The ferry usually sails daily from December to mid-January, and always at weekends and on public holidays, but there is a more restricted service at other times. Check sailing days when you book. The boat usually leaves Tiritiri Matangi at 15.30 arriving at Gulf Harbour at 16.00 and Auckland at 16.45. Advance booking is essential: see www.kawaukat.co.nz

Coffee, tea and soft drinks are available but not food, so bring your own supplies. To arrange to stay overnight at the bunkhouse (16 beds), contact the ranger on 09 476 0010: you need to book well in advance.

KEY SPECIES

Resident
Little Spotted Kiwi (plan to stay overnight), Brown Teal, Brown Quail

(I), Spotless Crake, Takahe (I), New Zealand Pigeon, Red-crowned Parakeet, Morepork, Fernbird,

Whitehead, New Zealand Robin, Stitchbird, New Zealand Bellbird, Kokako, Saddleback

Further Away

7 Tawharanui Regional Park (90km)

This is an Open Sanctuary with a beautiful mix of farm and forest, complemented by a wonderful coastline and surf beach, part of which is protected as a marine park. Mature native bush in the park contains puriri, kauri and nikau palms. Large stands of maturing kanuka and manuka are evidence of a major regeneration programme. The Red-breasted

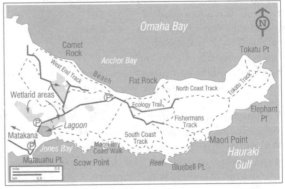

Dotterel nests in dunes on the northern coast. Bellbirds from Little Barrier Island have colonised the bush here and breed in the area called Ecology Bush. The park is surrounded by sea on all but one side and offers panoramic views of the Hauraki Gulf, and Kawau and Little Barrier Islands. Pacific Reef Egrets are often seen on the lagoon and Australasian Bitterns in the wetland areas.

From Auckland, head north on the **Northern Motorway** (State Highway 1) to Warkworth. At the second set of traffic lights, turn right and follow the signs to **Matakana**. After Matakana village, turn right at the Omaha intersection, head along **Takatu Road** and follow the signs to the park.

KEY SPECIES

Pacific Reef Egret, Australasian Bittern, Paradise Shelduck, Red-

breasted Dotterel, Variable Oystercatcher, White-fronted Tern,

Grey Gerygone, New Zealand Bellbird, Tui

8 Miranda (64km)

The Miranda Shorebird Centre is on the Firth of Thames, one hour southeast of Auckland and half an hour west of the Coromandel Peninsula. There are no public transport services to the Shorebird Centre from Miranda, however an on-demand shuttle service operates locally. The Firth of Thames includes 8,500ha of intertidal flats. These attract thousands of migratory waders. Extensive shell banks have formed along the coast and these provide excellent and safe roosting areas for the birds at high tide, which can be viewed from hides. Miranda is the most important wintering ground for Wrybills. This endemic species is unique among birds in having the bill curved to the side. The estimated total species population is only 5,300 birds, up to 40 per cent of which flock to Miranda at the end of the breeding season.

A successful visit to this site is really dependent on being there two hours before high tide until two hours after high tide (see website address for tide times on p. 185). The Miranda Shorebird Centre is a good starting point for information. There are two main birding options. Travel north from the centre, where the road crosses a small river, turn right and follow a track to a small parking area near the river mouth. This is often the best site for Wrybills. About 2km south of the centre there is an obvious lay-by and tracks lead across from here to the estuary edge. The shell banks here hold Red-breasted Dotterels, and the mangroves near the parking area are excellent for Banded Rails. Approach the pools in the field carefully as they often hold waders.

KEY SPECIES

Resident
White-faced Heron, Royal Spoonbill, Banded Rail, Purple Swamphen, Variable Oystercatcher, White-headed Stilt, Masked Lapwing, Red-breasted Dotterel, Kelp Gull, Black-billed Gull, White-fronted Tern

Summer
South Island Oystercatcher, Double-banded Plover, Pacific Golden Plover, Wrybill, Red Knot, Sharp-tailed Sandpiper, Terek Sandpiper, Red-necked Stint, Curlew Sandpiper, Marsh Sandpiper, Far Eastern Curlew, Whimbrel, Hudsonian Godwit, Black-tailed Godwit, Bar-tailed Godwit

(eastern race), Ruddy Turnstone, White-fronted Tern, Little Tern

Winter
Pied Oystercatcher, Double-banded Plover, Wrybill, Bar-tailed Godwit, Red Knot, Ruddy Turnstone (small numbers of juveniles overwinter)

9 Whangamarino Swamp (55km)

This is a huge area of marshland to the south of Auckland and is the second largest bog and swamp complex in the North Island, covering 5,900ha. The area is heavily hunted for waterfowl in May and June. The swamp holds both Spotless and Baillon's Crakes and both Australasian Bitterns and Fernbirds.

Take the **Southern Motorway** (State Highway 1) out of Auckland south towards Hamilton. The motorway now ends at **Mercer**. About 2km after Mercer, turn left into Island Block Road and follow this into the wetland. At the first major bridge, there is a car park on the right. Park here and follow the grassed track out into the wetland. Continue on around Island Block Road into Falls Road on the eastern side for a view across the swamp.

KEY SPECIES

Resident
Little Black Cormorant, Great

Cormorant, Australasian Bittern, Indian Peafowl (I), Wild Turkey (I),

Spotless Crake, Baillon's Crake, White-headed Stilt, Fernbird

10 Pelagic Trip in Hauraki Gulf

Regular day trips are run by Pterodroma Pelagics and sometimes by Wrybill Birding Tours (see websites below) and depart from **Sandspit Quay**, Leigh or Ti Point locations, a few kilometres east of **Warkworth** (about an hour's drive north of Auckland). These provide excellent birding and the chance to see a very rare bird, the New Zealand Storm-petrel. Other seabirds include the locally endemic Buller's Shearwaters, Parkinson's Petrels and Pycroft's Petrels. Important populations of Cook's Petrels, Fluttering Shearwaters, Little Shearwaters, Flesh-footed Shearwaters and White-faced Storm-petrels also occur in the Hauraki Gulf. A number of island species such as Kaka and Red-fronted Parakeet, and even Saddlebacks, can be seen when you moor for lunch and supper. Cetaceans including Common and Bottlenosed Dolphins, Short-finned Pilot Whales, Orcas and Bryde's Whales have all been recorded. From January through to April the full-day trips give you a good chance of seeing Grey Ternlets in small numbers on Maori Rocks, Mokohinau Islands. They arrive in New Zealand waters post-breeding from the tropical Pacific.

The New Zealand Storm-petrel was described from three specimens collected during the 19th century but was presumed extinct as there were no subsequent sightings. Then, in January 2003, one was seen off the Coromandel Peninsula and this was followed by ten that were filmed near Little Barrier Island in November 2003. There have since been regular sightings subsequently around the outer Hauraki Gulf between October and March.

KEY SPECIES

Little Penguin, Parkinson's Petrel, Cook's Petrel, Pycroft's Petrel, Black-winged Petrel, Great-winged Petrel, Buller's Shearwater, Sooty Shearwater, Fluttering Shearwater, Flesh-footed Shearwater, Little Shearwater, Common Diving Petrel, Fairy Prion, White-faced Storm-petrel, New Zealand Storm-petrel, Australian Gannet, Little Black Cormorant, Pacific Reef Egret, Variable Oystercatcher, Red-breasted Dotterel, Red-billed Gull, Kelp Gull, White-fronted Tern, Caspian Tern, Grey Ternlet, New Zealand Kaka, Red-fronted Parakeet, Saddleback

Useful contacts and websites

Ornithological Society of New Zealand
www.osnz.org.nz
PO Box 12397, Wellington, New Zealand.
Email: OSNZ@xtra.co.nz

Miranda Shorebird Centre
www.miranda-shorebird.org.nz
Firth of Thames, 283 East Coast Rd, R.D. 3, Pokeno, New Zealand.

Supporters of Tiritiri Matangi Inc.
www.tirimatangi.org.nz
Information about Tiritiri Matangi Island and conservation efforts.

Department of Conservation
www.doc.govt.nz
Government Department website with useful information about bird conservation in New Zealand.

Tawharanui Regional Park
www.arc.govt.nz/arc/auckland-regional-parks/northern-parks/tawharanui.cfm

Wenderholm Regional Park
www.arc.govt.nz/arc/auckland-regional-parks/northern-parks/wenderholm.cfm

Pterodroma Pelagics
www.nzseabirds.com
Pelagic birding trip organisers.

Wrybill Birding Tours
www.wrybill-tours.com
Birding tour operators.

Kiwi Wildlife Tours (New Zealand) Ltd
www.kiwi-wildlife.co.nz
Birding tour operators.

Birding NZ email group
http://groups.yahoo.com/group/BIRDING-NZ

MAXX Regional Transport
www.maxx.co.nz
Bus and train services in Auckland area.

Intercity Coachlines
www.intercitycoach.co.nz
Bus services outside Auckland.

Tide Tables www.hydro.linz.govt.nz/tides/
majports/index.asp

Books and publications

The Hand Guide to the Birds of New Zealand
Hugh Robertson and Barry Heather (2001) OUP
ISBN 019850831X

Birds of New Zealand: Locality Guide
Stuart Chambers (2000) Arun Books
ISBN 0473073277

Field Guide to New Zealand Seabirds
Brian Parkinson (2000) New Holland
ISBN 1877246328

The Reed Field Guide to New Zealand Birds
Geoff Moon (1992) Reed Publishing
ISBN 0790005042

● ●

TYPICAL SPECIES

Australasian Grebe, Great Cormorant, Little Black Cormorant, Little Pied Cormorant, White-faced Heron, Australian White Ibis, Straw-necked Ibis, Black Swan, Mallard (I), Pacific Black Duck, Maned Duck, Dusky Moorhen, Purple Swamphen, Eurasian Coot, Silver Gull, Sulphur-crested Cockatoo, Welcome Swallow, Eurasian Blackbird (I), Willie-wagtail, White-browed Scrubwren, Superb Fairywren, Silver-eye, Red Wattlebird, Little Wattlebird, White-plumed Honeyeater, Eastern Spinebill, Magpie-lark, European Starling (I), Common Myna (I), Australasian Magpie, Little Raven, European Greenfinch (I)

Musk Lorikeet is a garden bird in suburban areas of Melbourne

Melbourne

Although overshadowed by Sydney, there is a lot to like about Melbourne, Australia's second-largest city and also the world's most southerly metropolis. It has access to many birding areas that include marshes and coast, wooded hills and mallee – an Australian habitat that describes a semi-arid plain with scattered eucalyptus trees. The picturesque Dandenong ranges are 30km to the east and the vast flat basalt plains head off a long way to the west. With a population of 3.2 million, the city is actually less than 200 years old and covers an area of 8,800km². Perhaps the only negative thing about it is the notoriously changeable weather. A standing joke in Australia is that Melbourne can experience four seasons in a day! From December to February the hottest days can actually reach up to 40°C but from June to August the maximum is 14°C. However be aware that a southerly wind at any time of the year will quickly remind you that the Antarctic is not so far away!

GETTING AROUND

Some of the best sites in the Melbourne area are very distant from the city and a car is essential both to get there and to provide mobility when you arrive. Even though Melbourne is on the coast there is no good coastal birding site closer than 30km from the city and even there (Werribee Sewage Farm) you need a car to cover the large area, provided you are allowed access. For the few sites close to the city, the public transport system (bus, tram and train)

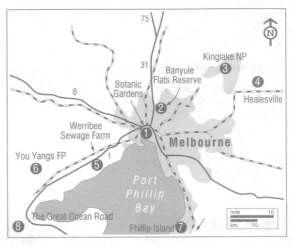

will work fine, so you really have to decide whether you will be content to visit just the city sites or will rent a car and do quite a bit of driving.

Locations

1 **Royal Botanic Gardens** (City Centre)

Birding in the city is really limited to parks and open spaces, but a visit to the Botanic Gardens is worth the trip – if only to get Blackbird and Song Thrush onto your Australia list! The gardens offer a combination of habitats including several watercourses. If travelling within the city you can ride there on the free Melbourne City Tourist Shuttle. The shuttle begins at the Melbourne Museum and runs every 15 minutes from 10.00–16.00 daily. The opening time is always 07.30, when the gardens are at their quietest.

KEY SPECIES

Resident	Rufous Night Heron, Rainbow	Fantail, Bell Miner, Grey Butcherbird
Hoary-headed Grebe, Grey Teal,	Lorikeet, Red-rumped Parrot, Song	
Chestnut Teal, Darter, Great Egret,	Thrush (I), Brown Thornbill, Grey	

2 **Banyule Flats Reserve** (16km)   *(Rosanna Railway Station) (Bus No. 517)*

This site is on the Yarra River at Heidelberg in the Melbourne suburbs. The habitat consists of swamp and billabong, paddock and riverside bush, enhanced by extensive indigenous plantings in recent years. Many unexpected sightings have occurred here, including Powerful Owl and Painted Snipe.

Approach from **Burgundy Street** near the junction with Lower Heidelberg Road; turn left into **Beverley Road** beside Heidelberg Park and follow it to the T-junction with **Banyule Road**. Turn right into Banyule Road. The park can be entered from

Banyule Road but it is better to turn right into **Somerset Drive** and enter the reserve at the bottom of the road: turn right into the car park. At the bottom of the car park there is a billabong overgrown with young trees. If you turn left and follow the track upstream, you can go for miles and see a large range of bush and open country birds (and Platypus in the river). The other option is to turn right (outside the billabong fence)

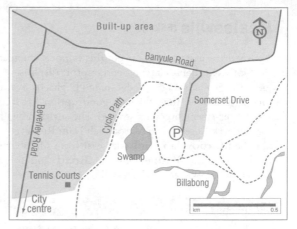

and walk west to the swamp, or walk along the circular bike path: there is a chance to see Buff-banded Rails in the drains. Return to the car park by crossing the paddocks from the tennis courts.

KEY SPECIES

Resident
Grey Teal, Chestnut Teal, Buff-banded Rail, Galah, Yellow-tailed Black Cockatoo, Eastern Rosella, Red-rumped Parrot, Laughing Kookaburra, Tree Martin, Black-faced Cuckoo-shrike, Brown Thornbill, Yellow Thornbill, Striated Thornbill, Grey Fantail, Yellow Robin, Crested Shrike-tit, Grey Shrike-thrush, White-throated Treecreeper, Spotted Pardalote, Noisy Miner, New Holland Honeyeater

Summer
Latham's Snipe

Winter
Cattle Egret

3 Kinglake National Park (65km)

This park is north of Melbourne and about an hour's drive via Whittlesea. The Mason's Falls picnic area is quite a reliable place for Superb Lyrebirds, particularly in damp weather when they are fairly approachable. Weekend and public holiday crowds send them back down the gullies: likewise dry spells of weather. The Crimson Rosellas can be so confiding as to be a nuisance. Another part of the park is Jehosaphat Gully: also good for Superb Lyrebirds and many other wet-forest birds.

Take the C727 north from **Heidelberg** in the northern suburbs of Melbourne towards **Whittlesea**. Turn right (east) in Whittlesea onto the **Whittlesea-Yea Road** (C725) and continue for about 10km, before turning right again on the **Kinglake Road**. The Mason's Falls picnic site is on a turnoff to the right after about 1km.

KEY SPECIES

Resident
Yellow-tailed Black Cockatoo, Superb Lyrebird, Laughing Kookaburra, Crimson Rosella, Yellow Robin, Brown Thornbill, Yellow Thornbill, Striated Thornbill, Grey Fantail, Crested Shrike-tit, Grey Shrike-thrush, White-throated Treecreeper, Spotted Pardalote, Noisy Miner, Bell Miner, New Holland Honeyeater, Yellow-faced Honeyeater, Brown-headed Honeyeater

4 Healesville (65km)

An hour's drive east of Melbourne, this famous sanctuary is a well-run zoo for local species, with excellent Platypus and raptor displays. A great many wild birds also frequent the grounds. Local Wedge-tailed Eagles fly over to display at captive birds. For visitors with limited time this is a good option for superb Lyrebirds (display closed during winter-spring breeding season). The surrounding hills are full of good sites such as Corranderrk Reserve and Maroondah Dam Reserve.

Take the **Maroondah Highway** east from Melbourne to **Healesville**. Turn right in Healesville onto **Badger Creek Road** and follow the signs for Healesville Sanctuary, about 4km from Healesville.

KEY SPECIES

Resident
Wedge-tailed Eagle, Superb Lyrebird, Laughing Kookaburra, Brown

Thornbill, Yellow Thornbill, Striated Thornbill, Grey Fantail, Yellow Robin, Crested Shrike-tit, Grey Shrike-thrush,

White-throated Treecreeper, Spotted Pardalote, Noisy Miner, New Holland Honeyeater

5 Werribee Sewage Farm (30km)   *(Werribee Train Station)*

This site west of Melbourne is particularly famous for hosting small numbers of the critically endangered Orange-bellied Parrot. Only around 180 adults survive in the wild and most of these overwinter in saltmarsh habitat along the central Victoria coast, while the remainder move west to the coast of South Australia. Up to 70 per cent of the entire population concentrates at three wintering sites around Port Phillip Bay and the Bellarine Peninsula. Unfortunately, while the overall numbers seem to be increasing slightly, their winter locations have become a mystery. They do get seen in winter at Werribee, but only very occasionally now. The site is important for passage and wintering birds and recent surveys have revealed the presence in winter of some 65,000 wintering waterfowl. It is also a major location for migratory waders in the Australian summer and a number of rarities turn up each year.

Access to the site (known as Melbourne Water Western Treatment Plant) is by permit only. Even if as a casual visitor you manage to obtain access you will really have great difficulty finding your way around, as there are over 200km of roads within the site. The best option is to visit with a local birder who both knows the area and has a key.

KEY SPECIES

Resident
Great Crested Grebe, Hoary-headed Grebe, Australian Pelican, Pied Cormorant, Great Egret, Little Egret, Royal Spoonbill, Yellow-billed Spoonbill, Cape Barren Goose, Freckled Duck, Musk Duck, Grey Teal, Chestnut Teal, Australasian Shoveler, Pink-eared Duck, Hardhead, Black-shouldered Kite, Whistling Kite, Swamp Harrier, Peregrine Falcon, Brown Falcon, Australian Crake, Black-tailed Native-hen, Pied

Oystercatcher, Masked Lapwing, Red-kneed Dotterel, Black-fronted Dotterel, White-headed Stilt, Red-necked Avocet, Pacific Gull, White Tern, Great Crested Tern, Blue-winged Parrot, Yellow-rumped Thornbill, Eurasian Skylark (I), Tree Martin, Australian Pipit, Little Grassbird, Golden-headed Cisticola, White-fronted Chat, European Starling (I), European Goldfinch (I), Red-browed Finch, House Sparrow (I)

Summer
Australian Shelduck, Banded Stilt, Sharp-tailed Sandpiper, Red-necked Stint, Curlew Sandpiper, Black-tailed Godwit, Common Greenshank, White-winged Black Tern, Fairy Martin

Winter
Cattle Egret, Double-banded Plover, Orange-bellied Parrot

6 You Yangs Forest Park (60km)

Situated just a few kilometres north of the Princes Highway, this 2000ha park includes an 800ha eucalypt plantation and is one of the better birding areas within an hour of the city. It has a good variety of birds at any time of the year. Over 200 bird species have been recorded within the park.

To reach it, leave the **Princes Highway** about 30km southwest of Melbourne at the **Little River** exit. Drive through Little River and go west for 4km to a T-junction. Turn left then right after 2km. After a further 5km, a road off to the right leads to the picnic area. Much of the surrounding area is dry open woodland but there are massive granite boulders and outcrops on the slopes. The best strategy is to walk along the roads and tracks, looking for flowering trees.

KEY SPECIES

Resident
Black-shouldered Kite, Brown Goshawk, Brown Falcon, Common Bronzewing, Purple-crowned Lorikeet, Eastern Rosella, Red-rumped Parrot, Horsfield's Bronze Cuckoo, Laughing Kookaburra, Tree Martin, Black-faced Cuckoo-shrike, Brown Thornbill, Buff-rumped Thornbill, Yellow-rumped Thornbill, Yellow Thornbill, Striated Thornbill, Restless Flycatcher, Grey Fantail, Scarlet Robin, Yellow Robin, Jacky Winter, Crested Shrike-tit, Grey Shrike-thrush, Varied Sittella, White-throated Treecreeper, Brown Treecreeper, Mistletoebird, Spotted Pardalote, Striated Pardalote, Yellow-faced Honeyeater, Brown-headed Honeyeater, White-naped Honeyeater, New Holland Honeyeater, Black-chinned Honeyeater, White-winged Chough, White-browed Woodswallow, Dusky Woodswallow, Red-browed Finch, Diamond Firetail

Summer
Pallid Cuckoo, Black-eared Cuckoo, Fan-tailed Cuckoo, Shining Bronze Cuckoo, Sacred Kingfisher, Rainbow Bee-eater, Fairy Martin, Rufous Whistler, Satin Flycatcher, Brown Songlark

Winter
Swift Parrot, Flame Robin, Golden Whistler

Further Away

7 Phillip Island (140km)

Phillip Island is on the east side of Port Phillip Bay and is world famous for the nightly 'Penguin Parade', a big tourist attraction. Every evening at dusk significant numbers of Little Penguins return to their burrows, much to the delight of the many tourists that pay to see the spectacle. For most people this is their first wild encounter with penguins and, despite some fairly annoying behaviour by non-birding tourists (flashlights and noise), the experience is memorable. The parade is open every day of the year. The best time to visit is at sunset/dusk if you wish to see the Little Penguins crossing Summerland Beach to return to their sand-dune burrows. The parade can be experienced from environmentally friendly observation boardwalks and viewing stands.

Travel from Melbourne on the **South Eastern Freeway** (toll way) for approximately 45 minutes. After you pass signs indicating Dandenong just keep going. After a few kilometres you will see large green freeway signs clearly indicating a turn off from the left lane to **Phillip Island**. Once you drive onto the island, the Phillip Island Information Centre will appear on your left, just past a roundabout. Here you can buy tickets to the Penguin Parade, and it is open till late in the evening.

KEY SPECIES

Resident
Little Penguin, Australian Gannet, Black-faced Cormorant, Great Egret, Cape Barren Goose, Australian Shelduck, Pied Oystercatcher, Great Crested Tern, Pacific Gull, Kelp Gull, Australian Pipit

8 The Great Ocean Road (110km)

Tourist sites such as the Twelve Apostles rock formation are very popular and many people visit the south coast from Geelong to Warrnambool known as the Great Ocean Road. If you are sightseeing here, then the coastal heath is worth searching for the near-threatened Rufous Bristlebird. This skulking species chooses cliff-top heathland, especially steep, densely-vegetated slopes where feral predators cannot penetrate. As it feeds on the ground and nests close to it, it is at the mercy of cats and foxes. It can be found from Point Addis in the east to at least Peterborough in the west. One of the best sites for it is The Arch, 5km west of Port Campbell. Other good sites are the London Bridge car park (3km further west), Loch Ard Gorge car park (just east of Port Campbell) and Airlies Inlet. Another good bird to look for is the Blue-winged Parrot.

TYPICAL SPECIES

Albatrosses (August to November), Blue-winged Parrot, Rufous Bristlebird

Useful contacts and websites

Birds Australia
www.birdsaustralia.com.au
Birds Australia, 415 Riversdale Rd, Hawthorn East, VIC 3123 — BirdLife Partner in Australia.

Victorian Branch of Birds Australia
www.babblersnest.com

Bird Observers' Club of Australia
www.birdobservers.org.au
Australian birding organisation.

BAYBOCA
www.cohsoft.com.au/bayboca/
BAYBOCA is the Bayside (Melbourne) Branch of the Bird Observers Club of Australia.

Birds Australia Rarities Committee
www.users.bigpond.net.au/palliser/barc/barc-home.html

Birding-Aus Mailing-List
www.shc.melb.catholic.edu.au/home/birding/index.html

Phillip Island Nature Park
www.penguins.org.au
Information about Penguin Parade at Phillip Island.

Werribee Sewage Farm
Information about access and bird checklist.
www.melbournewater.com.au/content/sewerage/western_treatment_plant/bird_watching_and_fishing.asp

Victoria National Parks
www.parkweb.vic.gov.au
Website with information about all National Parks in the State, including Kinglake Park.

Royal Botanic Gardens Melbourne
www.rbg.vic.gov.au

Australian Ornithological Services
www.philipmaher.com
Bird tour company specialising in inland trips.
Email: enquiries@philipmaher.com

Metlink
www.metlinkmelbourne.com.au
Information about public transport in Melbourne area.

Books and publications

Complete Guide to Finding the Birds of Australia Richard Thomas and Sarah Thomas (1996) Frogmouth Publications, ISBN 0952806509

A Bird Atlas of the Melbourne Region
Helen Aston and Rosemary Balmford. Victorian Ornithological Research Group

Field Guide to the Birds of Australia
Graham Pizzey & Frank Knight HarperCollins (1997) ISBN 0002201321

Field Guide to Australian Birds
Michael Morcombe (2000) Steve Parish Publishing ISBN 187628210X

Field Guide to the Birds of Australia
[7th Edition] Ken Simpson and Nicolas Day (2004) Christopher Helm, ISBN 0713669829

TYPICAL SPECIES

Resident
Little Pied Cormorant, Great Cormorant, White-faced Heron, Cattle Egret, Little Egret, Great Egret, Black Swan, Pacific Black Duck, Mallard (I), Grey Teal, Dusky Moorhen, Masked Lapwing, Silver Gull, Spotted Dove (I), Peaceful Dove, Rainbow Lorikeet, Sulphur-crested Cockatoo, Laughing Kookaburra, Welcome Swallow, Black-faced Cuckoo-shrike, Red-whiskered Bulbul (I), Eurasian Blackbird (I), Superb Fairywren, Willie-wagtail, Silvereye, Eastern Spinebill, New Holland Honeyeater, Red Wattlebird, Little Wattlebird, Noisy Miner, Magpie-lark, Grey Butcherbird, Pied Currawong, Australasian Magpie, Australian Raven, Green Figbird, European Starling (I), Common Myna (I), Red-browed Finch, House Sparrow (I)

Summer
Fairy Martin

Silver Gulls flying around Sydney Opera House

Sydney

Sydney is located on a narrow coastal plain about 60km wide, between the Pacific Ocean and the foothills of the Great Dividing Range. The city provides an excellent introduction to the Australian avifauna as it has both wet and dry forest and open mallee on its doorstep, as well as offering coastal and pelagic birding opportunities. It is particularly satisfying for a birder to experience the indigenous birdlife of the eucalyptus forests: eucalyptus is often encountered in so many other parts of the world where it is an invasive exotic and detrimental to the local species.

The resident bird populations are considerably augmented during summer (Oct to April) by winter migrants from the Palearctic, most notably Siberian waders, as well as by summer migrants from further north in Australia, which breed in the area. Mean temperatures range from 25°C from November to January to 16°C in June and July. It is almost always sunny with five to seven hours of daily sunshine in any month. Rainfall is fairly evenly spread, with a slight increase between January and May. Additionally, prolonged dry spells – on this the world's driest continent – result in nomadic behaviour for many birds so that some species can occur in unexpected numbers or not at all.

GETTING AROUND

Sydney has a population of almost 4 million and covers almost 1,120km², but it is a remarkably easy city to move around in. An excellent train, bus and ferry service covers most points of the greater metropolitan area. The roads, although congested at peak periods, are generally well planned and driving is fairly straightforward. Most of Sydney's tourist attractions are either around the central business district or at points around the harbour. Some of the best birding sites require a drive of two to three hours, although if you are lucky enough to join a pelagic trip from Sydney you can get a taxi to where the boat docks. Rail services can be used to visit the Royal National Park and Wollongong, but you need to plan your journey in advance to ensure you get enough time at these sites. There are a number of low-cost car-hire firms in addition to the usual large hire companies.

1 Royal Botanic Gardens (City Centre)

The Royal Botanic Gardens are just a short walk around the water's edge from the Sydney Opera House and are a good place to do some birding if you have just an hour to spare. They open daily at 06.30 and close at sunset. Entry is free. The gardens fill an area of land between the harbour and the eastern part of the central business district, and the harbour views from up on the hill are superb. A walkway skirts around the harbour front at the lower part of the gardens. As well as the vegetated areas, which provide cover for many birds and roosting Fruit Bats, the watercourses are also good places to get to close quarters with aquatic species. Green Figbirds are easy to spot here (and also at the Rose Bay Wharf, from where the Sydney pelagic departs).

KEY SPECIES

Resident
Little Black Cormorant, Australian White Ibis, Maned Duck, Buff-banded Rail, Tawny Frogmouth, Green Figbird

2 Warriewood Wetlands (25km)   (Bus Nos. 185/L85)

Situated an hour's bus ride north of the city, the wetlands are a small area close to a large shopping complex. Enter via Katoa Place and use the boardwalk to traverse the area. At the end of the path cross the road and continue along another path to a small waterfall. A stay of one or two hours would be suffi-cient at this site, but you can add a visit to the nearby Deep Creek and Long Reef if you have a vehicle. Deep

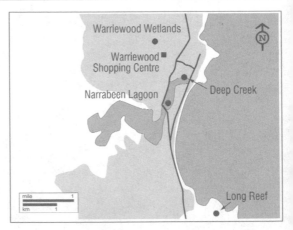

Creek is accessed from Wakehurst Parkway, on the north side of Narrabeen Lagoon, and Long Reef from Long Reef Golf Club car park.

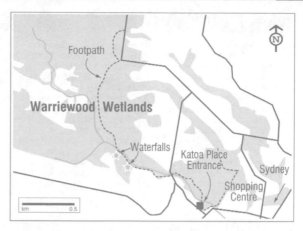

KEY SPECIES

Resident
Warriewood: Brown Goshawk, Purple Swamphen, Buff-banded Rail, Australian Crake, Crested Pigeon, Eastern Rosella, Fan-tailed Cuckoo, Horsfield's Bronze Cuckoo, Eastern Whipbird, Clamorous Reed Warbler, Tawny Grassbird, Little Grassbird, Superb Fairywren, Grey Fantail, Yellow Robin, Golden Whistler, Grey Shrike-thrush, White-browed Scrubwren, Brown Gerygone, Yellow Thornbill, White-throated Treecreeper, Spotted Pardalote, Noisy Friarbird, Lewin's

Honeyeater, Yellow-faced Honeyeater, Spangled Drongo

Deep Creek: Black Bittern, Rufous Night Heron, Whistling Kite, Crested Pigeon, Fan-tailed Cuckoo, Azure Kingfisher, Yellow Robin, Grey Butcherbird

Long Reef: Australian Gannet, Australian Pelican, Pacific Reef Heron, Australian Kestrel, Great Crested Tern, Australasian Pipit

Summer
Warriewood: Australian Koel, Sacred

Kingfisher, Dollarbird, Black-faced Monarch, Leaden Flycatcher

Deep Creek: Sacred Kingfisher, Rufous Whistler, Leaden Flycatcher, Olive-backed Oriole

Long Reef: Short-tailed Shearwater, Pacific Golden Plover, Bar-tailed Godwit, Grey-tailed Tattler, Red-necked Stint, Sharp-tailed Sandpiper, Ruddy Turnstone

Winter
Long Reef: Double-banded Plover

3 Royal National Park (32km)

(Engadine Railway Station)
(Heathcote Railway Station)

Established in 1879, Royal National Park is the world's second-oldest national park – after Yellowstone in the USA. It packs incredible natural diversity into a relatively small area, offering open woodland, beaches, cliff-top views, heathland and rainforest. It has an excellent selection of birds with a typical day total covering many habitats approaching 80 species. Over 270 species have been recorded in the park.

If driving, leave Sydney and head south on the **Princes Highway**. The park will eventually be signposted to the left. You have to pay a modest entrance fee and there are good park maps at the entrance and in the headquarters. You can also download maps from the Internet. One of the most profitable birding areas is Audley, although this gets busy with families at weekends. There is a café here also. From here a route called Lady Carrington Drive passes through great rainforest habitat: the trail can also be tackled from its southern end. For heathland habitat a walk on the Curra Moors Trail towards the coast is recommended, and this joins with a coastal track. As with most remote places worldwide it's not advisable to leave visible items in a parked vehicle.

Trains from Sydney stop at Engadine and Heathcote, both of which are next to the park.

KEY SPECIES

Resident
Australasian Grebe, Little Pied Cormorant, Darter, Pacific Reef Heron, Maned Duck, Brown Goshawk, Grey Goshawk, Australian Hobby, Swamp Harrier, Brown Quail, White-fronted Tern, Topknot Pigeon, Wonga Pigeon, Brown Cuckoo-Dove, Sulphur-crested Cockatoo, Yellow-tailed Black Cockatoo, Little Corella, Crimson Rosella, Eastern Rosella, Australian King Parrot, Fan-tailed Cuckoo, Azure Kingfisher, Bassian Thrush, Eastern Whipbird, Superb Fairywren,

Variegated Fairywren, Southern Emuwren, Rock Warbler, Pilotbird, White-browed Scrubwren, Large-billed Scrubwren, White-browed Scrubwren, Yellow-throated Scrubwren, Chestnut-rumped Hylacola, Striated Thornbill, Brown Gerygone, Superb Lyrebird, Grey Fantail, Yellow Robin, Golden Whistler, Crested Shrike-tit, Grey Shrike-thrush, White-throated Treecreeper, Spotted Pardalote, White-eared Honeyeater, White-naped Honeyeater, White-cheeked Honeyeater, Scarlet

Honeyeater, Lewin's Honeyeater, Yellow-faced Honeyeater, Yellow-winged Honeyeater, Tawny-crowned Honeyeater, Noisy Friarbird, Dusky Woodswallow, Pied Butcherbird, Green Catbird, Satin Bowerbird, Beautiful Firetail

Summer
Sacred Kingfisher, Dollarbird, Rufous Whistler, Black-faced Monarch, Leaden Flycatcher, Rufous Fantail, Rose Robin

Further Away

4 Capertee Valley/Glen Davis (170km)

This is a rather distant site but if you have a full day available, or happen to be visiting the Blue Mountains, then this is a fantastic birding area to the west of Sydney. It is a terrific valley covered by woodland, hills and farmland. It is possible to rent accommodation in the valley, but you should plan this in advance. Accommodation is also available at Lithgow, 30km south of Glen Davis. A wide variety of birds can be seen but most of all the Capertee Valley is famous as a breeding area for the endangered Regent Honey-eaters. These birds feed on nectar and insects within box-ironbark eucalypt forests. After the breeding season the birds roam widely although they have been recorded here in every month. It is also a stake-out for Plum-headed Finches which are often found at the end of the valley in the Glen Davis churchyard, while the fabulous Turquoise Parrots can be seen anywhere along the last two-thirds of the road (20km or so) before Glen Davis. Eastern Grey Kangaroos and Wallaroos are easily seen in the valley.

To get to the site, drive west from Sydney on the **Great Western Highway**. Turn north onto **Highway 86** to Capertee which is the **Mudgee road** just after Lithgow. When you reach **Capertee**, turn right to Glen Davis. The whole road is good for birding: just stop where the habitat looks good or more particularly whenever you hear a bird calling. The best area for Regent Honeyeaters is the short section of bitumen on a steep hill about 10km from Capertee. Further along from the bitumen section the road crosses a couple of creeks and these are the best areas for Plum-headed Finches. There is a museum at Glen Davis itself and the trees around here and just before it are good for Turquoise Parrots, while Regent Honeyeaters have also bred here.

KEY SPECIES

Resident
Australasian Grebe, White-necked Heron, Maned Duck, Black-shouldered Kite, Brown Goshawk, Wedge-tailed Eagle, Little Eagle, Australian Hobby, Australian Kestrel, Brown Falcon, Stubble Quail, Brown Quail, Painted Buttonquail, Black-fronted Dotterel, Common Bronzewing, Crested Pigeon, Gang-

gang Cockatoo, Galah, Glossy Black Cockatoo, Little Corella, Little Lorikeet, Musk Lorikeet, Crimson Rosella, Eastern Rosella, Red-rumped Parrot, Turquoise Parrot, Swift Parrot, Australian King Parrot, Brush Cuckoo, Fan-tailed Cuckoo, Black-eared Cuckoo, Horsfield's Bronze Cuckoo, Tree Martin, Australasian Pipit, White-bellied Cuckoo-shrike, Cicadabird,

Spotted Quail-thrush, Grey-crowned Babbler, White-browed Babbler, Golden-headed Cisticola, Rufous Songlark, Variegated Fairywren, Rock Warbler, Speckled Warbler, White-browed Scrubwren, Chestnut-rumped Hylacola, Weebill, Western Gerygone, White-throated Gerygone, Brown Thornbill, Buff-rumped Thornbill, Yellow-rumped Thornbill, Yellow

Thornbill, Southern Whiteface, Grey Fantail, Flame Robin, Scarlet Robin, Red-capped Robin, Hooded Robin, Yellow Robin, Jacky Winter, Crested Shrike-tit, Grey Shrike-thrush, Varied Sittella, White-throated Treecreeper, Brown Treecreeper, Red-browed Treecreeper, Mistletoebird, Spotted Pardalote, Striated Pardalote, Noisy Friarbird, Bell Miner, Fuscous

Honeyeater, Yellow-tufted Honeyeater, Black-chinned Honeyeater, Regent Honeyeater, Striped Honeyeater, Yellow-faced Honeyeater, White-plumed Honeyeater, Brown-headed Honeyeater, White-naped Honeyeater, Scarlet Honeyeater, White-winged Chough, Diamond Firetail, Plum-headed Finch, Red-browed Finch, Zebra Finch (I), Double-barred Finch

Summer
Sacred Kingfisher, Rainbow Bee-eater, Pallid Cuckoo, Singing Bush Lark, Restless Flycatcher, Rose Robin, Rufous Whistler, Dusky Woodswallow, White-browed Woodswallow, Olive-backed Oriole

5 Pelagic trips out of Sydney and Wollongong

The seas off Australia support more species of seabird than anywhere else in the world with over 120 species recorded. The Sydney and Wollongong pelagics have recorded over 75 species, which is over a third of the world's seabirds! To join one of these trips is a great birding experience. Trips are organised from Sydney on the second Saturday of every month and from Wollongong (70km to the south) on the fourth Saturday of every month. The boats head for the deep waters along the continental shelf. Here the cold waters attract hundreds – sometimes thousands – of seabirds. You get about nine hours out at sea and every trip is accompanied by experts who know the birds well. There is always plenty to see but the widest range of petrels, shearwaters and albatrosses are seen between August and November.

The *Halicat* leaves Rose Bay Public Ferry Wharf in Sydney at 07.00, returning at about 15.00. Additional trips are sometimes planned depending on demand. This is a fast boat allowing additional time in deeper waters.

The *Sandra K* leaves Wollongong Harbour at 07.00, returning around 16.00. Again, additional trips may also be planned during peak periods. Contact details about both pelagics are given below and overleaf.

KEY SPECIES

All Year
Fluttering Shearwater, Hutton's Shearwater, Great-winged Petrel

Summer
Black Petrel, Gould's Petrel, Flesh-footed Shearwater, Wedge-tailed

Shearwater, Short-tailed Shearwater, Buller's Shearwater, Sooty Shearwater

Winter
Cape Petrel, Providence Petrel, Southern Giant Petrel, White-headed Petrel, Wilson's Storm-petrel, White-faced Storm-petrel, Wandering Albatross, Black-browed Albatross, Yellow-nosed Albatross, Shy Albatross, Fairy Prion

Useful contacts and websites

Birds Australia
www.birdsaustralia.com.au
Birds Australia, 415 Riversdale Rd, Hawthorn East, VIC 3123 – BirdLife Partner in Australia.

Birds Australia Rarities Committee
www.users.bigpond.net.au/palliser/barc/barc-home.html

Bird Observers' Club of Australia
www.birdobservers.org.au
Australian birding organisation.

Birding NSW
www.birdingnsw.org.au
NSW Field Ornithologists' Club – local branch of Bird Observers' Club of Australia.

Cumberland Bird Observers' Club
www.cboc.org.au
Sydney area birding group website.

SOSSA
www.sossa-international.org
Southern Oceans Seabird Study Association Inc. Operators of pelagic trips off Wollongong.

Tony Palliser's Pelagic Web Page
http://users.bigpond.net.au/palliser/pelagic/
Details of pelagic trips off Sydney and Wollongong.
Contact: Tony Palliser (+61) 02 94911678 or
Hal Epstein (+61) 0 411 311236
Email: tonyp@bigpond.net.au or
hepstein@bigpond.net.au

Pelagic Birding in Southeastern Australia
www.oceanwanderers.com/Aust.html
Information about pelagic trips off Southeastern
Australia.

NSW Maritime
www.waterways.nsw.gov.au
Weather and tidal predictions for Sydney Harbour.

Transport Infoline
www.131500.info/realtime/default.asp
Information on buses, trains and ferries in Sydney
area.

Sydney Buses
www.sydneybuses.info
Bus services in Sydney area.

CityRail
www.cityrail.nsw.gov.au
Suburban trains in Sydney area.

RailCorp
www.railcorp.info
Train services in New South Wales.

Local Bird Guides
Alan McBride. Email: amcbride1@mac.com
Andy Burton's Bush Tours. Email: abbt@mac.com

Books and publications

**The Complete Guide to Finding the Birds of
Australia** Richard and Sarah Thomas (1996)
Frogmouth Publications, ISBN 0952806509

Birdwatchers Guide to the Sydney Region
Peter Roberts. Kangaroo Press.

Where to Find Birds in Australia
John Bransbury (1987) Hutchinson
ISBN 0091689414

**Birdwatching in Royal & Heathcote National
Parks** Steve Anyon-Smith. NSW National Parks &
Wildlife Service.

Field Guide to the Birds of Australia
Graham Pizzey & Frank Knight HarperCollins (1997)
ISBN 0002201321

Field Guide to Australian Birds
Michael Morcombe (2000) Steve Parish Publishing
ISBN 187628210X

Field Guide to the Birds of Australia
[7th Edition] by Ken Simpson and Nicolas Day (2004)
Christopher Helm, ISBN 0713669829

TYPICAL SPECIES

Resident
Great Crested Grebe, Great Cormorant, Grey Heron, Mute Swan, Greylag Goose, Mallard, Tufted Duck, Eurasian Sparrowhawk, Common Buzzard, Moorhen, Common Coot, Black-headed Gull, Herring Gull, Wood Pigeon, Collared Dove, Great Spotted Woodpecker, White Wagtail, Wren, Dunnock, Robin, Blackbird, Song Thrush, Blue Tit, Great Tit, Eurasian Jay, Magpie, Western Jackdaw, Carrion Crow, Common Starling, House Sparrow, Chaffinch, Greenfinch, Goldfinch, Bullfinch, Reed Bunting

Summer
Common Tern, Common Swift, Sand Martin, Barn Swallow, House Martin, Blue-headed Wagtail, Common Redstart, Reed Warbler, Blackcap, Common Whitethroat, Common Chiffchaff, Willow Warbler, Spotted Flycatcher

Great Crested Grebes inhabit the canals in Amsterdam

Amsterdam

The Netherlands is the most densely populated country in Europe and the area around Amsterdam is very built-up with suburbs, motorways, Schiphol Airport and huge industrial estates. Yet some of the best birding in Europe can be experienced in pockets of habitat that are quite close to the city, while the coast is only 25km away. What is most surprising is that most of the habitat is man-made, yet winter and summer these areas attract some of the largest concentrations of birds in Europe.

Amsterdam is situated at the hub of a very complex system of waterways, some natural and some man-made. The River Amstel and the North Sea Canal converge at the city centre and enter the IJmeer to the east of the city. The surrounding areas are flat reclaimed polders, which are intensively cultivated. To the east lies the large inland sea, the Markermeer. The whole region is several metres below sea-level.

Many rare vagrants are found in The Netherlands, and the country is so small that virtually anywhere can be easily reached in the course of a day, so if you wish to follow up a rare bird report, this is quite feasible from Amsterdam. The only exception would be the Friesian Islands which require a rather lengthy journey (over 100km) followed by a ferry crossing, but are still possible in a day.

Within the city there are several parks which support a selection of woodland species and are certainly worth a look, but the area around Amsterdam has some of the best wetland sites in Europe and it is for its breeding aquatic birds and its wintering wildfowl that The Netherlands is best known. The flat polders around the IJsselmeer provide breeding habitat for the typical avifauna of lowland wet meadows, a habitat which is fast disappearing elsewhere, and the marshes support some of northern Europe's rarest breeding herons, waders and wildfowl. In winter, these areas are inundated with wildfowl from eastern and northern Europe and provide some of the continent's top wildlife spectacles – Veel plezier!

GETTING AROUND

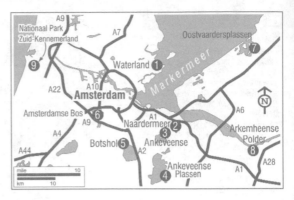

Amsterdam is a fairly small city, with narrow streets and canals seeming to criss-cross each other, making driving very confusing and often frustrating. The best advice is to forgo the use of a car in favour of renting a bicycle or using public transport within the city. The Netherlands being almost entirely flat, cycling is a popular method of getting about and the network of cycle paths makes it a safe, convenient way to get around, particularly in the summer. The railway system is quite comprehensive, especially south and west of Amsterdam, and most nature reserves are within a few kilometres of a railway station. The big advantage of a car is the extra flexibility it gives to visit several sites in a day and to cover the larger sites much more easily. Much of the birding in The Netherlands involves driving along roads that have been laid on the top of dykes and scanning the waterways and polders either side for raptors, herons and flocks of wildfowl. These roads are often narrow but busy, with very few stopping places, so drive carefully and try not to get too distracted by the birds! If you are going to rent a car it is better to pick it up at Schiphol Airport as the adjacent motorway makes it much easier to navigate to your destination. From some train stations in cities around Amsterdam you can use the Treintaxi (train taxi) to get to the reserves. This operates like a car pooling system in which you share the taxi with other travellers. Around Amsterdam, the Treintaxi is currently available at the stations of Alphen a/d Rijn, Hillegom, Hilversum, Naarden-Bussum and Weesp. In the Noord-Holland Province you can use the OV-taxi, which is a similar system.

As you will most likely be attracted to birding some of the wetland sites near Amsterdam, a telescope is desirable, particularly in winter, as the open, flat country-side doesn't permit a close approach to flocks of wildfowl. Most nature reserves have hides so that you can make do with binoculars. Also, since birding is such a popular pursuit in The Netherlands, you will probably run into fellow birders who can assist. It is a feature of many wetland nature reserves that the best access is by boat, and some-times there is a local boat service provided near the reserve. There will be restrictions on access during the breeding season so as not to disturb the birds. You will need a map that covers the western half of the country, including Noord-Holland, Zuid-Holland, Utrecht and Flevoland Provinces, in order to locate all the sites, and it would also be useful to get a city map of Amsterdam with a schematic diagram of the public transport system. These are available from any bookshop, tourist office (VVV), and bus and rail-way stations.

Locations

1 Waterland (14km)

GVB Bus No. 30

If you want to combine some sightseeing with birding, then the picturesque Waterland area just north of Amsterdam provides a perfect combination of canals, windmills and polders with access to the Markermeer, and it is also on the way to Edam. These low-land wet meadows provide breeding habitats for ducks, waders and passerines, and in winter the open waters of the Markermeer attract grebes, ducks and gulls.

From central Amsterdam cross the **North Sea Canal** by the **Schellingwouder Bridge**. Immediately turn right and follow the **Durgerdammerdijk**, which becomes the **Uitdammerdijk** as far as **Monnickendam**. From the dyke you can check the polders and Kinselmeer on the west side, and the Markermeer on the east side. Just south of Monnickendam the road runs past the Gouwzee which holds large numbers of wildfowl in winter. If you don't have your own transport (car or bicycle), there is a limited bus service (No. 30 only on weekdays) which operates as far as Holysloot on the Uitdammer Die and connects with buses Nos. 33 or 38 from Centraal Station.

KEY SPECIES

Resident
Great Bittern, Common Pheasant, Water Rail, Common Kingfisher, Bearded Tit, Penduline Tit, Tree Sparrow

Summer
Garganey, Northern Shoveler, Marsh Harrier, Oystercatcher, Northern

Lapwing, Ruff, Common Redshank, Common Snipe, Black-tailed Godwit, Common Cuckoo, Sky Lark, Meadow Pipit, Bluethroat, Sedge Warbler, Savi's Warbler, Marsh Warbler, Icterine Warbler

Winter
Whooper Swan, Greylag Goose,

Eurasian Wigeon, Greater Scaup, Goosander, Water Pipit, Fieldfare, Redwing, Brambling, Mealy Redpoll

Passage
Gadwall, Eurasian Teal, Red-crested Pochard, Osprey, Little Gull, Black Tern

2 Naardermeer (18km)   *Naarden Railway Station*

This lake is one of the closest wetland sites to Amsterdam, and also one of the best. However, it can be frustrating as the road does not run alongside the lakeshore and the only access is at the northeast corner where several small roads and footpaths run per-pendicular to the lakeshore. During the breeding season, you can take a boat trip on the lake to the Great Cormorant colony which provides excellent opportunities for close-up photography. This boat trip needs to be booked in advance, through Natuurmonumenten, the Dutch Nature Preservation Organisation (details on p. 205), but this is difficult to organise for a short-term visitor.

The lake, reedbeds and riparian woodland provide a mosaic of habitats that attract a wide range of aquatic birds and many woodland species. The area is best in late spring and summer when the breeding birds are on view but is worth a visit at any time of year.

The best way to get to Naardermeer is to take the **A1 Motorway** east from Amsterdam, exit south at **Exit 4 (Muiden)** and go east on **Zuidpolderweg** follow-ing the signs for **Naarden**. After about 2km turn south on **Googweg** and after passing over the motorway and the railway, turn east on **Meerkade Road**. Follow this road to a T-junction, where you can turn left or right to circle the lake. Although the road is often quite distant from the lakeshore, there are several tracks leading from the road down to the shore. Turn left at the T-junction and after about 1km, there is a hide erected on the northern shore near the De Machine House. Boat trips depart from here during the summer, and it is signposted from Meerkade Road.

An alternative is to hike through the wetlands directly south of the Naardermeer (Hilversumse Bovenmeent). In spring many waders and herons, including Eurasian Spoonbills and Great White Egrets, can be found here.

KEY SPECIES

Resident
Little Grebe, Great Cormorant, Great Bittern, Common Pochard, Common Pheasant, Water Rail, Northern Lapwing, Common Snipe, Bearded Tit, Rook, Tree Sparrow

Summer
Purple Heron, Northern Shoveler, Marsh Harrier, Black Tern, Common

Cuckoo, Bluethroat, Common Redstart, Savi's Warbler, Great Reed Warbler, Marsh Warbler, Sedge Warbler, Garden Warbler

Winter
Whooper Swan, Gadwall, Eurasian Wigeon, Eurasian Teal, Pintail, Common Goldeneye, Smew, Goosander, Hen Harrier, Common

Gull, Redwing, Goldcrest, Brambling, Siskin, Mealy Redpoll

Passage
Red-necked Grebe, Eurasian Spoonbill, Great White Egret, Common Crane, Osprey, Green Sandpiper, Little Gull

3 Ankeveense Plassen (22km)

Weesp Railway Station

Immediately south of Naardermeer is another large lake and marsh just west of Ankeveen. Unlike Naardermeer you don't need a boat to get to the best parts of the marsh and although the lake is larger than Naardermeer you need not walk around the whole circumference to see the birds. The area near the village of Ankeveen on the eastern shore is most productive. There is also a hide on the western shore (Kromme Googh).

Take the **A1** east from Amsterdam and then the **A9** south for about 2km before exiting east on the **N236**. After about 10km, turn south into **Ankeveen**. In the centre of the village there is a path – Dammerkade – on the west side of the road which goes down to the lakeshore, continues south to the hide, and then curves round to the southern shore. If using public transport you can get the train to Weesp which is about 7km away and from there cycle or take a train-taxi to Ankeveen.

KEY SPECIES

Resident
Little Grebe, Great Bittern, Common Pochard, Northern Goshawk, Common Pheasant, Water Rail, Northern Lapwing, Common Snipe, Bearded Tit, Tree Sparrow

Summer
Purple Heron, Northern Shoveler, Marsh Harrier, Spotted Crake, Black Tern, Common Cuckoo, Bluethroat, Common Redstart, Savi's Warbler, Great Reed Warbler, Marsh Warbler, Sedge Warbler, Garden Warbler

Winter
Whooper Swan, Gadwall, Eurasian Wigeon, Eurasian Teal, Common Goldeneye, Smew, Goosander, Hen Harrier, Common Gull, Fieldfare, Redwing, Brambling, Siskin, Mealy Redpoll

4 Tienhovense Plassen (32km)

Weesp Railway Station

There are a number of large wetlands along the course of the River Vecht between Amsterdam and Utrecht and the Tienhovense Plassen Reserve occupies the southeast corner of a large area of lakes and marshes called the Loosdrechtse Plassen. A road runs around the perimeter so you do not need a boat to access the reserve unless you want to explore the whole lake system. All of the typical marsh species can be seen here.

Take the **A2** south from Amsterdam and exit east at **Breukelen (Exit 5)**. In Breukelen, cross the River Vecht and

KEY SPECIES

Resident
Great Bittern, Common Pheasant, Water Rail, Common Kingfisher, Bearded Tit, Tree Sparrow

Summer
White Stork, Purple Heron, Great White Egret, Marsh Harrier, Hobby, Spotted Crake, Northern Lapwing, Ruff, Common Redshank, Common Snipe, Black-tailed Godwit, Black Tern, Common Cuckoo, Sky Lark,

Meadow Pipit, Bluethroat, Grasshopper Warbler, Sedge Warbler, Savi's Warbler, Marsh Warbler, Great Reed Warbler

Winter
Eurasian Teal, Eurasian Wigeon, Pintail, Common Pochard, Common Goldeneye, Smew, Goosander, Water Pipit, Fieldfare, Redwing, Brambling, Mealy Redpoll, Siskin

turn south on **Zandweg** and after about 2km turn east on **Nieuweweg** which runs alongside the southern shore of Loosdrechtse Plassen as far as **Tienhoven**. Just south and east of Tienhoven, the **Dwarsdijk** road runs along the perimeter of the reserve affording good views along the road. There is a train service to Hilversum, about 7km away and you could get a train-taxi to the reserve or rent a bike.

5 Botshol Nature Reserve (17km)   *Abcoude Railway Station*

Botshol is an area of marsh-land and watermeadows, to the south of Amsterdam, on the western shore of the Vinkeveense Plassen lake. There are many such areas in Utrecht Province, but Botshol is special because it is one of the few places in Western Europe where Red-crested Pochards breed. It also pro-vides habitat for breeding waders, which require low-

land wet meadows, and for a range of other marsh birds. There are colonies of Cormorants and Black-headed Gulls. The best time to see all these is in spring and summer but there are restrictions on access to certain areas, from 1 April to 16 June, in order to prevent disturbance of breeding birds. If water levels are high, you might need rubber boots.

Take the **A2** south from Amsterdam and exit at **Abcoude** (**Exit 3**). Go west on the **Winkeldijk Road** to **Gemaal**. Turn south at Gemaal on **Hoofdweg** and go about 1km until you come to a crossroads. Turn left (west) at the crossroads on **Botsholsedijk** and continue for about 2km until you come to the reserve entrance. For those travelling by public transport, there is a train station at Abcoude, where you can rent a bike.

KEY SPECIES

Resident
Great Cormorant, Great Bittern, Red-crested Pochard, Common Pheasant, Water Rail, Black-headed Gull, Common Kingfisher, Bearded Tit, Penduline Tit, Tree Sparrow

Summer
Eurasian Spoonbill, Purple Heron,

Marsh Harrier, Hobby, Spotted Crake, Northern Lapwing, Ruff, Common Redshank, Common Snipe, Black-tailed Godwit, Mediterranean Gull, Black Tern, Common Cuckoo, Sky Lark, Meadow Pipit, Bluethroat, Grasshopper Warbler, Sedge Warbler, Savi's Warbler, Marsh Warbler, Great

Reed Warbler, Icterine Warbler

Winter
Whooper Swan, Greylag Goose, Gadwall, Eurasian Wigeon, Northern Shoveler

6 Amsterdamse Bos (7km)   *Amstelveen Metro Station/ Bus No. 170*

This is a very large park which has extensive mature woodland and a large lake in the centre, and is located just south of the city, right beside Schiphol Airport. Although very popular with strollers and joggers, there are quieter areas towards the south end which provide habitats for woodland species in what is a very urban area. If you find yourself stuck in an airport hotel, then this is the place to go first thing in the morning, particularly in late spring and early summer. In winter, the park lake at the south end, Amstelveense Poel, holds a selection of wintering waterfowl.

The park is located east of Schiphol Airport adjacent to **Exit 6** on the **A9 Motorway**. The car park is on the east side of the motorway. The area is also well-served by Metro, bus and tram.

KEY SPECIES

Resident	Summer	Winter
Common Pheasant, Water Rail, Stock Dove, Green Woodpecker, Lesser Spotted Woodpecker, Mistle Thrush, Goldcrest, Long-tailed Tit, Marsh Tit, Willow Tit, Short-toed Treecreeper	Common Nightingale, Lesser Whitethroat, Garden Warbler, Wood Warbler, Pied Flycatcher, Golden Oriole	Eurasian Wigeon, Common Pochard, Common Goldeneye, Common Snipe, Common Gull, Great Black-backed Gull, Fieldfare, Redwing, Brambling, Siskin, Mealy Redpoll

---

7 Oostvaardersplassen (56km)   *Lelystad Railway Station*

The province of **Flevoland** lies east of Amsterdam and is a huge area of polders which were reclaimed from the **IJsselmeer** in the 1950s and 1960s. Most areas have been fully reclaimed and are now turned over to agriculture, but some areas on the western flanks of the reclaimed area did not dry out. These have become legendary sites for birds and although the vegetation is in succession, various species have moved in to exploit this new habitat. The reserve at Oostvaardersplassen is on the western flank of

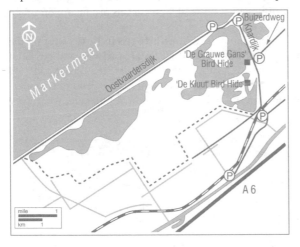

Flevoland and is very accessible from Amsterdam. This is probably the best site in the Amsterdam area and is well worth a visit at any time of the year. The numbers and variety of birds are amazing and in the case of several aquatic species, you would have to travel as far as Spain or Eastern Europe to find similar populations. Additionally, northeast of the reserve, there are some very good tracts of woodland, which support several scarce woodland species,

such as Northern Goshawk and Black Woodpecker, making this whole area a 'must-see' for any birder visiting Amsterdam. This is a site which attracts a lot of rarities: amongst the rarer, in recent times, being Greater Flamingo, Black Vulture and Pygmy Cormorant, so keep an eye on the Dutch Birding website, www.dutchbirding.nl, for the latest news.

Take the **A1** east from Amsterdam and turn north on the **A6**, going as far as **Lelystad**. Exit east on the **N302** for Lelystad and take the third turn to the left – **Buizerdweg**. This road runs through some very productive woodland and will bring you to the reserve. There are trails through the marsh and hides near the reserve HQ but, if you have a car, the best way to 'do' Oostvaardersplassen is to drive slowly along the perimeter roads: **Knardijk** which runs perpendicular to the **Markermeer** and allows you to scan the pools either side of the road, and the 20km **Oostvaardersdijk** which runs along the top of the dyke separating the polder from the Markermeer and provides height to overlook the open expanse of water. However, there are very few lay-bys so you may need to park the car and walk a bit. Also, the road along the southeast side of the reserve, the **Praamweg**, offers two of the best places to stop and have a view over the reserve: known as the **Grote** and **Kleine Praambult**. There are several hides in the area from which you have an excellent view over the wetlands and sometimes close encounters with species like Great White and Little Egrets.

KEY SPECIES

Resident
Little Grebe, Great Bittern, Greylag Goose, Eurasian Teal, Northern Shoveler, Northern Goshawk, Hen Harrier, Common Kestrel, Water Rail, Black Woodpecker, Sky Lark, Bearded Tit, Crested Tit, Penduline Tit, Tree Sparrow

Summer
Eurasian Spoonbill, Purple Heron, Great White Egret, Little Egret, Northern Shoveler, Garganey, Marsh Harrier, Montagu's Harrier, Hobby, Spotted Crake, Avocet, Kentish

Plover, Little Ringed Plover, Common Cuckoo, Bluethroat, Grasshopper Warbler, Savi's Warbler, Sedge Warbler, Marsh Warbler, Great Reed Warbler, Icterine Warbler, Golden Oriole

Winter
Tundra Swan, Whooper Swan, Bean Goose, White-fronted Goose, Barnacle Goose, Eurasian Wigeon, Common Pochard, Greater Scaup, Common Goldeneye, Smew, Goosander, Rough-legged Buzzard, White-tailed Eagle, Peregrine Falcon,

Water Pipit, Fieldfare, Redwing, Brambling, Twite

Passage
Gadwall, Common Crane, European Honey-buzzard, Osprey, European Golden Plover, Grey Plover, Red Knot, Little Stint, Dunlin, Curlew Sandpiper, Ruff, Black-tailed Godwit, Spotted Redshank, Common Sandpiper, Green Sandpiper, Wood Sandpiper, Little Gull, Black Tern, Tawny Pipit, Northern Wheatear, Whinchat

8 Arkemheense Polder (45km)

This area of polders straddles the border between Utrecht Province and Gelderland Province. It is important during the breeding season for its community of lowland wet meadow species and in winter it supports large flocks of geese and swans, being a very reliable site for Tundra Swans. It is quite a large area and the birds can move around over the course

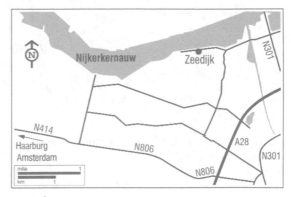

of the winter, so the best way to survey the area is to cruise the roads, scanning for the flocks wherever you can find a good vantage point.

Take the **A1** east from Amsterdam and exit north at **Exit 12** on the **N199**. Turn east at **Haarburg** on the **N806** and drive in a rectangle east as far as the A28 Motorway and north as far as the Nijkerkernauw Lake which separates Gelderland from Flevoland, then west along the lake as far as the drainage canal at **Zeedijk**. In early summer, this area along the lakeshore is good for displaying Ruff, although actual breeding numbers have declined in recent years. In late summer the lake acts as a moulting ground for almost all of Western Europe's feral Ruddy Shelducks. There is a train service to Baarn, which is about 10km from the polders, but without some transport (car or bicycle) it is not practical to reach here on foot.

KEY SPECIES

Resident
Great Bittern, Common Pheasant, Water Rail, Common Kingfisher, Bearded Tit, Tree Sparrow

Summer
Ruddy Shelduck, Garganey, Northern Shoveler, Marsh Harrier, Hobby, Oystercatcher, Northern Lapwing,

Ruff, Common Redshank, Common Snipe, Black-tailed Godwit, Common Cuckoo, Turtle Dove, Sky Lark, Meadow Pipit, Grey Wagtail, Bluethroat, Sedge Warbler

Winter
Whooper Swan, Tundra Swan, Bean Goose, Greylag Goose, Eurasian

Wigeon, Water Pipit, Fieldfare, Redwing, Brambling, Mealy Redpoll, Siskin

Passage
Common Crane, European Golden Plover, Green Sandpiper, Common Sandpiper, Little Gull, Black Tern

9 Kennemerland (25km)

Santpoort-Noord Railway Station
Connexxion Bus Nos. 70/75/82 (no winter service)

The North Sea coast is just 25km west of Amsterdam and the stretch just south of IJmuiden combines a number of different habitats, all of which offer good birding opportunities. Many rarities have been found along this stretch of coast but the local birds are just as interesting for the visitor. The largest area comprises the Zuid-Kennemerland National Park, an extensive sand-dune system with small lakes, emergent tree growth and mature forest. In spring and summer it is possible to see all of the typical woodland birds including the more difficult species such as Woodcocks,

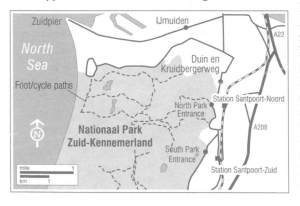

Black Woodpeckers and both Marsh and Willow Tits. At migration times the shoreline near the south pier (Zuidpier) of IJmuiden Harbour and, if water levels are low, the margins of the small lakes are good for waders. The coast offers opportunities to observe all of the typical seabirds of the North Sea as well as divers and sea-ducks in winter. Further north along the shore, the Zuidpier acts as a migrant trap for passerines in spring and autumn and can also be a good seawatch point during onshore winds. IJmuiden Harbour itself attracts large numbers of gulls.

To get to IJmuiden, take the **N202** west from Amsterdam, which basically runs alongside the **North Sea Canal** to Ijmuiden. Drive past the harbour area to get to the Zuidpier. The best way to access the Zuid-Kennemerland forest is to take the **A208** Motorway south from IJmuiden and exit west at **Santpoort-Noord**: the park entrance is on **Duin en Kruidbergerweg** at the west end of the town. It is also quite close to Schiphol Airport by motorway and is worth a look if you have a few hours to spare.

The area is very extensive and is difficult to cover without a bicycle. However, there is a train service to Santpoort-Noord and from there you could visit the national park and later take a bus to IJmuiden.

KEY SPECIES

Resident
Northern Goshawk, Grey Partridge, Water Rail, Oystercatcher, Northern Lapwing, Common Snipe, Woodcock, Common Redshank, Great Black-backed Gull, Common Kingfisher, Green Woodpecker, Black Woodpecker, Lesser Spotted Woodpecker, Sky Lark, Wood Lark, Meadow Pipit, Goldcrest, Bearded Tit, Long-tailed Tit, Marsh Tit, Willow Tit, Crested Tit, Short-toed Treecreeper

Summer
Common Shelduck, Northern Shoveler, Garganey, Marsh Harrier, Hobby, Avocet, Little Ringed Plover, Black-tailed Godwit, Eurasian Curlew,

Lesser Black-backed Gull, Turtle Dove, Common Cuckoo, Wryneck, Tree Pipit, Whinchat, Common Stonechat, Common Nightingale, Bluethroat, Northern Wheatear, Grasshopper Warbler, Savi's Warbler, Marsh Warbler, Icterine Warbler, Garden Warbler, Lesser Whitethroat, Firecrest

Winter
Red-throated Diver, Black-throated Diver, Gadwall, Eurasian Wigeon, Eurasian Teal, Common Goldeneye, Greater Scaup, Common Eider, Common Scoter, Velvet Scoter, Red-breasted Merganser, Smew, Turnstone, Purple Sandpiper,

Glaucous Gull, Common Gull, Mediterranean Gull, Guillemot, Rock Pipit, Fieldfare, Redwing, Twite, Snow Bunting

Passage
Fulmar, Northern Gannet, Brent Goose, Barnacle Goose, Pintail, Common Crane, Grey Plover, European Golden Plover, Red Knot, Sanderling, Dunlin, Bar-tailed Godwit, Whimbrel, Greenshank, Red-necked Phalarope, Grey Phalarope, Great Skua, Arctic Skua, Little Gull, Kittiwake, Sandwich Tern, Arctic Tern, Black Tern

Useful contacts and websites

Vogelbescherming – Dutch Bird Protection Society (English and Dutch)
www.vogelbescherming.nl
Vogelbescherming, PO Box 925, NL-3700 AX Zeist. Tel. (+ 31) 30 6937777.
The BirdLife International partner in Holland and publishers of the journal *Vogels*.

Nederlandse Ornithologische Unie (English and Dutch)
http://home.planet.nl/~boude112/main.htm
Netherlands Ornithologist's Union – publishers of the journals *Limosa* and *Ardea*.

Dutch Birding Association (English and Dutch)
www.dutchbirding.nl
Publishers of the journal *Dutch Birding*.

SOVON – Dutch Centre for Field Ornithology (English and Dutch) www.sovon.nl
Rijksstraatweg 178, 6573 DG, Beek-Ubbergen Tel. 024 6848111. Fax. 024 6848122.

Natuurbeleven (English and Dutch)
www.natuurbeleven.nl
NatuurBeleven, Oostermeerkade 6, 1184 TV Amstelveen. Tel. (+31) 20 4720777.
Guided birding tours for visitors to Amsterdam. Email: mark@natuurbeleven.nl

Natuurmonumenten – Dutch Nature Reserves (Dutch) www.natuurmonumenten.nl
Natuurmonumenten, Schaepp en Burgh, NL-1243 JJ 's-Graveland. Tel. 035 6559933. Fax. 035 6563174.
Access, boat-hire, maps and descriptions of every Dutch nature reserve.

Stichting Vogel- en Natuurwacht 'Zuid-Flevoland' (Dutch) www.degrauwegans.nl
South Flevoland Bird Study Group – surveys, excursions, bird-ringing activities.

Vogelwerkgroep Zuid-Kennemerland (Dutch) www.vwgzkl.nl

Maps, checklist and site details for IJmuiden and Zuid-Kennemerland National Park.

Dutch Birding Birdline
For information: 0900-BIRDING: Tel. 0900 2473464
To report sightings: Tel. 010 4281212.

GVB (English and Dutch) Public Transport in the Amsterdam Area: www.gvb.nl

Nederlandse Spoorwegen (English and Dutch) Dutch Railways www.ns.nl

Connexxion (Dutch) – Regional Bus Service outside Amsterdam www.connexxion.nl

Bever Zwerfsport, Stadhouderskade 4, 1054 ES Amsterdam www.bever.nl
Shop for outdoor equipment: Tel. 020 6894639.

Books and publications

Atlas van de Nederlandse Broedvogels 1998–2000 – Dutch Breeding Bird Atlas (2002) SOVON Vogelonderzoek Nederland ISBN 9050111610

Avifauna van Nederland 2 – common and scarce birds of the Netherlands R.G. Bijlsma, F. Hustings & C.J. Camphuysen (2001) GMB Uitgeverij/KNNV Uitgeverij, ISBN 9074345212

Rare Birds of the Netherlands
A. B. van den Berg & C. A. W. Bosman (1999) Pica Press, ISBN 1873403887

Where to Watch Birds in Holland, Belgium & Northern France
Arnoud van den Berg & Dominique Lafontaine (1996) Hamlyn, ISBN 0600579778

Collins Bird Guide
Lars Svensson *et al.* (1999) HarperCollins ISBN 0002197286

Birds of Europe
Lars Jonsson (2006) Christopher Helm ISBN 0713676000

The harsh calls of Sardinian Warbler can be heard above the crowds in Parc Guel

TYPICAL SPECIES

Resident
Grey Heron, Mallard, Common Buzzard, Common Kestrel, Peregrine Falcon, Moorhen, Common Coot, Yellow-legged Gull, Wood Pigeon, Collared Dove, Monk Parakeet (feral), Green Woodpecker, Crested Lark, White Wagtail, Wren, Black Redstart, Robin, Common Stonechat, Blackbird, Sardinian Warbler, Blackcap, Common Chiffchaff, Firecrest, Blue Tit, Great Tit, Short-toed Treecreeper, Eurasian Jay, Magpie, Carrion Crow, Common Starling, House Sparrow, Tree Sparrow, Greenfinch, Goldfinch, Serin

Summer
Black Kite, Hobby, Turtle Dove, Hoopoe, European Bee-eater, Common Swift, Barn Swallow, House Martin, Spotted Flycatcher, Golden Oriole

Winter
Great Cormorant, Cattle Egret, Black-headed Gull, Sky Lark, Meadow Pipit, Grey Wagtail, Dunnock, Song Thrush, Goldcrest, Chaffinch

Barcelona

Barcelona, the capital of Catalonia autonomous region, is in the northeast corner of Spain on the shores of the Mediterranean Sea. Its location gives the city an avifauna which is an interesting mix of northern European and Mediterranean birds. Rather like the Alps further east, the Pyrenees act as a dividing line between the Mediterranean and Continental European zones, and many species which are common in the Iberian Peninsula are rather uncommon north of the mountains. However, the littoral area between Barcelona and the French Riviera acts as a conduit for many birds, producing a wealth of wintering wildfowl in the wetlands of Catalonia as well as enabling extralimital vagrants to be recorded regularly. Spring is often the best time to observe the phenomenon of migration, as the prevailing winds at that time of year are from the north and can result in a build-up of many migrants in the Barcelona area, particularly in the marshes at the Delta del Llobregat.

For resident birds and summer visitors, the area around the city offers a variety of habitats. The city is virtually surrounded by a 'green ring' of low hills which are forested with holm oak and Aleppo pine, as well as having more open patches of stunted shrub growth – the typical Mediterranean garrigue vegetation. Further west, the foothills of the Pyrenees offer nesting sites for mountain and cliff-dwelling species.

Some of the best birding sites in northeast Spain are located about two hours drive from Barcelona but, as such, are beyond the scope of this book. The visiting birder with the use of a car could consider driving the extra distance involved to reach the steppes of western Catalonia near Lleida, which support a rich community of steppe birds, or the high peaks of the Pyrenees to the northwest. If your visit to Barcelona is en route to the Costa Dorada, south of the city, you have the opportunity to visit the Ebro Delta, one of western Europe's best wetlands. If you are going north to the Costa Brava you could visit the Aiguamolls de l'Empordà, another superb wetland site. However, even if you are restricted to the city and its immediate area, Barcelona has much to offer at any time of year and within easy reach of the centre.

GETTING AROUND

If your visit is short, then a car is unnecessary to visit the best birding sites near the city. You can use public transport to reach these locations with ease. Even the surrounding hills are served by cable cars and funicular railways so you don't have to exhaust yourself climbing up to the best habitats and, after a gruelling day's birding, you have an easy ride back to

town. All of the trains and metro lines converge on the main stations at Plaça de Catalunya or Sants, so if you have a choice, base yourself at a hotel near either station. For some of the outlying areas, public transport is a little more problematic but, provided you plan your route, there is a good suburban train service which can bring you out of town quite speedily. Outside peak periods you are allowed to bring a bicycle on the trains, so you could consider renting a bike to go birding, but this would be inadvisable for the mountain areas, unless you have the fitness of a mountain cyclist!

If you do have a car, you will have much more flexibility to visit several sites in the course of a day or to consider going further afield. However, bear in mind that the coast can get very congested during the summer, particularly on Sundays, as crowds of holidaymakers converge on the beaches. The roads are also congested with commuter traffic at peak periods on weekdays. In order to navigate through the city, make use of the two orbital 'Rondas', signposted 'green' anywhere in the city: the Ronda del Litoral (B10), which flanks the centre along the coast and the Ronda de Dalt (B20), which runs along the west side of the city through the foothills of Collserola Park. All of the main routes to the outlying areas fan out from this motorway.

Hiking in the surrounding hills is well catered for with designated trails, small kiosks and picnic sites but requires strong footwear. Remember to bring adequate water in summer.

Barcelona is very much a late-night city, so if you need to stock up with birding supplies, maps or books don't waste valuable birding time shopping during the day, as everywhere will be open until at least 20.00. A map including the whole of Barcelona Province is sufficient to cover the areas described. However a number of main roads in the area have been renumbered recently so, if purchasing a map, check that the road numbering is not the old system (in Catalan: antic or antiga).

Locations

---

1 Montjuïc Hill and Barcelona Port (City Centre)

(Parc de Montjuïc Funicular Station)

Dominating the southern part of the city centre and overlooking the port, Montjuïc Hill is a large, very popular public park which contains the Olympic Stadium, and many other public amenities. The park is extensively wooded and contains several botanic gardens, mainly along the southern flank of the hill. Needless to say it is very popular with tourists and local people, but the main problem is noise from the traffic through the park which can make it difficult to listen for bird calls. However, an early morning visit, away from the main thoroughfares when the area is quieter, can yield all of the typical birds of park and woodland. Typical Mediterranean species are to be found here: Sardinian Warblers, Short-toed Treecreepers and Serins for example, while the cliffs facing the sea, near the castle, hold breeding Yellow-legged Gulls, Common Kestrels and the occasional Peregrine Falcon. Flocks of Common Swifts should be checked for the occasional Alpine Swift.

In addition, the harbour which is at the foot of Montjuïc Hill, is worth a visit to check for gulls, cormorants and other seabirds. You can take a cable-car from the hilltop to the outer pier of the harbour. Although, generally-speaking, the Mediterranean Sea is rather poor for seabirds, the outer harbour is a good vantage point from which to seawatch when the winds are onshore, and it is possible to see all of the typical marine species.

KEY SPECIES

Barcelona Port

Winter
Great Crested Grebe, Northern Gannet, Shag, Mediterranean Gull, Lesser Black-backed Gull, Sandwich Tern, Razorbill

Passage
Cory's Shearwater, Balearic Shearwater, Common Scoter, Arctic Skua, Little Gull, Audouin's Gull, Black Tern

2 Delta del Llobregat (17km)

(El Prat del Llobregat Railway Station R1/R2)/(Bus L95)

This site can be the first stop for any visitor to Barcelona, as it is literally at the end of the runway of Barcelona Airport. Where the Llobregat River enters the Mediterranean just south of Barcelona, there were once extensive marshes and low-lying fields, prone to flooding, some of which still remain. There are now two main lagoons either side of the airport inter-

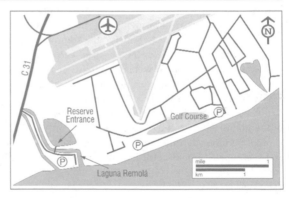

spersed with reedbeds, pinewoods, cultivation and scrub. The overall effect is a mosaic of different habitats which makes this area so attractive to a large variety of birds: the location list stands at 360 species. Much of this site is contained within the Reserves Naturals Delta del Llobregat. It is well worth a visit at any time of the year but is particularly good in winter, when large numbers of wildfowl are present, and during spring migration, when many migrants pass through on their way north. It is an important breeding site for some aquatic species and there is a Kentish Plover colony on the beach, which is off-limits during the breeding season.

This is one of Spain's rarity hotspots and is well-watched by local birders, frequently turning up rarities during migration (check www.rarebirdspain.net). However, most of the passerine rarities tend to be found through ringing activities and for the visiting birder the main focus will be on the abundant indigenous birdlife. There are two main areas of the reserve which should be looked at: the entrance to one lies northeast of the airport entrance and the other to the southwest. The northeastern area is almost permanently 'under construction' and large infrastructure works have been continuously carried out for the last five years, particularly around the airport. As a consequence, getting lost is very easy and prior contact is advisable. The southernmost area, the Remolar-Filipines Reserve, is, on the other hand, a well established and much recommended visit for the newcomer.

Take the **C-31** road southwest from the city centre towards the **airport**. When the road crosses the Llobregat River turn left (south) through the town of **Prat del Llobregat** on the **B250** and follow this road around the perimeter of the airport towards the beach. You can walk east along the beach into the reserve which brings you to the first lagoon (Laguna de la Ricarda). If you go west along the beach, the road takes you past the golf course to a viewing point in front of the Kentish Plover breeding area. This area is at its best in winter when the low-lying fields flood and attract many duck and egrets. However, if you are relying on public transport, getting to the reserve will involve a walk of about 4km from the railway station (except in summer when there is a bus service from the station to the beach). Moreover, this area is difficult to work, except for the beach for the Kentish Plovers, so it is advisable to go first to the Remolar-Filipines reserve.

The Remolar lagoon is reached by staying on the C-31 and driving past the airport. If coming from Barcelona, take the Gavà mar exit, turn back to Barcelona (north) and at km 188 there is a sign for RN Delta del Llobregat, pointing to a right-hand turn, which brings you to the coast. The reserve entrance is at the end of this road where there is a car park. From there, a bridge over the canal leads to a set of paths and two hides which overlook the lagoon and the marsh (Maresma de les Filipines). If travelling by public transport, this area is easier to cover, as you can get a bus to the turn-off on the C-31 which is within about 1km from the reserve entrance (Bus No. L-95 from Barcelona to Castelldefels).

KEY SPECIES

Resident
Little Grebe, Night Heron, Little Egret, Cattle Egret, Gadwall, Red-crested Pochard, Marsh Harrier, Kentish Plover, Purple Gallinule, Red-knobbed Coot (I), Water Rail, Sandwich Tern, Little Owl, Cetti's Warbler, Zitting Cisticola, Penduline Tit, Spotless Starling

Summer
Little Bittern, Purple Heron, Squacco Heron, Little Ringed Plover, Black-winged Stilt, Audouin's Gull,

Whiskered Tern, Gull-billed Tern, Great Spotted Cuckoo, Scops Owl, Sand Martin, Spanish Wagtail, Ashy-headed Wagtail, Savi's Warbler, Reed Warbler, Great Reed Warbler

Winter
Great Crested Grebe, Northern Gannet, Greater Flamingo, Gadwall, Eurasian Wigeon, Eurasian Teal, Northern Shoveler, Common Pochard, Tufted Duck, Northern Lapwing, European Golden Plover, Common Snipe, Jack Snipe,

Mediterranean Gull, Lesser Black-backed Gull, Common Kingfisher, Water Pipit, Moustached Warbler, Reed Bunting, Cirl Bunting

Passage
Balearic Shearwater, Glossy Ibis, Eurasian Spoonbill, Garganey, Osprey, Common Crane, Dunlin, Curlew Sandpiper, Sanderling, Ruff, Wood Sandpiper, Common Redshank, Spotted Redshank, Greenshank, Little Gull, Black Tern, Bluethroat, Sedge Warbler, Woodchat Shrike

3 Collserola Park (10km)   *(La Floresta Railway Station S1/S2/S5)*

Collserola Park is a large wooded area covering the hills just west of Barcelona. Although the area is rapidly becoming encircled by the growing conurbation of the city and its suburbs, there are a variety of habitats with pine forest, mixed woodland, garrigue scrub and even cultivated plots in the Llobregat Valley. However, apart from a few mountain streams, there is no large body of open water to provide habitats for aquatic birds. As a public park it is very popular but at least its elevation and steep slopes limit its appeal to cyclists and roller skaters so there is less disturbance from that quarter. At any rate, the park is so big that it is possible to find quiet areas well away from the crowds.

There are several possible routes through the park depending on how far you wish to hike. A suggestion would be to take the train to La Floresta and walk the road west to the church of St Bartomeu, a distance of about 2km. This route will take you through some riparian and pine woodland. You can either return to La Floresta or else, if you feel like hiking further, turn south at the church and walk for another 10km or so through some mixed woodland and open areas of garrigue to get to the park entrance at Peu de Funicular, where you can get trains back to the city centre.

KEY SPECIES

Resident	Summer	Passage
Eurasian Sparrowhawk, Northern Goshawk, Common Pheasant, Stock Dove, Mistle Thrush, Dartford Warbler, Firecrest, Long-tailed Tit, Coal Tit, Crested Tit, Southern Grey Shrike, Cirl Bunting (local)	Common Cuckoo, Wryneck, Common Nightingale, Subalpine Warbler, Melodious Warbler, Western Bonelli's Warbler, Woodchat Shrike (local)	European Honey-buzzard, Short-toed Eagle and other raptors
	Winter Siskin, Hawfinch (local)	

4 Garraf Hills (35km)   *(Garraf Railway Station R2)*

These are low limestone hills located on the coast just southwest of Barcelona. Much of the underlying limestone rock is exposed due to erosion. The vegetation consists of low bushes and stunted trees, the typical Mediterranean biome known as maquis or garrigue, supporting its characteristic avifauna.

Take the **A16 Motorway** southwest as far as **Garraf** and first check the harbour at Garraf for gulls and terns. In onshore winds you can seawatch from the harbour tip. Drive back towards Barcelona on the **C31** and turn north at **Port Ginesta**. This road ascends into the hills and runs along the southern slopes, towards **Vallgrassa**. There are several lay-bys and picnic sites along this road, all of which provide vantage points to scan the hillside.

You really need a car to make this visit as it would be a tough hike or cycle up the hillside. However, there is a train service from Barcelona if you just plan to visit the harbour at Garraf.

KEY SPECIES

Resident		
Bonelli's Eagle, Peregrine Falcon, Red-legged Partridge, Wryneck, Thekla Lark, Crag Martin, Blue Rock Thrush, Dartford Warbler, Crested Tit, Common Raven, Southern Grey	Shrike, Rock Sparrow, Cirl Bunting, Corn Bunting **Summer** Shag, Audouin's Gull, Yellow-legged Gull, Turtle Dove, Pallid Swift, Alpine	Swift, European Nightjar, Black Wheatear, Black-eared Wheatear, Rock Thrush, Subalpine Warbler, Bonelli's Warbler, Woodchat Shrike

5 Montserrat Mountains (50km)   *(Monistrol Railway Station R5)*

The Montserrat Mountains west of Barcelona offer the highest point (1,200m) within a 50km radius of the city, and support a varied avifauna typical of the Mediterranean coastal mountains, particularly those which require cliff faces and rocky precipices. The Monastery of Montserrat is a famous tourist and pilgrim site so there is ready access to the mountain at this point. From here there is a network of mountain trails, so this is a relatively easy way to get to the habitat of a number of montane species.

Take the **N11** northwest from the southwestern outskirts of Barcelona for about 27km to the town of **Abrera**. From Abrera turn north on the **C1411** towards **Monistrol de Montserrat**. Turn left (west) in Monistrol and take the **BP-1211** mountain road up to the monastery. This site is easy to get to by public transport as you can get a train to Monistrol and, from there, there is a funicular railway and a cable car up to the monastery. If possible try to avoid weekends in summer, as it can get very crowded.

KEY SPECIES

Resident	Summer	Winter
Bonelli's Eagle, Red-legged Partridge, Wryneck, Crag Martin, Blue Rock Thrush, Dartford Warbler, Crested Tit, Common Raven, Rock Sparrow (rare), Cirl Bunting, Rock Bunting	Turtle Dove, Common Cuckoo, Alpine Swift, European Nightjar, Rock Thrush, Subalpine Warbler, Western Bonelli's Warbler, Woodchat Shrike	Alpine Accentor, Wallcreeper (occasional)

6 El Corredor Nature Park (45km)   *(Sant Celoni Railway Station R2)*

A line of low wooded hills stretches northeastwards along the Costa Brava from a point just south of Barcelona to the Tordera River. A large area of these hillsides has been designated as a public amenity or otherwise protected from urban encroachment. Taken with the adjoining Montnegre Nature Park, El Corredor Nature Park represents a continuous stretch of forested slopes with a rich fauna. The geographical location of these woodlands at the interface of the Mediterranean and more Continental-type woodland communities means that there is an interesting mix of bird species, with some of the northern species at the extreme southern limit of their range. The north/south orientation of these hills, parallel to the coast, makes them good observation points for migrant raptors and storks.

Take the **A7 Autopista** northeast from Barcelona and after about 40km exit south at **Sant Celoni (Exit 11)** onto the B511 to **Vallgorguina**. The park entrance is about 3km on the right-hand (west) side of this road, and leads to the Visitor Centre from where there are a number of forest trails.

KEY SPECIES

Resident		
Eurasian Sparrowhawk, Northern Goshawk, Stock Dove, Woodcock, Grey Wagtail, Mistle Thrush, Dartford Warbler, Firecrest, Long-tailed Tit,	Coal Tit, Crested Tit, Cirl Bunting **Summer** European Honey-buzzard, Short-toed Eagle, Common Cuckoo, Scops Owl,	European Nightjar, Wryneck, Common Nightingale, Subalpine Warbler, Melodious Warbler, Woodchat Shrike

7 Estany de Sils (75km)   *(Sils Railway Station)*

This lake is one of the few substantial bodies of fresh water in the Barcelona area. It is rather distant from the city but quite near to Girona Airport so, if this is your point of arrival or departure, the lake and surrounding marshland are worth a look. The wetland has recently been restored for wildlife, and the water levels are controlled to maintain permanent water in the main lagoon. Another temporary lagoon has exposed muddy margins in the autumn, to provide the optimal habitat for migrating waders. The lakes are surrounded by marshland and riparian woodland and there is a set of footpaths which provide access to the lakeshores.

The site is very easy to get to. Take the **A7 Autopista** towards Girona and after about 70km exit east at **Sils (Exit 9)** onto the **N11** and follow the signs for Sils. The lake is located at the southwest end of the village. This is actually the best place in the area for Lesser Spotted Woodpeckers.

KEY SPECIES

Resident
Little Egret, Water Rail, Common Kingfisher, Little Owl, Lesser Spotted Woodpecker, Cetti's Warbler, Zitting Cisticola, Long-tailed Tit, Cirl Bunting

Summer
Hobby, Little Ringed Plover, Black-winged Stilt, Common Cuckoo, Sand Martin, Spanish Wagtail, Great Reed Warbler, Reed Warbler, Woodchat Shrike

Winter
Little Grebe, Eurasian Teal, Eurasian Wigeon, Tufted Duck, Common Pochard, Northern Lapwing, Common Sandpiper, Common Snipe, Lesser Black-backed Gull, Short-eared Owl, Water Pipit, Reed Bunting

Passage
Squacco Heron, Night Heron, Purple Heron, Great White Egret, Black Stork, White Stork, Garganey, Northern Shoveler, Ruff, Common Redshank, Spotted Redshank, Greenshank, Green Sandpiper, Wood Sandpiper, Whinchat, Bluethroat, Northern Wheatear, Pied Flycatcher

Further Away

8 **Aiguamolls de l'Empordà** (134km)

This site is rather far from Barcelona and would not normally be considered for inclusion in a city guide such as this, but is so good that it is worth making the extra effort to visit. It is one of the best birding sites on the east coast of Spain and if your visit to Barcelona takes you to the tourist resorts of the Costa Brava, then this site is actually very close by. It is

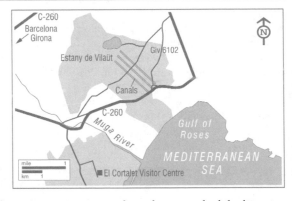

also less than an hour's drive from Girona Airport, making this a worthwhile diversion if that is your point of arrival or departure.

The site is a large expanse of coastal marshland, but the area between the Rivers Fluvia and Muga hold the greatest numbers and variety of birds. This is also where the Visitor Centre and several hides are located. The system of nature trails and hides provided brings you through several different wetland habitats: freshwater and saltwater lagoons, reedbeds, ricefields, riparian scrub and lowland wet meadows: habitats that are under severe threat in western Europe. The main trail provides access to the sand dunes and the beach and then loops back to the Visitor Centre. As a result of the protection afforded by the reserve several species have recolonised the area and it is by far the best place in Spain to see Little Crakes on migration. Both White Storks and Purple Gallinules have been reintroduced.

Take the **A7 Autopista** north to **Figueras** and exit east on the **C260** towards the holiday resort of **Roses**. Turn south after about 9km at Castelló d'Empúries towards **Sant Pere Pescador**. About 3km along this road there is a turn-off on the left-hand (east) side to the Visitor Centre at El Cortalet. Several paths depart from here and the Cortalet lagoon, close to the car park, has three hides conveniently overlooking the lagoon, while there are four hides and many towers along the paths to the beach. To reach the northern part of the reserve, return to Castelló d'Empúries and take the C260 towards Roses, but turn left (north), almost immediately, towards Palau-Saverdera. After the road crosses three irrigation canals in quick succession, there is a turn-off to the left which brings you to a hide overlooking a lagoon (Estany de Vilaüt). This dries out seasonally so it is best to check first at the information centre.

Unfortunately, this reserve is not easily reached by public transport. The nearest railway station is at Figueras, about 12km from the reserve. You could also consider renting a bike at the nearby resort of Roses.

KEY SPECIES

Resident
Little Grebe, Great Crested Grebe, Great Bittern, Night Heron, Little Egret, Cattle Egret, Gadwall, Purple Swamphen, Marsh Harrier, Stone Curlew, Kentish Plover, Common Kingfisher, Water Rail, Little Owl, Cetti's Warbler, Zitting Cisticola, Moustached Warbler, Penduline Tit, Reed Bunting

Summer
Little Bittern, Squacco Heron, Purple Heron, Little Ringed Plover, Black-winged Stilt, Whiskered Tern, Little

Tern, Great Spotted Cuckoo, European Roller, Sand Martin, Spanish Wagtail, Savi's Warbler, Reed Warbler, Great Reed Warbler, Subalpine Warbler, Woodchat Shrike

Winter
Great Cormorant, Greater Flamingo, Glossy Ibis, Greylag Goose, Gadwall, Eurasian Wigeon, Eurasian Teal, Northern Shoveler, Common Pochard, Tufted Duck, Northern Lapwing, European Golden Plover, Common Snipe, Mediterranean Gull, Lesser Black-backed Gull, Razorbill,

Sky Lark, Meadow Pipit, Water Pipit, Southern Grey Shrike

Passage
Spoonbill, Garganey, Osprey, Common Crane, Little Crake, Spotted Crake, Collared Pratincole, Dotterel, Dunlin, Curlew Sandpiper, Sanderling, Ruff, Wood Sandpiper, Marsh Sandpiper, Common Redshank, Spotted Redshank, Greenshank, Little Gull, Sandwich Tern, Common Tern, Black Tern, Bluethroat, Northern Wheatear, Sedge Warbler, Pied Flycatcher

Useful contacts and websites

Sociedad Española de Ornitologia (SEO)
(Spanish) www.seo.org
C/ Melquíades Biencinto, 34-28053 Madrid
Tel. (+34) 914 340910. Fax. (+34) 914 340911.
Spanish BirdLife International partner and publishers of the journal *Ardeola*.

Institut Català d'Ornitologia (ICO) www.ornitologia.org
Catalan Ringing Group which publishes the annual bird report *Anuari Ornitològic de Catalunya*. Address for bird records: ico@ornitologia.org.

Rare Birds in Spain www.rarebirdspain.net
Information on the latest sightings of rare birds in the whole of Spain, in English. Also includes a photographic section, information on rarities committee, and identification topics.

Delta del Llobregat (Catalan and Spanish)
www.gencat.net/mediamb/rndelta/cdll900.htm
Reserves Naturals Delta del Llobregat, Apartat de Correus 99 08840-Viladecans. Tel. 936 586761.
Reserve website with information about access, site diagrams and monthly reports. Some directions given in English.

Parc de Collserola (Catalan and English)
www.parccollserola.net/catalan/home/marcos.htm
Information about Collserola Nature Park.

Spanish Rarities Committee
www.seo.org/rarezas.cfm
José Ignacio Dies Jambrino, Delegación
SEO/BirdLife de Valencia, Av. Los Pinares 106, 46012 El Saler, Valencia.
Website of CR/SEO Spanish Rarities Committee.

Autoritat del Transport Metropolità (English)
www.atm-transmet.org/index_ang.htm
Web portal site for public transport services in Barcelona.

Transports Metropolitans de Barcelona
(English) www.tmb.net
Bus, tram and Metro service in Barcelona.

Ferrocarrils de la Generalitat de Catalunya
(English) www.fgc.es
Suburban train service for Barcelona. Tel. 932 051515.

RENFE (Spanish & English) www.renfe.es
Barcelona area train service. Tel. 932 240202.

Books and publications

Where to Watch Birds in North & East Spain
(2nd edition) Michael Rebane & Ernest Garcia
Due 2007. Christopher Helm
ISBN 0713647000

Where to Watch Birds in Spain & Portugal
Laurence Rose (1995) Hamlyn
ISBN 0600584046

Finding Birds in Northern Spain
Dave Gosney. BirdGuides, ISBN 1898110158

Guía de las Aves de España
Eduardo de Juana & Juan M. Varela (2005)
Lynx Edicions, Barcelona

Áreas Importantes para las Aves en España
C. Viada (1998) SEO

**Where to Watch Birds in Spain: The 100 best
sites** J. A. Montero (2006) Lynx Edicions
ISBN 8496553043

Where to Watch Birds in Catalonia
J. Sargatal and J. del Hoyo (1989) Lynx Edicions,
Barcelona

A Birdwatching Guide to the Pyrenees
Graham Hearl and Jacquie Crozier (1998)
Arlequin Publications, ISBN 1900159805

Collins Bird Guide
Lars Svensson *et al.* (1999) HarperCollins
ISBN 0002197286

Birds of Europe
Lars Jonsson (2006) Christopher Helm
ISBN 0713676000

• •

*A Black Tern flies over the River Spree with Oberbaumbrucke in the
background*

TYPICAL SPECIES

Resident
Little Grebe, Grey Heron, Mute Swan,
Mallard, Tufted Duck, Common
Pochard, Common Buzzard, Northern
Goshawk, Moorhen, Common Coot,
Black-headed Gull, Common Gull, Wood
Pigeon, Collared Dove, Great Spotted
Woodpecker, Wren, Dunnock, Robin,
Blackbird, Goldcrest, Marsh Tit, Coal Tit,
Blue Tit, Great Tit, Eurasian Nuthatch,
Eurasian Jay, Carrion Crow, Chaffinch,
Greenfinch, Bullfinch, Hawfinch,
Yellowhammer

Summer
Black Kite, Common Swift, Wryneck,
Sand Martin, Barn Swallow, House
Martin, White Wagtail, Black Redstart,
Common Redstart, Reed Warbler,
Blackcap, Barred Warbler, Common
Chiffchaff, Wood Warbler, Red-breasted
Flycatcher, Serin

Berlin

Situated in what was formerly East Germany and just 75km from the Polish
border, Berlin is a city with a real eastern European feel to its avifauna. Barred
Warblers and Red-breasted Flycatchers breed within the city limits and Great
Bustards and White-tailed Eagles do so just beyond. The flat surrounding countryside is
part of the North German Plain which extends right across northern Europe, with
barely any high ground between the North Sea and the Ural Mountains. Originally, this
area was a patchwork of open steppe, forests, lakes and marshes and, although it is now
mostly reclaimed for agriculture, there are still pockets of the original habitats.

The city is fortunate in having a number of important nature reserves nearby as well as some most impressive stretches of woodland in its parks, so a visitor can experience a very wide cross-section of Continental European birds. Being situated so far east, the city's extensive waterways are often the first stop for wintering wildfowl arriving from Siberia and, for as long as the city remains ice-free, the numbers of birds are maintained through the winter. A trip to the Baltic Sea coast, with its islands, wetlands and migrant traps, would be a must for any birder with more than a few days to spare. However, the coast is almost 150km away and beyond the scope of this book. Equally, there are some excellent sites, notably Owczary, just over the border in Poland which are also worth a visit. However, the birder restricted to the immediate vicinity of the city will not be disappointed.

GETTING AROUND

Although the largest city in Germany and covering a wide area, the large urban parks, cultural attractions and the airports are all within an orbital autobahn, the A10 or Berliner Ring, which is useful for orientation. Most of the best birding sites are just beyond the A10 but for sites within the ring-road, the best way to travel is by public transport: which is cheap, efficient and fully

integrated, so that one ticket will serve for buses, trams, subway and trains. Berlin's transport system is frequently represented as an example of the way all public transport should operate.

Rental cars are expensive in Germany but worth the cost if your time is limited. While not essential, a car would enable a much more efficient use of your time outside the city. A less pricey and healthier alternative is to rent a bicycle. Cycling is a very popular activity among Berliners and consequently there is a very well-developed infrastructure which makes getting around by bike easy, safe and secure. You may also take the bicycle on the U-Bahn and S-Bahn, so that you can cover much more distance when you reach your destination.

Birding in Berlin is very comfortable, with many areas landscaped for ease of access, and easy on your feet since it is so flat. At times it can be very cold in winter, but generally the weather is reasonably pleasant. A telescope is useful for winter birding, as birds such as ducks, geese and Great Bustards can be quite distant, but is less of a necessity in summer. If you need to purchase gloves, binoculars, books or any other birding requisites, bear in mind that Germany still has some antiquated retail opening hours, which means that large stores remain closed for much of the weekend and there are restricted opening times during the week. You will need a map that covers Brandenburg State in order to locate all the sites. It would also be useful to get a city map with a plan of the public transport system. These are available from any bookshop or tourist office, and at bus and railway stations.

Locations

1 Flughafensee Tegel (10km) *(Bus No. 133)*

Tegel Airport Lake is a disused flooded sandpit which is surrounded at its west end by Tegel Forest. It is on the north side of the perimeter fence of Tegel Airport, less than ten minutes by taxi from the terminal, and is well worth a visit if travelling to or from the airport. Although the lake is very deep, the margins support a variety of habitats including reedbeds and woodland. There is a bird reserve at the west end of the lake, with nature trails and observation towers.

From Tegel Airport, take **A111 Autobahn** north to the first exit at **Holzhauserstrasse**. Exit west onto Holzhauserstrasse and continue west when it becomes **Bernauerstrasse**. After about 500m, turn left (south) onto **Sterkraderstrasse** which leads to the reserve entrance. There is a bus service (No. 133) from Alt-Tegel U-Bahn Station.

KEY SPECIES
Resident
Mandarin Duck (I), Northern Goshawk, Common Kingfisher, Black Woodpecker, Long-tailed Tit
Summer
Little Bittern, Common Cuckoo, Tree Pipit, Song Thrush, Great Reed Warbler, Golden Oriole
Winter
Great Bittern, Siskin

2 Spandauer Forst (16km) *(Rathaus Spandau U-Bahn Station U7)*

This forest in the western suburbs is about the best location in Berlin for an opportunity to see some of the Continental European woodland species. It is probably best in early spring for woodpeckers, and in late spring for warblers and other songbirds, but even in winter there is much to interest a visitor to Europe. The forest offers a mosaic of mixed woodland, open fields and hedges as well as some ponds. It is popular with city-dwellers but is large enough to offer secluded areas for birding.

The main entrance lies along Schonwalderstrasse, just northwest of the Spandau District. From the U-Bahn station in Spandau go north on **Altstäter Ringasse** to **Neuendorferstrasse**. Turn left onto **Bismarckplatz** which continues onto **Schonwalderstrasse**. After about 1km this road enters the forest and becomes Schonwalder Allee. There are two small nature reserves lying to the west of Schonwalder Allee, with nature trails and ponds. It is also quite close to Tegel Airport, so is a convenient stop if you are collecting or dropping off a rental car there.

KEY SPECIES	
Resident	Willow Tit, Crested Tit, Long-tailed Tit, Common Raven
Eurasian Sparrowhawk, Water Rail, Woodcock, Stock Dove, Common Kingfisher, Green Woodpecker, Lesser Spotted Woodpecker, Middle Spotted Woodpecker, Black Woodpecker, Wood Lark,	**Summer**
	Common Crane, Hobby, Common Cuckoo, Common Nightingale, Song Thrush, Barred Warbler, Red-backed Shrike, Golden Oriole

3 Köppchensee NSG (14km) *(Bus No. 107)*

This nature reserve contains an interesting mix of habitats lying along the the River Tegel floodplain (Tegeler Fließ). It was 'protected' for many years on the eastern side of the Berlin Wall and consists of flooded peat-workings, drainage canals, abandoned farmland and scattered areas of woodland. This area is worth a visit in late spring and early summer as it offers an opportunity to see such birds as River and Barred Warblers, which are more typical of eastern Europe, and provides the right habitat for both Common and Thrush Nightingales in the very narrow zone of overlap in their breeding ranges.

The reserve is in **Blankenfeld** on the northern outskirts. Go north on **Schildowerstrasse** to the junction with Lübarser Weg; turn left (west) on **Lübarser Weg** and continue for about 1km. The entrance is on the right-hand side of the road. There is a bus service (No. 107) from Hermann-Hesse-Strasse Tram Stop.

KEY SPECIES

Resident
Water Rail, Common Kingfisher

Summer
Corn Crake, Spotted Crake, Common Snipe, Common

Cuckoo, Common Nightingale, Thrush Nightingale, River Warbler, Barred Warbler, Penduline Tit

---

4 Müggelsee (20km)   (Friederichshagen S-Bahn Station S3)

The Müggelsee is a rather large lake on the River Spree in the southeastern sector of the city, near the old city area of Köpenick. There is a breeding colony of Black Terns, and it is generally quite good for gulls in winter and on migration. The Müggelsee marks the western flank of a large sprawling forest park (Müggelsee-Spree) which stretches along the River Spree as far as Fürstenwalde and includes a huge expanse of forests, lakes and riverside walks. You could easily spend a week exploring the area. Luckily, some of the best birding areas are the north and south shores of the Müggelsee itself. Take any street south from Friederichshagen S-Bahn Station to get to the lakeshore.

KEY SPECIES

Resident
Great Crested Grebe, White-tailed Eagle, Water Rail, Common Kingfisher, Green Woodpecker, Lesser Spotted Woodpecker, Black Woodpecker, Reed Bunting

Summer
Osprey, Common Sandpiper, Black Tern, Turtle Dove, Yellow Wagtail, Great Reed Warbler

Winter
Red-necked Grebe, Common

Goldeneye, Goosander, Yellow-legged Gull

Passage
Black-throated Diver, Red-throated Diver, Black-necked Grebe, Common Sandpiper, Little Gull

---

5 River Havel (25km)   (Wannsee S-Bahn Station/Bus No. 216)

One of the best stretches of riparian habitat along the River Havel is the east bank between the western outskirts of Berlin and Potsdam, and it is only 20 minutes from the city centre. There is a riverside walk constructed between Großer Wannsee and Glienicker Brücke. This provides a good vantage point to scan this wide stretch of the Havel, and the adjoining broadleaved woodland is particularly good for some of the forest specialities.

The best starting point is the **Pfaueninsel ferry terminal** at **Pfaueninselchaussee**, which is a short bus ride from Wannsee S-Bahn Station. You could take the ferry onto the island, which is a popular tourist spot, and explore the nature trails through the woodland. Otherwise continue on the riverside walk south towards Glienicker Brücke and take the train back to Berlin from Potsdam.

KEY SPECIES

Resident
Mandarin Duck (I), Eurasian Sparrowhawk, Water Rail, Common Kingfisher, Green Woodpecker, Lesser Spotted Woodpecker, Middle Spotted Woodpecker, Black Woodpecker, Bearded Tit, Long-tailed Tit

Summer
Black Kite, Osprey, Great Reed Warbler, Red-breasted Flycatcher

Winter
Common Goldeneye, Goosander, White-tailed Eagle, Yellow-legged Gull

Passage
Common Sandpiper, Little Gull, Black Tern

6 Döberitzer Heide NSG (22km)

(Dallgow Railway Station)

The Döberitzer Heide is a large area of heathland which was cleared of forest in the 19th century and kept that way for military manoeuvres ever since. It has now been designated a bird protection area and supports a good selection of species typical of open habitats, including heath, grassland, young forest and copses of more mature woodland. The area is also scattered with marshes and small wetlands in the more low-lying patches. There are many hiking paths and it is a very popular area with walkers and joggers but, because of its large area, there is little disturbance of the birds. Some of Berlin's best birds breed here and it is easily worth half a day's visit in spring and early summer.

Go west on the **B5** road, which is basically an extension of the main east/west carriageway through central Berlin, for about 18km as far as **Dallgow**. The next stretch of the B5 from Dallgow to **Elstal** runs through the Döberitzer Heath and there are several turn-ins and hiking trails on the left-hand (south) side of the road. Turn left at the town of **Rohrbeck**, which is about 2km west of Dallgow, and drive for about 3km, checking likely spots on either side of the road.

If you are using public transport, you could take a bicycle on the train (RE4 Suburban Line) to Dallgow or Elstar and cycle from there.

KEY SPECIES

Resident
White-tailed Eagle, Black Woodpecker, Middle Spotted Woodpecker, Corn Bunting

Summer
Red Kite, European Honey-buzzard, Marsh Harrier, Hobby, Common Crane, Woodcock, European Nightjar, Hoopoe, Wood Lark, Tawny Pipit,

Northern Wheatear, Common Stonechat, Whinchat, Barred Warbler, Golden Oriole, Red-backed Shrike

Winter
Great Grey Shrike

7 Belziger Landschaftswiesen (70km)

(Brück Railway Station)

If, for any reason, you fail to find Great Bustards at Havelland, then the Belzig Reserve is worth a look, not least because it also hosts a fine selection of marsh and grassland species. This reserve is about 40km south of Havelland and managed by the same conservation organisation, Förderverein Großtrappenschutz e.V. They have erected an observation tower at the edge of the reserve to permit viewing over a wide area.

Take the **A9 Autobahn** southwest from Berlin and exit west on the **B246** towards **Brück**. At Brück turn north on the **L85**: the observation tower is located after about 2km on the west side of the road. If you are coming from Havelland, it is less complicated to drive back to the Berliner Ring and from there get onto the A9, rather than drive through Brandenburg and risk getting lost.

KEY SPECIES

Resident
Great Bustard, Grey Partridge

Summer
Black Stork, Montagu's Harrier, Red Kite, Common Quail, Common Snipe, Eurasian Curlew, Wood Lark, Tree Pipit

Winter
Bean Goose, White-fronted Goose, White-tailed Eagle, Rough-legged Buzzard, Hen Harrier, Merlin, Great Grey Shrike

Passage
Northern Lapwing, European Golden Plover, Black-tailed Godwit, Wood Sandpiper

8 Rietzer See NSG (68km)

(Bus No. 553 from Brandenburg)

Berlin is surrounded by lakes and waterways, particularly along the course of the River Havel southwest of the city. One of the best lakes for birding is the Rietzer See, which supports a good selection of breeding birds as well as wintering ducks and other waterfowl. Apart from the lake itself, the reedbeds and water meadows which surround it are part of a nature reserve and over 250 species have been recorded here.

It is very easy to get to by car. Go west on the **A2 Autobahn**, one of the autobahns which radiate out from the **Berliner Ring**, and take **Exit 79** at **Netzen** onto the **L88**. Follow the L88 through Netzen and then through the villages of **Prützke** and **Rietz**. This road follows the western shore of the lake, from which you can scan for waterfowl. There is also an observation tower in Rietz and, depending on water levels, you can check exposed mudbanks for waders, gulls and marsh terns. Returning to Netzen, there is an observation tower just east of the village (**Schulstrasse**) along a gravel track and bridge over the drainage canal, which enables you to scan the area of the smaller **Netzener See** and the river (**Streng**) which connects the two lakes. There is a bus service (**No. 553**) to Rietz from **Brandenburg**, the nearest railway station and about 7km away.

KEY SPECIES

Resident Greylag Goose, Gadwall, Northern Shoveler, Common Kingfisher	Bluethroat, Savi's Warbler, Sedge Warbler, Barred Warbler, Bearded Tit, Penduline Tit, Ortolan Bunting	**Passage** Great White Egret, Eurasian Wigeon, Eurasian Teal, Pintail, Ringed Plover, Northern Lapwing, Ruff, Common Redshank, Wood Sandpiper
Summer Red-necked Grebe, Black-necked Grebe, Great Bittern, Garganey, Spotted Crake, Little Ringed Plover, Common Snipe, Eurasian Curlew, Common Tern, Black Tern,	**Winter** Bean Goose, White-fronted Goose, Common Goldeneye, Smew, White-tailed Eagle, Hen Harrier, Great Grey Shrike	

9 Havelländisches Luch (60km)

(Nennhausen Railway Station)

The reserve at Havelländisches Luch is one of the most reliable sites for Great Bustards: the population includes some 30 breeding males. It is also an excellent site for other grassland species as well as aquatic birds. Winter is probably best for numbers of birds but this site is well worth a visit at any time of year. The whole area is flat grassland with wet meadows, which

flood in winter producing an enormous wetland of over 5,000 hectares. As well as Great Bustards, the reserve is a breeding site for White Storks, Common Cranes, Corn

Crakes and waders. Up to 20,000 geese winter here and there is a chance of seeing the rare Red-breasted Goose.

Take the **B5** northwest from Berlin and after about 40 km turn west on the **L99** towards **Retzow**. Pass through Retzow and turn west on to the **L991** for **Nennhausen**. In Nennhausen take the **L982** south towards **Garlitz**. Drive slowly along this road and scan for bustards: they are often seen in rapeseed fields. Five km along this road is Buckow, where the Information Centre for the bird reserve is located, and you can probably obtain more precise directions there. Just south of **Buckow** on the left-hand side of the road is a turn-off to an observation tower, which enables you to scan a larger area. Another observation tower can be reached by turning east at Garlitz. It is important not to encroach on the reserve, particularly during the bustards' courtship display period and in the breeding season.

KEY SPECIES

Resident
Great Bustard, Grey Partridge, Sky Lark, Common Raven, Corn Bunting

Summer
White Stork, Black Stork, Montagu's Harrier, Red Kite, Common Quail, Corn Crake, Spotted Crake, Common Crane, Northern Lapwing, Ruff, Black-tailed Godwit, Common Redshank,

Common Snipe, Eurasian Curlew, Turtle Dove, Wryneck, Common Cuckoo, Tree Pipit, Blue-headed Wagtail, Common Nightingale, Barred Warbler, Ortolan Bunting

Winter
Whooper Swan, Tundra Swan, Greylag Goose, Bean Goose, White-fronted Goose, Gadwall, Pintail,

Northern Shoveler, Eurasian Wigeon, Eurasian Teal, Smew, White-tailed Eagle, Rough-legged Buzzard, Hen Harrier, Merlin, Water Pipit, Great Grey Shrike, Brambling

Passage
European Golden Plover, Meadow Pipit

10 Gülper See (80km)

This is a large lake about 80km west of Berlin on the borders of the next state, Saxony-Anhalt. The site is at the northern end of the Naturpark WestHavelland, a large national park which extends as far south as Brandenburg. Although it is quite distant from Berlin, it is well worth the trip as it holds vast numbers of geese and other waterfowl in winter, as well as being one of the most reliable sites for Great White Egrets and White-tailed Eagles. The area has a very rich variety of breeding birds and at migration times it attracts good numbers of waders and Common Cranes. The south shore is the most well watched as well as the most accessible part of the lake. There is a nature trail with observation hides along the south shore from Prietzen to Gülpe.

Take the **B5** west from Berlin as far as **Friesack**, and from there, take the **L17** east to **Rhinow**. At Rhinow, go south on the **L175** to Prietzen, from where there is access to the lakeshore. If using public transport, take the train to **Rathenow**, about 15km south of the Gülper See, and get the bus to Prietzen.

KEY SPECIES

Resident
Great Cormorant, Greylag Goose, Gadwall, Water Rail, Common Kingfisher, Sky Lark, Meadow Pipit, Bearded Tit, Great Grey Shrike, Corn Bunting, Reed Bunting

Summer
Great Bittern, White Stork, Garganey, Northern Shoveler, Red Kite, Spotted Crake, Common Snipe, Eurasian Curlew, Turtle Dove, Common

Cuckoo, Wood Lark, Tree Pipit, Yellow Wagtail, Whinchat, Common Nightingale, Great Reed Warbler, Savi's Warbler, River Warbler, Grasshopper Warbler, Icterine Warbler, Penduline Tit, Golden Oriole

Winter
Whooper Swan, Tundra Swan, White-fronted Goose, Bean Goose, Common Goldeneye, Goosander, Smew, White-tailed Eagle, Hen Harrier, Rough-

legged Buzzard, Common Raven, Brambling

Passage
Black Stork, Eurasian Wigeon, Eurasian Teal, Pintail, Great White Egret, Common Crane, Northern Lapwing, Ruff, Common Redshank, Spotted Redshank, Green Sandpiper, Wood Sandpiper, Black Tern, Little Gull

11 Linum Fish Ponds (45km)

Linum is a village north of Berlin where large numbers of White Storks nest on the buildings. Just east of the village is a set of fish ponds with an observation tower. The area around Linum Fish Ponds is a well-known site for Common Cranes migrating through the region in late autumn: several thousand can be seen in October and November. The fish ponds themselves offer good habitat

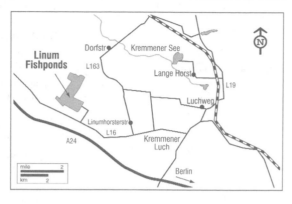

for aquatic birds and the fields between Linum and Kremmen support farmland birds as well as the flocks of Common Cranes.

The ponds are easy to find: take the **A24 Autobahn** to the northwest of the **Berliner Ring** and exit almost immediately north on the **B273**. After about 4km, turn west on the **L162** and follow this all the way into Linum. The fish ponds are accessed just east of the village by public footpath. If you return along the L162 and stay on this road all the way to Kremmen (approx. 10km) the fields on the left-hand (north) side of the road (known as the **Kremmener Luch**) hold the Common Crane flocks in autumn. There is a railway station in Kremmen quite close to the eastern edge of the Kremmener Luch, but the area is so extensive that a car is really the only practical way to 'do' this site.

KEY SPECIES

Resident
Greylag Goose, Red Kite, Common Snipe, Common Kingfisher, Bearded Tit, Reed Bunting

Summer
White Stork, Black Stork, Northern Shoveler, Hobby, Common Cuckoo, Tree Pipit, Whinchat, Common

Nightingale, Great Reed Warbler, Savi's Warbler, Icterine Warbler, Penduline Tit

Winter
Common Goldeneye, Goosander, White-tailed Eagle, Rough-legged Buzzard, Hen Harrier, Great Grey Shrike

Passage
Bean Goose, White-fronted Goose, Gadwall, Red-crested Pochard, Common Crane, Northern Lapwing, Ruff, Common Redshank, Spotted Redshank, Green Sandpiper, Wood Sandpiper, Eurasian Curlew, Black Tern, Little Gull

Useful contacts and websites

Berliner Ornithologische Arbeitsgemeinschaft (BOA) (German)
www.orniberlin.de
Berlin Ornithological Society – publishes the annual Berlin Bird Report: copies available from: Ludwig Schlottke, Nienkemperstr. 46c, 14167 Berlin. Tel. 0308177404.

Arbeitsgemeinschaft Berlin-Brandenburgischer Ornithologen
www.abbo-info.de
Berlin-Brandenburg Ornithological Group Publishers of *Otis* – Bird Report for Brandenburg and Berlin.

c/o Wolfgang Mädlow, Konrad-Wolf-Allee 53, 14480 Potsdam. Tel. 03316263488.

NABU RV Brandenburg/Havel e.V (German)
www.nabu-brandenburg-havel.de
Local Branch NABU – National Organisation for Nature Conservation (BirdLife partner in Germany)
Bruchstraße 60, 14778 Schenkenberg. Tel. 03320751271. Fax. 03320751271.

Vogelruf – general site for birding in Germany (German): www.vogelruf.de

Förderverein Großtrappenschutz e.V.
Havelland Great Bustard Reserve.
www.grosstrappe.de

Förderverein Großtrappenschutz e.V. Dorfstr. 34, 14715 Buckow bei Nennhausen. Tel. 03387860257.

Naturschutz-Förderverein Döberitzer Heide e.V (German) www.doeberitzerheide.de
Döberitzer Heath Nature Protection Organisation Information about nature reserve
Naturschutzzentrum 1, Ortsteil Elstal, 14641 Wustermark. Tel. 0332347080.

Naturpark Westhavelland (German) www.naturwacht.de

Die Vogelwelt des Naturparks Westhavelland (German) www.fedtke.de
Excellent website by Burghard Fédtke & Dr Rainer Warthold with site description, maps, winter wild-fowl counts, monthly bird reports and checklist.

German Birdnet www.birdnet.de

Berlin Verkehrsbertriebe (BVG). www.bvg.de Public Transport in Berlin Area

Deutsch Bahn
www.bahn.de/pv/uebersicht/die_bahn_ international_guests.shtml
Trains and feeder bus service in Germany. Tel. Berlin Region 01805194195.

Books and publications

Important Bird Areas (IBA) in Brandenburg and Berlin NABU Landesverband Brandenburg

Vogelparadiese, Band 1, Norddeutschland (North Germany and Berlin-West)
Michael Lohmann, Knut Haarman (1989) Blackwell Wissenschafts-Verlag ISBN 3826381513

Vogelparadiese, Band 3, Ost- und Mitteldeutschland (former GDR)
Michael Lohmann, Knut Haarman (1989) Blackwell Wissenschafts-Verlag ISBN 3826381505

Collins Bird Guide
Lars Svensson *et al.* (1999) HarperCollins ISBN 0002197286

• •

TYPICAL SPECIES

Resident
Great Crested Grebe, Great Cormorant, Grey Heron, Mute Swan, Greylag Goose, Mallard, Tufted Duck, Eurasian Sparrowhawk, Common Buzzard, Moorhen, Common Coot, Black-headed Gull, Herring Gull, Wood Pigeon, Collared Dove, Great Spotted Woodpecker, White Wagtail, Wren, Dunnock, Robin, Blackbird, Song Thrush, Goldcrest, Blue Tit, Great Tit, Eurasian Jay, Magpie, Jackdaw, Carrion Crow, Common Starling, House Sparrow, Chaffinch, Greenfinch, Goldfinch, Bullfinch, Reed Bunting

Summer
Common Swift, Sand Martin, Barn Swallow, House Martin, Black Redstart, Icterine Warbler, Garden Warbler, Blackcap, Common Whitethroat, Common Chiffchaff, Willow Warbler, Spotted Flycatcher

A White Wagtail inspects the lawns at Brussels' unique Atomium

Brussels

Brussels is located just about in the centre of Belgium and as such lies almost equidistant between the two best birding areas in the country: the North Sea coast and the forests of the Ardennes. Both locations are over 100km from the city and are beyond the scope of this book, although it would be possible to visit either in the course of a day. However, luckily for the city-bound birder, there are plenty of interesting birding sites in and around the city, some of which are close to each other and accessible by public transport.

The most outstanding feature of Brussels from a birder's point of view is the large belt of broadleaved woodland which stretches discontinuously across the southern suburbs and hosts a very rich community of woodland species. These areas are very much the focus of a birder's attention in late spring and early summer when the breeding birds are establishing territories.

In winter, attention turns to the network of waterways mainly to the north and east of the city, which remain largely ice-free and attract wintering wildfowl in varying numbers. Diurnal migration passes through Belgium on a fairly broad front and cranes, raptors, waders and hirundines may all be encountered as they overfly Brussels and its environs during spring and autumn.

GETTING AROUND

Getting around Brussels is quite easy using the public transport system, and for any site within the city limits this is the best way for the birder to travel. The bus, tram and metro systems are fully integrated under the municipal transportation authority (STIB/MIVB) so one ticket allows you to use all modes of transport. Beyond the city limits a car would definitely be an

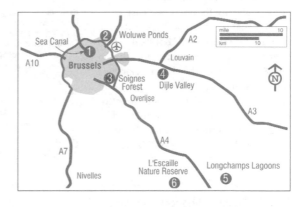

advantage as most sites are widely separated and a car enables you to visit more than one location in a day. However car rental is expensive in Belgium and if you are going to rent you would probably get more birding benefit by travelling as far as the Ardennes, the North Sea coast, or even to the southern provinces of The Netherlands, any of which involves a two-hour drive each way.

As well as the railway, there are two regional bus services which operate in the vicinity of Brussels: De LIJN which serves the Flemish Brabant region north and east of Brussels and TEC which operates in the French-speaking Walloon region to the south of Brussels. As Brussels is mainly a French-speaking city surrounded by a Flemish-speaking region, several sites have bilingual place names which appear on direction signs and public transport maps. For this reason, both languages are used here for the names of sites in Brabant.

Apart from the woodland sites, a telescope is useful in most cases: however the sites located in nature reserves normally have hides, permitting reasonably close approach. You will need a map that covers Brussels as well as the provinces of Brabant and Namur in order to locate all the sites, and it would also be useful to get a city map showing the public transport system. These are available from any bookshop and tourist office, and at bus and rail stations.

Locations

1 Sea Canal (3km)

(Gare du Nord & Schaerbeek Railway Stations)

The River Senne which flows through Brussels has been channelled and deepened to provide a navigable waterway to Antwerp. This stretch of canal is deeper than the lakes and ponds around the city and less likely to freeze over, so in winter it can hold good numbers of diving ducks, which are rare elsewhere in the Brussels area. The best area is adjacent to the Royal Gardens of Laeken (west bank), which has a large heronry, and on the opposite bank there is remnant marsh at Moeraske alongside some allotments which are good for migrating passerines.

The starting point for this area is the **Gare du Nord**. Exit west from the station along **Boulevard Simon Bolivar** to the **Bassin Vergote** and check for gulls. Walk north along the east bank of the canal until you are opposite the **Royal Gardens at Laeken**. This stretch is good in winter for diving ducks. Just past the Royal Gardens turn east at Rue de l'Avant-Port and continue southeast over Pont Albert to Rue Chaumontel and turn north. The Parc Walckiers and Moeraske Reserve are on the left-hand side of this street.

KEY SPECIES

Resident
Little Grebe, Water Rail, Common Kingfisher, Common Snipe, Green Woodpecker, Lesser Spotted Woodpecker, Long-tailed Tit, Tree Sparrow

Summer
Marsh Warbler, Reed Warbler, Lesser Whitethroat, Garden Warbler, Firecrest

Winter
Shelduck, Common Pochard,

Goldeneye, Smew, Goosander, Woodcock, Common Gull, Lesser Black-backed Gull, Fieldfare, Redwing, Brambling, Siskin, Mealy Redpoll

Passage
Northern Lapwing, Green Sandpiper, Common Sandpiper, Tree Pipit, Grey Wagtail, Northern Wheatear, Sedge Warbler

2 Woluwe Ponds (10km)

(STIB Bus No. 54; De LIJN Bus No. 281)

The Woluwe Ponds are two small reservoirs in northeastern Brussels which, in recent years, have become a popular birding site for Brussels-based birders. These ponds are important for wintering wildfowl, and waders are attracted to the muddy margins if water levels are low during migration periods. The site is also very close to Brussels Airport and if your flight is delayed by a couple of hours, the ponds are less than ten minutes away by bus (**No. 281**) or taxi.

From the **R0** orbital motorway, exit north on the **R22** (only possible from the southbound lane) towards **Machelen**. The Woluwe Ponds are on the east side of the R22, approximately 1.5km north of the R0 Exit.

KEY SPECIES

Resident
Little Grebe, Peregrine Falcon, Common Kestrel, Water Rail, Common Kingfisher, Willow Tit, Tree Sparrow, Linnet

Summer
Garganey, Little Ringed Plover, Common Cuckoo, Blue-headed Wagtail, Marsh Warbler, Lesser Whitethroat

Winter
Common Shelduck, Northern Shoveler, Eurasian Teal, Eurasian Wigeon, Common Pochard, Common Snipe, Jack Snipe, Common Gull, Lesser Black-backed Gull, Grey Wagtail, Redwing, Brambling

Passage
White Stork, Gadwall, Pintail, Common Sandpiper, Green Sandpiper, Greenshank, Sky Lark, Meadow Pipit, Northern Wheatear, Whinchat

3 Soignes Forest/ Zoniënwoud (10km)

(Herrmann-Debroux Sta. Metroline 1A)
(STIB Tram No. 44 ; STIB Bus No. 72)

The Soignes Forest is a large tract of woodland, more than 4,000 hectares, in the southeastern suburbs of Brussels. The forest is famous for its extensive beechwoods but because of this monoculture these areas are not the most attractive for birds. The greatest species diversity is in the pockets of mixed forest and riparian woodland alongside the many ponds and waterways. There are many access points to the forest across the southern suburbs of Brussels but one of the most convenient and productive sites is at the Abbey of Rouge-Cloître near Auderghem (Oudergem). A network of trails lead into the forest from here. A number of woodland lakes in this area add to the species diversity. The best time to visit is early morning in spring and early summer, but a good number of species can be found at any time of year.

Take the **N4 (Chaussée de Wavre / Waversesteeenweg)** southwest from the city centre to Auderghem, passing the **Auderghem Shopping Centre** on the right. Turn right onto **Drève du Rouge Cloître (Roodkloosterdreef)** about 1km past the junction with **Boulevard du Souverain**. The area is well served by public transport.

KEY SPECIES

Resident
Common Pheasant, Woodcock, Stock Dove, Common Kingfisher, Green Woodpecker, Lesser Spotted Woodpecker, Middle Spotted Woodpecker, Black Woodpecker, Mistle Thrush, Goldcrest, Eurasian Nuthatch, Long-tailed Tit, Willow Tit, Marsh Tit, Crested Tit, Short-toed Treecreeper, Hawfinch, Crossbill

Summer
European Honey-buzzard, Hobby, Tree Pipit, Common Nightingale, Reed Warbler, Lesser Whitethroat, Wood Warbler, Firecrest, Pied Flycatcher, Golden Oriole

Winter
Wigeon, Common Pochard, Goldeneye, Common Gull, Fieldfare, Redwing, Brambling, Siskin, Mealy Redpoll

4 Dijle Valley/Dyle Valley (20km)

(Oud-Heverlee Railway Station)
(Sint-Joris-Weert Railway Station)

The River Dijle is a tributary of the Senne which flows north and then east through Brabant approximately 20km east of Brussels. The stretch between Louvain and the Flemish / Walloon Border flows through a variety of habitats: including farmland, fish ponds, riparian woodland and mixed forest, and a day spent in this area can yield a good selection of the species found in this region.

The best way to visit is to drive a circular route south from Louvain along the west bank of the river and return north along the east bank, visiting the forest at Meerdaal. This area is worthwhile at any time of year: in late spring and summer there is a rich community of breeding birds in the forested areas and the lake margins, while in winter the fish ponds support good numbers of wildfowl and consequently raptors, including an occasional White-tailed Eagle. Water levels in the ponds are controlled by a system of dams and when levels are lowered in spring or autumn, this attracts migrating waders.

Take the **A3 Motorway** east from Brussels for about 15km, then exit southeast at the **N3 Interchange (Exit 22)** towards **Korbeek-Dijle**. From Korbeek-Dijle go east on **Stationstraat** to **Oud-Heverlee**. The first two ponds are located just on the western edge of Oud-Heverlee. There is a track leading to the ponds just before the railway station on **Bogaardenstraat**.

There is another pond (Grote Bron) just south of Korbeek-Dijle, which can be reached by returning to Korbeek and turning south towards Neerijse. Then turn left just after leaving Korbeek and go to the end of the road (**Kleinebroekstraat**). If water levels are low, this pond is good for waders on migration and also for Ospreys.

Return to the main road and continue towards **Neerijse**. From the village of Neerijse turn east towards **Sint-Agatha-Rode** on **Beekstraat**. There is another fish pond (**Grootbroek**) at Sint-Agatha-Rode which can be reached by taking **Leuvensebaan** north from the village towards **Sint-Joris-Weert**. A track to the pond is a short distance from the village on the right-hand side of Leuvensebaan. Although this is the largest of the ponds in the valley it holds fewer waterfowl. However, White-tailed Eagles are present occasionally in winter.

Continue north on Leuvensebaan through Sint-Joris-Weert and about 1km north of the village, turn left (west) onto **Reigersstraat**. This brings you to Doode Bemde Nature Reserve where there is a restored reedbed and a number of reedbed species breed here, including Bluethroat. There are 6km of trails through the reserve as well as a bird hide.

The final site on the circuit is Meerdaal Forest. Return to Sint-Joris-Weert and when you reach the village turn east on **Hollestraat** which becomes **Weertsedreef**. There are trails accessing the forest on the south side of this road. Five species of woodpecker breed in the forest as well as the other typical woodland species. Stay on Weertsedreef and it brings you to the **N25**: turn north and this connects with the A3.

Even if you don't have a car, you can cover some parts of the valley. There is a hiking trail from the Oud-Heverlee Ponds south through the Doode Bemde Nature Reserve to Sint-Joris-Weert, which you could walk comfortably in a day and return by train from Sint-Joris-Weert Station. Be aware that this area attracts wildfowlers during the game season and that there may be considerable disturbance to birds outside the Doode Bemde Nature Reserve, particularly at weekends.

KEY SPECIES

Resident
Little Grebe, Northern Shoveler, Common Pochard, Northern Goshawk, Common Pheasant, Water Rail, Common Snipe, Common Kingfisher, Green Woodpecker, Lesser Spotted Woodpecker, Middle Spotted Woodpecker, Black Woodpecker, Mistle Thrush, Willow Tit, Marsh Tit, Crested Tit, Bearded Tit, Rook, Tree Sparrow, Linnet

Summer
Garganey, European Honey-buzzard,
Marsh Harrier, Hobby, Turtle Dove, Common Cuckoo, Sky Lark, Wood Lark, Meadow Pipit, Bluethroat, Whinchat, Common Redstart, Common Nightingale, Grasshopper Warbler, Marsh Warbler, Sedge Warbler, Lesser Whitethroat, Garden Warbler, Golden Oriole

Winter
Great Bittern, Great White Egret, Gadwall, Eurasian Wigeon, Eurasian Teal, Smew, Hen Harrier, White-tailed Eagle, Common Snipe, Common Gull,
Lesser Black-backed Gull, Water Pipit, Redwing, Fieldfare, Goldcrest, Brambling, Siskin

Passage
Black-necked Grebe, Shelduck, Pintail, Osprey, Common Crane, Northern Lapwing, Golden Plover, Greenshank, Green Sandpiper, Common Sandpiper, Tree Pipit, Grey Wagtail, Common Stonechat, Northern Wheatear, Crossbill

5 Longchamps Lagoons (50km)

(De LIJN Bus Line E)

The Longchamps Lagoons are a set of settling ponds for discharges from a sugar factory and are southeast of the city, near the town of Éghezée. The muddy margins of these ponds are a magnet for significant numbers of migrating waders, making Longchamps one of the most popular birding sites for Brussels birders during spring and autumn. For an inland site, it has amassed an enviable list of rarities over the years. In winter, the lagoons attract many grebes, ducks and gulls. Even in summer, there are a number of interesting species breeding including Black-necked Grebes and Bluethroats, as well as a large Black-headed Gull colony. The premises are private property, but birders are permitted access provided they observe the directions of the company staff.

Take the **A4 Motorway** south from Brussels to **Namur** and after about 45km, exit east at **Exit 12** towards Éghezée on the N912. Longchamps is situated on this road, approximately 5km from the A4: the sugar refinery is just on the outskirts after you pass the village.

The sugar refinery grounds are fenced, but you can enter through several points: either through the main gate if it is open (it is best to park outside and walk in, in case the gate gets closed during your visit), or through a back gate that can be reached by taking the first road to the right before you pass the factory. Alternatively (if both gates are closed) enter via the garden of the small house just to the right of the main gate, walking along the left side of the house. (Yes, birders are allowed to do this without asking permission.)

You can walk the embankments that separate the ponds, but be careful not to stray off the gravel paths. The birds have become accustomed to human activity on the pathways but can be easily alarmed if you leave the path.

There is a bus service from Brussels to Éghezée, from where you can take a taxi to the lagoons. The bus departs hourly from the Gare du Luxembourg.

KEY SPECIES

Resident
Eurasian Teal, Common Pochard, Common Kestrel, Water Rail, Common Snipe, Common Kingfisher, Tree Sparrow, Reed Bunting

Summer
Little Grebe, Black-necked Grebe, Shoveler, Marsh Harrier, Hobby, Little Ringed Plover, Northern Lapwing, Lesser Black-backed Gull, Black-headed Gull, Common Cuckoo, Sky Lark, Meadow Pipit, Bluethroat, Grasshopper Warbler, Sedge Warbler, Marsh Warbler

Winter
Greylag Goose, Gadwall, Eurasian Wigeon, Hen Harrier, Jack Snipe, Water Pipit, Fieldfare, Brambling, Siskin

Passage
Garganey, Spotted Crake, Avocet, Ringed Plover, Dunlin, Little Stint, Temminck's Stint, Ruff, Common Redshank, Black-tailed Godwit, Greenshank, Green Sandpiper, Wood Sandpiper, Common Sandpiper, Little Gull, Common Tern, Black Tern, Grey Wagtail

6 L'Escaille Nature Reserve (40km)

(Gembloux Railway Station)

The elevated plain south and east of Brussels is given over to intensive cereal cultivation and as such is not very attractive for birds. However there are pockets of habitat around old gravel pits and industrial workings which can be attractive to aquatic birds and also act as resting points for passage migrants. One such site is L'Escaille Nature Reserve, just outside the town of Gembloux, about 40km southeast of Brussels. This small reserve was developed around an old mill on the River Orneau and its associated millpond and supports breeding and wintering waterfowl and other aquatic birds.

During migration it is also possible to witness some noteworthy diurnal passage of raptors, cranes and passerines.

Take the **A4 Motorway** south from Brussels. Exit west on the **N29** at **Exit 11** and go towards Gembloux. At the roundabout just before Gembloux, turn south on the **N4** and after about 200 metres turn east on **Rue de la Posterie**. The entrance to the reserve is on the right-hand side of this road. There is a trail around the pond and an observation hide.

KEY SPECIES

Resident
Common Pochard, Common Kestrel, Water Rail, Snipe, Common Kingfisher, Tree Sparrow, Reed Bunting

Summer
Little Grebe, Northern Shoveler, Marsh Harrier, Hobby, Little Ringed Plover, Lesser Black-backed Gull, Sky Lark, Meadow Pipit, Sedge Warbler, Marsh Warbler, Reed Warbler

Winter
Gadwall, Eurasian Wigeon, Eurasian Teal, Hen Harrier, Jack Snipe, Water Pipit, Fieldfare, Redwing, Siskin

Passage
Garganey, Common Crane, Northern Lapwing, European Golden Plover, Greenshank, Green Sandpiper, Common Sandpiper, Black Tern, Wood Lark, Grey Wagtail, Whinchat, Brambling, Linnet

Useful contacts and websites

AVES (French)
www.aves.be
Secretariat Aves, Maison Liegeoise de L'environnement, Rue Fusch, 3 – 4000 Liege
Tel. 04 250 9590. Fax. 04 222 1689.
Birding Organisation for French-speaking regions of Belgium and publishers of the journal *Aves*.

Les Reserves Naturelles et Ornithologiques de Belgique (RNOB) (French) www.rnob.be
Rue Royale, Sainte-Marie 105, B-1030 Brussels (BirdLife Partner). Tel. 02 245 4300.
Access, maps and descriptions of every nature reserve in French-speaking regions of Belgium.

Natuurpunt (Flemish) www.natuurpunt.be
Kardinaal Mercierplein 1, 2800 Mechelen.
Tel. 01 529 7220 (Birdlife Partner).
Access, maps and descriptions of every nature reserve in Flemish-speaking regions of Belgium.

Vogelwerkgroep Brussel (English)
http://perswww.kuleuven.ac.be/~u0017670/homeb.htm
Brussels Birding Work Group – information on birds in the Brussels area.

Instituut voor natuur – and bosonderzoek (Flemish and English) www.inbo.be
Research Institute for Nature and Forest – information on breeding bird surveys, wildfowl counts etc.

Belgian Rarities Committee, Flemish region (English) www.bahc.be
Secretary, c/o Willem de Zwijgerstraat 8, B-8020 Oostkamp.

Belgian Rarities Committee, French region (French and English)
http://users.skynet.be/ch-web/index_en.htm
Secretary c/o Rue de St-Hubert 518, B-5300 Vezin.

Erik Toorman's Birding Pages (English)
http://perswww.kuleuven.be/~u0017670/avifoto.html
Very helpful personal site with lots of information about birding in Belgium and links to other sites.

Belgian Birdline: Flemish 0900 00194.

STIB/MIVB (French/Flemish)
Public transport in the Brussels area
www.stib.be

SNCB/NMBS (English) – Belgian Railways
www.b-rail.be/E/index.php Tel. +32 (0)2 219 2640

De LIJN (Flemish) – regional bus service in Flemish Brabant
www.delijn.be

TEC (French) – regional bus service in Walloon
www.tec-wl.be
Place des Tramways 9, 6000 Charleroi
Info +32 (0)71 234 111 or +32 (0)71 234 115.

Books and publications

Where to Watch Birds in Holland, Belgium & Northern France
Arnoud van den Berg & Dominique Lafontaine (1996) Hamlyn, ISBN 0600579778

Collins Bird Guide
Lars Svensson *et al.* (1999) HarperCollins
ISBN 0002197286

Birds of Europe
Lars Jonsson (2006) Christopher Helm
ISBN 0713676000

Budapest

Syrian Woodpecker is resident in the woods of the Buda Hills

Unlike much of the rest of Hungary, which lies astride a flat grassland plain, Budapest has a variety of different habitats in its vicinity. This makes it possible, even on a short trip to the city, to see a range of species, some of which would be rare or difficult to find elsewhere in Europe.

The city is divided by the River Danube between the flat plains of Pest on the east side and the wooded hillsides of Buda on the west side. Many birders will pass through Budapest on their way to the birding hotspots of eastern Hungary, particularly during spring and early summer. The steppes, or puszta as they are known in Hungary, consist of a mosaic of natural grasslands and seasonal and permanent wetlands, as well as cultivated areas, which support large numbers of breeding, wintering and migrating birds. This habitat is somewhat fragmented closer to Budapest but, although numbers are smaller, it is possible to see the same variety. Additionally, the hills to the west and the lakes to the southwest add to the variety of habitats and species diversity, giving a range of birding possibilities at any time of year.

Spring and autumn are the best periods for birding as species diversity is high, and temperatures are very pleasant. In spring herons, waders and passerines pass through on migration or are displaying on their breeding grounds. In autumn the wetlands are staging areas for cranes, geese and other wildfowl. Midsummer and midwinter offer less variety, but the resident birds of the Budapest area still offer the visiting birder a glimpse of the rich birdlife of eastern Europe.

TYPICAL SPECIES

Resident
Little Grebe, Great Crested Grebe, Great Cormorant, Grey Heron, Great White Egret, Mute Swan, Mallard, Common Buzzard, Common Kestrel, Moorhen, Common Coot, Common Pheasant, Black-headed Gull, Wood Pigeon, Collared Dove, Great Spotted Woodpecker, Wren, Robin, Blackbird, Fieldfare, Marsh Tit, Blue Tit, Great Tit, Eurasian Nuthatch, Short-toed Treecreeper, Eurasian Jay, Magpie, Hooded Crow, House Sparrow, Tree Sparrow, Chaffinch, Greenfinch, Goldfinch, Linnet

Summer
White Stork, Common Swift, Sand Martin, Barn Swallow, House Martin, Sky Lark, White Wagtail, Common Stonechat, Common Redstart, Black Redstart, Song Thrush, Common Whitethroat, Lesser Whitethroat, Garden Warbler, Blackcap, Willow Warbler, Common Chiffchaff, Spotted Flycatcher, Western Jackdaw, Rook, Common Starling, European Serin

Winter
Common Pochard, Tufted Duck, Common Goldeneye, Yellow-legged Gull, Goldcrest, Coal Tit, Siskin

GETTING AROUND

Like many inland cities, the best birding sites will involve travel for quite some distance outside the urban limits, so a car is needed to make the most efficient use of a short visit. The best birding spots within the city limits are in the Buda Hills, which are easily accessed by public transport. There are chairlifts and a funicular railway, which make climbing into the hills relatively easy, and all are served by buses or trams which depart mainly from Moszkva Square, to the northwest of Buda's tourist area. Most of the other

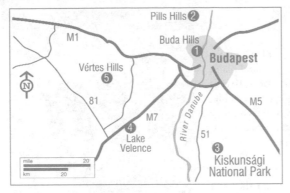

birding sites are served by train but, because of the open nature of the terrain, you would need to walk considerable distances to cover the best sites adequately. You could consider renting a bicycle and carrying this on the train, as Hungary's relatively flat countryside lends itself to cycling. There are a number of different railway stations in Budapest, each serving different parts of the country, so if you are travelling by train to a birding site, it is important to identify your departure point. All of the stations are interconnected by the Metro system.

If driving through the city, try to avoid peak hours as the streets become quite congested. There is a bypass motorway (M0) on the southern side of the city, which makes it relatively easy to reach the birding sites to the south and southwest from the airport. However, this is not a full orbital motorway and to get to the north of the city you need to travel through the city centre and cross the Danube. Beyond the city, driving is relatively stress-free. The rural roads around the birding areas are lightly travelled and there is little problem making roadside stops in order to watch birds, provided you take due care. In the open countryside of the puszta, overhead wires, electricity pylons and roadside fences all provide useful perches for raptors, shrikes and songbirds, so be prepared for unscheduled stops.

Many birding sites are nature reserves of some kind and are managed for the benefit of the wildlife. This means that some parts which are sensitive to disturbance may be off-limits or may be accessed only by permit holders. However, such reserves usually have some public areas which are served by landscaped footpaths and hides making it a little easier to observe the birds at closer quarters: important if you lack a telescope. In addition to a city map, you will also need a map which covers the neighbouring counties of Pest, Fejér and Bács-Kiskun, and you can also obtain trail maps for the national parks at the Park Headquarters.

Locations

1 **Buda Hills** (5km)   (Bus Nos. 21/22/56)

The west bank of the Danube is dominated by the Buda Hills, a north/south ridge of about 20km. They are extensively forested with deciduous woodlands but also have some cleared areas, which were originally vineyards and cultivated areas. The highest point in the hills, János-hegy (530m), is accessible by chairlift and from here there are several woodland trails fanning out to the lower elevations. The hills are popular with tourists, hikers and mountain-bikers but an early morning visit will be relatively undisturbed. Most of the typical deciduous woodland species of central Europe can be seen here in late spring and early summer, including seven species of woodpecker and both species of treecreeper.

There are several points of access to the hills. The **Szilagyi Erzsebet Avenue** going northwest from **Moszkva Square** cuts through the Buda Hills along **Budakeszi**

Road and **Nagykovácsi Road**, with access points to the forest along both roads. If you are using public transport, several bus routes originating in Moszkva Square serve a number of different points in the hills. Bus 158 brings you to the base of the chairlift which ascends up to János-hegy. You can then descend the western slope of János to the János-hegy Station on the so-called Children's Railway – a popular tourist attraction. From there, you may walk or take the railway north to Hüvösvölgy or south to Normafa: both have bus services back to Moszkva Square and both have good habitat in the vicinity.

There are a number of parks on the Pest side of the river, the most notable being the Városliget (City Park) and Népliget. These are not as good as the Buda Hills but still offer a chance to see some of the woodland species.

KEY SPECIES

Resident	Summer	Winter
Northern Goshawk, Eurasian Sparrowhawk, Stock Dove, Green Woodpecker, Black Woodpecker, Syrian Woodpecker, Middle Spotted Woodpecker, Lesser Spotted Woodpecker, Eurasian Treecreeper, Long-tailed Tit, Common Raven, Hawfinch	European Honey-buzzard, Hobby, Turtle Dove, Wryneck, Common Cuckoo, Wood Lark, Tree Pipit, Common Nightingale, Icterine Warbler, Barred Warbler, Wood Warbler, Collared Flycatcher, Pied Flycatcher, Golden Oriole, Red-backed Shrike	Meadow Pipit, Mistle Thrush, Fieldfare, Redwing, Great Grey Shrike, Brambling, Bullfinch, Siskin

2 Pilis Hills (30km)

(Pomáz Railway Station)

The Pilis Hills lie just north of Budapest and are a continuation of the Buda Hills. However, they have less human habitation than the Buda Hills and are contiguous with wider stretches of forest and open country. They provide some of the closest breeding sites to Budapest for Black Storks and Saker Falcons as well as an opportunity to see other raptors on migration. The Pilis Hills form part of the Danube-Ipoly National Park, which embraces large tracts of flood-meadows, forest and hillsides on either side of the Danube Bend. The hills are criss-crossed by hiking and cycling trails but the short-term visitor, with little time to spare, will find it easier to cover the area by car, as the start and/or end points for these trails are rather distant from public transport.

Take **Route 11** north to Pomáz and turn west towards **Esztergom**. This road which is about 30km long cuts right through the Pilis Hills and there are a number of stops along the way which are worth checking as they provide excellent vantage points to scan for raptors, or offer access into the forested areas. After about 4km, turn south towards **Csobánka**, where there is a lookout point, to scan the hillsides. Return to the Pomáz-Esztergom road and continue west towards **Dobogókő**, which at 700m is the highest point to which you can drive. Some of the best forest habitat is in this area and there is a hiking trail from the north side of Dobogókő to Dömös which passes through some good forest habitat including the Rám cliff face, worth a look for raptors and other cliff-dwelling species. You may obtain trail maps at the Visitor Centre in Dobogókő. For those relying on public transport, you can take the train from Batthyány Square in Budapest to Pomáz, and from Pomáz there is a bus service to Dobogókő.

KEY SPECIES

Resident		
Northern Goshawk, Eurasian Sparrowhawk, Stock Dove, Green	Woodpecker, Grey-headed Woodpecker, Black Woodpecker, Syrian Woodpecker, Middle Spotted	Woodpecker, Lesser Spotted Woodpecker, Eurasian Treecreeper, Long-tailed Tit, Common Raven,

Hawfinch, Rock Bunting, Corn Bunting

Summer
Black Stork, European Honey-buzzard, Short-toed Eagle, Hobby, Sober Falcon, Turtle Dove, European

Bee-eater, Wryneck, Common Cuckoo, Wood Lark, Tree Pipit, Northern Wheatear, Common Nightingale, Icterine Warbler, Barred Warbler, Lesser Whitethroat, Collared Flycatcher, Pied Flycatcher, Golden Oriole, Red-backed Shrike

Winter
Meadow Pipit, Mistle Thrush, Redwing, Great Grey Shrike, Brambling, Bullfinch, Common Crossbill

---

3 Kiskunsági National Park (50km)   (Dömsöd Railway Station)

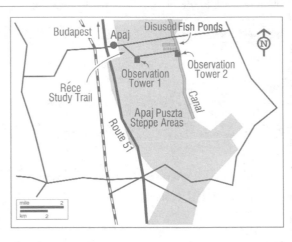

The level plains of the Kiskunsági National Park to the south of Budapest offer the best chance of seeing Great Bustards and other steppe species reasonably close to Budapest. The area is very large and the national park is made up of eight separate and non-contiguous units which stretch for about 75km along a north/south axis, between the Danube and Tisza Rivers. However the Upper-Kiskunság Plain at the northern end of the park is within an hour's drive of Budapest and can easily be done in a day. Each unit represents a unique habitat zone within the puszta supporting a distinct plant and animal community. Much remedial work is being carried out to restore habitats and species, and consequently large areas are either out of bounds or can be visited by permit only. Such permits are usually available only to groups accompanied by a guide. However there are a number of open access areas and nature trails which do not require a permit and these are of most interest to the visiting birder.

Take **Route 51** south from Budapest for about 45km to **Dömsöd**. Go east from Dömsöd to **Apaj** and then continue east from Apaj. There is a public nature trail (**Réce Study Trail**) located 2km east of Apaj on the right-hand (south) side of the road. This trail passes some disused fish ponds, reedbeds and open steppe, with observation towers at the start and finish. An alternative route is the road south from Apaj to **Kunszentmiklós**, which passes through a variety of wet and dry habitats and it is worth pulling over from time to time to check for birds. If you are relying on public transport, your options are more restricted. However, several villages within the national park area are served by train from Budapest-Józsefváros Station and, as the countryside is so flat, you could consider renting a bicycle and taking it on the train.

KEY SPECIES

Resident
Great Bittern, Greylag Goose, Gadwall, Ferruginous Duck, Grey Partridge, Water Rail, Great Bustard, Common Snipe, Common Kingfisher, Crested Lark, Bearded Tit, Penduline Tit,

Yellowhammer, Reed Bunting, Corn Bunting

Summer
Little Bittern, Purple Heron, Night Heron, White Stork, Eurasian

Spoonbill, Garganey, European Honey-buzzard, Black Kite, Red Kite, Marsh Harrier, Montagu's Harrier, Hobby, Common Quail, Spotted Crake, Little Crake, Corn Crake, Collared Pratincole, Black-winged

Stilt, Pied Avocet, Stone Curlew, Little Ringed Plover, Kentish Plover, Northern Lapwing, Ruff, Black-tailed Godwit, European Bee-eater, Hoopoe, European Roller, Tawny Pipit, Blue-headed Wagtail, Common Nightingale, Bluethroat, Grasshopper Warbler, Savi's Warbler, River Warbler, Sedge Warbler, Moustached Warbler, Marsh Warbler, Great Reed Warbler, Icterine Warbler, Golden Oriole, Red-backed Shrike, Lesser Grey Shrike

Winter
Bean Goose, White-fronted Goose, Eurasian Wigeon, Eurasian Teal, Northern Pintail, Northern Shoveler, White-tailed Eagle, Hen Harrier, Rough-legged Buzzard, Common Gull, Yellow-legged Gull, Caspian Gull, Meadow Pipit, Common Redpoll, Twite

Passage
Little Egret, Glossy Ibis, Red-footed Falcon, Common Crane, Dunlin, Little Stint, Temminck's Stint, Curlew Sandpiper, Common Redshank, Spotted Redshank, Greenshank, Eurasian Curlew, Wood Sandpiper, Common Sandpiper, Mediterranean Gull, Little Gull, Whiskered Tern, Black Tern, White-winged Black Tern

4 Lake Velence (50km)

(Velence Railway Station)
(Dinnyés Railway Station)

This is a medium-sized lake southwest of Budapest with extensive reedbeds at its southwestern end, which is a nature reserve. Much of the surface is dotted with reed islands. The lake is quite shallow and rich in nutrients, ideal habitat for dabbling duck, herons, waders and many other aquatic species. Normally, access to the nature reserve requires a permit, although

it may be possible for birders without permits to be admitted. There is a well-located observation tower in the reserve which permits a comprehensive scan over the reedbeds. Even if permission to enter cannot be obtained, much of the lake can be seen from the perimeter road along the east and south shorelines.

Take the **M7** motorway southwest from Budapest and after about 40km exit east at **Velence**. Follow the main road through Velence, keeping the lake on the right-hand side, to **Route 70**. Follow Route 70 along the eastern shore of the lake to **Dinnyés**. At Dinnyés take the road north towards **Pákozd**. The entrance to Velence Nature Reserve is about 2km on the right-hand side of this road. The area around Dinnyés is particularly good for birds: with marshes, fish ponds and flood meadows located between the village and the lake. If using public transport, you could take the train to Velence from Budapest-Déli Station and walk along the lakeshore to Dinnyés, a distance of about 8km, and take the train back to Budapest from there.

KEY SPECIES

Resident
Great Bittern, Greylag Goose, Gadwall, Water Rail, Common Kingfisher, Syrian Woodpecker, Bearded Tit, Penduline Tit, Yellowhammer, Reed Bunting, Corn Bunting

Summer
Little Bittern, Purple Heron, Night Heron, White Stork, Eurasian

Spoonbill, Garganey, Ferruginous Duck, European Honey-buzzard, Black Kite, Marsh Harrier, Montagu's Harrier, Hobby, Spotted Crake, Little Crake, Common Quail, Little Ringed Plover, Kentish Plover, Northern Lapwing, Ruff, Black-tailed Godwit, Common Tern, Turtle Dove, Common Cuckoo, European Bee-eater, Hoopoe,

Tree Pipit, Blue-headed Wagtail, Common Nightingale, Bluethroat, Grasshopper Warbler, Savi's Warbler, River Warbler, Sedge Warbler, Moustached Warbler, Marsh Warbler, Great Reed Warbler, Icterine Warbler, Barred Warbler, Golden Oriole, Red-backed Shrike, Lesser Grey Shrike

Winter	Passage	Redshank, Greenshank, Eurasian
Bean Goose, White-fronted Goose, Eurasian Wigeon, Eurasian Teal, Northern Shoveler, Smew, Common Gull, Caspian Gull, Water Pipit	Common Crane, Black-winged Stilt, Pied Avocet, Dunlin, Little Stint, Temminck's Stint, Curlew Sandpiper, Common Redshank, Spotted	Curlew, Wood Sandpiper, Common Sandpiper, Common Snipe, Little Gull, Black Tern

5 Vértes Hills (45km)

The level plains around Csákvár to the west of Budapest support a rich community of steppe birds, while the Vértes Hills to the north hold small breeding populations of Eastern Imperial Eagles, Lesser Spotted Eagles and Saker Falcons, all of which hunt over the Csákvár area. There is no easy way to 'do' this area, as the plains are extensive with no particular focal point. The best strategy is to drive along the roads, stopping occasionally to scan for birds, and particularly to check every electricity pylon for perched Sakers.

Take the **M1** west from Budapest and after about 35km exit south on **Route 811** towards Csákvár, stopping frequently to check the fields either side of the road. From Csákvár, take the road to **Gánt** up into the Vértes Hills, stopping at any vantage points which permit scanning across the forested slopes.

KEY SPECIES

Resident
Eastern Imperial Eagle, Saker Falcon, Grey Partridge, Crested Lark, Yellowhammer, Corn Bunting

Summer
White Stork, Black Kite, Red Kite, Montagu's Harrier, Lesser Spotted

Eagle, Hobby, Common Quail, Common Cuckoo, Tawny Pipit, Grey Wagtail, Whinchat, Lesser Grey Shrike

Winter
Bean Goose, White-tailed Eagle, Rough-legged Buzzard, Hen Harrier, Peregrine Falcon, Redwing, Great

Grey Shrike, Brambling

Passage
Black Stork, Red-footed Falcon, Eurasian Curlew, Tree Pipit, Meadow Pipit, Water Pipit, Northern Wheatear, Pied Flycatcher

Useful contacts and websites

Magyar Madártani és Természetvédelmi Egyesület MME (Hungarian) www.mme.hu
Hungarian Ornithological and Nature Conservation Society, Kolto u. 21, 391, H-1536, Budapest.
BirdLife partner in Hungary.

Birding Hungary (English)
www.birdinghungary.com
Birding website with information about birding tours, birding sites and species checklist.

Probirder (English) www.probirder.com
Birding tour company operating in Budapest area.

Kiskunsági National Park (Hungarian and English) www.knp.hu
Kiskunság National Park Management Centre, House of Nature, 6000 Kecskemét, Liszt F. u. 19.
National park website with site descriptions, access and information on natural history; and from whom visitor permits are available. Tel. 76 501596/76 500068.

Balaton Uplands National Park
(Hungarian and English) www.bfnpi.hu
Veszprém 8200, Vár u. 31. Tel. (+36) 88 577730
National park website with information about Lake Balaton area.

Hungarian Rarities Committee MME NB
(English) www.c3.hu/~mme/nb/hcrc.htm
Secretary, MME NB, Költö u. 2,1H – 1121
Budapest. Tel. (+36) 1 3957093.

Public Transport in Budapest (English)
publictransport-budapest.heliospanzio.hu

BKV Bus, tram and metro services in Budapest (English) www.bkv.hu

ELVIRA – Hungarian Railway Service (English)
www.elvira.hu

VOLÁN – Local bus services outside Budapest (Hungarian) www.volan.hu

Books and publications

A Guide to Birdwatching in Hungary
Gerard Gorman (1991) Corvina

An Annotated List of the Birds of Hungary
G. Magyar *et al.* (1998)

The Birds of Hungary
Gerard Gorman (1996) Christopher Helm
ISBN 0713642351

Where to Watch Birds in Eastern Europe
Gerard Gorman (1994) Hamlyn, ISBN 060057976X

Collins Bird Guide
Lars Svensson *et al.* (1999) HarperCollins
ISBN 0002197286

Birds of Europe
Lars Jonsson (2006) Christopher Helm
ISBN 0713676000

The Magpie is a familiar species in most European cities

TYPICAL SPECIES

Resident
Great Crested Grebe, Little Grebe, Great Cormorant, Grey Heron, Mute Swan, Mallard, Common Pochard, Tufted Duck, Eurasian Sparrowhawk, Common Buzzard, Common Kestrel, Moorhen, Common Coot, Black-headed Gull, Herring Gull, Common Gull, Great Black-backed Gull, Wood Pigeon, Collared Dove, Great Spotted Woodpecker, Wren, Dunnock, Robin, Blackbird, Fieldfare, Goldcrest, Blue Tit, Great Tit, Eurasian Jay, Magpie, Western Jackdaw, Hooded Crow, Common Starling, House Sparrow, Tree Sparrow, Chaffinch, Greenfinch, Goldfinch, Bullfinch

Summer
Greylag Goose, Common Tern, Common Swift, White Wagtail, Barn Swallow, House Martin, Song Thrush, Reed Warbler, Blackcap, Common Whitethroat, Lesser Whitethroat, Common Chiffchaff, Willow Warbler

Winter
Rough-legged Buzzard, Waxwing, Siskin, Brambling

Copenhagen

Any city on the coast of an island in the Baltic Sea is sure to arouse the interest of any birder. Also its position on the migration flightpath between Scandinavia and Continental Europe and its proximity to some good intact littoral habitat make Copenhagen a real birder's city. It is on the east coast of the island of Zealand (Sjælland), just 30km by the Øresund Bridge from Sweden. Indeed, because of the short distance, the more ambitious could consider visiting the neighbouring Swedish province of Scania, including a day trip to Falsterbo which is a Mecca for raptor enthusiasts and rarity hunters. However, the birder restricted to the immediate vicinity of the city will not be disappointed.

The city is quite small and there are lakes, woodlands and a fine coastal strip, which is rich in birdlife, within a short distance of the centre. Its strategic position for migrants means that it is possible to observe some spectacular diurnal migration within the city itself, and it is not infrequent for raptors or even Common Cranes to be seen migrating overhead during March–April and September–October. There are also good numbers of waterbirds in winter along the shorelines and in the neighbouring wetlands. Mid-summer is probably the least interesting time of year as then most breeding species are silent and difficult to observe. However, the first adult summer-plumaged shorebirds start arriving from their Arctic breeding grounds in July.

GETTING AROUND

Copenhagen is a rather small city, with a compact tourist area in the centre and the airport only 10km away. Like most Continental European cities, the public transport system is excellent and with a one-day or weekend ticket you can utilise buses, trains and the metro to get to the best birding locations. The train service reaches most parts of Zealand, including the coastline, and many

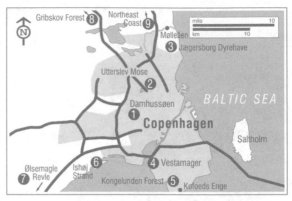

stations are served by a feeder bus system which will get you to your chosen birding site. This offers excellent possibilities to reach the main migration points within an hour and a half. Outside Copenhagen a rental car would offer the best flexibility, particularly if you want to visit two or more sites in the course of a day: car hire is rather expensive in Denmark, however.

Being a rather flat country, you might also consider renting a bicycle, as there is a very comprehensive network of cycle paths which makes cycling a safe, pleasant way to get around, particularly in the summer. Furthermore, it is possible to bring a bicycle on S-trains and Metro trains (check for restrictions during rush hours). Most English-speaking visitors to Denmark are unlikely to speak Danish. This is not too much of a problem since most Danes can converse in word-perfect English, but it is useful to familiarise yourself with some Danish terms for the purpose of following directions and locating sites: e.g. Skov = Forest; Sø = Lake; Vej = Road; Bro = Bridge; Ø (or Holm) = Island; Mose = Marsh or Bog; Fugletårn = Bird Observation Tower; Natursti = Walking trail; Fuglereservat = Bird Reserve; Ingen adgang = No trespassing.

Comfortable walking shoes are all that are required, unless you plan to stray from the footpaths. A telescope is almost a necessity for birdwatching at wetlands and coastal areas, as birds such as ducks, geese and shorebirds can be quite distant. You will need a map that covers the eastern half of Zealand in order to locate all the sites, and it would also be useful to get a city map with a diagram of the public transport system. These are available from any bookshop, public library and tourist office, and at bus and railway stations. Furthermore, Naturbutikken, the shop run by the Danish Ornithological Society (DOF) in the city centre at Vesterbrogade 140, can supply information and leaflets on many of the localities as well as provide binoculars, equipment and birding literature.

Locations

1 Damhussøen (6km)
(Vanløse S-Railway Station)

Damhussøen is a lake in the western suburbs of the city. Although it is partly surrounded by roads with heavy traffic, it holds very large numbers of wintering wildfowl such as Coot, Tufted Duck (up to several thousand) and Common Pochard as well as gulls. There is usually also a wintering flock of the attractive Smew. Another interesting species here is the Short-toed Treecreeper, a few of which have been present along the footpath on the northern shore of the lake in recent years.

Damhussøen is in Vanløse, west of the city centre. Go west on **Vesterbrogade** which becomes **Roskildevej** for about 5km to get to the lakeside. It is easily reached by S-train to Vanløse Station or by Metro to Flintholm Station. From either station there is only a 1km walk to the lake.

KEY SPECIES

Resident
Bearded Tit, Water Rail, Short-toed Treecreeper

Winter
Northern Shoveler, Tufted Duck, Common Pochard, Smew, Common Coot, Goosander, Redwing

2 Utterslev Mose (7km)
(Bus Nos. 2A and 6A)

There is a chain of small lakes and reedbeds across the northern suburbs of Copenhagen, many of which have been preserved for nature conservation. The largest of these is Utterslev Mose, a public park with three large lakes, reedbeds and riparian woodland. It is quite close to the city centre and constitutes a wild oasis in the heart of a built-up area. There is a large Black-headed Gull colony in summer and in spring the most conspicuous bird species is the Greylag Goose, which breeds everywhere along the lakeshores and walking trails. Although very tame, it is still a wild population which leaves the area in August for wintering grounds in Spain. The lakes attract wildfowl in winter, although in much lower numbers than Lake Damhussøen (see above), and a Bittern is a possibility where parts of the lakes remain unfrozen, particularly in severe winters. The area also experiences raptor and crane migration in spring and autumn. Utterslev Mose can be accessed at the west end from **Åkandevej**, served by Bus No. 2A and at the east end from **Frederiksborgvej**, served by Bus No. 6A. There are paved walking trails around the lakes over a distance of 8km.

KEY SPECIES

Resident
Bearded Tit, Water Rail

Summer
Greylag Goose, Black-head Gull, Common Cuckoo, Thrush Nightingale, Sedge Warbler, Marsh Warbler

Winter
Northern Shoveler, Goosander, Redwing

Passage
European Honey-buzzard, Osprey, Marsh Harrier, Common Crane, Sky Lark, Meadow Pipit, Northern Wheatear, Whinchat.

3 Jægersborg Dyrehave and Mølleåen (13km)
(Klampenborg S-Railway Station)
(Lyngby S-Railway Station/Bus Nos. 183/187)

Jægersborg Dyrehave (Deer Park) is a unique woodland area close to Copenhagen, consisting of patches of old forest and meadows, which was once the king's private deer park and hunting grounds. Today it has a large deer population of up to 2,000

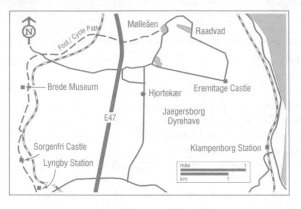

animals of three species. The Short-toed Treecreeper is a relatively scarce species outside southern Jutland but since 1989 has established a population of around 20 breeding pairs in Dyrehaven, making it the largest and most dense population in Denmark. It is mainly seen in the eastern part of the park, which offers an ideal habitat of old oak forest, and it is not unusual to see both European treecreeper species in the same tree. The best places are along the path leading north just after entering the park through the red wooden gate by Klampenborg Station. While the woodland areas provide Stock Doves and other hole-nesting birds, the open areas near Eremitage Castle can be good for watching diurnal migration during westerly winds. Dyrehave is a popular recreational area and if you plan a visit on weekends you should try to get there early.

North of the deer park runs Mølleåen (Mill Brook). The mills and old factory buildings are still visible near Raadvad. This idyllic area, with its many dammed mill-ponds, is the best place near Copenhagen for Dippers and Kingfishers in winter. The forest is mixed, with alder swamp supporting large flocks of Siskins in winter and old beech and oak trees on the slopes offering good roosting places for the resident Tawny Owls.

Jægersborg Dyrehave is reached very easily from Copenhagen by S-train to **Klampenborg Station** (20 min) from where you can walk the 4km north through the deer park to Raadvad and Mølleåen. Alternatively, you can take bus No. 187 or 183 from Lyngby Station to Raadvad or Hjortekær. The upper part of Mølleåen can also be visited on foot walking north from Lyngby, passing the park at Sorgenfri Castle and further along the stream to Brede Museum.

KEY SPECIES

Resident
Water Rail, Tawny Owl, Black Woodpecker, Long-tailed Tit, Marsh Tit, Coal Tit, Eurasian Nuthatch, Eurasian Treecreeper, Short-toed Treecreeper, Yellowhammer, Hawfinch, Common Raven

Summer
Stock Dove, Common Cuckoo, Sky Lark, Tree Pipit, Thrush Nightingale, Common Redstart, Icterine Warbler, Wood Warbler, Garden Warbler, Pied Flycatcher

Winter
Goosander, Northern Goshawk, Common Kingfisher, Dipper, Redwing, Great Grey Shrike, Mealy Redpoll, Siskin

Passage
European Honey-buzzard, Red Kite, Osprey, Merlin, Meadow Pipit, Spotted Flycatcher, Northern Wheatear

4 Vestamager (10km)

(Vestamager Metro Station)
(Bus Nos. 33 and 36)

The city of Copenhagen straddles two islands: Zealand, the 'mainland', and Amager where the airport is located. The southern and southwestern parts of Amager comprise a superb birding area called Vestamager. Within an area of almost 3,000 hectares are lakes, marshes, woodlands and meadows. It is easily accessible from the airport making it worth a visit for anyone on a stop-over at Copenhagen. This is a very large area which is closed to motorised traffic, so if you are taking the bus (No. 36) or a taxi from the airport you can only visit a limited selection of sites. Probably the best way to cover Vestamager and to reach the most bird-rich part, Klydesøen (6km south of Vestamager Metro Station), is by (rental) bike.

Take the **metro** to **Vestamager Station** or the **bus** (No. 31 or 36) to **Finderupvej**, which is a short walk from the Naturcenter Vestamager. It is possible to rent a bicycle here on weekends from 1 May until mid-October (phone 32 52 04 03). From here you can continue west along Finderupvej to the lake at Store Høj. This road continues for a further 3km until it reaches the coast at the Kalvebodbroen Bridge. At this point you can scan the harbour in winter for sea-ducks: in particular this area has good numbers of Smew in severe winters. The small lake just south of the bridge, the Birkedam, holds ducks and grebes. Vestamager also supports a large breeding population of Grasshopper Warblers and this species is easy to find singing during late May.

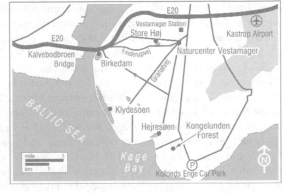

The large lake, Klydesøen, at the southwestern edge of Vestamager is excellent for breeding and wintering wildfowl, and this can be reached by road, Granatvej, from the Nature Centre. In 1997 a major restoration project was initiated, creating an even larger lake and wetland area. There are bird observation towers at the southern end of Kanalvej at the extreme southeast end of the lake, from where you can also check another small lake, Hejresøen (Natosøen), and on the north shore of the lake at Nihøjevej. The open areas north of Klydesøen along Granatvej are good for raptors in winter: including White-tailed Eagles and Peregrine Falcons occasionally, and also particularly during autumn migration. Klydesøen is a closed nature reserve and the birds are best viewed using a telescope from the observation towers or the southern dam.

KEY SPECIES

Resident
Peregrine Falcon, Common Snipe, Sky Lark, Meadow Pipit, Bearded Tit

Summer
Red-necked Grebe, Common Shelduck, Garganey, Gadwall, Marsh Harrier, Water Rail, Northern Lapwing, Oystercatcher, Ringed Plover, Common Redshank, Avocet, Ruff, Little Tern, Arctic Tern, Black Tern, Common Cuckoo, Penduline Tit,

Whinchat, Sky Lark, Meadow Pipit, Blue-headed Wagtail, Grasshopper Warbler, Sedge Warbler, Marsh Warbler, Red-backed Shrike, Linnet, Reed Bunting

Winter
Little Grebe, Whooper Swan, Greylag Goose, Canada Goose, Pintail, Northern Shoveler, Eurasian Wigeon, Smew, Red-breasted Merganser, White-tailed Eagle, Hen Harrier,

Redwing, Great Grey Shrike, Brambling, Twite

Passage
Barnacle Goose, Eurasian Teal, Common Crane, European Honey-buzzard, Osprey, Merlin, Hobby, Red Kite, Spotted Redshank, Greenshank, Sandwich Tern, Green Sandpiper, Wood Sandpiper, Red-throated Pipit, Northern Wheatear

5 Kongelunden Forest and Kofoeds Enge (13km)

(Bus No. 32)

Immediately southeast of Vestamager is an area of mixed woodland which supports a wide cross-section of European passerines and other woodland species. The best time to visit the forest is late spring/early summer, when all summer migrants have arrived and are establishing territories, but it is also worth checking in winter, particularly in good years for irruptive species such as Nutcrackers and Tengmalm's Owls, and Parrot Crossbills and other finches. It can also be a good site for Long-eared Owls, which often roost in pine trees inside the forest in winter. Another aspect of Kongelunden is that it offers a good opportunity to see diurnal migration. In spring, raptors and passerines are seen during west to northwesterly winds, while numbers of Common Cranes may reach up to 2,000 during easterlies in late March to mid-April (the majority in the afternoon).

This location is very accessible both from the city centre and from the airport. The forest entrance is located at the southern end of Kongelundsvej, and is served by Bus No. 32 from Vestamager Metro Station or from Ørestad train and metro station. South of the forest, at Kofoeds Enge, you can access the shoreline which is one of the best areas in the vicinity of Copenhagen for waders at any time of year. For many years, it has been a spot for systematic wildfowl counts and consequently several rarities have been found here. Autumn migration starts with waders in July and continues through September and October with raptors, seabirds and passerines, most being seen in westerly winds and especially during cold northwesterlies.

If you follow Kongelundsvej until it becomes **Skovvej**, then turn west at the intersection with **Kalvebodvej**. Go right to the end of Kalvebodvej where there is a car park and walk southeast along the shore. Bus No. 32 also serves Kalvebodvej. There are some restrictions on access during the breeding season and the area east of Kofoeds Enge is a military zone without public access (heed the warning signs).

KEY SPECIES

Resident
Kongelunden Forest: Common Pheasant, Woodcock, Eurasian Treecreeper, Long-tailed Tit, Coal Tit, Hawfinch, Mealy Redpoll, Siskin

Summer
Kongelunden Forest: Tree Pipit, Thrush Nightingale, Common Redstart, Icterine Warbler, Lesser Whitethroat, Garden Warbler, Wood Warbler, Spotted Flycatcher, Pied Flycatcher.
Kofoeds Enge: Common Redshank

(ssp. *totanus*), Sky Lark, Meadow Pipit, Little Tern, Arctic Tern

Winter
Kongelunden Forest: Long-eared Owl, Redwing, Brambling, Chaffinch
Kofoeds Enge: Canada Goose, Long-tailed Duck, Peregrine Falcon, Common Redshank (ssp. *robusta*)

Passage
Kofoeds Enge: Bewick's Swan, Barnacle Goose, Brent Goose, Common Eider, Common Crane,

Osprey, Merlin, Hobby, Marsh Harrier, Hen Harrier, Red Kite, Grey Plover, European Golden Plover, Jack Snipe, Common Snipe, Dunlin, Curlew Sandpiper, Red Knot, Spotted Redshank, Common Redshank, Greenshank, Bar-tailed Godwit, Eurasian Curlew, Little Gull, Sandwich Tern, Caspian Tern, Black Tern, Wood Pigeon, Stock Dove, Wood Lark, Sand Martin, Red-throated Pipit, Rock Pipit, Grey Wagtail, Northern Wheatear, Common Crossbill

6 Ishøj Strand (20km)

(Ishøj S-Railway Station)

Ishøj is on the coast southwest of Copenhagen and contains a number of interesting lagoons separated from the sea by a narrow sandy beach. The lakes have gull breeding colonies which include a few pairs of Mediterranean Gulls as well as Sandwich Terns.

They also attract migrating waders and wintering wildfowl. Westerly winds in autumn can produce heavy raptor and passerine migration. Although not having quite the numbers and variety of waders as Ølsemagle Revle it is still a good alternative which is closer and quicker to reach from the city centre.

This site can be reached by taking Route **E20** and exiting south at **Exit 26** (**Ishøj Stationvej**). Continue south on Ishøj Stationvej for about 3km until you reach the coast. The lakes are immediately to the west. For those relying on public transport, the easiest way to get here is by S-train to Ishøj Station, from where there is a 1.5km walk to the beach. If you are travelling with a non-birding companion, birdwatching can be combined with a visit to the Arken Museum of Modern Art, located right on Ishøj Strand.

KEY SPECIES

Summer	Winter	Passage
Oystercatcher, Ringed Plover, Northern Lapwing, Mediterranean Gull, Little Tern, Arctic Tern, Sandwich Tern, Meadow Pipit, Sky Lark, Blue-headed Wagtail, Reed Bunting	Red-throated Diver, Eurasian Wigeon, Common Goldeneye, Greater Scaup, Long-tailed Duck, Red-breasted Merganser, Common Redshank, Rock Pipit, Twite, Snow Bunting	Eurasian Teal, Pintail, Northern Shoveler, Grey Plover, European Golden Plover, Common Snipe, Dunlin, Common Redshank, Greenshank, Eurasian Curlew

Further Away

7 Ølsemagle Revle (35km)   (Ølby S-Railway Station/ Bus No. 121)

Ølsemagle is south of Copenhagen on the shores of Køge Bay. There are many beaches in the area, and it is a popular bathing spot, but just offshore is a line of low sand and shingle dunes, which joined to the coast by a causeway at Revlen. The lagoons either side of the causeway are excellent for a wide variety of waders and are a regular site for Broad-billed Sandpipers in July to early September. The best time to visit is late summer and autumn and if your time is limited, the lagoon north of the causeway is usually the best location. More than 20 species of wader can be seen at peak periods of autumn migration, which is very good for western Europe. The surface substrate is sandy rather than mud, so rubber boots are not essential, but a telescope is desirable as some of the birds are quite distant. Note that the more remote parts of the area are closed to the public during the breeding season: 1 April–15 July.

Take the **E20** Motorway south from Copenhagen and after about 20km exit east at **Exit 31**. Take **route 6** to **Jersie Strand** on the coast and turn south on **Københavnsvej**. After about 2km, turn left onto **Revlen** to bring you out to the lagoons. If using public transport, your options are Bus No. 121 from Copenhagen which serves Jersie Strand and Københavnsvej, or you could consider renting a bicycle and taking the S-train to Ølby Station and cycling from there.

KEY SPECIES

Resident	Winter	Passage
Common Eider, Common Shelduck	Canada Goose, Greater Scaup, Long-tailed Duck, Red-breasted Merganser, Hen Harrier, Common Redshank (ssp. *robusta*), Short-eared Owl, Rock Pipit, Twite, Snow Bunting	Osprey, Avocet, Grey Plover, European Golden Plover, Common Snipe, Ruddy Turnstone, Dunlin, Curlew Sandpiper, Red Knot, Spotted Redshank, Greenshank, Bar-tailed Godwit, Eurasian Curlew, Common Sandpiper, Little Gull, Sandwich Tern, Caspian Tern, Black Tern

Summer
Oystercatcher, Ringed Plover, Northern Lapwing, Common Redshank (ssp. *totanus*), Broad-billed Sandpiper, Little Tern, Arctic Tern, Meadow Pipit, Sky Lark, Blue-headed Wagtail, Reed Bunting

8 Gribskov Forest (45km)

(Mårum Railway Station)

This is a large belt of almost unbroken forest across the north of Zealand which holds populations of all typical woodland species as well as being among the best places to see Black Woodpeckers and breeding Common Goldeneye. The forest is very extensive but there is a main road, **Route 227**, running through the centre and it is permitted to hike through the forest from any lay-by.

You can also get there by train from Hillerød which has several stops (Gribsø, Kagerup and Mårum) on its way through the forest, all of which offer opportunities to explore in the vicinity of the station: e.g. the road (Gantekrogsvej) south from Mårum station.

KEY SPECIES

Resident
Northern Goshawk, Black Woodpecker, Lesser Spotted Woodpecker, Treecreeper, Long-tailed Tit, Coal Tit, Marsh Tit, Hawfinch,

Crossbill, Mealy Redpoll

Summer
Common Goldeneye, Hobby, European Honey-buzzard, Tree Pipit, Icterine Warbler, Garden Warbler,

Spotted Flycatcher, Pied Flycatcher, Golden Oriole, Red-backed Shrike

Winter
Fieldfare, Redwing

9 Northeast Coast of Zealand (45km)

(Helsingør Railway Station)
(Gilleleje Railway Station)

Although not so close to Copenhagen, the localities along a 20km stretch of coastline on the northeast coast of Zealand can offer some excellent birdwatching during migration and also give you a possibility of seeing a number of seabirds which are rarely seen closer to the capital. The volume of spring migration here is among the largest in northern Europe and should not be missed by the visiting birder. For the best result, it is very important to check wind direction before giving it a try. Although migration goes on throughout the entire spring, the best migration is seen in April. Apart from wind direction, the weather development to the south of Denmark is also important. If temperatures rise to 20°C in northern Germany in late March or April, this often results in a massive migration the following day in Zealand! Unfortunately the often heavy sea-fog, which occurs in spring over the Baltic area, results in poor visibility and very few birds passing over to Denmark.

Helsingør (Elsinore) has the shortest distance between the Danish and Swedish coasts (3.7km) and it is therefore a place where, in spring and autumn, you can watch both migrating raptors and passerines as well as seabirds passing between the waters of Øresund Sound and the Kattegat Sea. The birds are seen well from Kronborg Castle or, for seabirds, from the pier of the harbour just north of the castle (telescope recommended). The best time for seabirds is when strong northwesterly winds force the birds from Kattegat down towards the mouth of Øresund and this can produce species such as Northern Gannet and Fulmar. In early spring large numbers of Common Eider pass by Kronborg and Puffins, Little Auks and King Eiders are observed occasionally. Helsingør is very easily reached by direct train (50 min) from Copenhagen, and offers a good way of combining birdwatching with a visit to the famous castle in *Hamlet*.

Hellebæk 5km west of Helsingør is the main spot for raptor migration. Observations are from the hilly areas at Hellebæk Avlsgård which is 2 to 3km inland from the coastline. In general, Hellebæk offers the most reliable raptor migration in spring (northerly and westerly winds), even in light winds when migration is less apparent on the coast.

Gilbjerg Hoved is the most northerly point in Zealand and an important spring migration point for raptors and passerines (southeasterly winds). The volume of migrants can be extraordinary on occasions, with thousands of sparrowhawks and buzzards and tens of thousands of finches and thrushes. Both localities have White-tailed Eagles and Peregrine Falcons as regular migrants and, as for Kronborg, Gilbjerg Hoved and the western pier in the harbour of Gilleleje are good places to seawatch, in particular in autumn during strong northwesterly winds.

KEY SPECIES

Winter
Red-throated Diver, Black-throated Diver, Northern Gannet, Fulmar, Common Eider, Common Goldeneye, Common Scoter, Velvet Scoter,

Goosander, Purple Sandpiper, Rock Pipit

Passage
European Honey-buzzard, Hen Harrier, Marsh Harrier, Red Kite,

Merlin, Hobby, Osprey, Common Crane, Little Gull, Stock Dove, Wood Lark, Sand Martin, Grey Wagtail, Northern Wheatear, Great Grey Shrike, Common Crossbill

Useful contacts and websites

Dansk Ornitologisk Forening (DOF) – Danish Ornithological Society (Danish)
www.dof.dk
Vesterbrogade 140, DK-1620 Copenhagen.
Tel. (+45) 3331 4404.
Main Danish ornithological organisation and BirdLife International partner. Publishers of *Dansk Ornitologisk Forenings Tidsskrift* journal containing the Danish Bird Report with summary in English.

Dansk Ornitologisk Forening (DOF) Copenhagen Branch (Danish)   www.dofkbh.dk
Specific information about birds in the Copenhagen area.

Netfugl.dk (English and Danish) www.netfugl.dk
The best general site for birding in Denmark.

Birdwatch.dk (Danish)
www.birdwatch.dk
Another general site about birding in Denmark.

Copenhagen Bird Ringing Centre (English and Danish) www.zmuc.dk
Copenhagen Bird Ringing Centre, Zoological Museum, Universitetsparken 15, DK-2100 Copenhagen. Tel. (+45) 3532 1029

Kongelundens Fuglestation – (Danish)
www.dofkbh.dk/kongelunden/kongelunden.htm
Birding information about the Kongelunden area of Amager including map and checklist.

Gilbjerg Fuglestation – Gilbjerg Hoved Bird Observatory (Danish)
www.dofkbh.dk/gilbjerg
c/o Dansk Folkeferie, Tinkerup Strandvej 8A, DK-3250 Gillelej.

Birding Spots in Denmark (English)
www.linander.dk/stig/birds_e.htm

Erik Agertoft's personal website (Danish)
http://home24.inet.tele.dk/agni/
Personal website with reports, checklists and site descriptions for Gundsømagle Sø and for Kagsmosen (a small marsh in north Copenhagen).

Copenhagen Metro Service (English)
www.m.dk/en/welcome.htm

Greater Copenhagen Authority (English)
Public transport in the Copenhagen area.
www.hur.dk

Books and publications

Important Bird Areas (IBA) in the Baltic Sea
BirdLife International

Fuglenes Danmark Michael Borch Grell (DOF)
Danish Atlas of breeding, migratory and wintering birds.

Danmarks fugle – en oversigt
Klaus Malling Olsen (1992) (DOF) Copenhagen

Collins Bird Guide
Lars Svensson *et al.* (1999) HarperCollins
ISBN 0002197286

Birds of Europe
Lars Jonsson (2006) Christopher Helm
ISBN 0713676000

Pale-bellied Brent Geese flying past Dublin power station

TYPICAL SPECIES

Resident
Great Cormorant, Grey Heron, Mute Swan, Mallard, Common Kestrel, Moorhen, Common Coot, Oystercatcher, Black-headed Gull, Herring Gull, Great Black-backed Gull, Wood Pigeon, Collared Dove, Pied Wagtail, Wren, Dunnock, Robin, Blackbird, Song Thrush, Goldcrest, Blue Tit, Great Tit, Magpie, Western Jackdaw, Hooded Crow, Rook, Common Starling, House Sparrow, Chaffinch, Greenfinch, Goldfinch, Linnet

Summer
Common Tern, Sandwich Tern, Lesser Black-backed Gull, Common Swift, Barn Swallow, House Martin, Blackcap, Common Chiffchaff, Willow Warbler

Winter
Brent Goose, Eurasian Wigeon, Eurasian Teal, Common Gull, Eurasian Curlew, Common Redshank, Turnstone, Blackcap

Dublin

Although the island of Ireland has a rather poor avifauna in terms of species diversity, this seems to be compensated by the numbers of birds, and a visitor to Dublin at any time of the year will be immediately struck by this abundance. Even in the city centre, the skies seem to be filled with gulls, corvids, Wood Pigeons and Starlings. Dublin is at the mouth of the River Liffey, on the shores of a sweeping C-shaped bay with rather low mountains to the south and an extensive fertile plain to the north and west. Unlike many European capitals it is also a busy seaport and although much habitat has been consumed by the development of the port, there are still extensive mudflats in Dublin Bay which are easily accessible to the birder. These areas are the focus of attention for Dublin birders during the winter and at migration times. Summer is a quieter period but there are several large seabird colonies both north and south of Dublin Bay, which support populations of almost all of the breeding seabirds of the northeast Atlantic.

For the European birder it is the coast which holds the most attraction, as the terrestrial avifauna of Ireland is rather impoverished by Continental standards and there are no species endemic to the island. However visitors from North America and from the southern hemisphere should try to make some time to visit a city park, such as Phoenix Park mentioned on p. 245, to see some of the Palearctic landbirds, particularly if your trip will not include any other European destination.

GETTING AROUND

Although Dubliners constantly complain about it, the public transport system is both cheap and comprehensive and works quite well for the birder's purposes. Business and leisure travellers will most likely be staying in a city-centre hotel and, as the birding locations are out of town, even during rush hour you will be travelling against the flow so traffic should not be too great a hindrance. However if you are staying out of town, you will need to adjust your schedule so that any travelling through the centre is accomplished outside of peak hours. For crossing the city by car, it is best to use the M50 Motorway which runs around the perimeter, and avoid driving through the centre.

Although Dublin has no subway, the DART train system provides a fast and frequent service to the coastal strip along Dublin Bay from three city-centre stations and this can be used to access some of the best coastal birding sites. The bus system serves all parts of the city including all of the birding sites mentioned here. Entrance is free to all the locations included. Also bear in mind that Ireland is undergoing a long and tortuous transition from the Imperial System to the Metric System, so that old signposts indicate miles while new ones indicate kilometres, and all car odometers are still calibrated in miles.

Birding in Dublin is not particularly weather depend-

ent, but the state of the tide does make a big difference when birding at the coastal locations. Comfortable walking shoes are adequate for most of the locations described but, in some cases as indicated, rubber boots would be necessary to cover areas of salt-marsh and mudflat. Special care should be exercised when visiting any seabird colonies as the sea cliffs can be extremely treacherous, especially in wet weather.

Locations

1 Phoenix Park (3 miles)   (Bus Nos. 25, 66, 67)

The Phoenix Park is the largest enclosed park in Europe, with a variety of habitats including mixed woodland, freshwater ponds and grassland. It even has its own herd of Fallow Deer. The western end is probably the least disturbed and contains the widest variety of habitats in a relatively small area. By carefully walking through the park early in the morning, it should be possible to see all the usual woodland passerines, including Coal Tits and Eurasian Jays, which are both represented by a race endemic to Ireland. In a city which has surprisingly few freshwater habitats, the ponds in the park are locally important for breeding Little Grebes and ducks. The best plan is to take the bus to **Knockmaroon Gate** and walk through the park as far as the Furry Glen.

Other parks which host a similar selection of species, although in smaller numbers, are Marlay Park and Bushy Park on the south side of the city, and St Anne's Park on the north side adjacent to the North Bull Island.

KEY SPECIES

Resident
Little Grebe, Tufted Duck, Eurasian Sparrowhawk, Grey Wagtail, Mistle Thrush, Coal Tit, Long-tailed Tit, Eurasian Treecreeper, Eurasian Jay

Winter
Common Pochard, Common Kingfisher, Meadow Pipit, Redwing, Fieldfare, Siskin, Lesser Redpoll

2 North Bull Island (5 miles)

(Bus No. 30)

The North Bull Island and its lagoon were formed on the north shore of Dublin Bay in the 19th century when the Bull Wall was constructed to prevent silting in Dublin Port. The resultant coastal lagoon and mudflats are now one of Ireland's top bird sites and the oldest bird sanctuary in the country. Autumn, winter and spring provide a wonderful bird spectacle with large flocks of wildfowl, waders and gulls feeding on the rich mudflats. Such large numbers of birds attract Peregrine Falcons, Merlins and Short-eared Owls in the winter. Most of these birds depart for the summer months, although stragglers of any species may remain, and terns can be seen feeding in the lagoon: there is a breeding colony nearby in Dublin Port.

Most species can be seen by walking along the **James Larkin Road** at low tide and scanning the mudflats and channel. The island can be accessed by two routes, either the wooden bridge at the south end or the causeway at the centre of the island. At high tide the southern side of the wooden bridge and Bull Wall provide opportunities to see divers, grebes and sea ducks as well as several gull species. The sandy beach on the seaward side of the island adjacent to the Bull Wall affords the best views of Sanderlings, Common Ringed Plovers and Red Knots, and occasionally Snow Buntings.

KEY SPECIES

Resident
Shag, Common Shelduck, Ringed Plover, Sky Lark, Meadow Pipit, Common Stonechat, Reed Bunting

Winter
Red-throated Diver, Great Northern Diver, Great Crested Grebe, Little

Egret, Northern Shoveler, Pintail, Common Goldeneye, Red-breasted Merganser, Peregrine Falcon, Merlin, Northern Lapwing, Common Snipe, European Golden Plover, Grey Plover, Red Knot, Dunlin, Sanderling, Common Greenshank, Bar-tailed Godwit, Black-tailed Godwit, Short-

eared Owl, Snow Bunting

Passage
Whimbrel, Ruff, Curlew Sandpiper, Arctic Tern, Common Cuckoo, Northern Wheatear

3 Swords Estuary (10 miles)

(Bus Nos. 41 and 41C)
(Malahide DART Station)

The rather small and insignificant Broadmeadows River enters the sea just north of Swords where it has formed a very wide estuary. A railway causeway with just a narrow outlet was constructed across the outer estuary and the resulting constriction of tidal movements has created a large lagoon of brackish water. The upper end of the estuary has a small area of mudflats which is very attractive to passage waders: The birds are easily viewed from the road running alongside the south shore of the estuary

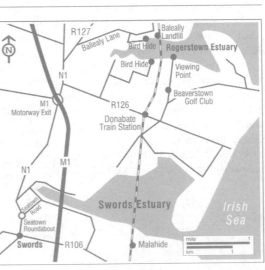

as far as Malahide. In many ways it is easier to cover this estuary than the other coastal sites as the birds are usually much closer to the road.

This site is less than ten minutes drive from Dublin Airport and is well worth a look if you have time at either the beginning or end of your visit. Take the **N1** north from Dublin to **Swords** and at Swords turn east on **Seatown Road** for about a mile. This road leads down to the south shore of the upper estuary. You can drive along the length of the south shore until you reach Malahide at the mouth of the lower estuary. If using public transport, bus routes 41 and 41C serve Swords via Dublin Airport. If you walk the length of the estuary to Malahide you can return to Dublin from Malahide by bus or DART train.

KEY SPECIES

Resident	Winter	Passage
Common Shelduck, Ringed Plover, Common Snipe, Common Kingfisher, Grey Wagtail, Tree Sparrow, Yellowhammer	Great Crested Grebe, Common Pochard, Common Goldeneye, Red-breasted Merganser, Northern Lapwing, European Golden Plover, Dunlin, Common Greenshank, Black-tailed Godwit, Stock Dove	Whimbrel, Ruff, Curlew Sandpiper, Little Stint, Spotted Redshank, Common Sandpiper, Mediterranean Gull, Little Gull

4 Rogerstown Estuary (15 miles)

(Bus Nos. 33 and 33X)
(Donabate Railway Station)

This is the most northerly of the Dublin estuaries and is located in an area of farmland, well clear of the urban sprawl of Dublin. Consequently it provides an opportunity to see a greater range of farmland and hedgerow species. There is a large landfill site on the north shore which attracts large numbers of gulls and it is the best site in Dublin for Yellow-legged Gulls as well as northern gulls. The main focus for birders is the upper estuary, east of the railway causeway, but the lower estuary is also worth a look as it can hold large numbers of geese and ducks. However, this is a very extensive mud-flat and a telescope is necessary.

Take the **M1** north from Dublin and exit east at the **Skerries** exit onto the **N1**. After about a mile turn right onto the **R127** towards Skerries but turn right again after about 100 yards onto **Baleally Lane** towards **Baleally Tiphead**. The entrance to the **Birdwatch Ireland Reserve** is about half a mile on the right-hand side of this road. (Drive slowly; it is easy to miss the entrance gate.) There is a hide here which enables you to check the north shore of the inner estuary. You can reach the south shore of the inner estuary by returning to the M1 and then turning east on the **R126** towards **Donabate**. Continue for about half a mile before coming to the sign pointing to the Birdwatch Ireland hide. Turn left at the sign and drive along an unsurfaced track through the fields and take the right-hand turn past the allotments. The gate closes at dusk so be sure to return before this time, which is indicated on the sign at the entrance. The hide can only be accessed by keyholders but it is still possible to view the estuary by standing alongside it.

You can then get to the outer estuary by returning to the R126 and continuing east-wards until you reach the village of Donabate. Turn left in the village and follow the signs for **Beaverstown Golf Club**. Drive past the golf club until the road reaches the shore of the estuary. There is a bus service (Route No. 33) to Baleally Lane and there is a train service to Dublin from Donabate Station.

5 Howth (10 miles)

(Howth DART Station)
(Bus No.31)

The Howth Peninsula rises to a height of 985ft and forms the northern boundary of Dublin Bay. It is surrounded on the northern and eastern perimeter by sea cliffs and the interior is a mixture of heath and woodland. During spring and summer, the cliffs hold large numbers of breeding seabirds, while onshore winds in late summer and autumn provide an opportunity to seawatch from the eastern side of the peninsula and Manx Shearwaters, terns and skuas can be seen. Howth Harbour is a busy fishing centre and in winter holds large flocks of gulls, offering the opportunity to see both Iceland and Glaucous Gulls. Ireland's Eye is a small island about a mile offshore. In addition to the seabirds on Howth Head, it also hosts a Northern Gannet colony and a small Atlantic Puffin colony. Boats to the island are available from the East Pier of Howth Harbour.

To reach the cliffs, follow **Balscadden Road** from the **East Pier** of the harbour for about a mile to a car park at the start of the footpath along the top of the sea cliffs. Great care should be exercised when walking along the cliff path as the cliffs are very treacherous.

6 Sandymount Strand (4 miles)

(Booterstown DART Station)

Vast sandflats are exposed at low tide on the southern shore of Dublin Bay. Although these are less productive than mudflats, they still support a wide variety of waders in autumn, winter and spring, as well as the ubiquitous Brent Geese. However the big attraction of Sandymount is the large tern roost in the autumn and large gull roosts in the winter and spring. The tern roost contains up to 5,000 birds in August and September and offers one of the best opportunities to see Roseate Terns. Approximately 500 of these rare terns roost among the much larger numbers of Common and Arctic Terns. The gull flock in winter can contain up to ten gull species and this is one of the few sites in Europe where Ring-billed Gulls can be guaranteed.

This location is easy to get to, by taking the DART to Booterstown Station and walking back towards the city along the shoreline at low tide. Ideally a telescope would offer a better opportunity to sort through the large number of birds, but the flocks are reasonably tolerant of close approach. However, walking out on to the sandflats would require rubber boots and care should be exercised when the tide is coming in so that you don't become marooned on a sandbar!

KEY SPECIES

Resident	European Golden Plover, Grey Plover,	Passage
Shag, Common Shelduck, Ringed Plover, Meadow Pipit	Red Knot, Dunlin, Sanderling, Common Greenshank, Bar-tailed Godwit, Black-tailed Godwit, Mediterranean Gull, Ring-billed Gull	Whimbrel, Ruff, Curlew Sandpiper, Arctic Skua, Common Tern, Arctic Tern, Roseate Tern, Black Tern
Winter		
Brent Goose, Northern Lapwing,		

7 South Dublin Bay (7 miles)

(Dun Laoghaire DART Station)
(Dalkey DART Station)

Dun Laoghaire Harbour is a deepwater harbour on the south shore of Dublin Bay and in winter supports a wide variety of divers, grebes and sea ducks. The best opportunity to see these birds is to walk along the West Pier, looking on the seaward side of the pier wall. Ideally, a telescope is required, but at high tide the birds can be observed at closer quarters. Purple Sandpipers may be seen at the tip of the pier and Black Redstarts and Snow Buntings are possible in winter. About half a mile further south is Sandycove, the best location in Dublin to see Mediterranean Gulls, and further south again at the southern tip of Dublin Bay lies Dalkey Island, about 100 yards offshore. In summer this supports a mixed tern colony, which includes small numbers of Roseate Terns. The tern population is augmented by migrants in late summer and autumn and although numbers are smaller here than on Sandymount Strand, the birds are often easier to see.

Dun Laoghaire, Sandycove and Dalkey are all served by bus and adjacent DART stations.

KEY SPECIES

Resident	Summer	Goldeneye, Long-tailed Duck,
Shag, Common Shelduck, Ringed Plover, Kittiwake, Black Guillemot, Meadow Pipit, Rock Pipit	Arctic Skua, Arctic Tern, Roseate Tern	Common Scoter, Red-breasted Merganser, Purple Sandpiper, Mediterranean Gull, Little Gull, Black Redstart, Snow Bunting
	Winter	
	Red-throated Diver, Great Northern Diver, Great Crested Grebe, Common	

8 Kilcoole Marsh (15 miles)

(Bus Nos. 84 and 84X)

The coast of north County Wicklow, just to the south of Dublin, is completely free of human dwellings for a distance of almost 12 miles between Greystones and Wicklow town. The low-lying land is prone to winter flooding. There is a shingle beach running the entire length of the coast which supports Ireland's largest Little Tern colony. The whole area forms the Murrough Wetlands Complex. Midway along this stretch of coastline, Kilcoole Marsh is an important wetland, part of which is a Birdwatch Ireland Reserve, and although numbers and variety of wintering wildfowl and waders are lower than the North Dublin estuaries, it is a locally important site for breeding ducks, waders and other aquatic species. Onshore winds during the autumn can produce a significant passage of seabirds.

Take the **N11** south from Dublin and after about ten miles turn east on the **R762** towards **Delgany**. Follow the signs for Kilcoole and at Kilcoole village turn east on **Sea Road** towards **Kilcoole Railway Station**, where there is a car park. You may walk south along the railway line as far as the Birdwatch Ireland Reserve at **Blackditch** which is a distance of about three miles, checking the marsh on the inland side and scanning the sea for divers, ducks and seabirds.

KEY SPECIES

Resident
Shag, Common Shelduck, Common Pheasant, Water Rail, Ringed Plover, Northern Lapwing, Common Snipe, Common Kingfisher, Sky Lark, Meadow Pipit, Rock Pipit, Common Stonechat, Linnet, Reed Bunting

Summer
Northern Gannet, Common Sandpiper, Little Tern, Common Cuckoo, Sand Martin, Northern Wheatear, Sedge Warbler

Winter
Red-throated Diver, Little Egret, Whooper Swan, Greylag Goose, Red-breasted Merganser, European

Golden Plover, Dunlin, Common Greenshank, Little Gull, Redwing, Fieldfare, Lesser Redpoll

Passage
Manx Shearwater, Common Scoter, Whimbrel, Ruff, Curlew Sandpiper, Arctic Skua, Great Skua, Common Guillemot, Razorbill, Whinchat, Grasshopper Warbler

---

9 **Killakee Mountain** (10 miles)   (Bus No.161)

The 'Dublin Mountains' are, in fact, the northern slopes of the Wicklow Mountains which are located in South County Dublin. They only rise to about 2,000ft but provide some coniferous forest and heather moorland habitat, as well as superb views of Dublin City and Bay. Killakee Mountain is one of the summits closest to Dublin, and the heather moor on the northern slope provides the best opportunity to see Red Grouse, a race of Willow Grouse which is endemic to Britain and Ireland.

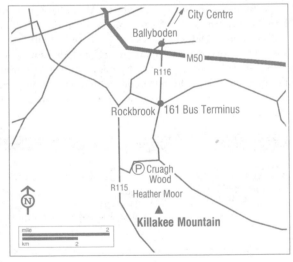

From **Ballyboden** in the southern suburbs of Dublin take the **R116** south for approximately 2.5 miles following the signs for Cruagh Wood. If using public transport, take Bus No. 161 to its terminus on the R116 at Rockbrook. Walk south on the R116 for about a mile until you arrive at a T-junction. Check any of the streams either side of this road for Dippers, which are of a race endemic to Ireland. At the T-junction, turn right and after about 300 yards you will arrive at the public entrance to **Cruagh Wood**. There are public footpaths within the forest. If you walk through in the late evening, you should be able to hear and see Eurasian Woodcock as they leave their roosts.

Continuing past Cruagh Wood onto the **R115**, the road veers to the left and after approximately a mile from this bend it emerges from the forest into open moorland carpeted in heather. Walk onto the moor on the eastern side of the road and you should be able to flush Red Grouse from among the heather. Overall, the distance from the bus stop to the open moor is about three miles and involves an ascent of over 1,000ft, so comfortable walking shoes are required as well as weatherproof clothing.

KEY SPECIES

Resident	Raven, Siskin, Lesser Redpoll,	Sedge Warbler, Common Whitethroat
Red Grouse, Woodcock, Sky Lark,	Common Crossbill	**Winter**
Meadow Pipit, Grey Wagtail, Common	**Summer**	Fieldfare, Redwing, Brambling
Stonechat, Mistle Thrush, Dipper,	Common Cuckoo, Whinchat, Northern	
Coal Tit, Long-tailed Tit, Eurasian	Wheatear, Grasshopper Warbler,	
Treecreeper, Eurasian Jay, Common		

10 Glendalough (30 miles)

Glendalough is one of many glens or valleys in the Wicklow Mountains, south of Dublin. There are two mountain lakes in the valley and the steep surrounding hillsides are forested with oak and birch as well as coniferous plantations. The sixth century monastic site at Glendalough is a key element of the tourist trail out of Dublin, particularly for tour buses. However, in early summer, the surrounding oakwoods are the stronghold in Ireland of the Wood Warbler and also support one or two pairs of Common Redstarts. There is a good selection of other woodland species and the Upper Lake is visited by small numbers of Goosanders. There are trails enabling you to walk through the woodland and around the lakes. However this is a very popular site with tourists and members of the public, so an early morning visit is essential. Late May and early June are the best times to ensure that all the summer migrants have arrived and are establishing territories. However, foliage growth is so advanced by then that you must rely on song to locate the birds.

Take the **M11** south from Dublin and turn west at **Kilmacanogue** onto the **R755**. Follow this road for about 16 miles through the village of **Roundwood** towards Glendalough, where there is a large car park between the two lakes. The woodland trails commence at the car park and follow along both sides of the Upper Lake. If you don't have a car, there is a bus service operated by the St Kevin Bus Company from Dublin and from Bray to Glendalough, but the buses are infrequent and there is no early morning service to get you there before the crowds build up.

KEY SPECIES

Resident	Treecreeper, Eurasian Jay, Common	Common Redstart, Northern
Goosander, Peregrine Falcon,	Raven, Siskin, Lesser Redpoll	Wheatear, Grasshopper Warbler,
Woodcock, Grey Wagtail, Common	**Summer**	Sedge Warbler, Common Whitethroat,
Stonechat, Mistle Thrush, Dipper,	Common Cuckoo, Sand Martin,	Wood Warbler, Spotted Flycatcher
Coal Tit, Long-tailed Tit, Eurasian		

11 Blessington Lake (17 miles) (Bus No. 65)

Blessington Lake is actually a large, rather deep, reservoir (also known as Poulaphouca Reservoir), formed by constructing a dam across the River Liffey, and is the largest body of fresh water in the Dublin area. Consequently, it is important for wintering diving ducks such as Common Pochard, although being inland, it tends to be neglected by Dublin birders. There is also a wintering flock of Greylag Geese which usually includes a small number of Greenland White-fronted Geese. The surrounding farmland and woodland support all of the typical passerine species and the beechwoods along the western shore are good for flocks of Bramblings in winter, particularly in years that produce a good crop of beechmast.

Take the **N81** southwest from the **M50** as far as Blessington village. Turn left (east) in the village to get to the lakeshore and take the road northeast along the lakeshore to get to the fields at **Threecastles**, which is normally where the geese feed. There is a regular bus service (Route 65) from Dublin to Blessington.

KEY SPECIES

Resident
Little Grebe, Great Crested Grebe, Tufted Duck, Common Pheasant, Ringed Plover, Common Snipe, Common Kingfisher, Grey Wagtail, Mistle Thrush, Coal Tit, Long-tailed

Tit, Eurasian Treecreeper, Common Raven, Eurasian Jay, Siskin, Lesser Redpoll

Winter
Whooper Swan, Greylag Goose,

White-fronted Goose, Tufted Duck, Common Pochard, Common Goldeneye, Northern Lapwing, European Golden Plover, Common Sandpiper, Lesser Black-backed Gull, Redwing, Fieldfare, Brambling

Useful contacts and websites

Birdwatch Ireland
www.birdwatchireland.ie
Birdwatch Ireland, Rockingham House, Newcastle, Co. Wicklow. Tel. 01 2804322. Fax. 01 2844407.
National bird conservation organisation. BirdLife International partner for Ireland.

Birdwatch Ireland – Fingal
www.bwifingal.ie
North Dublin Branch of Birdwatch Ireland. Website provides information about local sites and checklist.

Birdwatch Ireland – South Dublin
www.birdweb.net
Information about birding and bird conservation work in South Dublin.

Birdwatch Ireland – Tolka Branch
http://136.206.208.1/tolka/
Northwest Dublin Branch of Birdwatch Ireland. Information about local excursions.

Irish Rare Birds Committee
www.birdwatchireland.ie/bwi/irbc
Details of rare bird reporting in Ireland.

Irishbirding.com www.irishbirding.com
Website with topical information about birdwatching in Ireland with news and photographs of the past week's important sightings.

BirdsIreland.com www.birdsireland.com
Provides information about guided tours and recent reports. Operates the Birds of Ireland News Service.

Irish Bird Images
www.irishbirdimages.com
Photographic website with images of Irish birds.

WildlifeSnaps.com
www.wildlifesnaps.com
Photographic website with images of Irish birds, butterflies and dragonflies.

Birdline Birds of Ireland News Service
Tel. 155011700 for news. Tel. 01 8307364 to leave an update.

Local Recorder: Editor, East Coast Bird Report, c/o Birdwatch Ireland, Rockingham House, Newcastle, Co. Wicklow.

Dublin Bus – bus service for Dublin and suburbs. Tel. 01 8720000. www.dublinbus.ie

St Kevin's Bus Service – bus service to Glendalough. www.glendaloughbus.com/index.html

Iarnród Éireann – Dublin suburban train service. Tel. 01 8366222.
www.irishrail.ie

Tidal Information for Dublin Bay
www.ireland.com/weather/tides.htm

Books and publications

Where to Watch Birds in Ireland
Clive Hutchinson (1994) Christopher Helm
ISBN 0173638273

Birds of Ireland
Gordon D'Arcy (1986) Appletree Press
ISBN 0862811627

Checklist of the Birds of Ireland
IRBC (1998) Birdwatch Ireland
ISBN 1899204105

The Complete Guide to Ireland's Birds
Eric Dempsey and Michael O'Clery (1993) Gill &
Macmillan, ISBN 0717119734

Collins Bird Guide
Lars Svensson *et al.* (1999) HarperCollins
ISBN 0002197286

Birds of Europe
Lars Jonsson (2006) Christopher Helm
ISBN 0713676000

Frankfurt

The tinkling song of the Serin can be heard above the traffic in central Frankfurt

Frankfurt is near the confluence of the River Main and the River Rhine in central Germany. It is surrounded by a flat plain which is heavily populated and criss-crossed by motorways and railways, but considerable pockets of mixed and deciduous forest, and areas of intensive farmland, are located between the urban areas and surrounding Frankfurt Airport. Germany would not normally be considered a birding destination in its own right as its avifauna, although a rich and varied selection of Continental European species, is merely a subset of what can be found at the more popular destinations in eastern Europe and in the Mediterranean. That said, Frankfurt's location and ease of access to some good habitat makes it possible to find some really good birds with little effort.

In general, the spring and summer, when all summer migrants have arrived to complement the resident species, is best for landbirds. Winter is best for aquatic birds as resident numbers are swollen by winter arrivals. March and April are best for viewing woodpeckers as the birds are quite vocal at this time and foliage growth is still sparse, enabling you to see them.

If you have the time and the use of a car, you could also consider a trip to the Taunus Mountains northwest of Frankfurt. The extensive coniferous forests that clad these slopes support small populations of rarer European species such as Hazel Grouse and Tengmalm's Owl. However these birds are very thinly distributed and your chances of seeing them are very limited without local knowledge.

TYPICAL SPECIES

Resident
Little Grebe, Great Crested Grebe, Great Cormorant, Grey Heron, Mute Swan, Mallard, Tufted Duck, Common Pochard, Common Buzzard, Common Kestrel, Moorhen, Common Coot, Black-headed Gull, Common Gull, Wood Pigeon, Collared Dove, Green Woodpecker, Great Spotted Woodpecker, Wren, Dunnock, Robin, Blackbird, Fieldfare, Goldcrest, Marsh Tit, Coal Tit, Blue Tit, Great Tit, Eurasian Nuthatch, Short-toed Treecreeper, Eurasian Jay, Carrion Crow, Common Starling, Chaffinch, Greenfinch, Linnet, Bullfinch, Hawfinch, Yellowhammer

Summer	Black Redstart, Common Redstart,	Passage
Garganey, European Honey-buzzard, Black Kite, Marsh Harrier, Hobby, Common Swift, Sand Martin, Barn Swallow, House Martin, White Wagtail,	Reed Warbler, Blackcap, Common Whitethroat, Common Chiffchaff, Wood Warbler, Firecrest, Red-backed Shrike, Serin	White Stork, Common Crane, Osprey

GETTING AROUND

The city is very well served by an integrated bus, tram and subway system for which you can obtain an inexpensive visitor's day or weekend pass (Hotelgastkarten) at your hotel. The system works very well even at peak periods. Also, if you plan to use the train to go birding, bear in mind that it is possible to rent a bike at many railway stations in Germany or to carry one on the train, thus making many sites much more accessible.

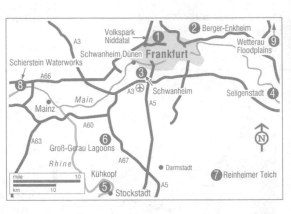

Many of the sites mentioned in this chapter are nature reserves and are designated by the letters NSG (Naturschutzgebiet). Unlike parks and nature reserves in other countries, in Germany these areas are used extensively by the public for recreation, particularly if they are close to an urban area. So expect substantial numbers of joggers, cyclists and dog-walkers even midweek and early in the morning. There really is no time when you can expect to have the area to yourself. Birding in Frankfurt is not really weather dependent. Comfortable walking shoes are adequate for all of the locations described below.

The best map to use in conjunction with those in this guide is the Falk Cityplan 1:16500 (www.falk.de) which illustrates bus and tramways and also includes a subway plan. It is available from bookshops, tourist offices, bus and rail stations.

Locations

1 Volkspark Niddatal (5km)

(Bus No. 67)

This is a large public park near Praunheim in the northwest of the city which provides an opportunity to see some woodland species if visited early in the morning. Take Bus No. 67 from the city centre and get off at the Geiselwiesen stop. There is a park entrance 100m east of the bus stop. There is a mixture of open grass and deciduous woodland but the absence of a lake in the park grounds limits the range of possible species.

KEY SPECIES

Resident	Woodpecker, White Wagtail, Song	Winter
Eurasian Sparrowhawk, Grey-headed	Thrush, Mistle Thrush, Long-tailed Tit	Brambling, Hawfinch

2 Berger-Enkheim (6km)

(Enkheim U-Bahn Station U7)

The suburb of Berger-Enkheim lies in the east of Frankfurt on the north bank of the Main. There are two nature reserves in the area: Enkheimer Marsh (NSG Enkheimer Ried) and Berger Hill (NSG Am Berger Hang). Both reserves complement each other as they support a different range of species, and this area could provide a whole day's birding since they are set amidst a larger area of forests and small lakes. The Lower Main Bird Observatory (VBU) is located in Berger Hill Reserve and information about the work of the observatory, including recent sightings, is available here.

The Enkheimer Ried is accessed at several points on the shoreline by a footpath, with good vantage points over the open water. Several aquatic species breed at the reserve and it is good for wintering waterfowl. The Berger Hill Reserve is an area of open parkland with fruit trees and hedges. The two reserves are adjacent to each other and can be accessed from the Sports Centre on **Fritz-Schubert–Ring Road** in east Enkheim. At the ramp for the **A66 Autobahn**, go east on **Borsigallee** and continue east on **Leuchte** until it curves north to Fritz-Schubert-Ring. The Sports Centre is on the right-hand side. The Enkheim U-Bahn Station is on Borsigallee.

KEY SPECIES

Resident	Summer	Winter
Northern Goshawk, Water Rail, Grey-headed Woodpecker, Lesser Spotted Woodpecker, Middle Spotted Woodpecker, Common Kingfisher, Sky Lark, Long-tailed Tit, Tree Sparrow, Reed Bunting	Wryneck, Common Nightingale, Marsh Warbler, Garden Warbler, Lesser Whitethroat, Willow Warbler, Pied Flycatcher, Golden Oriole, Penduline Tit	Eurasian Teal, Eurasian Wigeon, Northern Shoveler, Goosander, Siskin **Passage** Tree Pipit, Bluethroat, Grasshopper Warbler

3 Schwanheim (8km)

(Tram No. 21; Bus No. 51)

The Schwanheimer Wald is an extensive forest park to the southwest of the city, just north of Frankfurt Airport. It is probably the best area of mature mixed woodland that is easily accessible from Frankfurt. There is a park entrance on Schwanheimer Bahnstrasse just west of the tram terminus and from here you can take a circular forest walk of about 3km, which passes through mature oakwood on the north side of the forest and coniferous woodland on the south side. The oakwood contains most of the typical woodland species including Middle Spotted and Black Woodpeckers. The coniferous woodland is less interesting but still worth checking. It is also possible to see Wild Boar in the park.

Take **Tram** (Strassenbahn) **No. 21** from the City Centre to its terminus at **Rheinlandstrasse**. There is also a bus service (No. 62) from the tram terminus direct to Frankfurt Airport, but it is not as frequent as the trams.

Further west of the Schwanheimer Wald lies a system of sand dunes at a bend in the River Main. This is Schwanheim Dünen NSG, an area with some open habitat, pine woods, allotments and flooded pits. Small numbers of waterfowl occur on the ponds and Black Redstarts, Wood Larks, Golden Orioles and Wrynecks may be seen in the woods.

Take the No. 51 bus from the tram terminus at Rheinlandstrasse and get off at Völklinger Weg. Take the road that crosses over the Autobahn and on the other side take any of the right-hand turn-offs.

KEY SPECIES

Resident
Red Kite, Northern Goshawk, Eurasian Sparrowhawk, Grey-headed Woodpecker, Middle Spotted Woodpecker, Black Woodpecker, Song Thrush, Mistle Thrush, Long-tailed Tit, Common Raven, Siskin

Summer
Wryneck, Wood Lark, Golden Oriole

4 River Main near Seligenstadt (25km)   (Seligenstadt Railway Station)

To the east of Frankfurt, there is a footpath (Mainuferweg) along the bank of the River Main which extends for about 5km on either side of Seligenstadt. You could walk the entire length but the most interesting locations for the birder are at the start of the path opposite Stauding Power Station and at the end of the path at Bongsche Gravel Pits.

Take the **A66 Autobahn** east from Frankfurt and then the **A43** south to the **Mainhausen Exit**. Exit north on the **L2310** to Seligenstadt and from there take the **L3065** north to **Hainburg**. The start of the Mainuferweg is just before Hainburg, at the point where the road meets the river.

The power station and the locks at **Großkrotzenburg** are on the east bank of the river. Peregrines breed on one of the power station cooling towers and there is a cooling pond beside the power station which holds some waterfowl. However, it is fenced off, so it is probably not worth the trip across the river as the nearest bridge is about 1km downstream. The Großkrotzenburg Locks are good in winter for gulls and to scan the river for ducks.

South of Seligenstadt on the L2310, turn east on the **K185** to **Mainflingen**. Take the first turn on the right, **Am Sportplatz**, and park in the car park at the end of this road. The continuation of this road leads to an area of flooded gravel pits with good riparian woodland on the banks. Cormorants and herons breed here. South of the gravel pits on the south side of the **A43** Autobahn are the **Kleinostheim Locks** and wastewater lagoons which are also worth checking. If you don't have a car, there is a train service to Seligenstadt from Frankfurt, but it is probably more convenient to get the S-Bahn to Hanau (S8 or S9) and get the local bus to Hainburg and from there to Mainflingen.

KEY SPECIES

Resident
Great Bittern, Peregrine Falcon, Common Kestrel, Grey-headed Woodpecker

Summer
Black-necked Grebe, Turtle Dove,

Common Cuckoo, Blue-headed Wagtail, Garden Warbler, Penduline Tit

Winter
Eurasian Teal, Eurasian Wigeon, Common Goldeneye, Smew,

Goosander, Red Kite, Herring Gull, Bearded Tit

Passage
Wood Sandpiper, Black Tern, hirundines

5 Kühkopf (40km)   (Goddelau-Erfelden Railway Station)
(Stockstadt Railway Station)

The Kühkopf Reserve on the Rhine is one of the best birding areas in central Germany. Any species that you are likely to see in the Frankfurt area can be seen at Kühkopf and in larger numbers than anywhere else. Over 250 species have been recorded here so it is well worth a visit, winter or summer. Kühkopf and the contiguous reserve at Knoblochsaue NSG are about 40km south of Frankfurt on the east bank of the Rhine.

This is a large wetland, almost 2,500 hectares, which was formed by an oxbow at a curve in the river. The new course of the Rhine was cut in the 1800s creating an island bounded by the current Rhine and the 'old' river. This extensive area is now a mosaic of different habitats with woods, reedbeds, open water and wet meadows. Many aquatic species breed here in summer and winter birds include several duck species and large numbers of Bean Geese. There are 60km of trails through the reserve and several hides and viewing platforms.

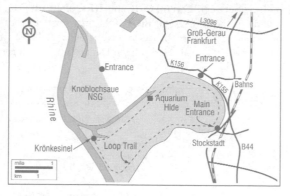

The main entrance is at Stockstadt where there is an information centre providing maps and guides to the reserve. A loop trail circles the island but perhaps the most productive areas are along the northern part of this trail, in particular the Aquarium Hide and the sandbars at Krönkesinsel, the northwest tip of the island, which can hold waders in the autumn. Another entrance at Erfelden provides the only other bridge onto the island. A third entrance west of Erfelden off the K156 provides access to Knoblochsaue NSG.

The **S-Bahn S7** runs as far as **Erfelden** and there is a regular train service from Frankfurt serving both Erfelden and Stockstadt. In both cases the railway station is within walking distance of the Erfelden and Stockstadt entrances.

KEY SPECIES

Resident
Great Bittern, Greylag Goose, Red Kite, Water Rail, Grey Partridge, Common Snipe, Eurasian Curlew, Grey-headed Woodpecker, Lesser Spotted Woodpecker, Middle Spotted Woodpecker, Black Woodpecker, Crested Tit, Sky Lark, Willow Tit, Reed Bunting

Summer
Black Stork, Common Quail, Spotted Crake, Little Ringed Plover, Common Sandpiper, Turtle Dove, Wryneck, Common Cuckoo, Tree Pipit, Blue-headed Wagtail, Common Nightingale, Bluethroat, Savi's Warbler, Marsh Warbler, Great Reed Warbler, Icterine Warbler, Penduline Tit

Winter
Whooper Swan, Bean Goose, White-fronted Goose, Gadwall, Pintail, Northern Shoveler, Eurasian Wigeon, Eurasian Teal, Smew, White-tailed Eagle, Hen Harrier, Merlin, Woodcock, Yellow-legged Gull, Water Pipit, Great Grey Shrike

Passage
Great White Egret, Purple Heron, Glossy Ibis, Montagu's Harrier, Red-footed Falcon, Northern Lapwing, Little Stint, Dunlin, Ruff, Spotted Redshank, Greenshank, Black-tailed Godwit, Green Sandpiper, Wood Sandpiper, Common Tern, Black Tern, Meadow Pipit

6 Groß-Gerau Lagoons (28km) *(Groß-Gerau Railway Station)*

These are waste-water settling ponds beside a sugar refinery just on the northwest outskirts of the town. Like all such lagoons, they contain a very rich organic soup which nourishes abundant insect life. The site seems to be very attractive to migrating waders and it is well worth a visit in autumn if you are returning to Frankfurt from Kühkopf. It is also just a short drive (20 minutes) from Frankfurt Airport so it would be worth a look if you were picking up or returning a rental car.

Groß-Gerau is just off the **A67 Autobahn**, southwest of Frankfurt Airport, and midway between **Darmstadt** and **Rüsselsheim**. The lagoons are west of the inter-section of the **L3482** and the **L3094**, just northwest of the town centre. You can turn off either of these roads onto gravel tracks leading to the lagoons. There is a regular train service to Groß-Gerau from Darmstadt, which is served by S-Bahn S3 and S4.

KEY SPECIES

Resident Water Rail, Grey Partridge, Common Kingfisher, Common Snipe, Reed Bunting **Summer** Little Ringed Plover, Common	Sandpiper, Turtle Dove, Blue-headed Wagtail **Winter** Gadwall, Pintail, Eurasian Wigeon, Eurasian Teal, Hen Harrier	**Passage** Northern Lapwing, Dunlin, Ruff, Spotted Redshank, Greenshank, Green Sandpiper, Wood Sandpiper, White Wagtail

7 Reinheimer Teich (40km)   *(Reinheim Railway Station)*

This is a marsh south of Frankfurt. It is less interesting than some of the other sites and probably not worth a trip out from the city, but it may be worth a look if your business takes you to Darmstadt or elsewhere on the southern outskirts of Frankfurt. It can also provide an alternative, at weekends, to the crowds at Kükhopf. For many years this was the only breeding site in Hessen for Marsh Harriers and it is still a good site for this species. There is also a Grey Heron colony here. In summer there is a good selection of breeding aquatic birds and it is used as an autumn roost by large numbers of hirundines.

Take the **B26** east from Darmstadt for about 8km and turn south onto the B38 towards **Spackbrücken**. At Spackbrücken, take the **L3413** east for about 3km. The reserve is on the south side of the road: follow the signs for the Gliding Club. The nearest railway station is at **Reinheim**, approximately 4km from the site, and trains operate from Darmstadt.

KEY SPECIES

Resident Grey Heron, Red Kite, Water Rail, Common Kingfisher, Willow Tit, Tree Sparrow, Reed Bunting **Summer** Marsh Harrier, Northern Lapwing,	Blue-headed Wagtail, Common Nightingale, Bluethroat, Marsh Warbler, Garden Warbler, Penduline Tit **Winter** Pintail, Eurasian Wigeon, Eurasian	Teal, Great Grey Shrike **Passage** Black-necked Grebe, Great White Egret, Black Tern

8 Schierstein Waterworks (38km)   *(Schierstein Railway Station)*

These are just outside Wiesbaden on the north shore of the Rhine. There is a harbour area and linear path along the north bank of the river, as well as the lagoons. The lagoons are surrounded by reedbeds and riparian woodland, and the whole area has been landscaped and designated as a nature reserve. The main ornithological interest is the White Stork colony in summer, but in winter the harbour area attracts gulls and the riverside path provides a good opportunity to scan the Rhine for ducks.

Take the **A66 Autobahn** west from Frankfurt and exit south on the L3441 (**Grorotherstrasse**). Continue on the L3441 until you come to a T-junction and turn left. Then take the next left which will bring you to a car park at **Schierstein**

Harbour. You can check the harbour here and then walk west to the waterworks whilst scanning the river. You can also get here easily by public transport. Take the S-Bahn to Wiesbaden from the airport or from Frankfurt Main Railway Station (S8 or S9) to Wiesbaden Main Railway Station. You can either get a taxi from here to Schierstein Harbour or get a local train to Schierstein Station.

Further Away

9 **Wetterau Floodplains** (50km)   *(Echzell Railway Station)*

This is an area northeast of Frankfurt which incorporates several tributaries of the River Main. There are a number of shallow lakes, as well as marshes and water meadows, some of which are nature reserves and are described below. Summer or winter, this area is worth a visit if you have a car, as you could visit several of these locations in the course of a day.

Bingenheimer Ried is a lake surrounded by reedbeds and water meadows which provide habitats for about 30 breeding species, including White Storks, and for flocks of wintering wildfowl. The whole area is very flat, but there is an observation tower at the entrance to the reserve and if you have a telescope you can scan the whole area quite easily. Pfaffensee and Teufelsee are two lakes just west of Bingenheimer Ried and are the result of abandoned lignite mines. As both lakes are deeper than Bingenheimer Ried they are better for grebes, rarer diving ducks and for the occasional diver. Pfaffensee is the better of the two lakes as it has better habitat for waders. There are hides on the shores of both lakes.

Take the **A66 Autobahn** west from Frankfurt and then the **A45** north. After about 20km, exit the A45 west at **Staden** on the **B275**. In Staden, turn north on the **L3188** to **Bingenheim**. The reserve is a short walk west of Bingenheim and there is parking in the town.

Also worth visiting are the lakes at Knappensee and the Kuhweide water meadows. Take the **L3188** north from **Echzell** for about 3km and you come to a T-junction. Turn right at the T-junction and then the next left turning. This will bring you to **Unterwiddersheim**. On the left-hand side of the road is a signpost directing you to an observation platform which offers a superb view over the Kuhweide meadows. In autumn, several thousand Common Cranes pass through this area, and there are flocks of geese in winter (mainly Greylags). Continue past Unterwiddersheim and take the next turn west towards **Utphe**. On either side of this road is access to the Upper and Lower Knappensee Lakes. It is a little more difficult to get here by public transport. The best way is to get the train to Echzell which lies at the centre of this area, a few kilometers from each site. It would be worth considering renting a bicycle and taking this on the train.

KEY SPECIES

Resident
Great Bittern, Greylag Goose, Red Kite, Water Rail, Grey Partridge, Common Snipe, Sky Lark, Meadow Pipit, Reed Bunting

Summer
Red-necked Grebe, Black-necked Grebe, White Stork, Spotted Crake,

Common Sandpiper, Common Cuckoo, Blue-headed Wagtail, Common Nightingale, Bluethroat, Northern Wheatear, Penduline Tit

Winter
Whooper Swan, Bean Goose, White-fronted Goose, Gadwall, Pintail, Northern Shoveler, Eurasian Wigeon,

Eurasian Teal, Smew, Hen Harrier, Merlin, Water Pipit, Great Grey Shrike

Passage
Great White Egret, Red-footed Falcon, Common Crane, Northern Lapwing, Ruff, Dunlin, Eurasian Curlew, Black-tailed Godwit, Green Sandpiper, Wood Sandpiper, Black Tern

Useful contacts and websites

Hessische Gesellschaft für Ornithologie und Naturschutz e.V. (German)
www.hgon.de
Hessen Ornithological Organisation, Lindenstraße 5, 61209 Echzell. Tel. (+49) 6008/1803.

Lower River Main Bird Observatory (German)
www.vogelkunde-untermain.de

Kühkopf-Knoblochsaue Nature Reserve
(German) www.rpda.de/kuehkopf

Norbert Kuehnberger Website
(German with some English content)
www.norbert-kuehnberger.de
Personal website with lots of information about birding in Hessen.

NABU – Naturschutzbund Deutschland
(BirdLife partner in Germany)
www.nabu.de
Herbert-Rabius-Str 26, D-53225 Bonn.
Tel: (+49) 2284036166.

German Birdnet www.birdnet.de

Rhein-Main-Verkehrsverbund
Public transport in Frankfurt area
www.rmv.de

Deutsch Bahn (German and English)
Trains and Feeder Bus services in Germany
www.bahn.de

Books and publications

Vögel beobachten in Süddeutschland
Christoph Moning and Christian Wagner

Avifauna von Hessen
(1997) Hessische Gesellschaft für Ornithologie und Naturschutz e.V. (Vols. 1 – IV) ISBN 3980109216

Vogelparadiese, Bd.2, Westdeutschland und Süddeutschland
Michael Lohmann and Knut Haarman (1989)
Blackwell Wissenschafts-Verlag
ISBN 3826381521

Collins Bird Guide
Lars Svensson et al. (1999) HarperCollins
ISBN 0002197286

Geneva

Geneva is at the southwest end of Lake Geneva, where the River Rhône begins its descent into the Rhône Valley. Over 300km inland and at about 1,000m elevation, the city can experience cold winters but the moderating influence of the lake on temperatures ensures that it stays ice-free and is a significant wintering area for diving ducks, grebes and even small numbers of divers. Indeed, Lake Geneva has accumulated a growing list of vagrant seabirds which have penetrated this far inland.

Another factor which makes Geneva such an interesting city for birds is its location at the neck of a natural 'funnel'. With the Jura Mountains to the northwest and the Alps to the south, southerly movements of migrants are channelled towards the southwest shores of Lake Geneva. In spring many migrants following the Rhône valley northwards pass through the Geneva area.

A drake Goosander rests on Lake Geneva with the Jet d'Eau in the distance

Geneva is bounded on three sides by France, and several of the locations described in this guide involve a short journey over the border. In fact the French Alps, which support the full range of Alpine specialities, are closer to Geneva than the Swiss Alps and they are much more accessible to the Geneva-based birder. For those willing to undertake a more distant journey into France, a visit to the Dombes wetland area north of Lyon would be well worthwhile. This involves a round-trip by car of about four hours but will yield much higher numbers and a greater variety of herons, wildfowl and waders than can be seen in the wetlands closer to Geneva.

TYPICAL SPECIES

Resident
Little Grebe, Great Crested Grebe, Grey Heron, Mute Swan, Mallard, Goosander, Common Buzzard, Common Kestrel, Moorhen, Common Coot, Wood Pigeon, Collared Dove, Green Woodpecker, Great Spotted Woodpecker, Wren, Dunnock, Robin, Blackbird, Marsh Tit, Coal Tit, Blue Tit, Great Tit, Eurasian Nuthatch,

Short-toed Treecreeper, Eurasian Jay, Carrion Crow, Common Starling, House Sparrow, Tree Sparrow, Chaffinch, Greenfinch, Bullfinch, Hawfinch, Yellowhammer, Reed Bunting

Summer
Black Kite, Common Swift, Sand Martin, Barn Swallow, House Martin, White Wagtail, Black Redstart,

Common Redstart, Song Thrush, Reed Warbler, Blackcap, Common Whitethroat, Common Chiffchaff, Wood Warbler, Spotted Flycatcher, European Serin, Linnet

Winter
Great Cormorant, Tufted Duck, Common Pochard, Black-headed Gull, Yellow-legged Gull, Goldcrest, Western Jackdaw, Rook, Siskin

GETTING AROUND

Geneva is a rather small city, with its central area adjacent to the lakeshore, so it is possible for the visiting birder to enjoy its birding attractions within walking distance of the hotel. For the more distant sites, the bus and train services, almost all of which operate from the main railway station (Gare Cornavin), are comprehensive and regular, although the train service to some of the neighbouring

stations in France is not very frequent. Some of the train services into France depart from the Gare Eaux-Vives, on the south side of the city. If you plan to use the train to go birding, bear in mind that it is possible to carry a bicycle with you, so you might consider renting one as it can be very useful for getting around the birding locations (at least the flat, level sites!).

If you are renting a car, make sure that you inform the rental company that you intend to drive into France: there is generally a small surcharge to cover the breakdown recovery service. The A1 Motorway links the airport and the western suburbs of the city with the interior of the country to the northwest and with the French autoroute system to the south. Be prepared, however, for the road numbering system to change when you drive across the border.

Birding in Geneva is not really weather dependent, although some winter days can be very cold. Comfortable walking shoes are adequate for all the locations described, except if you are planning a trip into the Alps, where warm clothing and strong footwear are required even during the summer, as the weather at altitude can be very changeable.

Choose a city map which includes the neighbouring regions of France, as some of the sites involve travelling into the French region of Rhône-Alpes.

Locations

1 Geneva Harbour (City Centre)

Situated as it is where the River Rhône exits Lake Geneva, the inner harbour of Geneva plays host to a steady stream of visible migration as birds migrate along the course of the river and, in winter, good numbers of wildfowl congregate here. The lakeshore at Geneva is landscaped by a series of contiguous parks from north to south, and all are within walking distance of the main business and hotel district. Several jetties which jut out into the lake on the north and south shores are perfect vantage points to scan the open waters. The north shore from the Jetée de Pâquis to the Perle du Lac Restaurant is the most productive, especially in winter, for gulls, ducks, grebes and, often, divers. In autumn, terns and small numbers of waders can be seen on the Jetée de Pâquis.

KEY SPECIES

Winter
Black-throated Diver, Black-necked Grebe, Red-necked Grebe, Gadwall, Common Goldeneye, Ferruginous Duck, Lesser Black-backed Gull, Common Kingfisher

Passage
Garganey, Eurasian Teal, Eurasian Wigeon, Pintail, Northern Shoveler, Common Eider, Velvet Scoter, Marsh Harrier, Osprey, Ringed Plover, Sanderling, Little Stint, Dunlin, Ruff,

Common Redshank, Greenshank, Common Sandpiper, Mediterranean Gull, Little Gull, Black Tern

2 Verbois Reservoir (8km)   (Russin Railway Station/Bus Route K)

A dam across the River Rhône near a bend in the river just east of Geneva has created a reservoir which is very important for wintering wildfowl and for some breeding aquatic birds. Additionally, there are several wetland sites in the vicinity of the reservoir which together make this area well worth a visit at any time of year.

From the **A1 Motorway**, exit west onto **Route de Chancy** and at **Bernex** turn northwest towards **Aire-la-Ville**. Take the **Route de Verbois** west from Aire-la-Ville and after about 1km you will cross the dam. You may scan the reservoir from the car parks at either end of the dam. There is a footpath leading north on the west side of the reservoir. Tens of thousands of diving ducks winter here: most are Tufted Ducks and Common Pochards but a number of scarcer species are found among the flocks.

From the dam take the footpath leading south along the east bank of the Rhône, which leads to the small reserve at Moulin-de-Vert. This is an area of small ponds, reedbeds and riparian woodland which supports a rich variety of breeding species: the highest breeding-bird diversity in the Geneva Canton. It also attracts a wide range of migrants, including waders. There is a hide at one of the ponds and several trails between the other ponds.

The footpath south from the dam on the west side of the Rhône (first left after the hydroelectric works) passes through the Teppes de Verbois. This is an area of gravel-banks, scattered woodland and some small ponds, presenting a mosaic of habitats which support a wide range of breeding species making it particularly interesting in late spring and summer.

If you are using public transport, there is a railway station at Russin on the west bank of the Rhône and just south of the dam.

KEY SPECIES

Resident
Common Pheasant, Water Rail, Common Snipe, Common Kingfisher, Green Woodpecker, Lesser Spotted Woodpecker, Sky Lark, Grey Wagtail, Long-tailed Tit, Cirl Bunting, Corn Bunting

Summer
Little Bittern, Spotted Crake, Little Ringed Plover, Common Sandpiper, Common Tern, Turtle Dove, Wryneck,

Common Cuckoo, Blue-headed Wagtail, Common Nightingale, Melodious Warbler, Red-backed Shrike

Winter
Great Bittern, Common Shelduck, Gadwall, Smew, Greater Scaup, Tufted Duck, Common Pochard, Peregrine Falcon, Lesser Black-backed Gull

Passage
Little Egret, Garganey, Pintail,

Northern Shoveler, Eurasian Wigeon, Eurasian Teal, Red Kite, European Honey-buzzard, Marsh Harrier, Osprey, Hobby, Northern Lapwing, Little Stint, Dunlin, Ruff, Common Redshank, Greenshank, Black-tailed Godwit, Eurasian Curlew, Green Sandpiper, Wood Sandpiper, Whiskered Tern, Black Tern, Willow Warbler, Pied Flycatcher, Penduline Tit

3 Rhône Valley at L'Étournel (25km) *(Pougny Railway Station/ Bus Route K)*

Just over the border in France, where the River Rhône widens and becomes more sluggish, it allows extensive reedbeds to grow, thus creating an extensive marshland habitat. Further downstream at the Fort L'Écluse Gorge, the river runs between two mountainous ridges at the southern end of the Jura Mountains. In autumn this is a migration watchpoint for migrating raptors (mainly buzzards and kites) and storks, as well as passerines. This whole stretch of the Rhône Valley is worth a visit at any time of the year but particularly in autumn when migration is in full swing.

This area is easily accessed from Geneva. Take the **Route de Chancy** west from Geneva, crossing the Rhône (and the border) at **Chancy** where the road becomes the French route **D928B** and follow this road to **Collonges**. In Collonges, turn left (south) on the **N206** to get to the **Carnot Bridge** over the Rhône where there is a car park. You can walk along the riverside path on the south bank of the river through an area of reedbeds, willow scrub and marsh.

You get to the raptor watchpoint by continuing south past the bridge on the N206 for about 1.5km until you come to a railway bridge. Just before the bridge turn right (west) and follow this road up the hill

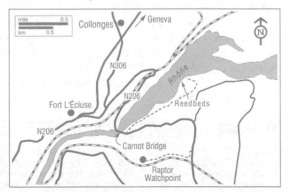

alongside the railway for about half a kilometre to a point overlooking the valley and facing north.

If you return to Collonges take the N206 west from the town, passing through a tunnel about 200m long. As you emerge from the tunnel, a fort is on the right-hand side. This is one of the most reliable sites, close to Geneva, for wintering Wallcreepers, and if this is your target species a visit here obviates the need for a strenuous climb into the mountains! Carefully check the walls of the fort for this species and the general area for wintering finches and buntings.

KEY SPECIES

Resident
Golden Eagle, Common Pheasant, Water Rail, Common Snipe, Common Kingfisher, Lesser Spotted Woodpecker, Black Woodpecker, Sky Lark, Grey Wagtail, Long-tailed Tit, Willow Tit, Crested Tit, Cirl Bunting, Corn Bunting

Summer
Little Bittern, Spotted Crake, Little Ringed Plover, Common Sandpiper, Common Tern, Turtle Dove, Alpine Swift, Wryneck, Common Cuckoo,

Blue-headed Wagtail, Crag Martin, Common Nightingale, Melodious Warbler, Western Bonelli's Warbler, Red-backed Shrike

Winter
Great Bittern, Common Shelduck, Gadwall, Northern Shoveler, Eurasian Wigeon, Eurasian Teal, Smew, Greater Scaup, Hen Harrier, Peregrine, Common Gull, Lesser Black-backed Gull, Wallcreeper, Rock Bunting

Passage
White Stork, Great White Egret, Night Heron, Garganey, Pintail, Red Kite, European Honey-buzzard, Marsh Harrier, Osprey, Eurasian Sparrow-hawk, Hobby, Common Kestrel, Common Crane, Northern Lapwing, Ruff, Common Redshank, Green-shank, Black-tailed Godwit, Eurasian Curlew, Green Sandpiper, Wood Sandpiper, Whiskered Tern, Black Tern, Wood Lark, Willow Warbler, Pied Flycatcher, Penduline Tit

4 Préverenges (52km)

(Morges Railway Station)

The Bay of Préverenges is on the north shore of Lake Geneva and right in the path of many migrating birds. In spring, when water levels are lower, exposed sand and gravel banks at the mouth of the Venoge River act as roosting sites for gulls, waders and terns. However in autumn the lake level is usually too high and covers the roosting sites. There is too much disturbance on the lakeshore for the birds so an artificial gravel bank has been constructed just offshore, which provides a roost and feeding site at all times of year. Flocks of waders are a most unusual sight in a land-locked country such as Switzerland, but the new island (l'Île aux Oiseaux) regularly attracts small flocks of even essentially maritime species, such as Red Knot and Sanderling. Over 30 species of wader have been recorded here and it has emerged as one of the best rarity sites in Switzerland. Common Terns have also commenced breeding on the island. The lake here is also productive in winter for waterfowl, including the scarcer species of ducks and also grebes and divers.

Take the **A1 motorway** northeast towards **Lausanne** and exit at **Morges** onto **Rue de Lausanne (Route 1)** towards Préverenges. On the east side of Préverenges, this road crosses over the River Venoge. After the bridge take the second turn right (**Chemin de la Venoge**). This road brings you to the lakeshore. The island lies about 100m offshore and small waders can be hard to discern on the stony substrate, so a telescope is desirable.

KEY SPECIES

Winter
Red-throated Diver, Black-throated Diver, Black-necked Grebe, Red-necked Grebe, Eurasian Teal, Eurasian

Wigeon, Pintail, Gadwall, Common Goldeneye, Red-crested Pochard, Velvet Scoter, Ferruginous Duck, Greater Scaup, Common Gull, Lesser

Black-backed Gull, Common Kingfisher

Passage		
Garganey, Northern Shoveler, Velvet Scoter, Red-breasted Merganser, Osprey, Ringed Plover, Little Ringed	Plover, Sanderling, Red Knot, Dunlin, Ruff, Common Redshank, Greenshank, Black-tailed Godwit, Wood Sandpiper, Green Sandpiper,	Common Sandpiper, Mediterranean Gull, Little Gull, Sandwich Tern, Common Tern, Whiskered Tern, Black Tern, Blue-headed Wagtail

5 Excenevex (28km)

The south shore of Lake Geneva, in France, provides some sheltered bays for wintering wildfowl. In spring, when water levels are at their lowest, sand and gravel banks are exposed, providing feeding and roosting sites for small numbers of waders. The bay at Excenevex is shallow and also has the only sandy beach on the lake, making it attractive to small waders. Northerly winds in winter (La Bise) can produce accumulations of wildfowl in the shelter of the Golfe de Coudrée. If you can endure the cold for long enough, careful scrutiny can turn up rarities such as a Slavonian Grebe or Long-tailed Duck.

From the south side of Geneva, take the **Route de Thonon** into France where it becomes the **N5 Route** and continue for about 12km to **Sciez**. Turn left (north) in Sciez onto the **D25** towards Excenevex. Just north of Excenevex turn right onto **Chemin Bellevue**, which leads to the beach. If you are using public transport, there is a regular though somewhat infrequent bus service to Sciex and other towns on the south shore of Lake Geneva, which operates from the bus station (Gare Routière) near the main railway station (Gare Cornavin) in Geneva.

KEY SPECIES

Winter	Passage	
Red-throated Diver, Black-throated Diver, Black-necked Grebe, Red-necked Grebe, Gadwall, Red-crested Pochard, Greater Scaup, Common Goldeneye, Velvet Scoter, Ferruginous Duck, Lesser Black-backed Gull, Common Kingfisher	Garganey, Eurasian Teal, Eurasian Wigeon, Pintail, Northern Shoveler, Common Scoter, Red-breasted Merganser, Osprey, Ringed Plover, Little Ringed Plover, Sanderling, Dunlin, Ruff, Common Redshank, Greenshank, Black-tailed Godwit,	Wood Sandpiper, Green Sandpiper, Common Sandpiper, Mediterranean Gull, Little Gull, Sandwich Tern, Whiskered Tern, Black Tern, Blue-headed Wagtail

Further Away

6 River Rhône at Motz (47km)   (Seyssel Railway Station)

Although Motz, in France, is only about 50km south of Geneva, the Rhône Valley at this point is at a lower elevation and not bounded by mountain ranges to the south. Consequently it supports several Mediterranean species which are scarce or absent in the Geneva area. The marsh at Motz is at the confluence of the River Fier and the Rhône. The Rhône widens at this point with a small island in the centre: the river level is controlled by a system of sluices and when water levels are low, this is a very reliable spot to see crakes walking out on the mud. The site is a watersports centre which has a nature trail alongside the riverbank and a hide which permits good undisturbed views of the reedbeds, open water and exposed mud.

Take the **A40** east from Geneva towards **Lyon** but after about 25km exit south at **Éloise (Exit 11)** onto the **N508**. After about 5km turn east onto the **D992** towards **Seyssel**. Pass through Seyssel and after about 2.5km the road passes over the River

Fier. Look for the **l'Espace Sport et Nature du Fier** on the right-hand side. Entrance is free for most of the year except the summer months. If you don't have a car you could consider renting a bicycle and bringing it on the train from Eaux-Vivres Station to Seyssel, which is only 2.5km from the site.

KEY SPECIES

Resident
Water Rail, Common Snipe, Common Kingfisher, Sky Lark, Grey Wagtail, Cetti's Warbler, Cirl Bunting, Corn Bunting

Summer
Little Bittern, Marsh Harrier, Spotted Crake, Little Crake, Little Ringed Plover, Common Tern, Turtle Dove, Wryneck, European Bee-eater, Hoopoe, Common Cuckoo, Wood Lark, Blue-headed Wagtail, Common

Nightingale, Great Reed Warbler, Melodious Warbler, Red-backed Shrike

Winter
Great Bittern, Common Shelduck, Gadwall, Pintail, Northern Shoveler, Eurasian Wigeon, Eurasian Teal, Peregrine Falcon, Lesser Black-backed Gull

Passage
Little Egret, Great White Egret, Cattle Egret, Purple Heron, Squacco Heron,

Eurasian Spoonbill, White Stork, Garganey, Red Kite, European Honey-buzzard, Short-toed Eagle, Osprey, Avocet, Northern Lapwing, Little Stint, Ruff, Common Redshank, Greenshank, Black-tailed Godwit, Eurasian Curlew, Green Sandpiper, Wood Sandpiper, Common Sandpiper, Whiskered Tern, Black Tern, Bluethroat, Penduline Tit, Willow Warbler, Pied Flycatcher, Ortolan Bunting

7 Réserve Naturelle de Sixt (60km) *(Cluses Railway Station)*

The French Alps are southeast of Geneva and are easier to get to than the Swiss Alps, which require a long drive along the shores of Lake Geneva into the interior of the country. They are higher than the Jura Mountains to the north of Geneva and offer a much better diversity of species due to their greater range of elevations and habitats. They are well served by road and rail, and by ski-lifts to the higher peaks, and so offer the best option for the Geneva-based birder to see some of the alpine specialities. These mountains are also a favoured haunt of Lammergeiers from the French reintroduction scheme.

Take the **A40 Autoroute** southeast from the French border and after about 30km exit north at **Cluses** onto the **D902** towards **Taninges**. At Taninges turn east onto the **D907** towards **Samoëns** and the ski resort at **Sixt-Fer-à-Cheval**. The road from Samoëns up to Sixt runs alongside the River Giffre with pine forests on each side of the valley offering several opportunities to stop and check for forest species. Once past the reserve headquarters at Sixt, the road ascends above the treeline towards the **Réserve Naturelle de Sixt** through an area of alpine meadows and bare rock. There is a mountain chalet at the end of this road and several trails over the mountains towards Switzerland. At these altitudes (2,500m) birds occur at low densities and it will often be necessary to walk significant distances in order to see just a few species.

If you are relying on public transport, there is a regular train service to Cluses from Eaux-Vives Station on the south side of Geneva, and a bus service from there to Sixt. There is also a bus service direct from Geneva Airport to Sixt. A number of ski-lifts at Sixt can bring you to the higher elevations but unfortunately they do not operate during May and June, which are the best months for seeing the alpine species on territory. In winter many birds move to lower elevations, so it is possible to see species such as Alpine Accentors, Alpine Choughs and Citril Finches around the village.

KEY SPECIES

Resident
Lammergeier (I), Golden Eagle,

Peregrine, Hazel Grouse, Ptarmigan, Black Grouse, Capercaillie, Rock

Partridge, Black Woodpecker, Grey-headed Woodpecker, Wallcreeper,

Alpine Accentor, Crested Tit, Willow Tit, Alpine Chough, Common Raven, Nutcracker, Crossbill, Citril Finch, Snow Finch, Rock Bunting	**Summer** Short-toed Eagle, Booted Eagle, Alpine Swift, Crag Martin, Water Pipit,	Common Stonechat, Fieldfare, Ring Ouzel, Western Bonelli's Warbler

8 Chavornay Clay-pits (80km)

(Chavornay Railway Station)

The Réserve Naturelle des Creux de Terre at Chavornay consists of two abandoned clay-pits which have flooded and become very attractive to waterfowl. This is a well-watched site, which frequently turns up rarities, with footpaths and an advantageously sited hide by one of the lakes. The reserve is on the floodplain of the River Orbe, which flows into Lake Neuchatel. The water levels are controlled in order to lower the water table to facilitate haymaking in the surrounding meadows during the summer. This preserves the unique flora and prevents more invasive plants from establishing a foothold but it also lowers the water levels in the lakes, resulting in muddy margins which attract small numbers of migrating waders. The site is strategically located for birds migrating along the valley between the Jura Mountains and the Alps, and the reedbeds and copses surrounding the lakes are good for migrant as well as breeding passerines.

Take the **A1** northeast towards **Lake Neuchatel** and after about 75km exit west at **Chavornay**. In Chavornay turn north alongside the railway line and the first lake is on the right-hand side after about 1km, where the road approaches the motorway. Stay on the road as it loops back towards Chavornay and after about 500m it passes between the two main lakes and the reserve entrance.

KEY SPECIES

Resident Common Pheasant, Water Rail, Common Snipe, Common Kingfisher, Sky Lark, Grey Wagtail, Long-tailed Tit, Yellowhammer, Cirl Bunting, Reed Bunting, Corn Bunting **Summer** Little Bittern, Marsh Harrier, Hobby, Spotted Crake, Little Ringed Plover, Common Sandpiper, Common Tern, Turtle Dove, Wryneck, Common	Cuckoo, Blue-headed Wagtail, Common Nightingale, Reed Warbler, Great Reed Warbler, Golden Oriole **Winter** Great Bittern, Gadwall, Eurasian Teal, Eurasian Wigeon, Hen Harrier, Peregrine Falcon, Common Gull, Lesser Black-backed Gull, Water Pipit **Passage** Great White Egret, Night Heron,	Purple Heron, Garganey, Pintail, Northern Shoveler, Red Kite, European Honey-buzzard, Osprey, Northern Lapwing, Dunlin, Ruff, Common Redshank, Greenshank, Bar-tailed Godwit, Eurasian Curlew, Green Sandpiper, Wood Sandpiper, Whiskered Tern, Black Tern, Bluethroat, Common Stonechat, Pied Flycatcher, Penduline Tit

Useful contacts and websites

Nos Oiseaux (French and English)
www.nosoiseaux.ch
Local birding group website with information about sites, recent reports and activities.

Cercle Ornithologique de Lausanne (French)
www.oiseau.ch
Lausanne birding group website with information about recent reports and activities.

ASPO – Birdlife Suisse (French and German)
www.birdlife.ch
ASPO, La Sauge, 1588 Cudrefin. Tel. 026 6770377.
BirdLife partner in Switzerland and publishers of the journal *Ornis*.

Schweizerische Vogelwarte (French and German)
www.vogelwarte.ch
Swiss Ornithological Institute, CH - 6204 Sempach.
Tel. 041 462 9700. Fax. 041 462 9710.

Ligue pour la Protection des Oiseaux – Haute Savoie (French) http://haute-savoie.lpo.fr
24 rue de la Grenette, 74370 Metz-Tessy.
Tel/fax. 04 50 27 17 74.
LPO group for neighbouring French Département – Haute-Savoie.

Swiss Rarities Committee (German) www.vogelwarte.ch/sak_e_f.html
Schweizerische Avifaunistische Kommission,
c/o Schweizerische Vogelwarte, CH-6204 Sempach.

Swiss Birdline (French and German)
Tel. 021 616 1222.

Transports Publics Genevois
(French and English) – Geneva public transport:
www.tpg.ch

Les Chemins de fer suisses CFF (French and
English) – Swiss Railways: www.cff.ch

SNCF (English) – French Railways
www.ter-sncf.com/uk/rhone-alpes/default.htm

Gare Routière (French and English)
www.gare-routiere.com
Bus services from Geneva to neighbouring areas of
France.

Books and publications

Where to Watch Birds in Switzerland
M. Sacchi *et al.* (1999) Christopher Helm
ISBN 0713651830

**Les Bons Coins Ornithologiques de Suisse
Romande** Groupes de Jeunes de Nos Oiseaux

Vögel in der Schweiz
Burkhardt & Schmid, Schweizerische Vogelwarte

Schweizer Brutvogelatlas
Schmid *et al.*, Schweizerische Vogelwarte

Collins Bird Guide
Lars Svensson *et al.* (1999) HarperCollins
ISBN 0002197286

TYPICAL SPECIES

Resident
Great Cormorant, Grey Heron, Mute Swan, Canada Goose, Mallard, Tufted Duck, Eurasian Sparrowhawk, Common Kestrel, Moorhen, Common Coot, Black-headed Gull, Lesser Black-backed Gull, Herring Gull, Tawny Owl, Wood Pigeon, Collared Dove, Great Spotted Woodpecker, Pied Wagtail, Wren, Dunnock, Robin, Blackbird, Song Thrush, Goldcrest, Blue Tit, Great Tit, Long-tailed Tit, Eurasian Jay, Magpie, Western Jackdaw, Carrion Crow, Common Starling, House Sparrow, Chaffinch, Greenfinch, Goldfinch, Bullfinch

Summer
Hobby, Little Ringed Plover, Common Tern, Common Swift, Barn Swallow, House Martin, Garden Warbler, Blackcap, Common Whitethroat, Common Chiffchaff, Willow Warbler

Winter
Common Pochard, Common Gull, Redwing, Fieldfare

*Tawny Owls can be found in most mature parks and gardens
but is strictly nocturnal*

London

A birder arriving in London has the comfort of visiting the most heavily-birded country in the world. Southeast England in particular probably has more birders per square mile than anywhere else on the planet. What this means for the visitor is that there is a wealth of up-to-the minute information about bird populations, distribution and movements in the London area. The main focus of many London birders may be on the birding hotspots of the south coast and East Anglia to the northeast. However, there is much for the visitor to see closer to London and the birding sites are easily accessed by public transport. Being part of an island, there are fewer breeding species in southeast England than on nearby parts of Continental Europe. However, enlightened conservation measures have ensured that many species that would have been lost to other urban areas can still be found in London, and the habitats thus created attract many rarities.

The River Thames is the most obvious physical feature of the London area, dividing the city between north and south and forming a large estuary at its mouth with extensive mudflats and saltmarshes. Sadly much of the river's natural hinterland has been built over and the estuarine habitats have now been largely reclaimed such that numbers of wintering wildfowl and waders are much reduced. It is in fact the tributaries of the Thames: the Lee and the Colne, which provide some of the best aquatic habitats in the London area. In the absence of large freshwater lakes, a densely populated urban area such as London needs to create many man-made water sources and storage facilities. The numerous reservoirs and sewage plants which have been constructed around the city, especially along these river valleys, provide excellent habitats for passage migrants and wintering wildfowl.

GETTING AROUND

By far the best way to get around is to use the London Underground system (or Tube), which actually runs above ground in the suburbs. The service is fast, frequent and fairly inexpensive. Although mainly serving areas north of the Thames, the system is very extensive, so you will nearly always be close to an underground station, and it connects with all

the major railway stations. If you are driving, the M25 orbital motorway is usually the best way to get from one side of the city to the other, although try to avoid using it during peak periods. Due to heavy traffic, scarce parking and congestion charges it is unwise to take a car into central London if you can avoid it.

You can use the train services for outlying areas of London and, although much-maligned by Londoners, outside peak periods the trains are quite reliable and adequately serve many of the birding sites. However, as a result of privatisation the rail system is fragmented among several different operators and it is not immediately obvious for a visitor which company operates on any given route. Also there are several main railway stations in London, none of which are interconnected. So it is important to identify which company serves your destination before setting out on your journey.

As, apparently, the most popular pastime in Britain, birding is very well-catered for in London, with many sites neatly landscaped and providing high quality access facilities. This also means that you will nearly always meet fellow-birders when out birding and they will be more than happy to assist you with directions and up-to-date information. Most large newsagents stock birding magazines, and there is a wealth of books, equipment and other birding aids available in general retail outlets. There are no special requirements for any of the sites in this guide other than comfortable walking shoes and perhaps a telescope for scanning the reservoirs and other wide expanses. A map of the Greater London area, which is widely available, should be adequate to locate all the sites and of course get a London Underground system map: probably the most familiar and ubiquitous graphic diagram in the world.

Locations

1 Regent's Park (City Centre)   *(Baker St Underground Station)*

There are several large and well-known parks in central London, all of which have a limited selection of resident birds as well as some wintering wildfowl. Of these, Regent's Park in the west end of the city is probably the best for both numbers of birds and diversity. The best areas are around the lake in the southwest section of the park and the northern perimeter along the banks of the Grand Union Canal. Also check Primrose Hill, just north of the park, an extension of the woodland habitat. The presence of London Zoo within the park attracts many visitors so it is best to visit the park on weekday mornings. The less specialised woodland species typical of southeast England breed here and in spring and autumn a number of passage migrants can be seen.

KEY SPECIES

Resident
Little Grebe, Great Crested Grebe, Common Pochard, Grey Wagtail, Mistle Thrush, Coal Tit

Winter
Siskin

Passage
Common Cuckoo, Sky Lark, Meadow Pipit, Tree Pipit, Yellow Wagtail, Whinchat, Common Stonechat, Northern Wheatear, Spotted Flycatcher

2 Hampstead Heath (4 miles)   *(Hampstead Underground Station)*

This famous area of high ground in north London is one of the best urban birding sites in the city. It supports a resident avifauna equal to many similar woodlands in the countryside, and due to its elevated position it has recorded many rare migrants. It is also worth visiting for the views over the city. The best areas for birds are along the western and northern fringes of the heath, either side of Spaniard's Road and the Highgate Ponds in the east. The heath is quite large so several hours are required to cover all of the relevant areas. Needless to say, the area is very popular with Londoners and an early morning visit is advised.

KEY SPECIES

Resident
Stock Dove, Green Woodpecker, Lesser Spotted Woodpecker, Mistle Thrush, Coal Tit, Eurasian Nuthatch, Eurasian Treecreeper, Reed Bunting

Summer
Reed Warbler, Lesser Whitethroat

Winter
Little Grebe, Northern Shoveler, Grey Wagtail, Linnet, Lesser Redpoll, Siskin

Passage

Yellow Wagtail, Northern Wheatear, Common Redstart, Common Stonechat, Spotted Flycatcher, Pied Flycatcher

3 London Wetland Centre (7 miles)   *(Hammersmith Underground Station)*

This area of wetland has been created on the site of a reservoir on the south bank of the Thames and is intensively managed by the Wildfowl and Wetlands Trust for the benefit of wildlife. Although only in existence since 2000 it has already built up a substantial list of records and has become a very valuable natural habitat in central London, attracting several species which would not normally encroach so far into the city.

The centre is in west London about 15 minutes walk from **Hammersmith Tube Station**. Go south on **Hammersmith Bridge Road** passing under the Hammersmith Flyover. Cross the Thames at Hammersmith Bridge and continue south on the **A306** for about a mile, following the signs for the Wetland Centre (also known as Barnes WWT), and turn left into **Queen Elizabeth Walk**. The entrance is on the left-hand side of the road. Several bus routes also cover the distance from the Tube station: look for the 'Duck Bus'.

KEY SPECIES

Resident
Little Grebe, Great Crested Grebe, Water Rail, Northern Lapwing, Common Redshank, Rose-ringed Parakeet (I), Black Redstart, Reed Bunting

Summer
Sand Martin, Sedge Warbler, Reed Warbler

Winter
Great Bittern, Common Shelduck, Gadwall, Northern Shoveler, Eurasian Wigeon, Eurasian Teal, Common Snipe, Jack Snipe, Common Kingfisher, Brambling

Passage
Garganey, Hobby, Ringed Plover, Black-tailed Godwit, Eurasian Curlew,

Greenshank, Green Sandpiper, Common Sandpiper, Little Gull, Sky Lark, Meadow Pipit, Northern Wheatear

---

4 Lee Valley Park (14 miles)

(Cheshunt Railway Station)
(Blackhorse Road Underground Station)

The River Lee is a tributary of the Thames which flows through the northeastern suburbs of London. Right along its length from Epping Forest to the Thames there is a string of reservoirs and flooded gravel pits which provide a complex chain of wetland sites which act as a conduit for migrants and provide important breeding and wintering grounds for wildfowl. There are many access points to the wetlands but Lee Valley Park is one of the best known and its Bittern watchpoint is probably the most reliable site in Britain for this species.

The park is just north of the **M25** at **Cheshunt**. Exit north from the M25 at **Exit 25** and take the **A10** north for about 1.5 miles. Turn east towards Cheshunt on the **B198** and continue on **Windmill Lane** towards **Cheshunt Railway Station**. There is a park entrance just beyond the railway station which gives you access to the west bank of the Lee and its associated flooded gravel pits. The Bittern watchpoint is on the east bank of the Lee and to get there you must leave the park at Cheshunt and take

the **B176** south and then the **A121** east in order to cross the river. Take the **B194** north for about 1.5 miles and enter the park on the left-hand side of the road at **Fisher's Green**. Trains to Cheshunt operate from Liverpool Street Station.

There are two large reservoirs on the south side of the M25 (King George and William Girling Reservoirs)

but although these sites support larger numbers of duck in winter and also large gull roosts, they are off-limits to visitors except for permit holders. However, further south, Walthamstow Reservoirs are home to a large heronry, breeding cormorants and many other waterbirds. From Blackhorse Road Station, head west along Forest Road for a quarter of a mile. Permits are available for a small fee at the office on the south side of Forest Road.

KEY SPECIES

Resident
Little Grebe, Great Crested Grebe, Greylag Goose, Gadwall, Northern Shoveler, Ruddy Duck, Common Buzzard, Little Owl, Green Woodpecker, Northern Lapwing, Common Redshank, Common Kingfisher, Sky Lark, Meadow Pipit, Grey Wagtail, Rook, Linnet, Reed Bunting, Yellowhammer

Summer
Common Shelduck, Ringed Plover, Common Sandpiper, Turtle Dove, Common Cuckoo, Sand Martin, Yellow Wagtail, Common Nightingale, Grasshopper Warbler, Sedge Warbler, Reed Warbler, Lesser Whitethroat

Winter
Great Bittern, Eurasian Wigeon, Eurasian Teal, Pintail, Common

Goldeneye, Goosander, Smew, Water Rail, European Golden Plover, Common Snipe, Great Black-backed Gull, Common Stonechat, Lesser Redpoll, Siskin

Passage
Garganey, Greenshank, Green Sandpiper, Whinchat, Common Stonechat, Northern Wheatear

5 Epping Forest (17 miles)

The forest is on the northern outskirts of London and, if you have time to spare, this is the best site for a good cross-section of the possible woodland species of southern England. This large, mainly broadleaved woodland provides a habitat which supports three species of woodpecker and a range of other interesting species. If you are flying into London via Stansted Airport, this site is on the way into London.

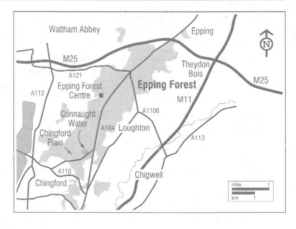

The best areas are at the southern end: Connaught Water for waterfowl and aquatic species, including Mandarin Duck; the wooded areas around this lake for forest species; and Chingford Plain in the southwest for migrants and for more open country species.

Take the **M11 Motorway** northeast from London and exit west at **Exit 5** towards **Loughton**. Drive through Loughton towards the **A104** and follow the signs for **Epping Forest Conservation Centre**, which has maps and information, or to **Queen Elizabeth's Hunting Lodge** on Chingford Plain. If travelling by public transport, take the train to Chingford (from Liverpool St Station) early in the morning in order to do Chingford Plain while it is least disturbed by joggers, model aircraft enthusiasts and dog-walkers. Then work your way northeastwards through the forest and return to London by Tube from Loughton.

KEY SPECIES

Resident
Mandarin Duck, Ruddy Duck, Woodcock, Green Woodpecker, Great Spotted Woodpecker, Lesser Spotted Woodpecker, Common Kingfisher, Sky Lark, Mistle Thrush, Eurasian Nuthatch, Eurasian Treecreeper, Reed Bunting

Summer
Common Cuckoo, Tree Pipit, Garden Warbler, Lesser Whitethroat

Winter
Northern Shoveler, Meadow Pipit, Grey Wagtail, Brambling, Siskin, Lesser Redpoll

Passage
Whinchat, Northern Wheatear, Common Redstart, Pied Flycatcher, Spotted Flycatcher

6 Rainham Marshes (14 miles) *(Purfleet Railway Station)*

Rainham Marshes are east of London on the north bank of the River Thames and represent one of the last stretches of saltmarsh on the upper Thames Estuary. The adjacent rubbish tip attracts many gulls and has become a regular site for Yellow-legged Gulls. Although much disturbed and surrounded by industrial development it has now been given protection and is managed for birds by the RSPB. For several species, this is the most reliable site in London. Much work has been done to encourage breeding waders and it also has good holding capacity for non-breeding waders on passage. Its location downstream from the city means that this site attracts several coastal species which would be extremely rare further inland. The reserve is very well watched at all times of the year and has developed a reputation as a rarity trap.

From the **M25**, exit west at **Exit 31** onto the **A1306** towards **Purfleet**. After about 1.5 miles turn south on the **A1090** (**Tank Hill Road**). The entrance to the reserve is a short distance on the right-hand (west) side of this road. Purfleet Railway Station is on Tank Hill Road, about a mile south of the entrance. Trains depart from Liverpool Street Station. You can then walk west along the seawall for access to the shoreline, where some derelict stone barges have been abandoned. This area is good for Short-eared Owls, Rock Pipits and Water Pipits in winter. If driving, you can reach this area by returning to the A13 and going west to the next exit (**Rainham**). Turn left (south) onto **Ferry Lane** and then left again into **Coldharbour Lane**. Park at the car park just east of the **Tilda Rice** factory.

KEY SPECIES

Resident
Little Grebe, Common Shelduck, Gadwall, Northern Shoveler, Northern Lapwing, Common Redshank, Common Snipe, Sky Lark, Meadow Pipit, Black Redstart, Common Stonechat, Linnet, Reed Bunting

Summer
Ringed Plover, Sand Martin, Yellow Wagtail, Sedge Warbler, Reed Warbler, Common Whitethroat, Lesser Whitethroat

Winter
Eurasian Teal, Pintail, European Golden Plover, Oystercatcher, Ruddy Turnstone, Great Black-backed Gull,

Caspian Gull, Yellow-legged Gull, Short-eared Owl, Water Pipit, Rock Pipit, Lesser Redpoll, Siskin

Passage
Garganey, Grey Plover, Dunlin, Ruff, Greenshank, Eurasian Curlew, Whimbrel, Common Sandpiper, Black-tailed Godwit

7 Staines Reservoirs (20 miles) *(Ashford Railway Station)* *(Bus No. 555 from LHR)*

If your plane approached London Heathrow Airport from the west, it is likely that you flew low enough over Staines Reservoirs to have seen flocks of gulls or swans from the window! Less than ten minutes from the airport, this is one of London's premier sites for wintering wildfowl. Additionally, if water levels are low in late summer or autumn, the muddy margins attract waders in varying numbers. The area is less attractive in summer, although a visit to nearby Staines Moor would produce some interesting passerines, and if you find yourself stuck in an airport hotel for a day, then this site is well worth a visit at any time of year.

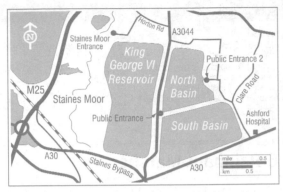

From the **M25 Motorway** take **Exit 13** and head east towards **Ashford** for about 1.5 miles on the **A30**, then turn north on the **A3044** for about half a mile. There is a public entrance to the reservoir on the right-hand side (east) of the road. The reservoir on the left-hand side (King George VI) is not open to the public. The path brings you up onto a causeway which separates the north and south basins of the reservoir and you can scan both basins from here. If you have time, you could return to the A3044 and go north for about a mile, then turn left on **Horton Road**, followed by the first turn left. This will bring you to the northern end of Staines Moor which is an area of wet meadow along the banks of the River Colne, with willow scrub and a flooded gravel pit. Take any of the footpaths and check for finches, buntings and, in summer, warblers.

If you are using public transport, there is a train service from London (Marylebone Station) to Ashford which is about a mile walk from the reservoir. From the railway station follow the B378 (Church Road) north, crossing over the A30 and continuing on the B378 which becomes Town Lane. The public entrance on the left-hand side of Town Lane brings you to the east end of the causeway between the north and south basins of the reservoir. There is also a bus service (No. 555) from Heathrow Airport to Ashford.

KEY SPECIES

Resident
Great Crested Grebe, Northern Lapwing, Sky Lark, Meadow Pipit, Rook, Linnet, Reed Bunting

Summer
Little Grebe, Ringed Plover, Common Redshank, Sand Martin, Yellow Wagtail, Reed Warbler, Common

Whitethroat, Lesser Whitethroat

Winter
Eurasian Wigeon, Eurasian Teal, Pintail, Northern Shoveler, Common Goldeneye, Goosander, Ruddy Duck, Common Snipe, European Golden Plover, Great Black-backed Gull, Water Pipit, Black Redstart, Lesser

Redpoll, Siskin

Passage
Black-necked Grebe, Garganey, Dunlin, Ruff, Greenshank, Common Sandpiper, Green Sandpiper, Little Gull, Black Tern, Whinchat, Common Stonechat, Northern Wheatear

Useful contacts and websites

Royal Society for the Protection of Birds
www.rspb.org.uk
The Lodge, Sandy, Bedfordshire SG19 2DL.
Tel. 01767 680551.
The UK's leading bird conservation organisation and also Europe's largest wildlife conservation charity.

RSPB – Central London Group
www.janja.dircon.co.uk/rspb
Local Branch of RSPB: Membership Secretary
Tel. 0208 2080981.

RSPB – Rainham Marshes
www.rspb.org.uk/reserves/guide/r/rainhammarshes
Information about Rainham Marshes RSPB reserve.

East London Birders' Forum www.elbf.co.uk
Information about recent sightings, sites, guided walks and much more.

Marylebone Birdwatching Society
www.geocities.com/birdsmbs
North London birdwatching group which organises local birding excursions.

London Natural History Society
www.users.globalnet.co.uk/~lnhsweb/orni.htm
Website dealing with ornithology of London including checklist and list of sites with grid references.

London's Birding
website.lineone.net/~andrewself/Londons birding.htm
Information about local sites, recent reports and checklist of London area.

Birding in London
myweb.tiscali.co.uk/calidris
Information about local sites, recent reports and news about birding events in London area.

Fatbirder – London
www.fatbirder.com/links_geo/europe/england_greater_london.html
Web page from the excellent fatbirder.com website devoted just to London.

Lee Valley Park
www.leevalleypark.org.uk
Information about natural history of Lee Valley with maps and access information.

London Wetland Centre (Barnes)
www.wwt.org.uk/visit/wetlandcentre
Information about reserve and about the work of the Wildfowl and Wetlands Trust.

Epping Forest
www.cityoflondon.gov.uk/living_environment/open_spaces/epping_forest.htm
Information about forest park with maps and general information.

Hampstead Heath
www.cityoflondon.gov.uk/living_environment/open_spaces/hampstead_heath.htm
Information about park with maps and general information.

London Area Recorder
Email: andrewself@lineone.net
Andrew Self, 16 Harp Island Close, Neasden, London NW10 0DD. Tel. 0208 2082139.

Rare Bird Alert: Birdline Southeast
www.southeastbirdnews.co.uk
Tel. 0906 8700240 for information.
Freephone 0800 0377240 to report sightings.

Transport for London www.tfl.gov.uk/tfl
Web portal site with links to information about train, Tube and bus transport in London area.

London Travel Information:
Tel. 0207 2221234 (24 hours a day).

Books and Publications
Where to Watch Birds in the London Area
Dominic Mitchell (1997) Christopher Helm
ISBN 0713638680

The Breeding Birds of the London Area
Jan Hewlett (2002) ISBN 0901009121

Birds in London
William Henry Hudson (1972) ISBN 0715346903

Collins Bird Guide
Lars Svensson *et al.* (1999) HarperCollins
ISBN 0002197286

Madrid

Most European birders have been to southern Spain on a birding trip at some stage in their lives in order to relish its outstanding avifauna, yet would consider Madrid to be too distant from this birding paradise. Moreover it is far inland and its birding sites less well known than those of the south. However, Madrid lies

Spanish Imperial Eagle can sometimes be seen soaring above the city centre in Madrid

at the centre of one of the most bird-rich regions in western Europe. The statistics for central Spain are astounding: over 90 per cent of western Europe's vultures, bustards and wintering cranes all inhabit this area, and breeding populations number in the thousands of pairs for species which would barely exceed double figures elsewhere. All of these birds are literally on Madrid's doorstep. Europe's only endemic raptor, the Spanish Imperial Eagle, may also be seen near the city

At almost 700 metres above sea-level, Madrid is one of the highest capital cities in Europe and its proximity to the high peaks of the Sierra de Guadarrama means that within an hour's drive of the city it is possible to see many specialist montane species for which you would normally need to plan a special birding trip. Bear in mind, however, that these birds exist at fairly low densities in this extensive habitat making birding at altitude quite hard work, especially if there is bad weather in winter.

Spring and early summer are the best times to visit as summer migrants will have arrived and the resident birds will be establishing territories. Temperatures are very pleasant and make for easy hiking in the surrounding mountains and the high plains. Winter can also be interesting as, although the area around Madrid is rather dry with few natural wetland areas, there are newly created wetlands formed as a result of gravel extraction and these abandoned sites are rapidly developing into wintering sites for ducks and gulls.

Many birds migrate to Africa through the Iberian Peninsula but most move on a rather broad front and there are few physical features in the area that would tend to concentrate migrants. Again it is the wetland areas which offer the best prospects as these attract migrating waders, terns and passerines.

TYPICAL SPECIES

Resident
Great Crested Grebe, Cattle Egret, White Stork, Grey Heron, Mallard, Common Buzzard, Common Kestrel, Moorhen, Common Coot, Wood Pigeon, Collared Dove, Little Owl, Green Woodpecker, Great Spotted Woodpecker, Crested Lark, Wood Lark, Grey Wagtail, White Wagtail, Wren, Black Redstart, Common Stonechat, Robin, Blackbird, Mistle Thrush, Zitting Cisticola, Sardinian

Warbler, Blackcap, Coal Tit, Blue Tit, Great Tit, Eurasian Nuthatch, Short-toed Treecreeper, Eurasian Jay, Magpie, Western Jackdaw, Carrion Crow, Spotless Starling, House Sparrow, Tree Sparrow, Rock Sparrow, Chaffinch, Greenfinch, Goldfinch, Linnet, Hawfinch, European Serin, Corn Bunting

Summer
Black Kite, Hobby, Turtle Dove, Scops

Owl, European Bee-eater, Hoopoe, Common Swift, Barn Swallow, House Martin, Common Nightingale, Melodious Warbler, Garden Warbler, Spotted Flycatcher, Golden Oriole

Winter
Black-headed Gull, Lesser Black-backed Gull, Sky Lark, Meadow Pipit, Song Thrush, Fieldfare, Goldcrest

GETTING AROUND

Although public transport is fine for travelling within the city and for some of the outlying areas, a car is essential in order to visit some of the best sites. Some of Madrid's most interesting and sought-after birds inhabit the remotest parts of the mountains to the north of the city and obviously these areas are not served by public transport. Even diehard ecotourists who would prefer to bird by bike, will be forced to use a car as, unless you have the fitness and stamina of a top-class professional cyclist, the mountains will present a lung-burning endurance trial. Fortunately car-hire is not too expensive, especially in spring and early summer when mountain birding is at its best. The road north to the mountains actually commences right in the city centre (Paseo de la Castellana) and goes directly to the heart of the Sierra de Guadarrama without deviating so you can't get lost! From anywhere else in the city, or from the airport, the road north and to all other points can be accessed easily from the M30 inner ring motorway or the M40 outer ring.

It is possible to reach the foothills of the Sierra and several other interesting sites by public transport. The Metro de Madrid is the fastest way to get around the city and to some of the birding sites in the northwest. The suburban train service known as the Cercanías is operated by the national railway company RENFE and has quite an extensive network serving the outer suburbs operating from both main railway stations at Chamartín and Atocha.

The regional buses provide a fairly comprehensive service to the main towns and villages in the mountains but you need to be prepared to hike several kilometres in order to get into good birding habitat. These buses are operated by several different companies and depart from either of two terminals in the city centre: Plaza de Castilla in the north of the city and Estación Sur near Méndez Álvaro Metro Station in the south.

Hiking in the mountains takes place along designated trails and there is no climbing required so a good pair of walking shoes is all that is required. However, bear in mind that in early spring, the weather at altitude can be quite cold and there is always the chance of thick mist forming, so the usual precautions about hiking in the mountains should be observed. The blistering heat of summer, on the other hand, makes birding during the day an arduous and rather unrewarding experience as most passerines remain quiet and concealed during the heat of the day. Early morning and evening will be much more productive.

Most shops in Madrid still observe the siesta period from 13.00 to 16.00 or thereabouts so make sure that anything you need is obtained beforehand. On the other hand, since most shops stay open until 20.00 or later this means that you don't have to waste birding time in order to stock up. Any map of the Madrid Autonomous Region will include all the sites mentioned in this chapter.

Locations

1 Casa de Campo (3km) (Lago Metro Station)

The Casa de Campo is a large suburban park in the western suburbs. The lake attracts some wildfowl but more importantly the park preserves an area of dehesa or grazing woodland, a habitat which is unique to the Iberian Peninsula comprising extensive pastures with scattered cork oaks and holm oaks. These areas support unique bird communities but are threatened by agricultural intensification. Despite its relative isolation from any contiguous stretch of similar habitat, it manages to support a selection of typical species. This is a popular park with strollers and picnickers, particularly the southeast end around the lake, so early morning is the best time to visit.

KEY SPECIES

Resident
Stock Dove, Monk Parakeet (I), Cetti's Warbler, Firecrest, Long-tailed Tit

Summer
White Stork, Subalpine Warbler, Woodchat Shrike

2 Monte de El Pardo (15km) (Bus No. 601)

The foothills of the Sierra de Guadarrama reach into the northern suburbs of Madrid at El Pardo. This is a large area of forest and dehesa which was protected as a hunting preserve and, although access is still restricted, the Spanish Ornithological Society (SEO) has created four birding trails through the area and some of the SEO members lead bird walks on Saturdays and Sundays.

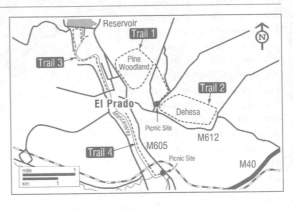

The village of El Pardo is northwest of Madrid and can be reached from the **M40** by exiting north at **Exit 51** onto the **M605** and continuing north on this road for 4.5km. There are two linear trails which follow the valley of the **Rio Manzanares** in opposite directions. One leads south to El Pardo from the football pitch at **Mingorubbio** and passes the large reservoir **Embalse de El Pardo**, an important wintering ground for wildfowl but not open to the public. The second leads north to El Pardo from the picnic site at the park entrance on the **M601**. The habitat is mainly riparian willow and ash, and this habitat is favoured by waterfowl, pipits, wagtails and Cetti's Warblers.

Two loop trails start from the picnic site, just southeast of El Pardo on the road to **Fuencarral (M612)** and one goes through an area of replanted pine forest which is good for woodpeckers and several tit species, while the second passes south through an area of more open dehesa. The bus serving this area (No. 601) departs from Moncloa Metro Station.

KEY SPECIES

Resident
Griffon Vulture, Eurasian Black Vulture, Spanish Imperial Eagle, Stock Dove, Common Kingfisher, Thekla Lark, Crag Martin, Dipper, Blue Rock Thrush, Cetti's Warbler, Dartford Warbler, Azure-winged Magpie, Red-billed Chough, Common Raven,
Southern Grey Shrike, Spanish Sparrow, Cirl Bunting

Summer
Booted Eagle, Common Cuckoo, Great Spotted Cuckoo, Red-necked Nightjar, Red-rumped Swallow, Black-eared Wheatear, Rock Thrush,
Subalpine Warbler, Orphean Warbler, Woodchat Shrike, Ortolan Bunting

Winter
Great Cormorant, Grey Heron, Gadwall, Eurasian Teal, Northern Shoveler, Common Pochard, Tufted Duck

3 Embalse de Santillana (45km) (Bus Nos. 725/726)

This is a fairly large reservoir north of Madrid set in an area of farmland and scattered clumps of woodland. As one of the few large waterbodies in the area, it attracts a significant number of wintering wildfowl as well as passage migrants, particularly storks, herons and cranes, and it is one of the most important winter wildfowl sites in Spain. There are parts of the reservoir, particularly at the east end, which are shallower and have more abundant plant growth and these are the most attractive areas for ducks and waders.

Take the **M607** north from Madrid to the town of **Soto del Real**. In Soto, take the **M608** west for about 3km. From this point on, the road runs along the north shore of the reservoir as far as **Manzanares el Real**. There are several lay-bys on the road where you can pull in and scan the open water and shore. However, also check the fields and open country on the opposite side of the road as this habitat is favoured by storks, bustards and larks. There is a regular bus service to Soto del Real from Plaza de Castilla.

KEY SPECIES

Resident	Summer	
Little Grebe, Great Crested Grebe, Gadwall, Northern Shoveler, Common Pochard, Water Rail, White Stork, Little Egret, Thekla Lark, Dartford Warbler, Southern Grey Shrike, Azure-winged Magpie, Common Raven	Little Bustard, Cattle Egret, Black-winged Stilt, Stone-curlew, Great Spotted Cuckoo, European Bee-eater, Short-toed Lark, Spanish Wagtail, Subalpine Warbler, Spectacled Warbler	Cormorant, Grey Heron, Eurasian Wigeon, Eurasian Teal, Pintail, Red-crested Pochard, Black-headed Gull
	Winter Black-necked Grebe, Great	**Passage** Purple Heron, Black Stork, Montagu's Harrier, Osprey, Avocet, Black-tailed Godwit, Bar-tailed Godwit, Green-shank, Black Tern, Whiskered Tern

4 La Pedriza del Manzanares (50km)   (Bus No. 724)

The Natural Park at La Cuenca Alta del Manzanares just north of Madrid contains very dramatic geological formations brought about through the erosion of the granite substrate. The area known as La Pedriza is famous for these rock formations and consequently attracts many tourists. However this is also an excellent site to see the birdlife of this dry rocky habitat, including Griffon Vultures which nest rather obviously on the granite outcrops. Its popularity with tourists means that there is good access but you would be advised to avoid the area at weekends as it can be very crowded, which makes for rather unsatisfying birding.

Take the **M607** north from Madrid to **Soto del Real** and turn west at Soto towards **Manzanares el Real** on the **M608**. This road runs alongside the north shore of **Embalse de Santillana** which is also worth a look. About 1km beyond Manzanares turn at the sign for La Pedriza and follow the road to a car park which overlooks the Manzanares River. There are several hiking trails which fan out from the car park, any of which will bring you into the habitat for rock thrushes, wheatears and other mountain passerines. Scan for raptors from any lookout points. There is a regular bus service to Manzanares from Plaza de Castilla.

KEY SPECIES

Resident	Summer
Griffon Vulture, Golden Eagle, Peregrine Falcon, Eagle Owl, Thekla Lark, Crag Martin, Water Pipit, Dipper, Black Wheatear, Blue Rock Thrush, Dartford Warbler, Azure-winged Magpie, Red-billed Chough, Common Raven, Rock Bunting	Booted Eagle, Tawny Pipit, Bluethroat, Black-eared Wheatear, Rock Thrush, Subalpine Warbler, Orphean Warbler, Woodchat Shrike, Ortolan Bunting

5 Puerto de la Morcuera (60km)

The Sierra de Guadarrama is just north of Madrid and at its highest elevation rises to an altitude of almost 2,500m. It is possible to visit a number of sites in this mountain range using the town of Miraflores de la Sierra as a base. The mountain pass at Puerto de la Morcuera is at 1,800m and the road from Miraflores to this pass offers a number of opportunities to explore the forest habitat and the more open scrub areas. The Puerto de Canencia is a mountain pass at a lower elevation which is forested and may offer alternative possibilities particularly if the higher pass is shrouded in mist.

Take the **M607** north from Madrid towards Miraflores de la Sierra. At Miraflores, turn west on the **M611** towards Puerto de la Morcuera, which is approximately 8km. There are a number of places to pull in along this road and it is worth checking any wooded areas and also scanning the valleys and hillsides for birds of prey. At the pass, there are trails across the mountain leading off from both sides of the road. If the weather at la Morcuera is bad, Puerto de Canencia is reached by taking the road (**M629**) north from Miraflores for about 15km. The elevation here is lower and there is an extensive forest of Scots Pines. Several trails lead off from the car park into the woods.

KEY SPECIES

Resident
Griffon Vulture, Eurasian Black Vulture, Northern Goshawk, Golden Eagle, Spanish Imperial Eagle, Dunnock, Alpine Accentor, Blue Rock Thrush, Firecrest, Crested Tit, Red-

billed Chough, Common Raven, Azure-winged Magpie, Rock Sparrow, Citril Finch, Crossbill, Rock Bunting

Summer
Black Stork, Short-toed Eagle, Booted

Eagle, Crag Martin, Water Pipit, Northern Wheatear, Bluethroat, Rock Thrush, Common Whitethroat, Western Bonelli's Warbler, Pied Flycatcher, Ortolan Bunting

6 Jarama and Henares Steppes (30km)

To the east of Madrid, between the Rivers Jarama and Henares, there is a large area of cereal-growing steppes which supports a diverse range of open-country species and is one of the best areas near the city for Great Bustards. It is also the favoured hunting territory of many of the raptors which breed in the adjacent Sierra de Guadarrama. However, this is a challenging location to explore because the area is so extensive and the birds are present in low densities. Moreover the habitat is under threat of fragmentation and depletion through motorway construction and industrial development.

Paradoxically, however, using a car is the only realistic way of seeing the birds. The area is best considered as a large triangle with one corner at **Madrid Airport**, the east corner at **Alcalá de Henares** and the third corner in the north at **Talamanca de Jarama**. Driving the road from Alcalá de Henares to **Daganzo de Arriba** (**M100**) and then north from **Cobeña** to Talamanca de Jarama (**M103**) takes you through the middle of this area. Preferably have one person driving and one or more observers free to scan the fields, skies, overhead wires and fenceposts. If possible you should stop and walk through any stubble fields, as these are attractive to larks and buntings.

KEY SPECIES

Resident
Red Kite, Eurasian Black Vulture, Hen Harrier, Golden Eagle, Spanish Imperial Eagle, Bonelli's Eagle, Red-legged Partridge, Little Bustard, Great Bustard, Stone-curlew, Black-bellied

Sandgrouse, Calandra Lark, Southern Grey Shrike, Cirl Bunting, Corn Bunting

Summer
Common Quail, Booted Eagle,

Montagu's Harrier, Lesser Kestrel, European Roller, Short-toed Lark, Tawny Pipit, Black-eared Wheatear, Woodchat Shrike, Rock Sparrow, Ortolan Bunting

7 Rivas-Vaciamadrid Lagoons (20km) *(Rivas-Vaciamadrid Metro Station)*

At the confluence of the Manzanares and Jarama Rivers there are a large number of abandoned gravel pits which have flooded and formed lagoons. This whole area has now been opened up as part of the Parque Regional del Sureste and bike trails, hiking trails and bird hides have been developed. There are more than 30 lagoons in the area extending along the river

banks north and south of Rivas-Vaciamadrid and this is now one of the most important wetland sites in the Madrid region. The surrounding countryside offers a number of different habitats including scrub vegetation, riparian woodland and rather impressive rocky cliff faces: formed as a result of erosion by the rivers. All are good for the typical species of these habitats.

Take the **A3** southeast from Madrid to Rivas-Vaciamadrid and exit north at **Exit 20**. You can park at the Metro station and then walk south alongside the metro line to the Laguna El Campillo. There is a circular path around the lagoon with several watch points along the way. You can return to the **A3** and continue south to the next exit (**Exit 21**) and exit south onto the **M832**. Continue south for about 1km and turn west for **Finca El Porcal**. There is a gravel track which runs between the lagoons. If on foot, you can walk from the metro station at Rivas-Vaciamadrid south through Finca El Porcal and along the banks of the Jarama as far as the SEO Reserve at Los Albardales. You could get the train back to Madrid from San Martín de la Vega.

KEY SPECIES

Resident
Little Grebe, Black-necked Grebe, Little Bittern, Purple Heron, Little Egret, Red-crested Pochard, Marsh Harrier, Peregrine Falcon, Purple Gallinule, Monk Parakeet (I), Eagle Owl, Common Kingfisher, Water Rail, Cetti's Warbler, Dartford Warbler, Black Wheatear, Blue Rock Thrush, Red-billed Chough, Penduline Tit, Rock Sparrow

Summer
Short-toed Eagle, Montagu's Harrier, Little Ringed Plover, Black-winged Stilt, Red-necked Nightjar, Sand Martin, Spanish Wagtail, Savi's Warbler, Reed Warbler, Great Reed Warbler, Subalpine Warbler

Winter
Great Cormorant, Greylag Goose, Gadwall, Eurasian Wigeon, Eurasian Teal, Pintail, Common Pochard, Tufted Duck, Ferruginous Duck, Ringed Plover, Common Snipe, Common Sandpiper, Green Sandpiper, Brambling, Reed Bunting, Rock Bunting

Passage
Osprey, Common Crane, Northern Lapwing, Common Redshank, Greenshank, Black Tern, Whiskered Tern, European Nightjar, Water Pipit, Bluethroat, Sedge Warbler

---

8 Los Albardales (25km) *(San Martín de la Vega Railway Station)*

Between its confluence with the River Manzanares and its confluence with the River Tagus, the River Jarama flows through an area of wetlands which are contained in the Southeast Regional Park (Parque Regional del Sureste). This large linear park is an important wetland for resident aquatic birds and wintering wildfowl. Within the park the SEO has created a small nature reserve near San Martín de la Vega called Los Albardales which provides an easy focal point to bird this area. There is a lagoon and reedbed at the reserve and a riverside walk along the west bank of the Río Jarama. This area is worth a visit at any time of year, as apart from the resident aquatic birds, it is a significant roost site for winter flocks of passerines as well as wintering wildfowl.

Take the **E5 Motorway** south as far as **Pinto** and exit east onto the **M506** to San Martín de la Vega. At San Martín, take the **M302** north to the northern outskirts of the town and the reserve is on the right-hand side of the road. If you are using public transport, the railway station at San Martín is beside the M506 and a short walk from the reserve.

KEY SPECIES

Resident
Little Grebe, Great Crested Grebe, Night Heron, White Stork, Little Egret, Northern Shoveler, Common Pochard, Marsh Harrier, Water Rail, Cetti's Warbler, Bearded Tit, Penduline Tit, Spanish Sparrow, Red Avadavat (I)

Summer
Little Bittern, Lesser Kestrel, Black-winged Stilt, Spanish Wagtail, Bluethroat, Great Reed Warbler, Spectacled Warbler, Ortolan Bunting

Winter
Great Cormorant, Gadwall, Eurasian Wigeon, Eurasian Teal, Pintail, Red-crested Pochard, Linnet, Reed Bunting

Passage
Purple Heron, Black Stork, Montagu's Harrier, Osprey, Black Tern, Whiskered Tern

Useful contacts and websites

SEO/BirdLife (Sociedad Española de Ornitologia) (Spanish) www.seo.org
C/ Melquíades Biencinto, 34-28053 Madrid
Tel. (+34) 914 340910. Fax. (+34) 914 340911.
Spanish BirdLife International partner and publishers of the journal *Ardeola*.

SEO – Monticola (Spanish)
www.uam.es/otros/monticola
Unidad de Zoología, Facultad de Ciencias, Universidad Autónoma de Madrid, 28049 Madrid.
Madrid University Branch of SEO – organises bird-walks in Valdelatas and elsewhere.

Spainbirds (English)
www.spainbirds.com
Information about organised birding trips from Madrid.

El Parque Regional Cuenca Alta del Manzanares (Spanish)
http://greenfield.fortunecity.com/deercreek/592/pedriza/parqueregional/parqueregional.htm
Leo Fernandez's personal website about Cuenca Alta del Manzanares Regional Park with lots of useful information about birds and habitat for La Pedriza del Manzanares and also the Embalse de Santillana.

Parque Regional del Sureste (Spanish)
www.elsoto.org
Official website of Parque del Sureste with details of access, natural history and bird checklist.

Spanish Rarities Committee (English)
www.seo.org and www.rarebirdspain.net/arbsh000.htm#English
Website of CR/SEO Spanish Rarities Committee with recent reports and other information. Contact: J. I. Dies. Delegación de Valencia de SEO/BirdLife. Avda. de los Pinares 106, 46012 El Saler (Valencia). Email. rarezas@seo.org

Consorcio Transportes Madrid (English and Spanish) www.ctm-madrid.es
Web portal site for public transport services in Madrid.

Metro de Madrid – Metro service in Madrid
www.metromadrid.es

EMT – (English and Spanish)
www.emtmadrid.es
Bus service for Madrid and suburbs. Tel. 914 068810.

RENFE (Spanish)
www.renfe.es/cercanias/madrid
Madrid suburban train service. Tel. 91 506 70 67 or 91 328 90 20.

Books and publications

Where to Watch Birds in North & East Spain
Michael Rebane & Ernest Garcia (2nd edition). Due 2007. Christopher Helm, ISBN 0713647000

Where to Watch Birds in Spain & Portugal
Laurence Rose (1995) Hamlyn
ISBN 0600584046

Where to Watch Birds in Spain: The 100 Best Sites J. A. Montero (2006) Lynx Edicions
ISBN 8496553043

Aves de España
Eduardo de Juana & Juan M. Varela (2005)
Lynx Edicions, ISBN 8487334881

Atlas de las Aves Nidificantes en Madrid
Mario Díaz, Ramón Martí, Ángel Gómez
Manzaneque & Alejandro Sánchez (1994)
SEO/BirdLife, ISBN 8445108026

Atlas de las Aves Invernantes de Madrid
Juan Carlos del Moral, Blas Molina, Javier de la
Puente y Javier Pérez-Tris (2002) SEO-Monticola
ISBN 8445122622

Collins Bird Guide
Lars Svensson *et al.* (1999) HarperCollins
ISBN 0002197286

Birds of Europe
Lars Jonsson (2006) Christopher Helm
ISBN 0713676000

TYPICAL SPECIES

Resident
Great Cormorant, Grey Heron, Mallard, Common Buzzard, Common Kestrel, Common Pheasant, Moorhen, Common Coot, Northern Lapwing, Wood Pigeon, Collared Dove, Great Spotted Woodpecker, Sky Lark, White Wagtail, Grey Wagtail, Wren, Black Redstart, Robin, Common Stonechat, Song Thrush, Blackbird, Cetti's Warbler, Blackcap, Common Chiffchaff, Long-tailed Tit, Coal Tit, Blue Tit, Great Tit, Short-toed Treecreeper, Eurasian Nuthatch, Eurasian Jay, Magpie, Western Jackdaw, Hooded Crow, Common Starling, House Sparrow, Tree Sparrow, Chaffinch, Greenfinch, Goldfinch, Serin

Summer
Little Egret, Black Kite, Hobby, Turtle Dove, Hoopoe, Common Swift, Pallid Swift, Barn Swallow, House Martin, Ashy-headed Wagtail, Common Whitethroat, Spotted Flycatcher, Golden Oriole

Winter
Black-headed Gull, Meadow Pipit, Dunnock, Fieldfare, Goldcrest, Brambling, Siskin

The gaudy Golden Oriole can be surprisingly elusive and is best located by its fluty song

Milan

Milan is situated at the western end of the Lombardy Plain. Although the area is very densely populated, there are pockets of habitat which have been preserved or, indeed, have been reclaimed from abandoned industrial sites, giving the visiting birder a reasonable opportunity to see some of the typical species of this region. The climate of Lombardy is more Continental in character than Mediterranean and many of the species which are common in peninsular Italy, or even two hours' drive away on the coast, are rare or absent.

Like many inland cities you need to travel some distance outside the city limits to encounter productive habitats, and in the Milan area the best opportunities tend to be wetland sites which offer aquatic species as well as the typical woodland species in the adjacent tree and shrub cover. Much of the open countryside is given over to intensive agriculture and offers poor habitat for birds, but the cultivation of rice creates artificial wetlands which provide rich feeding for herons.

The Italian peninsula is an important conduit for many migrant birds particularly raptors in spring and cranes and passerines in autumn. The river systems which link the Alps with the Lombardy Plain, such as the Ticino and the Sesia, provide migration pathways for these birds and offer the best birding opportunities during migration. Unfortunately hunting and bird-trapping in the area exact a severe toll on the populations of these birds, despite many efforts to curtail these activities.

The foothills of the Alps begin about 70km north of Milan but these mountains are too low to support all of the alpine specialities and it would require a two-hour drive, probably into Switzerland, to get to an elevation where the high altitude species can be seen. The coast is more than 100km further south and beyond the scope of this book. However, a birder might consider a trip to the Val Lerone, just west of Genoa, during spring migration, when large numbers of raptors and passerines can be observed migrating through this coastal gap in the Apennine Mountains.

GETTING AROUND

There is a very good system of public transport within Milan, encompassing the Metro, buses and trams: unfortunately it is of little use to the birder as the best birding sites are located some distance out of town. A car is the best option to get to these areas as they are rarely located within a short walk of a railway station or bus stop. There is a very dense and rather complex

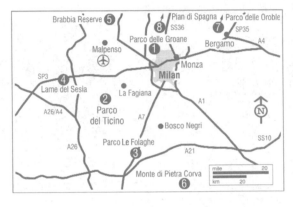

road system around Milan but always try to orientate from the orbital Autostrada A50/A51, which is the best way to navigate the city's road system. Fanning out from the orbital autostrada are several other motorways and the numbered minor roads (Strada Statale).

For those without a car, there is a very comprehensive suburban railway network but, with only a few notable exceptions, you need to be prepared to walk a few kilometres from the railway station to the birding sites. An easier option would be to rent a bicycle, which you can take on the train for a small fee or hire at your destination, and ride it between sites. There are several railway stations in Milan and although most trains depart from Stazione Centrale, for certain destinations and at certain times of the day, they depart from one of the other stations. It is thus important to familiarise yourself with the locations of these stations and to obtain a schematic map of the metro system, which interconnects all of them.

Milan experiences high temperatures in summer and the heat can be exacerbated by biting insects, which can be a problem when birding close to the rivers or in any wetland, so use some insect repellent. Late autumn and winter are mild, but the whole region is susceptible to thick fog at this time, which can linger for several days. A detailed map of the Lombardy Region is required to locate all of the sites mentioned. Most of the sites included here are either LIPU Reserves or state or regional parks and entry is almost always free.

Locations

1 Parco delle Groane (16km)

(Cesano Maderno Railway Station)
(Bollate Railway Station)

The Parco delle Groane is a large urban park located in a heavily industrialised area north of Milan. Although used for a variety of leisure activities, there are areas set aside for nature conservation which support all of the typical species of this area. Extensive pasture and moorland, formed through forest clearance by burning, are among the typical habitats of this region but they have been all but destroyed by development and urbanisation. However, attempts have been made to restore once-degraded moorland at the LIPU Reserve at Cesano Maderno. This reserve is located within the park and is a short train ride from the city centre. There are 4km of nature trails and a raptor rehabilitation centre in the reserve.

Take the **S44** north from Milan for about 15km to **Cesano Maderno**, to the point where this road runs alongside the Parco delle Groane, and turn west at **Corso Libertà**. Take the second turn on the left (**via Don Orione**) to the entrance to the LIPU reserve. The railway station is on Corso Libertà, about 1km from the reserve. There are several areas which are particularly interesting for a birder. The Caloggio WWF Reserve is at the extreme southern tip near Bollate. This is a newly reforested area adjacent to a disused canal and is very attractive to passerines. The Cesate Woods are in the centre of the park and consist of mature mixed woodlands of pine and oak with a small woodland pond, Il Laghetto Manuè. The Lentate sul Seveso Reserve at the extreme north of the park is a disused claypit which now attracts ducks and other aquatic species.

KEY SPECIES

Resident	Summer
Little Grebe, Water Rail, Little Owl, Common Kingfisher, Green Woodpecker, Lesser Spotted Woodpecker, Long-tailed Tit, Penduline Tit, Hawfinch, Reed Bunting	Little Bittern, Night Heron, Woodcock, Wryneck, Common Nightingale, Reed Warbler, Marsh Warbler, Great Reed Warbler, Melodious Warbler, Red-backed Shrike

2 Parco del Ticino (60km)

The River Ticino is a tributary of the River Po located about 25km southwest of Milan. A linear park of riparian woodland along both banks of the river forms a continuous belt of parkland with the Parco Agricolo Sud Milano, across the southern suburbs of Milan. The park covers a huge area stretching from Lake Maggiore in the north to the confluence with the River Po in the south, and you could easily spend a week in the area. For anyone with time to spare, it is worth renting a bicycle as there are many bike trails through the park. It is possible to see most of the woodland species typical of northern Italy as well as aquatic breeding birds and small numbers of wintering wildfowl, which frequent the broader stretches of the river. From any of the bridges across the Ticino, it is worth scanning either side for ducks and herons wading in the shallower sections. The river is used as a flyway during migration by cranes, geese and raptors. There are several routes to the park, all of which will yield a good selection of birds. The three routes described here are chosen on the basis that two are easily accessed by public transport from Milan, and the other is adjacent to Malpensa Airport and would be convenient for anyone who has time to spare while at the airport.

2a La Fagiana

(Magenta Railway Station)

Take the **A4** Autostrada west from Milan and after about 25km exit south at **Boffalora** on the **SP117** towards **Ponte-vecchio di Magenta**. Turn west at Pontevecchio following the signs for the **LIPU** reserve at **La Fagiana**. This reserve is mainly involved with the rehabilitation of raptors which have been injured, often by gunshot sad to say. Before reaching the LIPU reserve, you can park at the

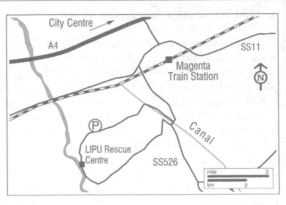

Park Visitor Centre and follow the trail which runs south through forest. This is a circular route of about 15km which eventually returns to the visitor centre via the village of Robecco and the old Grand Canal. It passes through a number of different habitats, including oak forest, rice fields, orchards and wetland areas.

2b Parco dei Fontanili

(Malpenso Railway Station)
(Cavaria Railway Station)

Malpenso Airport is actually within the northern sector of the park and there are a number of trails which start near the airport terminal at **Vizzola Ticino**. However, a

better trail is located near the village of **Premezzo** in the Parco dei Fontanili. This is on the **northeast side** of the **A26 Autostrada** and you will probably need to get a taxi from the airport. The entrance to the park is on the west side of the village, near the school soccer pitch, and the trail leads through the woods past a small lake with a bird hide.

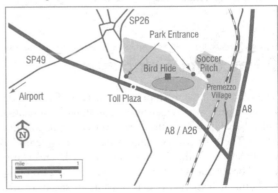

2c Bosco Negri

(Pavia Railway Station)

If you are sightseeing at the historic city of Pavia, just south of Milan, you could consider visiting the LIPU reserve at Bosco Negri. However, bear in mind that this reserve is only open at weekends between March and November and is closed throughout August. This is an area of forest and marsh, enclosed by the old Gravellone Canal system, and supports a good selection of woodland birds. To reach the reserve, cross the

Ticino from the old city on the **Ponte della Libertà** and continue south on the **S35** to **via Bramante**, where it crosses the canal, and turn right (west).

KEY SPECIES

Resident
Little Grebe, Mute Swan, Stock Dove, Marsh Harrier, Northern Bobwhite (I), Common Kingfisher, Green Woodpecker, Lesser Spotted Woodpecker, Mistle Thrush, Firecrest, Long-tailed Tit, Hawfinch, Yellowhammer

Summer
Little Bittern, Night Heron, European Honey-buzzard, Hobby, Little Ringed Plover, Common Sandpiper, Common Tern, Little Tern, Common Cuckoo, European Nightjar, European Bee-eater, Wryneck, Wood Lark, Common Nightingale, Common Redstart, Melodious Warbler, Red-backed Shrike

Winter
Eurasian Bittern, Great White Egret, Eurasian Wigeon, Pintail, Yellow-legged Gull

Passage
White Stork, Garganey, Black-winged Stilt, Tree Pipit

---

3 Parco Le Folaghe (53km)

Parco Le Folaghe lies just off the A7 Milan-Genoa Autostrada, only a 45-minute drive from the city centre. With a site list nearing 240 species, it is probably second only to Pian di Spagna as the best birding site within a 90-minute drive of Milan. It is an area of flooded gravel pits and reedbeds that holds many interesting species, especially during migration, but is worth visiting at any time of year. The condition of this site is dependent on water levels: some of the best areas may dry up in some summers and numbers of autumn migrants will be low, but water levels recover quickly after rain. As it is not a LIPU reserve, there are no restrictions on opening times.

Take the **A7 Autostrada** south towards Genoa, and after about 50km, exit east onto the **SP206** at **Casei Gerola**. In Casei Gerola go north on the **SP12** towards **Silvano Pietra** and after about 500m turn right onto the gravel road which accesses the park. If you are using public transport, the nearest railway station is at Voghera, about 8km east of Casei Gerola.

KEY SPECIES

Resident
Little Grebe, Great Crested Grebe, Mute Swan, Marsh Harrier, Water Rail, Yellow-legged Gull, Common Kingfisher, Zitting Cisticola, Reed Bunting

Summer
Purple Heron, Little Bittern, Montagu's Harrier, Little Ringed

Plover, Black-winged Stilt, Common Tern, Black Tern, European Bee-eater, Ashy-headed Wagtail, Great Reed Warbler, Melodious Warbler, Red-backed Shrike, Ortolan Bunting

Winter
Eurasian Bittern, Gadwall, Northern Shoveler, Eurasian Teal, Eurasian Wigeon, Red-crested Pochard,

Common Pochard, Tufted Duck, Hen Harrier, Common Gull, Water Pipit, Meadow Pipit, Penduline Tit, Great Grey Shrike, Rook

Passage
Garganey, Red-footed Falcon, Spotted Crake, Little Crake, Ruff, Black-tailed Godwit, Greenshank, Wood Sandpiper, Alpine Swift, Bluethroat

---

4 Lame del Sesia (65km)   *(Vercelli Railway Station)*

The River Sesia is a tributary of the Po which flows south through Piedmont about 60km west of Milan. Just south of the A4 Autostrada, the river broadens and forms a network of marshes and backwaters which flood occasionally and then become stagnant. The resulting wetland is like a small version of the Florida Everglades and is a haven for many aquatic species. There are several heronries in the area and a small breeding population of Sacred Ibises which have established a feral colony. As well as the breeding birds, this is an important migration flyway and a winter refuge for ducks. Most of the important sites are included in the Parco Naturale delle Lame de Sesia, and there is a LIPU reserve at Isolone di Oldenico

Take the **A4 Autostrada** west from Milan and after about 60km turn south at **Greggio** onto the S594. Turn east in Greggio onto the access road to the park entrance. Alternatively, you can continue on the S594 to the villages of **Albano Versellese** and to **Oldenico**, where there are access roads to the park on the east side of the villages. The path from the Oldenico entrance leads to a hide which overlooks the Isolone di Oldenico.

There are walking and biking trails along both banks of the river for almost 8km. However, to reach the east bank you must return to the A4, turn right and take the next turn south onto the **A26 Autostrada**. Then take the first turn west to **San Nazzaro**, where the road continues through the village to the park. The nearest railway station is in Vercelli, about 8km south of Oldenico on the S594 and a convenient cycle run away.

KEY SPECIES

Resident
Little Grebe, Eurasian Bittern, White Stork, Sacred Ibis (I), Marsh Harrier, Water Rail, Northern Bobwhite (I), Common Kingfisher, Penduline Tit, Reed Bunting

Summer
Little Bittern, Night Heron, Squacco

Heron, Purple Heron, Eurasian Spoonbill, Little Ringed Plover, Common Tern, Marsh Warbler, Great Reed Warbler

Winter
Gadwall, Northern Shoveler, Eurasian Teal, Tufted Duck, Common Pochard, Hen Harrier

Passage
Great White Egret, Glossy Ibis, Garganey, Black-winged Stilt, Ruff, Eurasian Curlew, Black-tailed Godwit, Greenshank, Wood Sandpiper, Black Tern, Whiskered Tern

5 Brabbia Reserve (50km)   (Bus No. 65/Varese Railway Station)

The Brabbia Marshes are an area of flooded peat workings south of Lake Varese, part of which are a LIPU reserve. The site is largely undisturbed and a variety of habitats including reedbeds, wet meadows, riparian woodland and open waterways encourage a diverse range of species, making this one of the top-ranking wetland sites in northwest Italy. As well as wintering wildfowl, it is also an important breeding site for several duck species and other aquatic birds. Its location on the migration path for birds crossing the Alps and the abundance of willow and alder make it an attractive spot for migrant passerines, as well as offering protection from bird-trappers.

Take the **A8 Autostrada** north from Milan and exit west at **Buguggiate**. From Buguggiate take the **S36** west towards **Cazzago Brabbia**, which runs alongside Lake Varese: after about 6km turn south to **Inarzo** on the **S53**. The visitor centre and car park are located in the centre of Inarzo. There are several trails leading to hides located at key observation points in the marsh. There is a regular bus service (No. 65) to Inarzo from Varese.

KEY SPECIES

Resident
Little Grebe, White Stork, Ferruginous Duck, Marsh Harrier, Water Rail, Penduline Tit, Reed Bunting

Summer
Little Bittern, Night Heron, Purple Heron, Hobby, Sand Martin, Ashy-headed Wagtail, Savi's Warbler, Marsh Warbler, Great Reed Warbler

Winter
Grey Heron, Gadwall, Northern Shoveler, Eurasian Teal, Eurasian Wigeon, Hen Harrier, Water Pipit, Meadow Pipit, Great Grey Shrike

Passage
Little Egret, Great White Egret, Glossy Ibis, Garganey, Osprey, Black-tailed Godwit, Greenshank, Wood Sandpiper, Black Tern, Whiskered Tern, Bluethroat

Further Away

6 Monte di Pietra Corva (80km)

The Apennine Mountains which form the backbone of the Italian peninsula are closest to Milan in the south of Pavia Province. An hour's drive from Milan would enable you to get to an elevation of about 1,000m. The area around Romagnese provides a variety of habitats, including cultivated fields, vineyards, pine forest, boulder-strewn slopes and an arboretum at the Giardino Alpino. This combination of habitats in a relatively small area provides for quite a productive day's birding in spring and early summer. The Giardino Alpino will be of interest to botanists and horticulturalists but, for birders, the main attraction is the track leading from the south side of the garden entrance to the summit of Monte di Pietra Corva. The track traverses some 2km of mixed forest, pine forest and ultimately scrub vegetation as you ascend the mountain.

From Milan, take the **A7 Autostrada** south to Pavia and from Pavia take the **S617** south to **Broni**. From Broni take the tortuous **S198** south, up into the mountains at **Zavattarello** and then take the S412 to Romagnese.

KEY SPECIES

Resident	Summer	
Red-legged Partridge, Green Woodpecker, Dipper, Mistle Thrush, Crested Tit, Common Raven, Bullfinch, Cirl Bunting, Rock Bunting	Short-toed Eagle, Alpine Swift, European Nightjar, Wryneck, Wood Lark, Tree Pipit, Tawny Pipit, Common Redstart, Northern	Wheatear, Subalpine Warbler (ssp. *moltonii*), Melodious Warbler, Wood Warbler, Western Bonelli's Warbler, Red-backed Shrike

7 Parco delle Orobie (80km)   (SAB Bus No. E32)

The Bergamo Alps which are about 80km northeast of Milan are foothills of the High Alps, separated from the main massif by the Adda River Valley. These mountains rise to over 2,000m and although they do not support the full range of alpine birds they present the best opportunity to see some of the montane species within a reasonable distance of Milan. However, in general, birding the Alps in Lombardy is rather tough going as most species occur at low densities and it can be very hard work finding them.

The Parco delle Orobie is an extensive national park which covers most of the Bergamo Alps and comprises mixed and coniferous forests at the lower levels, alpine meadows and rocky slopes. The most convenient starting point to explore this area is the Brembana Valley in the west of the range. This is a tourist area with ski resorts and many hiking and trekking trails through the mountains. A car is best for access to the area as, although there is a bus service from Milan, the ability to stop and look for birds along the road up into the mountains would add considerably to the range of species found. Whatever way you travel, be prepared for a significant amount of hiking through the forests and alpine meadows. The best time to visit is in late spring and early summer, when all of the birds are on territory and the long days permit an extended exploration. During winter most of the birds move to lower elevations but bad weather can make the viewing conditions very poor.

Take the **S470** north from **Bergamo** to **Lenna** and turn right onto the **SP2**, following this road all the way up to **Foppolo**. At Foppolo, there are several hiking trails which enable access to the higher peaks. The trail from Foppolo to Lago Moro and

Corno Stella takes you from 1,600m to 2,600m, above the tree line and into an area of boulders and rock scree. At this altitude, Ptarmigan, Rock Partridges and Alpine Accentors are all possible.

There is a bus service, provided by SAB, from Milan (Piazza Castello) to Foppolo, also serving Bergamo.

KEY SPECIES

Resident
Golden Eagle, Northern Goshawk, Hazel Grouse, Ptarmigan, Black Grouse, Capercaillie, Rock Partridge, Black Woodpecker, Dipper, Alpine Accentor, Crested Tit, Nutcracker,

Common Raven, Citril Finch, Common Crossbill

Summer
Alpine Swift, Crag Martin, Tree Pipit, Water Pipit, Common Redstart, Rock

Thrush, Northern Wheatear, Ring Ouzel, Wood Warbler, Western Bonelli's Warbler, Lesser Whitethroat, Goldcrest, Yellowhammer

8 Pian di Spagna (90km)　　　　　　　　　(Dubino Railway Station)

Lake Como is the closest of the large northern Italian Lakes to Milan, and it is a very popular tourist area. One of the best wetlands in Italy is the Pian di Spagna, right at the northern tip of Lake Como. This site is approximately 90km from Milan but if Lake Como is part of your itinerary then it is definitely worth visiting at any time of year. The Pian di

Spagna is a low-lying plain between the rivers Mera and Adda but is part of a larger wetland area between Lake Como and Lake Mezzola and constitutes the largest nature reserve in Lombardy. It is of prime importance for breeding, migratory and wintering aquatic birds, but its location deep in the foothills of the Alps means that many migrating birds get funnelled into the area. The list of rarities found here is increasing.

Take **S36** north from Milan. Just north of **Colico**, the road forks: S36 continues towards **Chiavenna** and **Madesimo** (left), while **S38** heads towards **Sondrio** and the **Valtellina** (right). Stay on S36 towards Chiavenna and Madesimo. About 1km from the fork you will come to a bridge over the river Adda: take the first left turn (signposted Pian di Spagna) immediately after this bridge. Follow the dirt road past the quarry, then park and explore the meadows, tree lines, ditches and marshy spots between here, the mouth of the river Adda, and the shore of Lake Como.

KEY SPECIES

Resident
Little Grebe, Great Crested Grebe, Mute Swan, Marsh Harrier, Water Rail, Yellow-legged Gull, Common Kingfisher, Crag Martin, Common Raven, Reed Bunting

Summer
Little Bittern, Night Heron, Common Sandpiper, Sand Martin, Marsh Warbler, Reed Warbler, Great Reed Warbler, Barred Warbler, Red-backed Shrike

Winter
Black-necked Grebe, Gadwall, Northern Shoveler, Eurasian Teal, Eurasian Wigeon, Red-crested Pochard, Common Pochard, Tufted Duck, Hen Harrier, Common Gull, Stock Dove, Water Pipit, Meadow

Pipit, Great Grey Shrike, Rook
Passage
Little Egret, Great White Egret, Purple Heron, Garganey, Spotted Crake,

Black-tailed Godwit, Greenshank, Wood Sandpiper, Black Tern, Whiskered Tern, Tree Pipit, Tawny Pipit, Blue-headed Wagtail, Grey-

headed Wagtail, Whinchat, Northern Wheatear, Bluethroat, Grasshopper Warbler, Sedge Warbler, Pied Flycatcher

Useful contacts and websites

Lega Italiana Protezione Uccelli (Italian)
www.lipu.it
Via Trento 49, 43100 Parma.
Tel. (+39) 0521 273043. Fax. (+39) 0521 273419.
Information about policies, activities, reserves and branches of LIPU – Italian BirdLife partner.

LIPU – UK (English)
www.lipu-uk.org
c/o David Lingard (LIPU-UK), Fernwood, Doddington Road, Whisby, Lincs, LN6 9BX, UK.

Il Birdwatching Italiano (Italian and English)
www.ebnitalia.it
Website for Italian birding with monthly reports, trip reports and site descriptions.

National Parks of Italy (Italian and English)
www.parks.it
Official website of Italian national parks.

Parchi Naturale Italiani (Italian)
www.amicianimali.it/itinerari
Site descriptions, directions and general information about Italian nature reserves.

Valle Brembana (Italian)
www.valbrembanaweb.com
Brembana Valley website.

Italian Rarities Committee (CIR) (English)
www.storianaturale.org/cir
c/o Andrea Corso / C.I.R. (Comitato Italiano Rarità), Via Camastra 10, 96100 Siracusa.
For records of species recorded more than ten times.

Italian Rarities Committee (COI) www.ciso-coi.org
Commissione Ornitologica Italiana, c/o Pierandrea Brichetti, Via V. Veneto 30, 25029 Verolavecchia. For records of species recorded fewer than ten times.

Azienda Trasporti Milanesi s.p.a
(English and Italian) www.atm-mi.it
Bus, tram and metro service for Milan.

Trenitalia (English and Italian)
National railway system
www.trenitalia.com

RegioneLombardia (Italian)
www.trasporti.regione.lombardia.it/trl_index.htm
Web portal site for regional travel (buses and trains) in Lombardy.

Books and publications

Where to Watch Birds in Italy
(LIPU) Lega Italiana Protezione Uccelli (1994)
Christopher Helm, ISBN 0713638672

Atlante degli uccelli svernanti in Lombardia.
L. Fornasari *et al.* (1992) Regione Lombardia e Università degli Studi di Milano

Check-list degli uccelli della Lombardia
P. Brichetti P. & D. Cambi (1987). *Sitta* 1: 57-71.

L'Atlante ornitologico della provincia di Pavia F. Barbieri *et al.* (1977)
Atti VII Simposio Naz. Conservazione Natura, *Bari*: 87 99.

Atlante degli uccelli nidificanti a Pavia.
Comune di Pavia F. Bernini, *et al.* (1998). LIPU

Atlante degli uccelli nidificanti in Provincia di Varese (Lombardia) 1983–1987.
(1998) W. Guenzani & F. Saporetti Edizioni Lativa, Varese.

Tra Cielo e Acqua: Migratori in volo sul Pian di Spagna
Lucio Bordignon & Walter Corti

Collins Bird Guide
Lars Svensson *et al.* (1999) HarperCollins
ISBN 0002197286

Birds of Europe
Lars Jonsson (2006) Christopher Helm
ISBN 0713676000

A pair of Jackdaws at home in the familiar surroundings of Red Square

TYPICAL SPECIES

Resident
Ruddy Shelduck (I), Mallard, Common Gull, Black Woodpecker, Lesser Spotted Woodpecker, Fieldfare, Goldcrest, Willow Tit, Coal Tit, Blue Tit, Great Tit, Long-tailed Tit, Eurasian Nuthatch, Eurasian Treecreeper, Magpie, Western Jackdaw, Hooded Crow, Greenfinch, Bullfinch, Goldfinch, Siskin, Common Crossbill

Summer
Grey Heron, Garganey, Black Kite, Common Kestrel, Hobby, Moorhen, Common Coot, Northern Lapwing, Black-headed Gull, Common Tern, Turtle Dove, Common Swift, Sand Martin, Barn Swallow, House Martin, Sky Lark, Tree Pipit, White Wagtail, Yellow Wagtail, Wren, Dunnock, Robin, Blackbird, Song Thrush, Redwing, Common Whitethroat, Blackcap, Garden Warbler, Common Chiffchaff, Wood Warbler, Common Starling, Eurasian Jay, Rook, House Sparrow, Tree Sparrow, Chaffinch, Linnet, Common Rosefinch

Winter
Rough-legged Buzzard, Herring Gull, Waxwing, Common Redpoll

Passage
Whooper Swan, Eurasian Wigeon, Eurasian Teal, Pintail, Northern Shoveler, Tufted Duck, Little Gull, Meadow Pipit, Whinchat, Northern Wheatear, Willow Warbler, Brambling

Moscow

In little more than a decade, Moscow has emerged as a popular destination in its own right for all types of traveller. Prior to that it may have been a transit point for birders on cheap flights to the Far East or for carefully controlled tour groups with little opportunity to depart from their programmes. However, this is no longer the situation and, apart from acting as a starting point for bird tour groups to Siberia and Central Asia, it has much to interest the visiting birder. Moscow lies right at the heart of the distributions of many eastern European species which are much sought-after rarities in western Europe and it is also at the edge of the ranges of several Siberian species which are absent further west.

Its main physical feature is the Moscow River, which meanders rather sluggishly through the city and is fed by several tributaries. These tributaries flow through a number of wooded parks and wetlands and also drain some sizeable lakes, all of which provide habitats for birds. A number of large reservoirs north of the city support significant numbers of breeding and migrant wildfowl.

The centralised planning of the Soviet era produced a city with very well-defined blocks of residential, commercial and industrial zones, which can be avoided, but also some large, well-forested parks set aside for recreational purposes, which are the main focus of a birder's attention.

Naturally, the bitter cold of winter militates against birding activity then and there is little to see anyway as all waterways are frozen, apart from some kept open for boat traffic. Spring arrives rather late but very suddenly and this produces a surge of migrant warblers, thrushes and other passerines, almost rivalling the similar kind of passage that can be observed in North American cities. The best time to visit is May and June when the woods are alive with singing birds establishing territories and the mosquitoes have not yet reached a density or a ferocity that would keep an elephant at bay.

GETTING AROUND

Although Moscow is a very large expansive city with a population of almost nine million, it has developed as a metropolis that caters for very low rates of private car ownership. Consequently the public transport system is very comprehensive and can be used to good effect to visit the birding sites within the Moscow Ring Road (MKAD). The Moscow metro system is world-renowned for its ornate

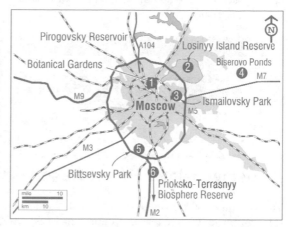

underground stations but it is also the world's most extensive network. There is a dense network of stations in the city centre, placing everywhere within easy walking distance of the metro, but in the suburbs the stations are much more sparsely distributed and there are long stretches of line without a station. Bus services interconnect with the metro stations and serve most of the sites.

Outside the ring road, a car is definitely an advantage and drivers used to western levels of traffic congestion will be pleasantly surprised by driving conditions in Russia. There are extensive forest reserves beyond the MKAD on all sides of the city. Taxis are a relatively expensive option in Moscow and should only be considered if no other choices exist.

Moscow lies in a flat, low-lying basin, with low hills to the south and southeast, so comfortable walking shoes are adequate. However the woodland paths can be very muddy just after the spring thaw and boots or perhaps an old pair of shoes are needed then. In winter, snow lies on the ground and piles up as the winter progresses, making it very difficult, if not dangerous to stray off the cleared tracks. If you need to purchase gloves, binoculars, books or any other birding requisites, bear in mind that the main shopping areas are in the city centre and there are few out of town shopping malls. You will need a map that covers Moscow City as well as Moscow Province (Oblast), which borders it, in order to locate all of the sites. It would also be useful to get a city map with a diagram of the public transport system. These are available from any bookshop or tourist office, and at bus and railway stations.

Locations

1 Botanical Gardens (9km)   *(Botanichesky Sad Metro Station)*

The Botanical Gardens and the nearby Economic Progress Park are popular urban parks, but a visit early in the morning can produce a good selection of woodland birds. The best and least crowded areas are towards the northwest end, between the Yauza and Likhoborka Rivers.

Take the **M8** (Prospekt Mira) **north** from **Moscow Garden Ring** until you reach the **Botanichesky Sad Metro Station** (about 4km). The park entrance is on the left side (west) of the road. A path leads from the entrance through some woodland to an open grassland meadow and continues west along the Likhoborka River to the Botanical Gardens.

KEY SPECIES

Resident	Summer	
Northern Goshawk, Eurasian Sparrowhawk, Green Woodpecker, Middle Spotted Woodpecker, White-backed Woodpecker	Corn Crake, Water Rail, Woodcock, Stock Dove, Common Kingfisher, Wryneck, Common Cuckoo, Thrush Nightingale, Song Thrush, Grasshopper Warbler, Marsh Warbler,	Blyth's Reed Warbler, River Warbler, Icterine Warbler, Booted Warbler, Lesser Whitethroat, Greenish Warbler, Red-breasted Flycatcher, Golden Oriole, Hawfinch, Reed Bunting

2 Losinyy Island Forest Reserve (17km) _(Mytishchi Railway Station)_

This is another of Moscow's parks and is located in the northeast of the city. Losinyy (Moose) Island Reserve forms a continuous forest with Sokolniki Park, which is close to the Leningradsky Railway Terminal. It would be possible to spend more than a week exploring this large forest. In addition to large tracts of conifers, the habitats include mixed forest, lakes, reedbeds, water meadows and old peat-workings, which support a wide variety of birds.

Take the **M8** (Yaroslavskoye Shosse) **northeast** from **Moscow Garden Ring** until you reach the **Outer Ring Motorway** (MKAD). Continue past the MKAD on the M8 for about 4km until you reach the bridge over the **Yauza River**. Turn right just past the bridge, and then turn right again after about 200m. This leads to the start of a trail which is about 6km long and leads through a large area of marsh, forest, cultivated allotments and peat bogs. Mytishchi Railway Station is about 1km northwest of the M8 slip-road.

KEY SPECIES

Resident		
Northern Goshawk, Eurasian Sparrowhawk, Hazel Grouse, Green Woodpecker, Grey-headed Woodpecker, Middle Spotted Woodpecker, Great Spotted Woodpecker, White-backed Woodpecker, Crested Tit, Common	Raven, Yellowhammer **Summer** European Honey-buzzard, Red Kite, Common Buzzard, Corn Crake, Stock Dove, Wryneck, Common Cuckoo, Citrine Wagtail, Common Redstart, Bluethroat, Thrush Nightingale, Mistle	Thrush, River Warbler, Sedge Warbler, Marsh Warbler, Blyth's Reed Warbler, Great Reed Warbler, Icterine Warbler, Barred Warbler, Greenish Warbler, Spotted Flycatcher, Red-breasted Flycatcher, Pied Flycatcher, Golden Oriole, Hawfinch, Reed Bunting

3 Ismailovsky Park (11km) _(Izmailovskaja Metro Station)_

Although the dominant woodland in the Moscow area is coniferous forest, Ismailovsky Park is an area of broadleaved woodland east of the city. The woodland is interspersed with open grassland, as well as marsh and open bodies of water, and so the park supports good numbers of all of the typical species of the Moscow area. The marshy areas near the banks of the Serebryanka River are overgrown with willows and alders. These areas provide excellent cover for Corn Crakes and waterside passerine species such as _Acrocephalus_ warblers and both White-spotted and Red-spotted subspecies of Bluethroat.

Take the **M7** (Entuziastov Schosse) **west** from **Moscow Garden Ring** for about 9km. The park entrance is on the left (north) side of the road.

KEY SPECIES

Resident
Eurasian Sparrowhawk, Grey-headed Woodpecker, Green Woodpecker, Great Spotted Woodpecker, White-backed Woodpecker

Summer
Common Goldeneye, Corn Crake,

Woodcock, Wryneck, Common Cuckoo, Thrush Nightingale, Common Redstart, Bluethroat, Mistle Thrush, Marsh Warbler, Blyth's Reed Warbler, Great Reed Warbler, River Warbler, Icterine Warbler, Barred Warbler, Lesser Whitethroat, Greenish Warbler,

Wood Warbler, Spotted Flycatcher, Red-breasted Flycatcher, Pied Flycatcher, Red-backed Shrike, Golden Oriole, Hawfinch, Common Rosefinch, Reed Bunting, Yellowhammer

---

4 Biserovo Ponds (34km) *(Kupavna Railway Station)*

This is an extensive area of fish ponds, some of which have been abandoned, providing a variety of habitats. The margins and reedbeds support breeding populations of several aquatic species, including Little Bitterns which are quite scarce in the Moscow region. The larger ponds hold concentrations of ducks during migration and, when drained, the muddy areas are very attractive to waders and gulls. The best time to visit is from April to October.

Take the **M7 east** from Moscow to **Kupavna**. The ponds are approximately 1km north of Kupavna. You may also take a train from Moscow (Kursk Station) to Kupavna.

KEY SPECIES

Summer
Little Grebe, Great Crested Grebe, Black-necked Grebe, Eurasian Bittern, Little Bittern, Greylag Goose, Marsh Harrier, Common Buzzard, Corn Crake, Spotted Crake, Water Rail, Little Ringed Plover, Green Sandpiper, Common Sandpiper, Common Redshank, Spotted Redshank,

Common Snipe, Little Gull, Black Tern, White-winged Black Tern, Common Kingfisher, Sedge Warbler, Marsh Warbler, Blyth's Reed Warbler, Great Reed Warbler, Icterine Warbler, Booted Warbler, Red-backed Shrike, Reed Bunting

Passage
Red-throated Diver, Black-throated

Diver, White Stork, Whooper Swan, Gadwall, Goosander, Osprey, Red-footed Falcon, Common Crane, Ringed Plover, Grey Plover, European Golden Plover, Ruff, Wood Sandpiper, Greenshank, Bar-tailed Godwit, Eurasian Curlew, Whimbrel, Mediterranean Gull, Grasshopper Warbler, Penduline Tit

---

5 Bittsevsky Park (22km) *(Bittsevsky Metro Station)*

Bittsevsky Park is a large wooded area in the south of the city which contains a huge belt of coniferous forest with many trails running through it. Virtually all of Europe's temperate zone forest species can be seen in this park although the dense woodland ensures that only a fraction of birds present will be seen. This area would repay several visits during the May/June period.

Take the **M2** (Varschavskoye Shosse) **south** from **Moscow Garden Ring** until you reach the **Outer Ring Motorway** (MKAD). Turn **west** and after about 2km the park entrance is on the right-hand side.

KEY SPECIES

Resident
Northern Goshawk, Eurasian Sparrowhawk, Green Woodpecker,

Grey-headed Woodpecker, Middle Spotted Woodpecker, Common Raven

Summer
European Honey-buzzard, Red Kite, Common Buzzard, Corn Crake, Stock

| Dove, Wryneck, Common Cuckoo, Thrush Nightingale, Mistle Thrush, Icterine Warbler, Lesser Whitethroat, | Greenish Warbler, Spotted Flycatcher, Red-breasted Flycatcher, Pied Flycatcher, Red-backed Shrike, | Golden Oriole, Hawfinch |

Further Away

6 Priokso-Terrasnyy Biosphere Reserve (100km)

(Serpukhov Railway Station)

This is one of the oldest reserves in the Moscow region and is designated a Zapovednik or Strict Nature Reserve. It is south of Moscow on the banks of the River Oka and although it is rather distant (100km) it is such a superb site that it is worth the effort to get there. The main forest habitat is mixed: pine, spruce, birch and lime. A feature of this area are the terraced banks of the River Oka. The low-lying terraces by the river are subject to seasonal flooding and are carpeted by sphagnum bogs. The higher terraces are drier and support relict meadow-steppe vegetation. The reserve also contains a herd of European Bison.

Take the **M2 south** to **Serpukhov** and **turn east** at the exit for Serpukhov (Serpukhov is west). The reserve stretches for several square kilometres on the east side of the motorway. Continue east for about 7km to the reserve entrance. You can also take the train to Serpukhov (Kursk Station) and take a taxi from there to the reserve.

KEY SPECIES

Resident	Summer	
White-tailed Eagle, Northern Goshawk, Eurasian Sparrowhawk, Hazel Grouse, Capercaillie, Green Woodpecker, Grey-headed Woodpecker, Great Spotted Woodpecker, Middle Spotted Woodpecker, White-backed Woodpecker, Willow Tit, Great Grey Shrike, Common Raven	European Honey-buzzard, Red Kite, Pallid Harrier, Montagu's Harrier, Marsh Harrier, Common Buzzard, Common Quail, Corn Crake, Little Gull, Black Tern, White-winged Black Tern, Stock Dove, Wryneck, Common Cuckoo, Citrine Wagtail, Thrush Nightingale, Mistle Thrush, Grasshopper Warbler, Sedge Warbler,	Marsh Warbler, Blyth's Reed Warbler, Great Reed Warbler, Icterine Warbler, Booted Warbler, Barred Warbler, Greenish Warbler, Spotted Flycatcher, Red-breasted Flycatcher, Pied Flycatcher, Red-backed Shrike, Lesser Grey Shrike, Golden Oriole, Hawfinch

Useful contacts and websites

Russian Bird Conservation Union
(Russian and English) www.rbcu.ru/en
Sojuz Ohrany Ptits Rossii, d. 60, korp. 1, Shosse Entuziastov, Moscow 111123.
Tel. (+7) 095 176 10 63.
Russian bird protection organisation, BirdLife partner for Russia, and publishers of the journal *BirdWorld*.

Moscow Birds (Russian)
www.mosbirds.narod.ru
Website of Moscow branch of Russian Bird Conservation Union (RBCU).

Russian Nature Press Information Service
(English) www.rusnatpress.org.uk
Information about Russian natural history publications, nature reserves and organised tours.

BirdWatching.ru (Russian)
http://birdwatching.boom.ru/
Information about birding sites in Moscow area.

Winter-birds (Russian)
www.winter-birds.narod.ru
Website on wintering birds of Moscow region.

Water-birds (Russian)
www.water-birds.narod.ru
Website on waterbirds of Moscow region.

Ciconii-birds (Russian)
www.ciconii-birds.narod.ru
Website on herons and storks of Moscow region.

Zoological Museum of Moscow University
(English) http://zmmu.msu.ru/eng/
Collection of birds – http://zmmu.msu.ru/eng/r_o_m.htm

Moscow Zoo (English) www.moscowzoo.ru

Botanical Garden of Russian Academy of Sciences (Russian) www.hortus.ru

State Darwin Museum (English) www.darwin.museum.ru

Working Group on Raptors of Northern Eurasia (English) www.raptors.ru

Working Group on Waders (English) www.waders.ru

Discussion group (mailing list) 'Birds in Russia and adjacent countries' (English) http://groups.yahoo.com/group/BirdsinRussia/

Moscow Metro (Russian) www.mosmetro.ru Russian only, but with downloadable metro system diagram.

Books and publications

Atlas: Birds of Moscow City and the Moscow Region M. V. Kalyakin & O. V. Voltzit (eds) (2006) Pensoft Publishers, ISBN 9526422622

Ptitsy Moskvy i Podmoskov'ya [Birds of Moscow city and surroundings] V. D. Il'ichev, V. T. But'yev, V. M. Konstantinov (1987) Edited by V.E.Sokolov. Nauka Press.

Collins Bird Guide Lars Svensson et al. (1999) HarperCollins ISBN 0002197286

Birds of Europe Lars Jonsson (2006) Christopher Helm ISBN 0713676000

Common Pochards on the River Isar in Munich

TYPICAL SPECIES

Resident
Little Grebe, Great Crested Grebe, Great Cormorant, Grey Heron, Mute Swan, Mallard, Tufted Duck, Common Pochard, Common Buzzard, Common Kestrel, Moorhen, Common Coot, Black-headed Gull, Common Gull, Wood Pigeon, Collared Dove, Green Woodpecker, Great Spotted Woodpecker, Wren, Dunnock, Robin, Blackbird, Fieldfare, Goldcrest, Marsh Tit, Coal Tit, Blue Tit, Great Tit, Eurasian Nuthatch, Short-toed Treecreeper, Eurasian Jay, Carrion Crow, Chaffinch, Greenfinch, Linnet, Bullfinch, Hawfinch, Yellowhammer

Summer
Garganey, European Honey-buzzard, Black Kite, Marsh Harrier, Hobby, Common Swift, Sand Martin, Barn Swallow, House Martin, White Wagtail, Black Redstart, Common Redstart, Reed Warbler, Blackcap, Common Whitethroat, Common Chiffchaff, Wood Warbler, Firecrest, Red-backed Shrike, Serin

Passage
White Stork, Osprey, Common Crane

Munich

Munich is on the River Isar about 100km north of the Bavarian Alps. Like all inland cities, there are few birding opportunities close to the centre that would yield a large variety of birds and it is necessary to travel some distance to seek out the habitats that support more specialised species. The city is surrounded by a flat plain, with farmland interspersed with mixed woodlands. However, there are some lakes and reservoirs within an hour's travel and, for those who are prepared to travel a little further, the Bavarian Alps are accessible in a day trip. For migrant birds, the lakes south of Munich are often a stop-over prior to crossing the Alps in autumn and also the first stop in spring after crossing the mountains. The Alps also act as a kind of barrier to vagrants which have wandered too far south, so the lakes can attract some surprising winter visitors.

For the birder, all of these factors offer opportunities within reach of the city so, despite its inland location, a rich variety of birds can be seen in the vicinity. Early spring is the best time to look for woodpeckers and other resident forest species as they will be establishing territories and are quite vocal at this time. In late spring and summer, the emphasis is on the montane species, while at other times of the year the best birding is offered by the lakes, reservoirs and wetlands.

GETTING AROUND

For the visitor without a car, the public transport system is excellent within the city and extends quite far out into the surrounding countryside, making many sites accessible. The underground railway or U-Bahn is the best way of getting around within the

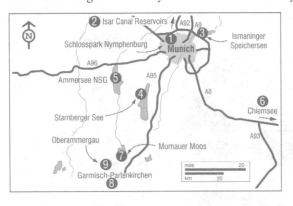

city, while the suburban railway or S-Bahn reaches the outlying areas. You can purchase a one-day or three-day ticket which is valid on the U-Bahn, S-Bahn, buses and trams, including the airport service, and so is really good value.

When time is short, the visiting birder needs to decide whether to restrict himself to the city's immediate suburbs and parks or whether to plan a day trip to one of the more outlying areas. The advantage of the latter is that, although some planning and preparation are required, a wider range of species will be seen. For those with little time and flexibility, it is still possible to see a selection of birds within easy reach of the city centre or airport. However, a car is necessary for those wishing to visit several of the more distant sites over one or two days.

Parks and nature reserves in Germany are used extensively by the public for recreation purposes, unlike in some other countries. So even midweek and early in the morning, expect substantial numbers of joggers, cyclists and dog-walkers. There really is no time when you can expect to have the area to yourself.

Birding in Munich itself is not really weather dependent. Comfortable walking shoes are adequate for all the locations mentioned. However, travel in the Bavarian Alps in winter is usually not possible because of heavy snowfall, and even in late spring there can be restrictions due to bad weather. Cheap but adequate binoculars can be purchased in any camera store, sporting goods store and major department stores. The best maps to use in conjunction with those in this guide are Sussman's Map of Munich (1:17000) which illustrates the bus and U-Bahn network and also includes a handy city guide book. Because several sites are quite distant from Munich, you will also need a map of Bavaria, such as the Michelin Road Map of Southern Germany. Both are available from bookshops, tourist offices and bus and railway stations.

Locations

1 Schlosspark Nymphenburg (6km)   *(Tram No.17; Bus No. 41)*

Of all the city parks in Munich, this is probably the most productive for birders because of its size and the more secluded areas at the western end, which don't attract the crowds. The lakes in the park are good for waterfowl. For a birder who has no time or opportunity to travel beyond the city centre this is about the best site to see the common birds of Bavaria. Even shy species such as Goshawk and Black Woodpecker are possible here. Culture vultures will also want to visit Nymphenburg Castle, which is quite impressive. The park is just west of the city centre within easy reach of several U-Bahn stations and bus routes.

KEY SPECIES

Resident	Grey-headed Woodpecker, Black	**Winter**
Greylag Goose, Red-crested Pochard,	Woodpecker, Song Thrush, Mistle	Gadwall, Eurasian Wigeon, Northern
Goosander, Eurasian Sparrowhawk,	Thrush, Long-tailed Tit	Goshawk

2 Isar Canal Reservoirs (60km)   *(Moosburg Railway Station)*

There are two reservoirs situated between the towns of Moosburg and Eching, about 60km northeast of Munich which are good for waterfowl and a variety of other species. The Moosburg Reservoir has very little shoreline, and the main attraction here are the large flocks of wildfowl. The Eching Reservoir, on the other hand, is shallower and has considerable mudflats, reedbeds and riparian woodland which are good for waders and passerines, particularly at migration times.

Take the **A92** north from Munich and exit west at **Moosburg Nord Exit** onto the **B11**. There are two lay-bys on the B11 which enable you to scan the Moosburg Reservoir. Going east on the B11, about 4km from the A92 will bring you to the Eching Reservoir. Two exits on the B11 will bring you to the shores of the reservoir: one at **Kronwinkl** and one at **Viecht**. The nearest railway station is at Moosburg, but you can rent a bike at the station and there is a bicycle track connecting the two reservoirs along the Isar Canal.

KEY SPECIES

Resident	Warbler, Great Reed Warbler,	Yellow-legged Gull, Siskin
Great White Egret, Greylag Goose,	Penduline Tit	**Passage**
Gadwall, Goosander, Northern		Grey Plover, European Golden Plover,
Goshawk, Water Rail, Common Snipe,	**Winter**	Little Stint, Temminck's Stint, Curlew
Water Pipit, Willow Tit	Red-necked Grebe, Whooper Swan,	Sandpiper, Dunlin, Common
	White-fronted Goose, Eurasian	Redshank, Spotted Redshank, Little
Summer	Wigeon, Pintail, Northern Shoveler,	Gull, Whiskered Tern, Black Tern
Black-necked Grebe, Spotted Crake,	Red-crested Pochard, Common	
Common Tern, Bluethroat, Marsh	Goldeneye, Smew, Hen Harrier,	

3 Ismaninger Speichersee (16km) *(Ismaning S-Bahn Station S8)*

This is a reservoir site which is quite close to Munich and also on the way to the rather distant Munich Airport, so it makes a worthwhile stop off on your way to or from the airport, or even if you have a few hours to kill before your flight. The site is best in winter, when it attracts flocks of ducks, but the surrounding woods hold all the typical species. It is also important in late summer as a moulting refuge for up to 50,000 waterfowl. Unfortunately, the fish ponds on the south shore of the reservoir are not open to the public.

This site can be accessed from the **A99 Autobahn** which is the semi-orbital motorway which encircles most of the city. Exit north on the **B471** which crosses the west end of the reservoir and take the next right (**Nordlicher Speicherseeweg**). Follow this road, which basically skirts the shore of the reservoir, turning inland for a couple of kilometres to pass the perimeter of the BMW Test Track but turning south again to cross the central dam, from where the open water can be scanned. The inland fields should be checked in winter for flocks of Greylag and White-fronted Geese. There is an S-Bahn Station in Ismaning which is about 2km on the B471 from the reservoir.

KEY SPECIES

Resident
Black-necked Grebe, Great White Egret, Greylag Goose, Gadwall, Eurasian Teal, Red-crested Pochard, Goosander, Northern Goshawk, Common Kingfisher, Long-tailed Tit, Willow Tit

Summer
Ferruginous Duck, Common Sandpiper, Common Tern, Yellow

Wagtail, Golden Oriole, Sedge Warbler, Icterine Warbler, Spotted Flycatcher, Reed Bunting

Winter
Red-necked Grebe, White-fronted Goose, Bean Goose, Eurasian Wigeon, Pintail, Northern Shoveler, Common Goldeneye, Smew, Yellow-legged Gull, Brambling, Siskin

Passage
Marsh Harrier, Hen Harrier, Northern Lapwing, Ruff, Green Sandpiper, Wood Sandpiper, Little Gull, Whiskered Tern, Black Tern, Sky Lark, Water Pipit, Meadow Pipit

4 Starnberger See (35km) *(Tutzing S-Bahn Station S6)*

The Starnberger See is a superb winter birding site southwest of Munich. It attracts up to 20,000 wildfowl in winter and is the best inland site in Germany for divers, grebes and seaducks, which is quite surprising given its distance from the coast. The lake is about 21km long and about 5km wide at its widest point. The best way to 'do' the lake is to drive a complete circuit of the

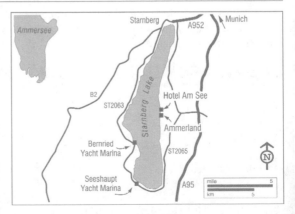

lakeshore in the course of a day, stopping at likely points to scan the water. The best locations are at **Starnberg** at the north end, **Bernried Yacht Marina** on the west shore, **Seeshaupt Yacht Marina** at the south end and the Hotel am See at **Ammerland** on the east shore. For those reliant on public transport, the S-Bahn operates along the west side of the lake serving the towns as far as Tutzing, and from there you can catch a train as far as Seeshaupt at the south end of the lake.

KEY SPECIES

Resident
Black-necked Grebe, Greylag Goose, Gadwall, Eurasian Teal, Red-crested Pochard, Goosander

Summer
Common Sandpiper, Common Tern, Yellow Wagtail, Golden Oriole, Spotted Flycatcher, Reed Bunting

Winter
Black-throated Diver, Red-throated Diver, Red-necked Grebe, Whooper Swan, Eurasian Wigeon, Pintail, Northern Shoveler, Greater Scaup, Common Goldeneye, Velvet Scoter, Common Eider, Smew, Yellow-legged Gull, Great Grey Shrike, Nutcracker, Brambling, Siskin

Passage
Marsh Harrier, Hen Harrier, Mediterranean Gull, Little Gull, Whiskered Tern, Black Tern, Water Pipit

5 Ammersee NSG (45km)

(Hersching S-Bahn Station S7)
(Diessen Railway Station)

The Ammersee is another lake southwest of Munich, not too far from Starnberger See. It is similar in size to Starnberger See and attracts similar flocks of wildfowl in winter. However, it also has rather extensive mudflats at the southern end, where the River Ammer enters the lake, and this is a good spot for waders at migration time. The area around the Ammer River also includes marshes

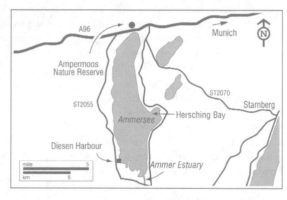

and water meadows, as well as riparian woodland which supports a variety of land-birds, making this the best location on the lake. Other locations worth a look are Hersching Bay on the east shore and the reedbeds along the River Ammer at the north-ern extremity, where it flows out of the lake and which comprise the Ampermoos Nature Reserve.

Ammersee can be reached by taking the **A36 Autobahn** west from Munich and exiting south at the **Inning** Exit onto the **ST 2067**. Turn right in the village of Inning onto **Landsbergerstrasse**, where the nature reserve office is located in an old brewery. Continuing on Landsbergerstrasse under the autobahn brings you to the Ampermoos Nature Reserve. Continue south on the **ST2067** to Hersching, where you can scan the lake for ducks and grebes. From Hersching, take the **ST2068** to **Fischen** and turn west onto the **ST2056**. Take an immediate right turn onto **Ammerweg** which leads to the east shore of the Ammer Estuary. You can also access the western side of the Ammer Estuary from the village of **Dießen** which lies further west along the ST2056. The S-Bahn serves Hersching and there is a train service to Dießen.

KEY SPECIES

Resident
Great White Egret, Greylag Goose, Gadwall, Northern Shoveler, Red-crested Pochard, Goosander, Water Rail, Common Snipe, Eurasian Curlew, Yellow-legged Gull, Lesser Spotted Woodpecker

Summer
Black-necked Grebe, Eurasian Bittern, Little Ringed Plover, Common Sandpiper, Common Tern, Corn Crake, Blue-headed Wagtail, Bluethroat, Whinchat, Sedge Warbler, Marsh Warbler, River Warbler, Great Reed Warbler, Icterine Warbler, Penduline Tit, Golden Oriole

Winter
Red-necked Grebe, Whooper Swan, White-fronted Goose, Bean Goose, Eurasian Wigeon, Pintail, Common Goldeneye, Velvet Scoter, Smew, Hen Harrier, Herring Gull, Great Grey Shrike, Siskin, Common Redpoll

Passage
Red-footed Falcon, Northern Lapwing, Whimbrel, Little Stint, Temminck's Stint, Ruff, Common Redshank, Spotted Redshank, Greenshank, Wood Sandpiper, Little Gull, Whiskered Tern, Black Tern, White-winged Black Tern, Meadow Pipit

Further Away

6 Chiemsee (80km)

(Übersee Railway Station)

Chiemsee is the largest lake in Bavaria and a RAMSAR-designated Wetland of International Importance. Although the lake itself supports large numbers of breeding and wintering waterfowl, it is the surrounding marsh, woodland and reedbeds, providing habitats for a very diverse range of species, which make this site well worth a visit at any time of the year, although it is rather distant (about 80km) from Munich. For gull enthusiasts, winter flocks containing Herring, Yellow-legged, Caspian and Lesser Black-backed Gulls in all plumages can be seen. However, unlike at coastal sites, these flocks are usually just a hundred or so birds in total, so it is relatively easy to examine each individual.

It is easy to get to by taking the **A8 southwest** from Munich, which runs along the south shore of the lake. There are exits at **Bernau** and at **Übersee**. There is a Conservation Centre at Übersee where you can get more detailed maps and information. The most interesting location is at the mouth of the Tiroler Achen River near Übersee where the sheltered Hirschauer Bay supports large numbers of gulls and waterfowl and, in winter, White-tailed Eagles. There is an observation tower on the lakeshore east of the river mouth, near Baumgarten. Other productive sites are: the southwest corner near Bernau (waders), Seebruck Harbour on the north shore of the lake (grebes and diving ducks) and the reedbeds on the Sassau Peninsula at the northwest corner (marshland species). This is a very large area and impossible to cover completely without a car, but if you are relying on public transport there are railway stations at Prien, Bernau and Übersee, so that you could visit one or two locations on the lakeshore in a day.

KEY SPECIES

Resident
Great White Egret, Greylag Goose, Gadwall, Northern Shoveler, Red-crested Pochard, Goosander, Water Rail, Common Snipe, Eurasian Curlew, Yellow-legged Gull, Common Kingfisher, Grey-headed Woodpecker, Lesser Spotted Woodpecker, Dipper, Water Pipit

Summer
Black-necked Grebe, Eurasian Bittern, Black Stork, Spotted Crake, Little

Ringed Plover, Common Sandpiper, Common Tern, Yellow Wagtail, Bluethroat, Sedge Warbler, Marsh Warbler, Savi's Warbler, Great Reed Warbler, Penduline Tit, Golden Oriole, Common Rosefinch

Winter
Red-throated Diver, Red-necked Grebe, Whooper Swan, White-fronted Goose, Bean Goose, Eurasian Wigeon, Pintail, Common Goldeneye, Velvet Scoter, Smew, White-tailed Eagle, Hen

Harrier, Herring Gull, Caspian Gull, Lesser Black-backed Gull, Siskin

Passage
Red-footed Falcon, Little Crake, Whimbrel, Little Stint, Temminck's Stint, Ruff, Common Redshank, Spotted Redshank, Greenshank, Wood Sandpiper, Little Gull, Whiskered Tern, Black Tern, White-winged Black Tern, Tree Pipit, Meadow Pipit, Ring Ouzel

7 Murnauer Moos (65km)   *(Murnau Railway Station)*

Murnau is a very picturesque Bavarian town about 60km southwest of Munich. It is a popular tourist spot and so can be a good compromise destination for a day trip for any birder who is travelling with a non-birder because of the nearby Murnauer Moos Nature Reserve. This is an area of alpine meadows and moorland at the foothills of the Bavarian Alps which supports a distinctive community of open-country grassland species in a habitat which is fast disappearing in western Europe. The best time to visit is late May and June, when the birds are establishing territories and also the flora is at its best. This is also a favoured location for many dragonfly species.

Take the **A35** southwest and exit onto the **ST2062** following the signs for **Murnau**. At Murnau take the **B2** south for about 1km and turn west at **Ramsachstrasse**. When you get to the small church at **Ähndl** (St Georg) you can park and follow the signs for the **Moosrundweg** (hiking and cycling trail). The trail is about 6km long and so is quite a long walk, although you don't have to do the full circuit as the best birding locations are along the first 3km: especially the trees along the banks of the Ramsach River, and the surrounding meadows. Many passerines use the trees as song-perches.

KEY SPECIES

Resident
Red Kite, Common Snipe, Eurasian Curlew, Grey-headed Woodpecker, White-backed Woodpecker, Sky Lark, Wood Lark, Meadow Pipit, Common Raven

Summer
Spotted Crake, Water Rail, Corn Crake, Quail, Tree Pipit, Meadow Pipit, Yellow Wagtail, Bluethroat, Whinchat, Common Stonechat, Grasshopper Warbler, Western Bonelli's Warbler, Pied Flycatcher, Penduline Tit,

Nutcracker, Common Crossbill, Common Rosefinch

Winter
Hen Harrier, Alpine Accentor, Redwing, Great Grey Shrike

8 Garmisch-Partenkirchen (90km)   *(Garmisch-Partenkirchen Railway Station)*

Garmisch-Partenkirchen is a ski resort close to the Austrian border where the 1936 Winter Olympics were held. It is about two hours drive south of Munich and might not seem a practical proposition for someone who has only a day or two to spare. However, the area around Garmisch-Partenkirchen is the closest site to Munich for the specialist avifauna of the High Alps, and therefore a must-see if you want to get to grips with these species. Also, bear in mind that the best time of year to see these birds is late

May and June when daylight is at its longest, so a lot can be achieved in a day. Even for those who are restricted to public transport, there is an hourly train service from Munich and a system of cable cars and mountain railways, operated by Bayerische Zugspitzbahn, to get you into the mountains. The usual precautions apply to anyone birding in the mountains even in summer. Be sure to take adequate footwear for hiking and warm clothing as the weather can be very changeable.

Any of the cable cars will bring you into good birding areas, but perhaps the best tactic is to drive or take the **Zugspitz Mountain Railway** as far as **Eibsee**. The forest on the north shore of the lake is a good spot for both Three-toed and White-backed Woodpeckers, as well as Hazel Grouse. Then take the railway up to **Zugspitz** (the highest peak in Germany at 2,962m) where you can get the alpine species. A less strenuous trip would be to take the Alpspitz cable car to **Osterfelderkopf Ski Resort** (2,050m), which is just above the tree line and hike the trail to **Kreuzeck** (1,651m). From here you can return to Garmisch-Partenkirchen by another cable car.

The area is less interesting in winter; the resident birds will still be present but will have moved to lower elevations. Deep snow will prevent you from travelling far from the ski lifts then but it should be possible to see Snow Finches, Alpine Choughs and Nutcrackers around the restaurants.

KEY SPECIES

Resident
Golden Eagle, Peregrine Falcon, Hazel Grouse, Ptarmigan, Black Grouse, Capercaillie, Grey-headed Woodpecker, White-backed Woodpecker, Three-toed Woodpecker,

Black Woodpecker, Wallcreeper, Alpine Accentor, Crested Tit, Alpine Chough, Common Raven, Nutcracker, Common Crossbill, Citril Finch, Snow Finch

Summer
Crag Martin, Water Pipit, Common Stonechat, Ring Ouzel, Western Bonelli's Warbler, Red-breasted Flycatcher

---

9 Oberammergau (85km)   (Oberammergau Railway Station)

If you missed some species at Garmisch-Partenkirchen then the cliffs at Oberammergau are worth a look on your way back to Munich. Although not as extensive an area as Garmisch-Partenkirchen, it is possible to get all of the alpine species here and there is a quite reliable site for Crag Martins and Wallcreepers.

Head back to Munich from Garmisch-Partenkirchen on the main road (**E533**) and after about 7km, turn west at **Oberau** onto the **B23**. This road takes you through **Ettal**. After about 1km past Ettal turn west onto the **ST2060**, which runs along the valley of the River Ammer. Proceed along this road for about 1.7km and take the first turn right (**ST2560**) which continues on towards Oberammergau. About 500m along this road on the western side you can scan the cliffs for breeding Crag Martins and, with patience, Wallcreepers.

If it is evening as you return to the main road at Oberau, it

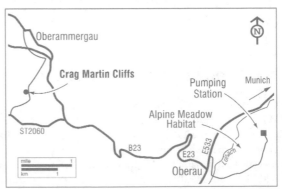

is worth checking the meadows east of Oberau for Corn Crakes. Cross the bridge over the **River Loisach** in central Oberau and follow this road east for about 2.5km until it comes to a dead end at a pumping station. Listen here for Corn Crakes, and also keep an eye out for Tawny and Water Pipits. There is a train from Munich to Oberammergau but the station is about 3km from the Crag Martin colony.

KEY SPECIES

Resident		Summer
Golden Eagle, Peregrine Falcon, Ptarmigan, Capercaillie, Grey-headed Woodpecker, White-backed Woodpecker, Three-toed Woodpecker, Black Woodpecker, Wallcreeper,	Alpine Accentor, Crested Tit, Alpine Chough, Common Raven, Nutcracker, Common Crossbill, Citril Finch, Snow Finch	Corn Crake, Wryneck, Crag Martin, Tawny Pipit, Water Pipit, Common Stonechat, Ring Ouzel, Western Bonelli's Warbler, Red-breasted Flycatcher

Useful contacts and websites

Bavarian Birds – local birding website (in English) www.bavarianbirds.de

Birding Germany – Bavarian webpages (German with some English content) www.birdinggermany.de/bayern.htm

Ornithological Society in Bavaria e.V. www.og-bayern.de Zoologische Staatssammlung, Münchhausenstraße 21, 81247 München.

Landesbund für Vogelschutz in Bayern e.V. (LBV) Bavarian Society for Habitat Protection Eisvogelweg 1, 91161 Hilpoltstein. Tel. 09174/47750. www.lbv.de

NABU – Naturschutzbund Deutschland BirdLife partner in Germany www.nabu.de Herbert-Rabius-Str 26, D-53225 Bonn. Tel. (+49) 2284036166

German Birdnet www.birdnet.de

Deutsch Bahn – railway travel in Germany: www.bahn.de/pv/uebersicht/die_bahn_ international_guests.shtml

Münchener Verkehrsverbund – public transport in Munich: www.mvv-muenchen.de

Bayerische Zugspitzbahn – rail and cable cars in Bavarian Alps: www.zugspitze.de

Books and publications

Vogelparadiese, Bd.2, Westdeutschland und Süddeutschland Michael Lohmann & Knut Haarman (1989) Blackwell Wissenschafts-Verlag ISBN 3826381521

Atlas der Brutvögel Bayerns 1979–1983 G. Nitsche; H. Plachter (1987) Ornithologische Gesellschaft Bayern/Bayerisches Landesamt für Umweltschutz

Avifauna Bavariae – Die Vogelwelt Bayerns im Wandel der Zeit – Volume I and II W. Wüst (1986) Ornithologische Gesellschaft Bayern

Collins Bird Guide Lars Svensson *et al.* (1999) HarperCollins ISBN 0002197286

Birds of Europe Lars Jonsson (2006) Christopher Helm ISBN 0713676000

• •

Paris

P aris is almost in the centre of the Île-de-France region and, situated as it is in northern France, a birder's expectations of the city would be much lower than for any of the cities in southern France. The city is almost 150km from the nearest stretch of coastline, and there are few natural features in its hinterland to give rise to the mosaic of habitats which is essential for supporting a diversity of species. However, its location is just far enough north for it to lie within the wintering range of several species of northern waterfowl as well as being far enough south for some southern landbirds to breed.

Although much of northern France is utilised for agriculture, given its regal history there are very significant belts of forest which were preserved for hunting in the Paris area. Even within the central parts of the city these forests provide urban habitats for woodland species and are well worth visiting, particularly in spring and early summer. The River Seine is the dominant natural feature of the Paris region and the course of the river over the centuries has created large sand and gravel deposits along its watershed. Exploitation of these deposits has created new wetland areas which are attractive to birds. Protection and management of these wetlands has created breeding habitats for reedbed species and winter habitats for wildfowl. These sites are well watched by Parisian birders and are productive at any time of the year.

Eurasian Jay is common in the central city parks of Paris

KEY SPECIES

Resident
Great Crested Grebe, Grey Heron, Mute Swan, Mallard, Eurasian Sparrowhawk, Common Kestrel, Moorhen, Black-headed Gull, Yellow-legged Gull, Wood Pigeon, Stock Dove, Collared Dove, Great Spotted Woodpecker, White Wagtail, Wren, Dunnock, Robin, Black Redstart, Blackbird, Song Thrush, Goldcrest, Blue Tit, Great Tit, Short-toed Treecreeper, Eurasian Nuthatch, Eurasian Jay, Magpie, Western Jackdaw, Carrion Crow, Common Starling, House Sparrow, Chaffinch, Greenfinch, Goldfinch, Bullfinch, Yellowhammer

Summer
Common Swift, Barn Swallow, House Martin, Garden Warbler, Blackcap, Common Whitethroat, Common Chiffchaff, Willow Warbler, Spotted Flycatcher

GETTING AROUND

The best way to get around Paris is to use the metro, which is fast, convenient and simple to use. There is a very dense network of subway stations so that it is always just a short walk to the metro, which is fully integrated with the suburban train service (RER) so you can quickly navigate through the city and out to the suburbs. The metro is definitely the best mode of transport within the inner ring-road, the Périphérique, as driving a car is stressful, and parking is difficult to find and expensive.

Outside the Périphérique a car enables much more efficient use of your time as, although you don't have to go too far outside the city to find good birding locations, the sites may be extensive and distant from the nearest bus or railway station. The outer orbital ring, the A86 Autoroute, enables you to reach the autoroutes which fan out from the city with-

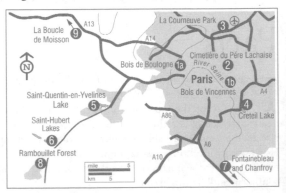

out having to cross through the city centre. However, be advised that driving around Paris is not an enjoyable experience for the faint-hearted or the hesitant and, for some, the more leisurely pace of birding by public transport is a small price to pay to avoid the dangers of Parisian traffic. As an alternative, the railway company (SNCF) has a service by which you can rent a bicycle which you pick up at your destination and this can be a pleasant way to reach the birding sites, particularly in summer. There are six main railway stations in Paris and it is important to identify which one serves the railway line for your destination.

For birding at the wetland sites, you will probably need a telescope as many French nature reserves exist for the benefit of the birds, not the birders, and access may be restricted or there may not always be a bird hide available, so you need to be able to scan the waterways from a distance. You will need a map that covers the whole of Île-de-France in order to locate all the sites, and it would also be useful to get a city map with a diagram of the public transport system. These are available from any bookshop or tourist office, and at bus and rail stations.

Locations

1 City Parks

There are two parks in the Paris urban area which hold most interest for birders: the Bois de Boulogne (1a) in the west of the city and the Bois de Vincennes (1b) in the east. Like all city parks they are heavily utilised by the general public and being surrounded by the city are isolated from a hinterland that might permit colonisation by a wider range of species. Nevertheless both parks are a superb birding resource located right in the heart of this beautiful city.

The Bois de Boulogne is just beyond the Périphérique and is a remnant of the vast forests which once extended around Paris. Luckily, this particular forest was preserved for hunting and the mature oakwoods support a reasonable diversity of woodland species, despite a recent (1999) major winter storm, which caused severe damage to the old trees. This park is large and to some extent rather crowded with tourists and strollers, but it is possible to find some less-disturbed locations particularly in early morning. The most productive areas are the northeast quadrant between the Allée de Longchamp and the Lac Inférieur, and the southeast quadrant, south of the Route de la Grande Cascade. The lakes themselves tend to be disturbed by boaters and picnickers but in winter they hold some wildfowl, although mainly introduced exotic species. There are bird walks organised by local birders every Sunday at 09.00, departing from the Porte d'Auteuil Metro, located at the southeast corner of the park.

The Bois de Vincennes holds a similar range of woodland species to that of the Bois de Boulogne. Its location near the confluence of the Seine and the Marne attracts some wildfowl in the winter to the park lakes, and a hide has been constructed near a recently created pond. This park is well known among Parisian birders as good for Tawny Owls (up to 20 pairs), so with some local assistance you stand a good chance of seeing a roosting bird. There are bird walks organised every Saturday at 09.00, departing from the Musée des Arts Africains et Océaniens which is at the west end of the park. Both parks are well served by public transport.

KEY SPECIES

Resident
Common Pheasant, Tawny Owl, Common Kingfisher, Black Woodpecker, Green Woodpecker, Lesser Spotted Woodpecker, Common Stonechat, Mistle Thrush, Firecrest, Marsh Tit, Long-tailed Tit, Tree Sparrow, Cirl Bunting (Bois de Vincennes only)

Summer
Common Nightingale, Common Whitethroat, Melodious Warbler, Wood Warbler, European Serin

Winter
Pochard, Tufted Duck, Common Gull, Herring Gull, Meadow Pipit, Grey Wagtail, Redwing, Siskin, Common Redpoll, Hawfinch

Passage
Common Cuckoo, Blue-headed Wagtail, Common Redstart, Whinchat, Northern Wheatear, Ring Ouzel, Pied Flycatcher, Golden Oriole

2 Cimetière du Père Lachaise   *(Philippe-Auguste Metro Station)*

This central site may well be on your tourist itinerary as it contains the graves of such notables as Chopin, Molière, Oscar Wilde and Jim Morrison. If so, bring your binoculars as it is also a migration point with many migrants passing through both in spring and in autumn, as well as having an interesting resident avifauna. The site owes its attraction to its elevated position in the east of the city and the extensive cover afforded by the shrubs in the cemetery. The slightly more reverential visitors to the cemetery mean there is less disturbance to the birds than in the city parks and consequently easier birding. It is also quite a reliable site for Tawny Owls so pay close attention to any agitated flocks of tits.

KEY SPECIES

Resident
Tawny Owl, Lesser Spotted Woodpecker, Firecrest, Marsh Tit, Coal Tit, Crested Tit, Long-tailed Tit

Winter
Grey Wagtail, Redwing, Linnet, Siskin

Passage
Common Redstart, Mistle Thrush, Pied Flycatcher, European Serin

3 La Courneuve Park (10km)   *(La Courneuve RER Railway Station)*
(Bus 249)

This is a rather good forest park, with some open waterways, located in the north of Paris. It is en route to Charles de Gaulle Airport, so you might consider stopping off on the way. The best areas are to the north and east of the park and include mixed woodland, wet meadows and some open farmland at the eastern end. Little Bitterns breed here and the semi-natural areas attract migrants in spring and autumn.

Take the **A1 Autoroute** north from Paris and exit north at **Le Bourget (Exit 5)** towards **Dugny**. At Dugny turn left on the **D114** going southwest: the park entrance is on the right-hand side of the road. The nearest railway station is at La Courneuve about 15 minutes' walk (five minutes by bus) from the park.

KEY SPECIES

Resident
Common Coot, Green Woodpecker, Meadow Pipit, Mistle Thrush, Firecrest, Long-tailed Tit, Crested Tit, Tree Sparrow

Summer
Little Bittern, Little Ringed Plover,

Turtle Dove, Sky Lark, Common Stonechat, Common Nightingale, Reed Warbler, Melodious Warbler, Pied Flycatcher, European Serin

Winter
Great Cormorant, Common Pochard, Common Kingfisher, Fieldfare,

Redwing, Siskin, Common Redpoll

Passage
Green Sandpiper, Common Sandpiper, Grey Wagtail, Blue-headed Wagtail, Northern Wheatear, Ring Ouzel, Pied Flycatcher

4 Créteil Lake (13km)

(Créteil-Préfecture Metro Station)

This is an artificial lake in the suburbs of southeast Paris which was created as a result of gravel extraction works. It is not particularly rich in resident birdlife but it occasionally attracts interesting winter visitors and is only 15 minutes from Orly Airport, making it worth a stop on your way to or from there. You could even visit the lake if your flight got delayed for a couple of hours.

From the **A86** outer ring Autoroute, exit east for **Créteil** at **Exit 22** and follow the **D1** south to the **Centre Commercial Créteil**. There is a car park at the commercial centre and a footpath which circles the lake, with a nature centre on the west shore.

KEY SPECIES

Resident
Great Cormorant, Common Pochard, Tufted Duck, Water Rail, Common Coot, Common Snipe, Crested Lark, Linnet

Summer
Little Grebe, Lesser Black-backed

Gull, Common Cuckoo, Sand Martin, Reed Warbler, European Serin

Winter
Gadwall, Eurasian Wigeon, Eurasian Teal, Common Goldeneye, Goosander, Common Kingfisher, Redwing

Passage
Garganey, Green Sandpiper, Common Sandpiper, Common Tern, Black Tern, Grey Wagtail, Blue-headed Wagtail

5 Saint-Quentin-en-Yvelines Lake (25km)

(Trappes RER Railway Station)

This, the largest lake in Île-de-France, is southwest of Paris, not far from the Palace of Versailles: and a worthwhile diversion for anyone visiting the palace. However try to avoid sunny weekends, when the area is very crowded! There are extensive reedbeds and when water levels are low there is a considerable expanse of mud exposed at the southeast end where it is shallowest, which can attract waders in variable numbers in autumn. This area is a nature reserve and there are two hides and a viewing point on the lakeshore. Most of the lake is surrounded by a public park which contains some areas of interest, including a small farm and tree nurseries, and a small pond with breeding Little Bitterns.

Take the **A12 Autoroute** southwest from Paris to the exit at **Trappes**. Turn west on the **D912** and after about 1km you will come to a roundabout. Turn right into a car park at the lake (this is actually the golf course car park). Access is restricted but there is an external perimeter path, 7km long, and with a telescope you can look into the reserve from outside. The reserve is open only for half a day (Saturday or Sunday morning) every fortnight: in summer it is also sometimes opened in the evening. It is not an LPO reserve and is free to everyone: telephone 01 30 164 440, or email: resnat.sqy@wanadoo.fr for information on opening times. There is a train service to Trappes (from Gare Montparnasse and La Défense). The station is about 1km from the lakeshore.

KEY SPECIES

Resident
Great Cormorant, Canada Goose, Northern Shoveler, Common Pochard, Tufted Duck, Water Rail, Common Coot, Northern Lapwing,

Common Snipe, Tree Sparrow, Reed Bunting

Summer
Little Grebe, Little Bittern, Garganey,

Marsh Harrier, Little Ringed Plover, Mediterranean Gull, Common Cuckoo, Sky Lark, Meadow Pipit, Grasshopper Warbler, Sedge Warbler, Marsh Warbler, Golden Oriole

Winter	Passage	
Eurasian Bittern, Common Shelduck, Gadwall, Eurasian Wigeon, Eurasian Teal, Ferruginous Duck, Short-eared Owl, Common Kingfisher, Water Pipit, Fieldfare, Redwing	Black Kite, Osprey, Spotted Crake, Ringed Plover, Ruff, Common Redshank, Black-tailed Godwit, Greenshank, Green Sandpiper, Wood Sandpiper, Dunlin, Little Stint,	Common Sandpiper, Little Gull, Common Tern, Whiskered Tern, Black Tern, Tawny Pipit, Sand Martin, Bluethroat, Ring Ouzel

6 Saint-Hubert Lakes (45km)   *(Le Perray-en-Yvelines RER Railway Station)*

These are a chain of lakes just north of Rambouillet Forest which are important for wintering wildfowl and for a range of breeding aquatic species. There is no access to the lakeshores but each lake is traversed at either end by embankments which enable you to scan the lake surface and the reedbeds. This would take rather too long if trying to negotiate the site on foot

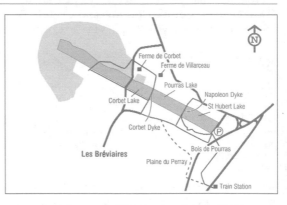

but, if driving, you can cover the site quite adequately. This is one of the best sites near Paris for Little Bitterns, particularly in spring.

Take the **A12 Autoroute** southwest from Paris to the exit at **Trappes**. Follow the **N10** towards Rambouillet, but turn west onto the **D191** at **Saint-Hubert**. This road runs along the northern shore of the lakes and there are a series of turn-offs to the left which bring you down to the embankments. To the south of the lakes lie the plains of Les Bréviaires and Le Perray, which are good for Hen Harriers and Golden Plovers in winter.

KEY SPECIES

Resident
Great Cormorant, Northern Shoveler, Common Pochard, Tufted Duck, Common Buzzard, Water Rail, Common Coot, Northern Lapwing, Common Snipe, Little Owl, Cetti's Warbler, Tree Sparrow

Summer
Little Grebe, Little Bittern, Garganey, Marsh Harrier, European Honey-buzzard, Hobby, Sand Martin, Grasshopper Warbler, Sedge Warbler, Marsh Warbler, Melodious Warbler, Golden Oriole

Winter
Eurasian Bittern, Gadwall, Eurasian Wigeon, Eurasian Teal, Hen Harrier, European Golden Plover, Common Kingfisher, Water Pipit, Fieldfare

Passage
Great White Egret, White Stork, Black Kite, Osprey, Spotted Crake, Ringed Plover, Ruff, Common Redshank, Black-tailed Godwit, Greenshank, Green Sandpiper, Wood Sandpiper, Common Sandpiper, Mediterranean Gull, Little Gull, Whiskered Tern, Black Tern, Sky Lark, Meadow Pipit, Bluethroat

Further Away

7 Fontainebleau Forest and Chanfroy Plain (60km)   *(Fontainebleau Railway Station)*

If you can only visit one forest in the Paris area, then this is the one to see. Although it is fairly distant from the city, it cannot be beaten for habitats, numbers of birds and species diversity! This is a very extensive forest and a birder could spend weeks roaming the trails, but it is worth mentioning two sites which are well-known for their birds, at least one of which can be accessed by public transport. The area itself was originally preserved for the purposes of hunting, but over the years forest fires and forestry operations have

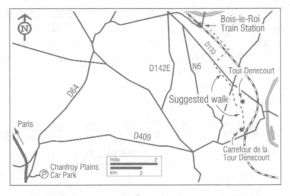

meant that the forest consists of old growth, secondary growth and new plantations, as well as open sandstone heaths, thus creating a mosaic of habitats that supports a variety of species. It is also sufficiently far south for a number of more southerly species such as Bee-eaters and Hoopoes to be present at their most northerly limit.

The Chanfroy Plain is at the western edge of the forest and is an area of open heath interspersed with small wooded areas and shallow ponds. Several heathland species can be found here, in particular Dartford Warblers, Wood Larks and Nightjars. Take the **A6 Autoroute** south from Paris and after about 50km, exit east at **Arbonne-la-Forêt**. At Arbonne, take the D64 south for about 1.5km and turn left at the sign for **Carrière des Fusillés**. There is a car park at the end of this track and you can walk the trail heading east from the car park.

The nearest railway station to Chanfroy Plain is at Fontainebleau which is about 10km away, so if you are reliant on public transport this is not an easy option. However there are several trails through the forest which originate near the railway station. A suggestion is to take the Carrefour de la Tour Denecourt which passes through the forest for about 7km terminating at Bois-le-Roi, from where you can get the train back to Paris (Gare de Lyon). In spring and early summer, any trails off this road can yield up to six species of woodpecker as well as other woodland species, including some interesting birds of prey.

KEY SPECIES

Resident
Northern Goshawk, Common Pheasant, Woodcock, Green Woodpecker, Grey-headed Woodpecker, Great Spotted Woodpecker, Middle Spotted Woodpecker, Lesser Spotted Woodpecker, Black Woodpecker, Crested Lark, Mistle Thrush, Dartford Warbler, Firecrest, Long-tailed Tit, Willow Tit, Marsh Tit, Crested Tit, Tree Sparrow, Linnet, Common Crossbill, Hawfinch, Yellowhammer, Cirl Bunting

Summer
Short-toed Eagle, Booted Eagle, European Honey-buzzard, Hobby, Common Cuckoo, European Nightjar, European Bee-eater, Wryneck, Wood Lark, Tree Pipit, Common Nightingale, Common Redstart, Reed Warbler, Melodious Warbler, Lesser Whitethroat, Western Bonelli's Warbler, Wood Warbler, Pied Flycatcher, Red-backed Shrike, Golden Oriole

Winter
Fieldfare, Redwing, Siskin, Common Redpoll

Passage
Hoopoe, Tawny Pipit, Grey Wagtail, Whinchat, Ring Ouzel

8 Rambouillet Forest (50km)   (Rambouillet RER Railway Station)

The Rambouillet Forest is southwest of Paris and, together with the nearby Saint-Hubert Lakes, makes a worthwhile trip at any time of year. This forest supports a much more diverse range of species than the parks within the city. It is possible to see five woodpecker species here, particularly in early spring when the birds are establishing territories and the trees are still largely free of foliage.

Take the **A12 Autoroute** southwest from Paris to the exit at **Trappes**. Follow the N10 to **Rambouillet**: the entrance to the forest and chateau is at the west end of the town centre. The forest trails in the northwest sector of the forest are the least disturbed and most productive for birds.

KEY SPECIES

Resident
Common Pheasant, Woodcock, Green Woodpecker, Great Spotted Woodpecker, Middle Spotted Woodpecker, Lesser Spotted Woodpecker, Black Woodpecker, Mistle Thrush, Firecrest, Long-tailed

Tit, Willow Tit, Marsh Tit, Crested Tit, Hawfinch

Summer
European Honey-buzzard, Hobby, European Nightjar, Tree Pipit, Common Nightingale, Reed Warbler,

Lesser Whitethroat, Western Bonelli's Warbler, Wood Warbler, Pied Flycatcher, Golden Oriole

Winter
Fieldfare, Redwing, Brambling, Siskin, Common Redpoll

9 La Boucle de Moisson (50km)   *(Rosny-sur-Seine RER Railway Station)*

Northwest of Paris the River Seine meanders through a number of oxbow bends in the course of its journey to the English Channel. The resultant sand deposits have been excavated for sand and gravel over the centuries and have created a number of gravel extraction pits, which have flooded and are attractive places for birds. One of the best sites is at Moisson to the

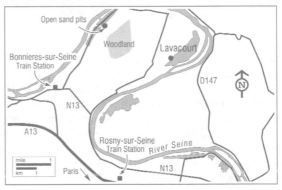

northwest of the city on the south bank (Left Bank) of the Seine. There are several different habitats in this area including forest, open areas and deep-water gravel pits, as well as the banks of the Seine itself.

Take the **A13 Autoroute** northwest from Paris and exit north at **Gassicourt (Exit 13)**, following the **N13** as far as **Rolleboise** before turning right onto the **D124**. This road runs largely parallel to the Seine towards Moisson. However before reaching Moisson turn right on the **D125** at **Lavacourt**. The large lake at Lavacourt is rather deep and, in addition to the usual diving ducks, it also attracts the occasional diver and rare grebe in winter as well as having a large gull roost. Continue on the D124 towards the forest at Moisson which supports a rich population of woodland species including Black Woodpeckers. Further east from the forest is an open area of sand pits and more flooded extraction sites where Stone-curlews, Wrynecks and Red-backed Shrikes breed. The full circuit of the D124 is about 20km and is best suited to those travelling by car. If reliant on public transport, there is a train service to Rosny-sur-Seine, which is near Rolleboise on the N13, from Gare St Lazare and you could consider renting a bicycle to make the same trip.

KEY SPECIES

Northern Shoveler, Common Pochard, Tufted Duck, Common Buzzard, Water Rail, Common Coot, Northern Lapwing, Common Snipe, Black Woodpecker, Crested Lark, Wood Lark, Sky Lark, Meadow Pipit, Mistle Thrush, Firecrest, Long-tailed Tit, Willow Tit, Marsh Tit, Crested Tit, Bearded Tit, Linnet, Common Crossbill, Tree Sparrow

Summer
Little Grebe, European Honey-buzzard, Marsh Harrier, Hobby, Stone-curlew, Little Ringed Plover, Common Tern, European Nightjar, Wryneck, Sand Martin, Grasshopper Warbler, Sedge Warbler, Marsh Warbler, Melodious Warbler, Red-backed Shrike, Golden Oriole, European Serin

Winter
Gadwall, Eurasian Wigeon, Pintail, Eurasian Teal, Common Goldeneye, Velvet Scoter, Goosander, Hen Harrier, Golden Plover, Common Gull, Lesser Black-backed Gull, Common Kingfisher, Water Pipit, Fieldfare, Siskin, Common Redpoll

Passage
Garganey, Black Kite, Ringed Plover, Ruff, Common Redshank, Greenshank, Green Sandpiper, Wood Sandpiper, Common Sandpiper, Mediterranean Gull, Little Gull, Whiskered Tern, Black Tern, Tawny Pipit, Northern Wheatear

Useful contacts and websites

Ligue pour la Protection des Oiseaux (LPO)
(French) www.lpo.fr
La Corderie Royale, BP 263, F-17305, Rochefort Cedex, France.
Leading French bird conservation organisation and rarities committee; publishers of the journal *Ornithos*.

LPO Île-de-France (French)
ile-de-france.lpo.fr
Local Branch of LPO: 62 rue Bargue, 75015 Paris.
Tel. 0153 585 838. Fax. 0153 585 839.

LPO – UK Branch (English)
www.kjhall.org.uk/lpo.htm
Ken Hall, LPO (UK), The Anchorage, The Chalks, Chew Magna, Bristol BS40 8SN.
Tel. (+44) (0)1275 332980.

CORIF Centre Ornithologique Île-de-France, (French) www.corif.net
18 Rue Alexis Lepère, 93100 Montreuil-Sous-Bois.
Tel. 0148 519200. Fax. 0148 519240. Local ornithological group; information about conservation of sites and species in Paris area.

La Courneuve Park (French and English)
www.parcs93.info/fr/parcs/courneuve/
Map and general information about La Courneuve Park.

Association des Naturalistes de la Vallée du Loing – anvl (French)
http://perso.club-internet.fr/anvl/
General information about geology, flora and fauna of Fontainebleau Forest.

Ornithologie en Île-de-France (French)
http://ornitho.idf.free.fr
Stéphane Boitel's personal website with information about local birding sites.

Des Oiseaux et des Plumes (French)
jpmaist.club.fr Jean-Paul Maistre's personal website about birding in the parks of Paris.

French Rarities Committee (French)
www.lpo.fr/homologation
Secretary, c/o LPO, La Corderie Royale, BP 263, F-17305, Rochefort Cedex, France.

Paris Birdline: CORIF/Ile-de-France/ Tel. 0149 840790.

RATP (English) – public transport in the Paris area www.ratp.fr

SNCF (English) – French Railways. Tel. 0891 362020 http://idf.sncf.fr/GB/default.htm

Books and publications

Where to Watch Birds in Holland, Belgium & Northern France
Arnoud van den Berg & Dominique Lafontaine (1996) Hamlyn, ISBN 0600579778

Where to Watch Birds in France
Philippe J. Dubois (2006) Christopher Helm
ISBN 0713669802

Finding Birds in Northern France
Dave Gosney. BirdGuides, ISBN 1898110115

Guide des Oiseaux de Paris
Guilhem Lesaffre & Jean Chevalier
ISBN 2840960842

Birdwatching Guide to France
Jacquie Crozier (2000) Arlequin Publications
ISBN 1900159368

L'Inventaire des Oiseaux de France Métropolitaine
Philippe J. Dubois, Pierre Le Maréchal, Georges Olioso & Pierre Yésou (2000) Nathan, Paris

Oiseaux et Forêts en Île-de-France CORIF
Map, checklists and trails guide to birding in Île de France.

Les Oiseaux d'Île-de-France
Pierre le Maréchal, Guilhem Lesaffre & Jean Chevallier (2000) Delachaux et Niestlé
ISBN 2603011774

Collins Bird Guide
Lars Svensson *et al.* (1999) HarperCollins
ISBN 0002197286

Birds of Europe
Lars Jonsson (2006) Christopher Helm
ISBN 0713676000

Prague

Prague is a popular tourist city which many birders may visit with the sole intention of sightseeing and enjoying the impressive cultural attractions. However be sure to carry your binoculars with you as you walk around as there are a number of birding opportunities among the tourist areas. The city sits astride the Vltava River with high ground on either bank which is dotted with parks and small woodlands. The river itself attracts a surprising variety of migrant and wintering waterfowl, while the green spaces provide habitats for many passerines and even some of the more specialised woodpecker species. There is, however, no high quality wetland near the city and like most inland cities you need to travel some distance out of town in order to reach a significant birding site. Increasing numbers of birders visit the Czech Republic to sample its superb forest habitats for woodpeckers and owls, or its outstanding wetlands in South Bohemia. However, these areas are beyond the scope of a city-bound birder and moreover, due to the secretive nature of the forest species, they would probably require several days on-site to stand a reasonable chance of finding these birds.

A pair of Tufted ducks on the River Vitava with Prague's famous Charles Bridge in the background

There are some good stretches of forest within easy reach of central Prague and it is possible to see at least six woodpecker species, as well as some eastern European specialities such as River Warblers and both Red-breasted and Collared Flycatchers. Late spring offers the best birding opportunities when the greatest diversity of species is present and summer visitors are establishing territories. Later in summer will be disappointing as most birds fall silent at this time and small birds are practically impossible to find in the dense foliage. Brehynský Fish Pond to the north of the city makes a worthwhile day trip at any time of the year, as its aquatic habitats support a different range of birds from that of the forest habitats around Prague.

TYPICAL SPECIES

Resident
Grey Heron, Mute Swan, Mallard, Tufted Duck, Common Pochard, Common Buzzard, Eurasian Sparrowhawk, Common Kestrel, Moorhen, Common Coot, Black-headed Gull, Wood Pigeon, Collared Dove, Great Spotted Woodpecker, White Wagtail, Wren, Dunnock, Robin, Blackbird, Goldcrest, Marsh Tit, Coal Tit, Blue Tit, Great Tit, Eurasian Nuthatch, Short-toed Treecreeper, Eurasian Jay, Magpie, Carrion Crow, House Sparrow, Tree Sparrow, Chaffinch, Hawfinch, Greenfinch, Goldfinch, Linnet

Summer
Black Kite, Common Swift, Sand Martin, Barn Swallow, House Martin, Common Redstart, Black Redstart, Song Thrush, Garden Warbler, Blackcap, Common Chiffchaff, Firecrest, Spotted Flycatcher, Common Starling, European Serin

Winter
Great Cormorant, Yellow-legged Gull, Goldcrest, Western Jackdaw, Siskin

GETTING AROUND

If your visit is confined within the city limits, then the public transport system is perfectly adequate for the sites which are close to the city. The metro, tram and bus systems are all integrated so the journey to any site rarely involves more than one interchange and the service is frequent and reliable. If travelling during peak periods, expect the trains and buses to be very crowded, and it may not be a comfortable trip if you are carrying a lot of birding gear. Given the number of hills in Prague, renting a bicycle is not as practical as in other European cities and you will find that whether you are sightseeing or birding you will do a lot of walking up hills! If you do have a car, you would be better off leaving it at any of the park and ride facilities on the outskirts of the city and

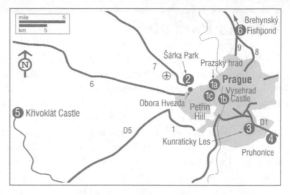

taking the metro into town. Driving across the city can be problematic as the ring road system is discontinuous so that you may need to drive in close to the city centre, particularly if your journey involves crossing the Vltava. However, once you are out on the open road, traffic is generally quite light and journey times are fairly brisk.

For most of the year, birding in and around Prague is generally pleasant. There can be periods of severe cold during midwinter but you are unlikely to be standing stationary for long periods, gazing through a telescope, as there are few large areas of open water.

Locations

1 Parks in Central Prague

As well as being a great city for sightseeing, Prague has some very productive bird habitat in its many parks, so that even if your visit is short and restricted to walking round the centre there are plenty of birding opportunities. The River Vltava itself can also be quite good for wintering waterfowl and many of the tourist sites such as Prague Castle (Prazsky hrad, 1a) and Vysehrad Castle and Cemetery (1b) are surrounded by wooded parks and gardens which support a surprising range of species. The most productive of the parks is Petřín Hill (1c) on the west side, which has several woodland trails and provides a superb view over the city. There is a funicular railway to the top of the hill, so you don't have to climb. In general, be sure to carry your binoculars even when walking round the city sightseeing, as you can also scan the river for wildfowl from any of the bridges. However, to avoid most of the crowds take one of the riverside walks located near the zoo (Bus No. 112).

KEY SPECIES

Resident
Peregrine Falcon, Stock Dove, Green Woodpecker, Grey-headed Woodpecker, Middle Spotted Woodpecker, Lesser Spotted Woodpecker, Marsh Tit, Long-tailed Tit

Summer
Hobby, Turtle Dove, Wryneck, Common Cuckoo, Tree Pipit, Common Nightingale, Willow Warbler, Icterine Warbler, Barred Warbler, Lesser Whitethroat, Golden Oriole,

Red-backed Shrike

Winter
Greater Scaup, Common Goldeneye, Velvet Scoter, Smew, Fieldfare, Redwing, Brambling, Bullfinch

2 Šárka Park (7km)

(Tram 20 and 26/Bus 119)

There are several large parks on the outskirts of Prague and this is one of the largest and one of the best in terms of habitat. The large limestone cliff-faces set amidst the woods, with streams and waterfalls running down the sides, are the main feature. There is also a large lake which attracts some wildfowl, and overall the park offers much more wild and rugged terrain than is normally associated with a city park. Despite this, it is well laid out with a network of footpaths giving easy access to all areas. If you have time, there is another park, Obora Hvezda, just south of Šárka Park in the same area, that has some good woodland. It is adjacent to Bílá Hora, the highest point in Prague (380m).

Šárka Park is in the northwest of the city, just about midway between the centre and the airport on the main route to the airport (Highway 7). If you are using public transport, take the tram (20 or 26) to the terminus which is right by the main entrance. If you are going to the airport, Bus 119 departs from the tram stop.

KEY SPECIES

Resident
Little Grebe, Northern Goshawk, Peregrine Falcon, Stock Dove, Green Woodpecker, Grey-headed Woodpecker, Black Woodpecker, Middle Spotted Woodpecker, Lesser Spotted Woodpecker, Mistle Thrush, Fieldfare, Willow Tit, Long-tailed Tit

Summer
European Honey-buzzard, Hobby, Turtle Dove, Wryneck, Common Cuckoo, Tree Pipit, Grey Wagtail, Blue-headed Wagtail, Common Nightingale, Icterine Warbler, Barred Warbler, Lesser Whitethroat, Willow Warbler, Collared Flycatcher, Red-breasted Flycatcher, Golden Oriole, Red-backed Shrike

Winter
Redwing, Brambling, Bullfinch

3 Kunraticky Les (10km)

(Roztyly Metro Station)

There are large tracts of mixed forest on the southeastern outskirts of the city, which can be accessed at several points by metro. One of the best areas is Kunraticky Les, a mature mixed woodland, bordered by the Kunraticky River, a tributary of the Vltava, and the D1 Motorway. There are many trails here leading through woods, open parkland and some riparian habitat. Although popular with the public this park is large enough to offer quiet corners where birds are most active.

KEY SPECIES

Resident
Red Kite, Northern Goshawk, Stock Dove, Green Woodpecker, Grey-headed Woodpecker, Black Woodpecker, Middle Spotted Woodpecker, Lesser Spotted Woodpecker, Crested Lark, Willow Tit, Long-tailed Tit

Summer
European Honey-buzzard, Hobby, Turtle Dove, Wryneck, Common Cuckoo, Tree Pipit, Common Nightingale, Marsh Warbler, Icterine Warbler, Barred Warbler, Lesser Whitethroat, Willow Warbler, Collared Flycatcher, Red-breasted Flycatcher, Golden Oriole, Red-backed Shrike

Winter
Mistle Thrush, Fieldfare, Redwing, Brambling, Bullfinch

4 Pruhonice (14km)

(Opatov Metro Station)

Your visit to Prague may take in this site, as Pruhonice Castle is a popular tourist and conference centre. Adjacent to the castle is a large forest park with almost 40km of forest paths, and a Botanical Park. Several small lakes are dotted throughout the park and the Botic River snakes its way through the woodlands. This site supports larger numbers and a better diversity of species than the urban parks as it is outside the city, with only open countryside between it and larger tracts of forest habitat stretching further south and east.

Take the **Brno Highway (D1)** southeast from Prague and exit east at **Pruhonice (Exit 6).** The castle and forest are at the southwest end of the town.

KEY SPECIES

Resident Northern Goshawk, Eurasian Sparrowhawk, Stock Dove, Green Woodpecker, Grey-headed Woodpecker, Black Woodpecker, Middle Spotted Woodpecker, Lesser Spotted Woodpecker, Crested Lark, Mistle	Thrush, Fieldfare, Willow Tit, Long-tailed Tit **Summer** European Honey-buzzard, Hobby, Woodcock, Turtle Dove, Wryneck, Common Cuckoo, Tree Pipit,	Common Nightingale, Marsh Warbler, Icterine Warbler, Barred Warbler, Lesser Whitethroat, Willow Warbler, Collared Flycatcher, Red-breasted Flycatcher, Golden Oriole, Red-backed Shrike

Further Away

5 Křivoklát Castle (46km)   *(Krivoklat Railway Station)*

Křivoklát Castle is a historic and rather impressive Gothic castle which is the object of many excursions out of Prague. It is also set on the banks of the River Rakovnicky and amidst one of the largest areas of lowland mixed forest in the Czech Republic. This huge forest was originally a hunting reserve and is now protected as a UNESCO Biosphere Reserve. The mixed woodlands of oak, beech and coniferous trees support a rich community of woodland birds and if you are visiting the castle, take some time to explore them. There are several paths from the castle grounds into the adjacent woodland. The best time to visit is between early spring and early summer; expect to see all of the typical woodland species of central Europe including seven species each of woodpecker and tit. If you are new to the European avifauna such challenging species pairs as Chiffchaff/Willow Warbler, Eurasian/Short-toed Treecreeper and Marsh Tit/Willow Tit can be seen and heard in good numbers.

Take the **D5 Motorway** southwest from Prague to **Beroun**. Exit north at Beroun on the **116**, and after 16km turn right (north) on the **236**. The castle is approximately 6km on the left-hand side of this road just past the town of **Křivoklát**.

KEY SPECIES

Resident Red Kite, Northern Goshawk, Stock Dove, Common Kingfisher, Green Woodpecker, Grey-headed Woodpecker, Black Woodpecker, Middle Spotted Woodpecker, Lesser Spotted Woodpecker, White-backed Woodpecker, Eurasian Treecreeper, Mistle Thrush, Fieldfare, Willow Tit,	Crested Tit, Long-tailed Tit, Common Raven, Common Crossbill, Yellowhammer, Reed Bunting **Summer** Little Grebe, Black Stork, European Honey-buzzard, Hobby, Turtle Dove, Wryneck, Common Cuckoo, Tree Pipit, Common Nightingale, Whinchat, River	Warbler, Grasshopper Warbler, Icterine Warbler, Barred Warbler, Lesser Whitethroat, Wood Warbler, Willow Warbler, Collared Flycatcher, Red-breasted Flycatcher, Golden Oriole, Red-backed Shrike

6 Brehynský Fish Pond (70km)

(Doksy Railway Station)

The practice of fish-farming in central Europe has resulted in the creation of excellent aquatic habitats at many localities. The technique requires the construction of several large ponds to accommodate the fish at various stages of their life cycle, and the nutrient rich waters promote the growth of abundant vegetation. The fish pond at Brehynský is one of the closest such habitats to Prague and is accessible in a day trip from the city. This pond has been rehabilitated in order to produce optimum conditions for wildlife. Apart from the open waters which attract wildfowl on passage, the littoral areas consist of reedbeds and water meadows, and the lagoon is surrounded by coniferous woodland.

Take **Highway 9** north from Prague for about 60km and turn right (east) onto **Route 38** towards the town of **Doksy**. At Doksy, turn north on **Route 270** towards **Brehyne**. The fish pond is about 2km along this road on the right-hand side.

KEY SPECIES

Resident
Little Grebe, Eurasian Bittern, Greylag Goose, Gadwall, Red-crested Pochard, Common Pheasant, Water Rail, Northern Lapwing, Common Kingfisher, Bearded Tit, Penduline Tit, Yellowhammer, Reed Bunting, Corn Bunting

Summer
Little Bittern, Night Heron, Great White Egret, White Stork, Black Stork, Garganey, European Honey-buzzard, Black Kite, Red Kite, Marsh Harrier,

Hobby, Common Crane, Spotted Crake, Little Crake, Little Ringed Plover, Common Tern, Turtle Dove, Common Cuckoo, Tree Pipit, Blue-headed Wagtail, Common Nightingale, Bluethroat, Grasshopper Warbler, Savi's Warbler, River Warbler, Sedge Warbler, Marsh Warbler, Great Reed Warbler, Icterine Warbler, Barred Warbler, Golden Oriole, Red-backed Shrike

Winter
White-tailed Eagle, Common

Goldeneye, Smew, Goosander, Common Gull, Water Pipit

Passage
Black-necked Grebe, Osprey, Corn Crake, Ruff, Common Redshank, Spotted Redshank, Greenshank, Eurasian Curlew, Black-tailed Godwit, Wood Sandpiper, Common Sandpiper, Common Snipe, Little Gull, Black Tern, Wryneck, Sky Lark, Meadow Pipit, Fieldfare, Redwing, Pied Flycatcher

Useful contacts and websites

Ceska spolecnost ornitologicka (CSO)
(English and Czech) www.birdlife.cz
Czech Society for Ornithology, Hornomecholupska 34, 102 00 Praha 10. Tel./fax. 274 866 700.
BirdLife partner in Czech Republic.

Czech Rarities Committee
(English and Czech)
aix.upol.cz/~vavrik/rarcom/czech-rc.htm
Dr Martin Vavrik, Sobotin 54, 788 16
Czech Republic

Probirder www.probirder.com
Birding tour company operating in Czech Republic.

Dopravní podnik hl. m. Prahy
(English and Czech)
www.dp-praha.cz
Public transport for Prague area.
Tel. (+420) 296 191 111. E-mail: dpp@r.dpp.cz

Books and publications

Where to Watch Birds in Eastern Europe
Gerard Gorman (1994) Hamlyn, ISBN 060057976X

Birds of the Czech Republic
Joseph Kren (2000) Christopher Helm
ISBN 0713647841

Collins Bird Guide
Lars Svensson *et al.* (1999) HarperCollins
ISBN 0002197286

Birds of Europe
Lars Jonsson (2006) Christopher Helm
ISBN 0713676000

Rome

Given its location midway along the Italian Peninsula in the Lazio region, Rome offers an excellent opportunity to see a wide range of Mediterranean birds. Although an inland city, lying astride the River Tiber, Rome is only 25km from the coast and a number of productive coastal wetlands. The entire Italian Peninsula acts as a conduit for many trans-Saharan migrants and the coastal marshes of Lazio are important feeding and resting areas. Spring and autumn are the best times to visit this area as, apart from the fact that the temperatures are very equable and make for a pleasant day in the field, the greatest diversity of species can be found. As well as the standard north/south migration, because of its location in the centre of the Mediterranean, Italy also experiences west/east movements. Several species, such as Red-footed Falcons and White-winged Black Terns breed in large numbers in eastern Europe and winter in west Africa, passing through Italy in both directions.

A Blue Rock Thrush finds the ruins of the Roman Forum to be an ideal habitat

The extensive coastal marshes of Lazio were drained even in ancient times and today's remnants are a tiny fraction of what was once a huge wetland ecosystem. A few have been preserved and, indeed, attempts are now being made to restore former wetlands, as is happening at the Centro Habitat Mediterraneo near Ostia. The native Mediterranean lowland woodlands and maquis habitats have also long since been cleared for agriculture but, where pockets remain, the unique bird communities of this ecosystem can be found within easy reach of the city.

The foothills of the Appenine Mountains extend to within 50km of the western suburbs of Rome, but the high peaks which support the specialised montane species are further west and would involve a journey of two to three hours, beyond the scope of this book. Likewise, the extensive coastal marshes of Circeo National Park are over 100km and a two-hour drive south of Rome, but are well worth a two-day trip for the birder with some extra time to spend in the area.

Finally, of course, the fascinating history, art and archaeology of Rome will be a constant distraction for any visitor especially if you are visiting with a non-birding partner or spouse. However there need not be any conflict as many of the historical sites, particularly those around Ostia, are surrounded by good birding habitat. *Veni, Vidi, Observavi!*

TYPICAL SPECIES

Resident
Grey Heron, Mallard, Common Buzzard, Common Kestrel, Moorhen, Common Coot, Yellow-legged Gull, Collared Dove, White Wagtail, Wren, Black Redstart, Common Stonechat, Blackbird, Sardinian Warbler, Blackcap, Common Chiffchaff, Blue Tit, Great Tit, Short-toed Treecreeper, Magpie, Hooded Crow, Common Starling, Italian Sparrow, Tree Sparrow, Greenfinch, Goldfinch, European Serin

Summer
Black Kite, Turtle Dove, European Bee-eater, Common Swift, Barn Swallow, House Martin, Spotted Flycatcher

Winter
Black-headed Gull, Sky Lark, Meadow Pipit, Robin, Song Thrush, Goldcrest

GETTING AROUND

The Ancient Romans ensured that all roads lead to Rome, so getting in and out of the city is quite easy as all the old thoroughfares radiate out from the city centre. The modern addition of the orbital motorway, the Grande Raccordo Anulare (GRA), means that wherever you are located in the city or on the outskirts, you can navigate easily to the autostradas. It is inside the GRA that driving is

much more complicated and, at times, terrifying. You are best advised to avoid driving within the GRA and to use the public transport system, which comprehensively serves both Rome and Ostia by bus, tram, metro and suburban train.

Driving on the autostradas and provincial roads is much less taxing than in the city and offers the opportunity to visit several sites in the course of a day. If you do not have a car, then there is a train service which operates along the coast north of Ostia. You can hire a bicycle to take on the train and use this to ride between the stations and birding sites. Almost all trains operate from the central railway station (Termini), which, of course, is also a hub for many bus, tram and metro routes. However the suburban train to Ostia departs from Piramide Metro Station in the south of the city centre. There is also a regional bus service between Rome and Civitavecchia along the Via Aurelia which also serves these locations and operates from the bus terminal adjacent to Lepanto Metro Station. Make sure that you obtain a schematic map of the transport system in order to navigate your way through all of these modes of transport.

Apart from excessive heat in the summer months, there is little discomfort involved in birding around Rome. Winters are mild, with most of the rain falling between September and December, while spring is blessed with a perfect climate, although you can expect some showers. A detailed map of the Lazio region is required to locate all of the sites. Some of the sites mentioned are either LIPU Reserves or regional parks and entry is almost always free. However LIPU Reserves operate on a limited access basis and often only open on one or two days a week, and even then, only for a few hours, making them unsuitable for early-morning birding. So if you are planning to visit LIPU Reserve, check the opening hours on their website or phone ahead.

Locations

1 **Parco dell'Appia Antica** (5km)

(Parco Appia Antica Metro Station)
(ATAC Bus No. 118/660)

The Appian Way led from the Roman Forum to southern Italy in ancient times and a linear park has now been created along the old route. This is one of the best birding sites within the GRA and provides a superb mix of birding and Roman archaeology for anyone interested in both. The park begins at the southeast end of the Ancient Forum and continues along the Appian Way as far as the GRA. However the best area for birds is the Caffarella Valley and there is a footpath (Via della Caffarella) which begins about

500m south of the visitor centre. This path passes through an area of undulating hills, grasslands, and riparian habitat along the Caffarella stream. Other habitats within the park include cultivated plots and oak groves, as well as the crumbling ancient structures themselves which provide niches for cliff-nesting species. The winter flocks of Starlings, which roost on the Forum buildings, always attract Peregrines and occasionally Lanner Falcons. The park is well served, internally, by public transport and there is also a hop-on/hop-off tourist bus circulating through it.

Several other city centre parks, including the Villa Borghese and Villa Ada east of the Tiber and the Villa Doria Pamphili west of the Tiber near the Vatican City, also support limited populations of typical woodland species and are worth a look in the early morning. The banks of the Tiber itself, from the Villa Borghese northwards, support a good variety of aquatic species including Kingfishers, Cetti's Warblers and, in winter, Penduline Tits. For the truly lazy, relaxing with a coffee at the Piazza Termini is a good way to look for Pallid Swifts. Keep a check on the hordes of swifts overhead and as soon as you see one that resembles the colour of your caffè latte – you've got one!

KEY SPECIES

Resident	Summer	Winter
Common Kingfisher, Little Owl, Green Woodpecker, Crested Lark, Common Stonechat, Cetti's Warbler, Great Reed Warbler, Long-tailed Tit, Corn Bunting	Common Cuckoo, Pallid Swift, Hoopoe, Wryneck, Common Nightingale, Subalpine Warbler, Melodious Warbler, Red-backed Shrike, Woodchat Shrike, Cirl Bunting	Peregrine Falcon, Larner Falcon, Common Snipe, Grey Wagtail, Blue Rock Thrush, Penduline Tit

2 Pineta di Castel Fusano (25km)   *(Castel Fusano Railway Station)*

The Roman Coast State Nature Reserve contains some good tracts of typical Mediterranean coastal woodland and one of the most extensive is located at Pineta di Castel Fusano, just south of Ostia. These Mediterranean woods are dominated by holm oaks and maritime pines. Although they cover a sizeable area they are not like the dense forests found in wetter climates: they form an open canopy and there are many clearings, making the birding relatively easy. The forest is bounded on its northern perimeter by the Canale del Pescatori (Canale dello Stagno), a waterway which attracts several aquatic species, and is worth checking. There are several footpaths through the park, making it very easy to cover the area, and its proximity to several other good sites in the Roman Coast State Nature Reserve make this whole area well worth a visit.

From the **GRA**, exit south on the **Via de Mare (Exit 28)** to **Ostia Antica**. At Ostia turn left (south) onto **Via di Castel Fusano**, and follow this road for about 3km to the park entrance, crossing a bridge over the Canale del Pescatori. You may park at the entrance and walk along the Via Severiana footpath which runs parallel to the coast through some of

the best habitat. If you are using public transport you can take the train to Lido Castel Fusano: the station is at the northwest corner of the park. If returning via Ostia, you might consider visiting the newly created Centro Habitat Mediterraneo (CHM), which is a very promising wetland habitat developed on an old dump-site and which has already built up an impressive species list. This is on Via Carlo Avegno, just behind the new harbour development at Ostia. Like most LIPU Reserves, visiting hours are restricted so check ahead with LIPU.

KEY SPECIES

Resident
Common Pheasant, Little Owl, Green Woodpecker, Crested Lark, Common Stonechat, Cetti's Warbler, Sardinian Warbler, Firecrest, Eurasian Nuthatch, Long-tailed Tit, Coal Tit

Summer
Little Bittern, Black Kite, Turtle Dove, Common Cuckoo, Little Owl, European Bee-eater, Wryneck, Reed Warbler, Great Reed Warbler, Subalpine Warbler

Passage
Eurasian Bittern, Eurasian Spoonbill, Osprey, Hobby, Whinchat, Garden Warbler, Common Whitethroat, Wood Warbler, Pied Flycatcher

---

3 **Maccarese Ponds** (25km)   *(Maccarese-Fregene Railway Station)*

There are a number of wetland sites on the Lazio Coast to the west of Rome. Many of these are included in the Roman Coast State Nature Reserve, an extensive although non-contiguous protected area containing isolated sites of environmental importance, antiquities and wetland habitats. The Maccarese Ponds are artificially created but have matured into excellent freshwater ponds with reedbeds, willows and larger stands of eucalyptus trees. They are quite shallow in places and are frequented by long-legged

waders such as redshanks, godwits and stilts. The ponds are studded with small reed-fringed islets which are important for breeding aquatic birds. The surrounding farmland is subject to flooding in winter and attracts flocks of geese. This is a popular site, well-watched by local birders and many rarities are found here, particularly in spring and autumn.

Take the **A12 Autostrada** north and about 5km past **Leonardo da Vinci Airport** perimeter, exit west at the **Maccarese Exit** and drive into the village. The ponds are about 500m south of the village. Although the area is a nature reserve, it is not developed for visitors so birding here involves crossing fields and ditches. It is also worth checking the beach near Maccarese, taking any of the roads down to the coast. In winter and on passage, seabirds can usually be seen. In fall conditions – the first two hours after heavy showers in spring or autumn – the beach can be littered with herons and waders and the surrounding bushes on the ponds filled with warblers.

Several other wetland sites in the vicinity of Leonardo da Vinci Airport are worth a look if you have time before departure or after arriving. Macchiagrande WWF Reserve is a brackish marsh surrounded by remnant Mediterranean coastal woodland just at the northwest end of the airport. This is an important site for migrants as well as a breeding and wintering site for wildfowl. There is also a lake and reedbed, at the Coccia di Morto Estate, at the western edge of the airport perimeter between Viale Coccia di Morto and the seashore. While all these places are worth a check, Maccarese is the place to concentrate on if time is limited.

KEY SPECIES

Resident
Little Egret, Water Rail, Common Kingfisher, Crested Lark, Cetti's Warbler, Zitting Cisticola, Penduline Tit

Summer
Little Bittern, Marsh Harrier, Hobby, Little Ringed Plover, Black-winged Stilt, Common Cuckoo, Short-toed Lark, Sand Martin, Ashy-headed Wagtail, Common Nightingale, Great Reed Warbler, Reed Warbler

Winter
Little Grebe, Eurasian Bittern, Great

Crested Grebe, Great White Egret, Eurasian Teal, Gadwall, Eurasian Wigeon, Northern Shoveler, Common Pochard, Ferruginous Duck, Hen Harrier, Peregrine Falcon, Northern Lapwing, Common Snipe, Water Pipit, Common Chiffchaff, Moustached Warbler, Corn Bunting, Reed Bunting

Passage
Squacco Heron, Night Heron, Purple Heron, Eurasian Spoonbill, Glossy Ibis, White Stork, Garganey, European Honey-buzzard, Common Crane, Little Crake, Avocet, Ringed Plover,

Curlew Sandpiper, Ruff, Common Redshank, Spotted Redshank, Greenshank, Green Sandpiper, Wood Sandpiper, Marsh Sandpiper, Little Stint, Temminck's Stint, Black-tailed Godwit, Little Gull, Mediterranean Gull, Gull-billed Tern, Whiskered Tern, Black Tern, White-winged Black Tern, Wryneck, European Bee-eater, Yellow Wagtail (three or four races including Black-headed), Whinchat, Pied Flycatcher, Wood Warbler, Subalpine Warbler, Northern Wheatear

4 Castel di Guido (17km)

Much of the natural habitats of the countryside surrounding Rome has been lost to intensive agriculture and urban developments. However an area of typical Mediterranean woodland and more open maquis has been preserved at the Castel di Guido LIPU Reserve. This is also a ringing station and is set within an agricultural research facility which conducts extensive agricultural

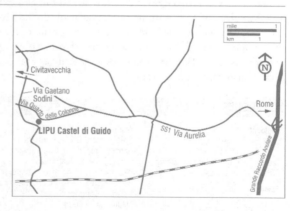

practices. These are more sympathetic to the environment and provide optimum conditions for many birds threatened by intensification of agriculture. There is a campsite and two nature trails through the reserve. The latter allow you to explore this area of open countryside without entering private property. Unlike other LIPU reserves, access is possible at all times.

Take the **Via Aurelia (SS1)** northwest from Rome and about 7.5km past the **GRA** turn left (south) at the flyover exit onto **Via Gaetano Sodini**. After about 500m, take the next turn on the left (**Via Quarto delle Colonne**). Follow the signs for the **Azienda Agricola di Castel di Guido**. The reserve is located 2km inside the research facility. Several regional bus routes serving Civitavecchia and the North Lazio coast operate along the Via Aurelia and pass by the turn-off to the reserve.

KEY SPECIES

Resident
Common Pheasant, Great Spotted Woodpecker, Green Woodpecker, Crested Lark, Common Stonechat, Zitting Cisticola, Sardinian Warbler, Long-tailed Tit, Cirl Bunting, Corn Bunting

Summer
European Honey-buzzard, Turtle Dove, Common Cuckoo, Little Owl,

Wryneck, Tawny Pipit, Ashy-headed Wagtail, Melodious Warbler, Subalpine Warbler, Red-backed Shrike, Woodchat Shrike

Winter
Hen Harrier, Northern Lapwing, European Golden Plover, Woodcock, Common Snipe, Wood Pigeon, Stock Dove, Sky Lark, Meadow Pipit, Siskin

Passage
Hobby, Short-toed Eagle, Red-footed Falcon, Common Crane, European Roller, Whinchat, Icterine Warbler, Garden Warbler

5 Macchiatonda WWF Reserve (45km)   *(Santa Severa Railway Station)*

The WWF reserve at Macchiatonda is a freshwater wetland right on the coast near Civitavecchia. A mixture of habitats includes artificial freshwater ponds, wet meadows, a sand-dune system and a small remnant of woodland. The reserve provides unhindered access to the beach so you can seawatch when the conditions are right and also look for waders on the shore. The freshwater ponds are attractive to roosting gulls, terns and waders when water levels are low in late summer and autumn and this is one of the most reliable sites on the Lazio coast for Audouin's Gulls. In winter this stretch of coast is good for sea ducks and wintering seabirds.

Take the **A12 Autostrada** north from Rome and exit west at **Ladispoli** onto the **Via Aurelia (SS1)**. Continue northwards on the SS1 for about 9km and the signs for the reserve are on the left-hand (west) side of the road. Unlike the LIPU reserves, this one is open every day for several hours. If you are using public transport, the reserve entrance is about 1km from Santa Severa Railway Station. The Rome/Civitavecchia bus service also passes the entrance.

KEY SPECIES

Resident
Little Egret, Water Rail, Common Kingfisher, Little Owl, Calandra Lark, Crested Lark, Cetti's Warbler, Zitting Cisticola, Corn Bunting

Summer
Marsh Harrier, Hobby, Common Quail, Little Ringed Plover, Common Cuckoo, Short-toed Lark, Sand Martin, Ashy-headed Wagtail, Great Reed Warbler, Reed Warbler

Winter
Great Crested Grebe, Northern Gannet, Great White Egret, Greylag Goose, Eurasian Teal, Eurasian Wigeon, Pintail, Common Pochard, Red-breasted Merganser, Northern Lapwing, Common Snipe, Mediterranean Gull

Passage
Yelkouan Shearwater, Cory's Shearwater, Squacco Heron, Night Heron, Purple Heron, Greater Flamingo, Garganey, Northern Shoveler, Avocet, Ruff, Eurasian Curlew, Common Redshank, Spotted Redshank, Greenshank, Curlew Sandpiper, Wood Sandpiper, Audouin's Gull

6 Tolfa Hills (60km)   *(Santa Severa Railway Station)*

Although just a little over 600m high, the Tolfa Hills provide an area of high ground on the otherwise flat coastal plain of Lazio and offer a relatively undisturbed refuge for many species in this rather built-up coastal zone. Moreover, their location on a promontory on the Tyrrhenian Sea coast means that they act as a migrant trap for many species, especially raptors in spring. The habitat at the lower elevations is typical Mediterranean maquis with scrub vegetation. Higher up, the hillsides are more exposed and barren but the valleys are well wooded, mostly with beech. The whole area supports a rich community of Mediterranean birds, some of which are difficult to find closer to Rome.

Take the **A12 Autostrada** north towards **Civitavecchia** and exit north at **Santa Severa** towards Tolfa. This road (**SP3B**) loops through the hills passing close to the summit (**Mt Tolfaccia**) and returns to the A12 near Civitavecchia. The best way to do this area is to drive the loop road, stopping at likely vantage points and at areas of good habitat. It is difficult to access this area without a car, as it would require walking a considerable distance from the railway station, and the hills would make for rather difficult cycling.

KEY SPECIES

Resident	Spring/Summer	
Red Kite, Green Woodpecker, Wood Lark, Crested Lark, Blue Rock Thrush, Long-tailed Tit, Linnet, Cirl Bunting, Yellowhammer, Corn Bunting	Eurasian Sparrowhawk, European Honey-buzzard, Short-toed Eagle, Black Kite, Montagu's Harrier, Hobby, Common Quail, European Roller, Turtle Dove, Alpine Swift, Wryneck,	Tree Pipit, Tawny Pipit, Short-toed Lark, Common Nightingale, Subalpine Warbler, Melodious Warbler, Red-backed Shrike, Woodchat Shrike, Lesser Grey Shrike

7 Lago di Vico (55km)   *(Capranica Railway Station)*

There are three large volcanic lakes northwest of Rome: Lakes Martignano, Bracciano and Vico. Although Lake Vico is the furthest from the city, it is surrounded on its northern side by very productive deciduous woodland. In spring and early summer at least, it is worth making the longer trip in order to visit this habitat, which supports three woodpecker species and several hole-nesting passerines. The northern shores of the lake and surrounding marsh and woodland are enclosed within the Lago di Vico Nature Reserve. Several foot and cycle paths pass through the more interesting areas and there is a hide overlooking the marshes at the northwestern bay. You may hike through the beechwoods to the summit of Mount Venere or take the less demanding trail through the woods at the base of Mount Fogliano on the western shore. As well as being an important winter site for wildfowl, significant numbers of Great Crested Grebes breed in the large reedbeds at the north end.

Take the **Via Cassia (SS2)** northwest from Rome and turn north at **Monterosi** towards **Ronciglione** on the **SP1 (Via Cimina)**. Follow this road through Ronciglione and after it swings northwards, take the second turn left (west) which is about 2.5km after Ronciglione. This road brings you to the reserve entrance. Follow the road, which encircles the lake, passing close to the shore at several points and leading to the woodland trails on the western shore. If you are relying on public transport the best option is to bring a rented bicycle on the train to Capranica, about 6km south of Ronciglione, and ride it from there.

KEY SPECIES

Resident	Summer	
Great Crested Grebe, Little Egret, Red Kite, Peregrine Falcon, Water Rail, Common Kingfisher, Little Owl, Green Woodpecker, Lesser Spotted Woodpecker, Cetti's Warbler, Zitting Cisticola, Marsh Tit, Coal Tit, Penduline Tit, Reed Bunting, Corn Bunting	Little Bittern, Marsh Harrier, Hobby, Little Ringed Plover, Black-winged Stilt, Common Cuckoo, Sand Martin, Ashy-headed Wagtail, Common Redstart, Great Reed Warbler, Reed Warbler, Spotted Flycatcher, Pied Flycatcher, Red-backed Shrike	Eurasian Teal, Eurasian Wigeon, Gadwall, Northern Shoveler, Northern Pintail, Common Pochard, Ferruginous Duck, Red-crested Pochard, Hen Harrier, Northern Lapwing, Common Snipe
	Winter Little Grebe, Black-necked Grebe,	**Passage** Squacco Heron, Garganey, Common Redshank, Greenshank, Wood Sandpiper, Black-tailed Godwit

Useful contacts and websites

Lega Italiana Protezione Uccelli (Italian)
www.lipu.it
Via Trento 49, 43100 Parma.
Tel. (+39) 0521 273043. Fax. (+39) 0521 273419.
Information about policies, activities, reserves and
branches of LIPU – Italian BirdLife partner.

LIPU – Roman Branch (Italian and English)
www.lipu.it/sezione/roma
Vicolo Silvestri 57/59, 00164 Roma.
Tel. 06 66166833. Fax. 06 66167140. Information
about local branch, local reserves and activities.

LIPU – Ostia Branch (Italian and English)
www.lipuostia.it
c/o ATI – Lungomare Duca degli Abbruzzi 84, 00121
Ostia Lido. Tel. 06 56339770. Fax. 06 56339770.
Information about local branch and local reserves,
including opening hours.

LIPU – Lazio – Regional Branch of LIPU (Italian)
www.lipu.it/sezione/roma/lipulazio.asp
Via Aldrovandi 2 (adiacente Bioparco – Villa
Borghese), 00197 Roma. Tel/fax. 06 32110752.

LIPU – UK (English) www.lipu-uk.org/
c/o David Lingard (LIPU-UK), Fernwood,
Doddington Road, Whisby, Lincs, LN6 9BX, UK.

Il Birdwatching Italiano (Italian and English)
www.ebnitalia.it
Website for Italian birding with monthly reports, trip
reports and site descriptions.

National Parks of Italy (Italian and English)
www.parks.it
Official website of Italian National Parks with information about access, facilities and natural history.

Riserva naturale statale del Litorale Romano
(English) www.romacivica.net/cyberia/riserva/
ereserve.htm
Site descriptions, directions and general information
about the Roman Coastal State Nature Reserve.

Parco dell'Appia Antica (English)
www.parcoappiaantica.it
Site description, directions and general information
about Parco dell'Appia Antica.

Riserva Naturale Lago di Vico (Italian)
www.parks.it/riserva.lago.vico
Site description, directions and general information
about Lago di Vico.

Italian Rarities Committee (CIR) (English)
www.storianaturale.org/cir
c/o Andrea Corso / C.I.R. (Comitato Italiano
Rarità), Via Camastra 10, 96100 Siracusa.
For records of species recorded more than ten times.

Italian Rarities Committee (COI)
www.ciso-coi.org
Commissione Ornitologica Italiana,
c/o Pierandrea Brichetti, Via V. Veneto 30, 25029
Verolavecchia. For records of species recorded fewer
than ten times.

ATAC – public transport in Rome (Italian and
English) www.atac.roma.it
Portal site with information about bus, tram and
metro services for Rome.

Trenitalia (Italian and English) National Railway
System: www.trenitalia.com

COTRAL – regional bus services in Lazio (Italian
and English): www.cotralspa.it

Books and publications

Where to Watch Birds in Italy (LIPU)
Lega Italiana Protezione Uccelli (1994)
Christopher Helm, ISBN 0713638672

Guida alla natura dei Monti della Tolfa
Elisabetta Faraglia and Francesco Riga (1997)
Franco Muzzio, ISBN 8870218074

Ornitologia Italiana Pierandrea Brichetti and
Giancarlo Fracasso (2003-2006) Alberto Perdisa
Editore.

Collins Bird Guide
Lars Svensson et al. (1999) HarperCollins
ISBN 0002197286

Birds of Europe
Lars Jonsson (2006) Christopher Helm
ISBN 0713676000

Collared Flycatcher and Middle Spotted Woodpecker are two key species in Vienna Woods

Vienna

Vienna is very favourably located ornithologically as it is far enough east to be within range of many eastern European breeding species but also far enough south and west for its waterways to remain ice-free through the winter and support wintering populations of wildfowl. Added to that it also lies on the western edge of the Hungarian Plains, with the foothills of the Alps to the southwest and with the Danube flowing right through the city. This provides an interesting mix of habitats and there are few inland cities in Europe to rival Vienna in terms of species diversity.

Late spring and early summer attract many birders to the hotspots of eastern Austria, all of which are within easy reach of Vienna. The wetlands of Burgenland, the state to the east of the city, are absolutely full of birds at this time of year, both migrants and breeders, while alpine species and the shyer forest species are more easily found at this time of year. However, autumn also produces many good birds as migrants travel down the Danube and March river valleys. In late autumn and winter the Danube and the artificial 'New Danube' hold a broad diversity of waterfowl and gulls right in the heart of the city.

Hungary, Slovakia and the Czech Republic are all within an hour or two by road from Vienna and all offer some excellent birding sites just over their borders. You may be tempted to take the time to visit some of these sites also, as some species occur in higher numbers or are easier to see. Nevertheless, for most birders, Vienna and its environs will offer more than enough to fill any spare time available.

GETTING AROUND

Vienna has a very comprehensive public transport system which integrates metro (U-Bahn), trams, buses and trains (S-Bahn and R-Bahn), and there is no corner of the city without a frequent and convenient service. For birding in the city parks just purchase a day ticket, which permits the use of all modes of transport, and leave the

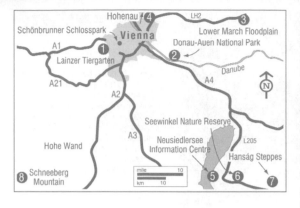

car behind. This saves the hassle of navigating the city streets and looking for parking spaces. Even for the outlying birding sites, there are comprehensive train services to most towns and feeder buses which connect with the trains serve most of the sites. So you can still access these superb locations even without a car. Of course, a car gives much more mobility and enables you to cover several sites in a day. It is also essential for birding the Hanság Steppes as you need to cover quite a distance and the car will afford some concealment when scanning for bustards and geese. Even if you haven't ridden one for years, you could also consider renting a bike and taking it on the train. This would be a practical and healthy way to commute between sites in the Neusiedl area, although beware the strong and unremitting winds which blow across the steppes!

Many birding sites are nature reserves of some kind, managed either as national parks or by conservation organisations such as WWF, and there may be a small entrance fee payable. However, such reserves are usually well provided with landscaped footpaths and hides, making it a little easier to observe the birds at closer quarters if you lack a telescope. If you are planning a trip into the mountains, bring warm clothing and strong footwear even during the summer, as the weather at altitude can be very changeable. In addition to a city map, you will also need one covering the whole of Niederösterreich State and perhaps also a detailed map of the Neusiedl area, which you can obtain at the Information Centre at Illmitz.

Locations

1 Lainzer Tiergarten (10km)   (Hütteldorf U-Bahn Station U4)

Like all former imperial cities, Vienna's rulers ensured that large tracts of forest near the city were preserved for hunting. The Lainzer Tiergarten is a forest park which is part of the well-known Vienna Woods (Wienerwald) which extend across the western suburbs of the city. Happily for the deer and wild boar, this park is now a nature reserve and it supports a rich birdlife. It is, however, very popular with tourists and members of the public, so a visit is best confined to weekdays and early mornings. Nevertheless it offers the city-bound birder the opportunity to sample the full range of woodland species which are typical of the lowland deciduous and mixed forests of central Europe. Although there are a few woodland streams and ponds the park lacks a substantial waterbody, thus limiting the overall range of species.

There are several entrances: the main entrance, which has a car park nearby, is the Lainzer Gate. If using public transport, the Nikolai Gate at the northeast corner is nearest to the U-Bahn station. In all cases, concentrate on the western side of the park which is least disturbed and has the largest stretches of continuous woodland.

The Schönbrunner Schlosspark is another city park with good woodland habitat and is closer to the city centre. However it is smaller than Lainzer Tiergarten and isolated from the larger forests of the Vienna Woods, so its species diversity is poorer, as is that of the other city parks.

KEY SPECIES

Resident
Northern Goshawk, Eurasian Sparrowhawk, Stock Dove, Grey-headed Woodpecker, Black Woodpecker, Middle Spotted Woodpecker, Lesser Spotted

Woodpecker, Mistle Thrush, Long-tailed Tit, Hawfinch

Summer
European Honey-buzzard, Turtle Dove, Wryneck, Common Cuckoo, Tree Pipit, Common Nightingale,

Icterine Warbler, Firecrest, Collared Flycatcher, Red-breasted Flycatcher, Golden Oriole, Red-backed Shrike

Winter
Woodcock, Fieldfare, Redwing, Brambling, Bullfinch

2 Donau-Auen National Park (30km)

(Bus No. 391)
(Maria Ellend S-Bahn Station S7)
(Haslau S-Bahn Station S7)

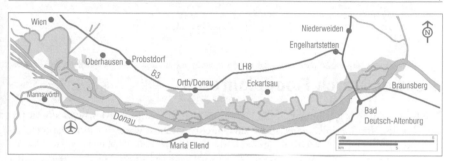

The floodplains of the River Danube between Vienna and Bratislava in Slovakia contain one of the longest stretches of uninterrupted riparian woodland and water meadows in Europe. Almost 40km of riverbank on either side has been preserved as a linear park extending from the eastern suburbs of Vienna to the Slovak border. There are several habitats within the park, including riparian woodland, broadleaved woodland on the higher ground, lowland wet meadows, oxbows and reedbeds. This is a very rich eco-system for woodland and aquatic breeding birds and is a tremendous natural resource right on Vienna's doorstep. For the birder visiting Europe – and indeed for European birders! – it provides a valuable opportunity to become familiar with such tricky species pairs as Chiffchaff / Willow Warbler, Reed Warbler / Marsh Warbler and Eurasian / Short-toed Treecreeper, all of which occur in sizeable numbers. There is an extensive network of footpaths, mostly on the north bank, and access points to the park at several towns along the riverbanks. It is also adjacent to Vienna Airport, making it convenient for a visiting birder at the point of arrival.

To access the south bank, take the **A4 Autobahn** east from Vienna and exit east at **Exit 19** onto the **E58**. This road runs parallel to the river as far as Slovakia. The first access points to the park that you will encounter are at **Maria Ellend** and at **Haslau**. You can walk along the river trail between the two towns, a distance of about 2km. To access the north bank, take the **B3** west from the north side of Vienna. This road also runs parallel with the river and you can access the extensive network of trails from any of the towns along the way, in particular **Groß-Enzersdorf** near the beginning of the park and **Orth an der Donau** midway along the park. There is an S-Bahn service (S7) running along the south bank, serving the airport as well as the towns on the south-bank. The north bank is served by bus from Kagran U-Bahn Station in Vienna

(No. 391). You can also take a guided boat trip down the river, which provides an opportunity for both birding and photography. The boat departs from Salztorbrücke on the Danube Canal in the city centre (Schottenring U-Bahn Station).

KEY SPECIES

Resident
Northern Goshawk, Common Pheasant, Common Kingfisher, Stock Dove, Black Woodpecker, Middle Spotted Woodpecker, Lesser Spotted Woodpecker, Long-tailed Tit, Eurasian Treecreeper, Hawfinch, Yellowhammer, Reed Bunting

Summer
White Stork, European Honey-buzzard, Black Kite, Water Rail, Little Ringed Plover, Common Sandpiper, Turtle Dove, Wryneck, Common Cuckoo, Sky Lark, Tree Pipit,

Common Nightingale, Common Stonechat, Grasshopper Warbler, Savi's Warbler, River Warbler, Sedge Warbler, Marsh Warbler, Great Reed Warbler, Icterine Warbler, Barred Warbler, Willow Warbler, Collared Flycatcher, Penduline Tit, Golden Oriole, Red-backed Shrike

Winter
Gadwall, Eurasian Wigeon, Eurasian Teal, Red-crested Pochard, Common Goldeneye, Goosander, Smew, White-tailed Eagle, Eurasian Sparrowhawk, Common Gull, Yellow-legged Gull,

Caspian Gull, Water Pipit, Mistle Thrush, Fieldfare, Redwing, Goldcrest, Great Grey Shrike, Brambling, Bullfinch

Passage
Great White Egret, Black Stork, Marsh Harrier, Osprey, Greenshank, Wood Sandpiper, Green Sandpiper, Common Tern, Caspian Tern, Black Tern, Meadow Pipit, Grey Wagtail, Blue-headed Wagtail, Whinchat, Pied Flycatcher

3 Lower March Floodplain (40km) (Marchegg Railway Station)

The River March is a tributary of the Danube which forms the border between Austria and Slovakia. Its floodplain consists of permanent marshes, water meadows and riparian woodland which is regularly inundated. The area around Marchegg is a WWF Reserve and supports a well-known colony of White Storks as well as many other aquatic species. Apart from the Stork colony there are two trails within the reserve. For birders the more interesting is the shorter trail, which runs north along the flood retention embankment, taking in several habitats and permitting views over a wide area. The second trail leads further downstream to a small wetland, Breitensee, and is also worth visiting if time permits. This trail is accessed from the Hungarian Gate (Ungartor) on the east side of Marchegg. This area is also a good spot for raptors and occasionally records rarer species such as White-tailed Eagles and Saker Falcons. Bring plenty of insect-repellent if visiting in summer as the mosquitoes are quite relentless.

Take the **A4** and then **B9** east towards Bratislava, cross the Danube on the bridge between Bad Deutsch-Altenburg and Hainburg and proceed straight northwards on the B49 via **Groissenbrun** to Marchegg. The entrance to the reserve is by the castle on the northern side of the village. If you are taking the train, bear in mind that the station is about 3km south of the village. There is, however, a Postbus service into Marchegg. The trains depart from Vienna – Südbahnhof.

KEY SPECIES

Resident
Great Cormorant, Greylag Goose, White-tailed Eagle, Northern Goshawk, Eurasian Sparrowhawk, Common Pheasant, Stock Dove, Common Kingfisher, Grey-headed Woodpecker, Black Woodpecker, Middle Spotted Woodpecker, Lesser Spotted Woodpecker, Long-tailed Tit, Yellowhammer, Reed Bunting, Corn Bunting

Summer
White Stork, Black Stork, Garganey, European Honey-buzzard, Black Kite, Red Kite, Marsh Harrier, Common Quail, Water Rail, Spotted Crake, Corn Crake, Northern Lapwing, Little Ringed Plover, Turtle Dove, Wryneck, Common Cuckoo, Sky Lark, Tree Pipit, Blue-headed Wagtail, Common Nightingale, Grasshopper Warbler, Savi's Warbler, River Warbler, Sedge

Warbler, Marsh Warbler, Great Reed Warbler, Icterine Warbler, Collared Flycatcher, Penduline Tit, Golden Oriole, Red-backed Shrike, Common Raven

Winter
Gadwall, Eurasian Wigeon, Eurasian Teal, Goosander, Smew, Common Gull, Water Pipit, Bearded Tit, Great Grey Shrike

Passage	Redshank, Spotted Redshank,	Common Snipe, Black Tern, Meadow
Great White Egret, Purple Heron,	Greenshank, Wood Sandpiper, Green	Pipit, Whinchat, Fieldfare, Redwing
Night Heron, Osprey, Ruff, Common	Sandpiper, Common Sandpiper,	Pied Flycatcher

Further Away

4 Hohenau (75km)

(Hohenau Railway Station)

This is another site along the River March at the border with Slovakia, which has recently become one of Austria's top birding sites. The alluvial forests of the March floodplain contain very old stands of oaks which hold a variety of woodland birds, especially woodpeckers. These forests are also a nesting site for Black Storks and several species of birds of prey, which can be seen hunting on the adjacent meadows and over arable land. The main attraction is several settling ponds of a sugar beet factory (closing down in 2006) on the edge of the forest. These ponds are partly managed for conservation and hold a variety of waterbirds, and are only matched in importance by the much more famous Neusiedlersee area. An observation tower at the largest pond and a hide at one of the smaller ones offer perfect viewing conditions. There is also a ringing station which is open to the public. Guided birding tours are carried out by Auring, the local birding and conservation organisation. When birding around Hohenau it is also worth visiting the alluvial forests and adjacent open landscapes around the next villages: Rabensburg and Bernhardsthal or nearby sites in Slovakia (Adamov) and the Czech Republic.

From Vienna take the **B8** to **Gänserndorf** and on to **Angern**. From there, proceed northwards on the **B49** until you reach Hohenau. The large pond is directly next to the road leading towards the Slovakian border crossing. The smaller ponds can be reached from there on a dirt road.

KEY SPECIES

Resident
Great Cormorant, Greylag Goose, Red Kite, White-tailed Eagle, Northern Goshawk, Eurasian Sparrowhawk, Imperial Eagle, Saker Falcon, Grey Partridge, Black-headed Gull, Stock Dove, Common Kingfisher, Grey-headed Woodpecker, Black Woodpecker, Middle Spotted Woodpecker, Lesser Spotted Woodpecker, Great Grey Shrike, Hawfinch, Yellowhammer, Reed Bunting, Corn Bunting

Summer
Black-necked Grebe, Night Heron, White Stork, Black Stork, Garganey, European Honey-buzzard, Black Kite, Marsh Harrier, Common Quail, Water

Rail, Spotted Crake, Corn Crake, Little Ringed Plover, Northern Lapwing, Common Redshank, Common Tern, Turtle Dove, Common Cuckoo, Wryneck, Sky Lark, Sand Martin, Tree Pipit, Blue-headed Wagtail, Common Nightingale, Bluethroat, Common Stonechat, Grasshopper Warbler, Savi's Warbler, River Warbler, Sedge Warbler, Marsh Warbler, Great Reed Warbler, Icterine Warbler, Collared Flycatcher, Penduline Tit, Golden Oriole, Red-backed Shrike

Winter
Great White Egret, Bean Goose, White-fronted Goose, Smew, Hen Harrier, Rough-legged Buzzard, Merlin, Mistle Thrush, Brambling

Passage
Eurasian Spoonbill, Eurasian Wigeon, Gadwall, Eurasian Teal, Northern Shoveler, Red-crested Pochard, Montagu's Harrier, Lesser Spotted Eagle, Osprey, Peregrine Falcon, Little Crake, Ringed Plover, European Golden Plover, Dunlin, Little Stint, Temminck's Stint, Curlew Sandpiper, Ruff, Common Snipe, Black-tailed Godwit, Spotted Redshank, Greenshank, Wood Sandpiper, Green Sandpiper, Common Sandpiper, Black Tern, Meadow Pipit, Water Pipit, Whinchat, Northern Wheatear, Fieldfare, Redwing, Pied Flycatcher

5 Neusiedlersee (50km)

This large, shallow, reed-fringed lake is one of Europe's best wetland sites, supporting diverse populations of herons, wildfowl and waders. Since it is close to several other top-class sites for birds this makes this whole area of eastern Austria, part of the Neusiedlersee – Seewinkel National Park, a very popular location for birding trips.

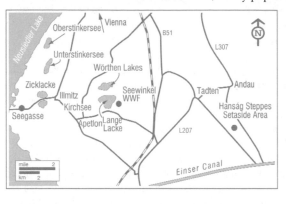

This lake is very shallow and low-lying and, with no outlet other than through evaporation (and an artificial overflow), it acts as a kind of giant 'sump' for the adjoining steppes. It is virtually surrounded by dense reedbeds but there are a number of points on the shoreline where access is easier and you can scan the open waters. As it is so shallow, small fluctuations in water level can inundate small islands or expose mudbanks, making for easier viewing of waders and even crakes. Late spring provides the best variety of species, with both breeding birds and passage migrants present, but winter is also good for wintering wildfowl and autumn turns up many rarities. Whatever time of year you are in Vienna, a visit to this general area is a must. There is an established 'circuit' of the lake and nearby Seewinkel, which the birding trips usually follow, but you could easily spend a week in spring or early summer birding these sites and still not see everything. However, by concentrating on the east shore of the lake, you will probably find the best selection of birds.

Take the **A4 Autobahn** east from Vienna and exit south at the Gols/Weiden Exit (**Exit 51**) towards **Illmitz** at the southeast end of the lake. At Illmitz, where the Park Information Centre (bird checklist available) is located, there is a road (Seegasse) down to the lakeshore, with shallow saline ponds (Kirchsee and Zicklacke) on either side. When crossing the reedbed to the Illmitz bathing area, the large open pools within the reeds on both sides of the dam road are worth checking. This road and the lagoons are among the best locations in the reserve. The wooded areas around Illmitz hold Syrian Woodpeckers and the Seegasse road has substantial areas of roadside scrub which gives way to reedbeds: good for warblers and Penduline Tits. Between the main road from Illmitz back to **Podersdorf** (**L205**) and the lakeshore there is a series of small shallow saline ponds. All are worth checking depending on water levels, wind direction and general disposition of the birds. The ponds, reedbeds and the lakeshore at Illmitz are interlinked with footpaths and there are observation towers at each pond.

KEY SPECIES

Resident
Great Cormorant, Great White Egret, Greylag Goose, Black-headed Gull, Syrian Woodpecker, Bearded Tit, Reed Bunting, Corn Bunting

Summer
Great Crested Grebe, Purple Heron,

Little Bittern, Eurasian Bittern, Night Heron, White Stork, Eurasian Spoonbill, Garganey, Northern Shoveler, Red-crested Pochard, Common Pochard, Ferruginous Duck, Marsh Harrier, Water Rail, Little Crake, Black-winged Stilt, Avocet, Little

Ringed Plover, Kentish Plover, Northern Lapwing, Black-tailed Godwit, Common Redshank, Common Tern, Turtle Dove, Common Cuckoo, European Bee-eater, Hoopoe, Sky Lark, Blue-headed Wagtail, Common Nightingale, Bluethroat,

Common Stonechat, Grasshopper Warbler, Savi's Warbler, River Warbler, Sedge Warbler, Moustached Warbler, Marsh Warbler, Great Reed Warbler, Barred Warbler, Willow Warbler, Penduline Tit, Golden Oriole, Red-backed Shrike

Winter
White-tailed Eagle

Passage
Eurasian Wigeon, Gadwall, Eurasian Teal, European Honey-buzzard, Ringed Plover, Dunlin, Little Stint, Temminck's Stint, Curlew Sandpiper,

Ruff, Spotted Redshank, Greenshank, Wood Sandpiper, Green Sandpiper, Common Sandpiper, Common Snipe, Little Gull, Black Tern, Common Kingfisher, Wryneck, Sand Martin, Tree Pipit, Meadow Pipit, Water Pipit, Fieldfare, Redwing, Icterine Warbler, Pied Flycatcher

6 Seewinkel Nature Reserve (70km)   *(Neusiedler S-Bahn Station S60/ Bus 871)*

Seewinkel is a vast wetland area just east of Neusiedlersee. There are a large number of shallow saline lakes interspersed with water meadows, marshes and small woodland copses. The wet meadows are important breeding zones for waders and other species that depend on this specialised and increasingly threatened habitat. The lakes change in size according to fluctuations in the water table and create feeding zones for migrating waders. Winter floods create habitat for wildfowl concentrations including thousands of geese. With about 45 lakes in total, it would be impossible to cover all in the course of a short visit, but the best developed for birders is the Lange Lacke, the largest lake, with a footpath around three sides, several observation towers and the WWF Education Centre nearby at Seewinkelhof. The paths around the lake link up with the Wörthen Lakes further north and lead through some varied habitats including reedbeds, pasture and some cultivation.

From **Illmitz** go west on the **L205**, passing through **Apetlon** and after about 6km in total, the entrance to the reserve is on the left-hand (north) side of the road.

KEY SPECIES

Resident
Great White Egret, Greylag Goose, White-tailed Eagle, Grey Partridge, Black-headed Gull, Long-eared Owl, Crested Lark, Bearded Tit, Rook, Reed Bunting, Corn Bunting

Summer
Black-necked Grebe, Little Egret, Night Heron, Little Bittern, Eurasian Bittern, Eurasian Spoonbill, White Stork, Common Shelduck, Garganey, Northern Shoveler, Red-crested Pochard, Marsh Harrier, Montagu's Harrier, Common Quail, Water Rail, Spotted Crake, Black-winged Stilt, Avocet, Little Ringed Plover, Kentish

Plover, Northern Lapwing, Eurasian Curlew, Black-tailed Godwit, Common Redshank, Mediterranean Gull, Common Tern, Turtle Dove, Short-eared Owl, Common Cuckoo, Sky Lark, Tree Pipit, Blue-headed Wagtail, Common Nightingale, Common Stonechat, Grasshopper Warbler, Savi's Warbler, Sedge Warbler, Marsh Warbler, Great Reed Warbler, Icterine Warbler, Barred Warbler, Penduline Tit, Golden Oriole, Red-backed Shrike

Winter
Bean Goose, White-fronted Goose, Hen Harrier, Eurasian Sparrowhawk, Merlin, Fieldfare, Great Grey Shrike

Passage
Eurasian Wigeon, Gadwall, Eurasian Teal, Pintail, Osprey, Red-footed Falcon, Saker Falcon, Peregrine Falcon, Common Crane, Ringed Plover, European Golden Plover, Grey Plover, Dunlin, Little Stint, Temminck's Stint, Curlew Sandpiper, Ruff, Spotted Redshank, Greenshank, Marsh Sandpiper, Wood Sandpiper, Common Sandpiper, Common Snipe, Little Gull, Lesser Black-backed Gull, Caspian Tern, Whiskered Tern, Black Tern, White-winged Black Tern, European Bee-eater, Sand Martin, Meadow Pipit, Whinchat, Northern Wheatear, Redwing

7 Hanság Steppes (75km)   *(Neusiedler S-Bahn Station S60/Bus 572)*

From Neusiedlersee east to the Hungarian border, the terrain takes on an aspect of the steppes further east. The area was originally a much wetter habitat which was drained for agriculture and it is now quite intensively cultivated. However a large portion has been set aside for preservation of the steppe flora and fauna and the area is now home to a small number of Great Bustards as well as other steppe species. The set-aside area is bounded by two roads running from Tadten and Andau to the Hungarian border. There are observation towers on both roads to help birders to scan the surrounding fields. The best time to look for bustards is in spring, when the males are displaying and the vegetation has not grown high enough to obscure the birds. In winter this area is frequented by several raptor species of open country.

From **Illmitz** go west on the **L205** to **Wallern**, and turn north on the **L207** towards Tadten and Andau. It is best to go as far as Andau and then turn south towards Hungary, cruising the eastern flank of the set-aside area. Drive as far as the border, where you can cross a bridge over the **Einser Canal**. The canal banks support several species of warbler, and the bridge is a good point to scan the skies for storks and raptors, especially above the treetops of the forest on the Hungarian side. You can then return to Tadten along the road running west of the set-aside area. The whole circuit is a drive of about 18km.

KEY SPECIES

Resident
Great Bustard, Grey Partridge, Yellowhammer, Reed Bunting, Corn Bunting

Summer
White Stork, Marsh Harrier, Montagu's Harrier, Common Quail, Northern Lapwing, Eurasian Curlew, Turtle Dove, Short-eared Owl, Common Cuckoo, Sky Lark, Tree Pipit, Blue-headed Wagtail, Common

Nightingale, Common Stonechat, Whinchat, Grasshopper Warbler, River Warbler, Sedge Warbler, Marsh Warbler, Great Reed Warbler, Icterine Warbler, Barred Warbler, Golden Oriole, Red-backed Shrike

Winter
Greylag Goose, Bean Goose, White-fronted Goose, White-tailed Eagle, Rough-legged Buzzard, Hen Harrier, Merlin, Peregrine Falcon, Fieldfare,

Redwing, Great Grey Shrike, Brambling

Passage
Black Stork, European Honey-buzzard, Eurasian Sparrowhawk, Imperial Eagle, Red-footed Falcon, Saker Falcon, Meadow Pipit, Water Pipit, Northern Wheatear, Pied Flycatcher

8 Schneeberg Mountain (85km)   (Schneeberg Mtn Railway Station)

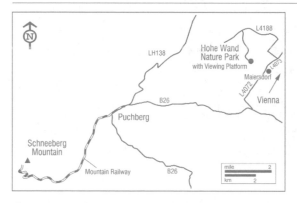

At over 2,000m high, Schneeberg is the closest mountain to Vienna with sufficient altitude to support alpine species. Although it is rather distant from the city centre, it is relatively easy to get to even if you don't have a car, as there is a train service to Puchberg at the foot of the mountain and from there a mountain railway will bring you up through the spruce forests to the summit. Also bear in mind that the best time to see the alpine birds is in May and June, when days are longest, so you can accomplish quite a lot in the course of a day.

Take the mountain railway from Puchberg to the summit and from here there are several trails across the mountain ridge. Keep your binoculars handy even in the train as it goes so slowly in places that it is perfectly feasible to watch birds from the train. Although most of the alpine species are possible in this area, you are unlikely to 'clean-up' as the birds are thinly distributed. If you do not see a Wallcreeper, a better place to look is at the Hohe Wand on the way back to Vienna although, even here, they are reportedly no longer seen, but you can still look for Rock Buntings. (Wallcreepers are now best seen in winter at various quarries south of Vienna, notably Gumpoldskirchen and Bad Vöslau.)

Take the **A2 Autobahn** south from Vienna and exit west onto the B26 at **Wiener Neustadt (Exit 44)**. Stay on the B26 all the way to Puchberg, a distance of about 23 km. From here, you take the mountain railway to Schneeberg. If you wish to visit the Hohe Wand cliff-face, take the B26 back towards Vienna, but after about 10km, turn north at **Unterhöflein** onto the **L4072**. Hohe Wand is located about 4km along this road on the left-hand (west) side. Trains to Puchberg depart from Vienna – Südbahnhof.

KEY SPECIES

Resident
Golden Eagle, Peregrine Falcon, Ptarmigan, Black Woodpecker, Grey Wagtail, Alpine Accentor, Dipper, Crested Tit, Willow Tit, Wallcreeper,

Alpine Chough, Common Raven, Nutcracker, Snow Finch, Common Redpoll, Common Crossbill, Rock Bunting

Summer
Water Pipit, Ring Ouzel, Western Bonelli's Warbler, Firecrest

Useful contacts and websites

BirdLife Austria (English and German)
www.birdlife.at
BirdLife Austria – Austrian Ornithological Society, Museumsplatz 1/10/8, A-1070 Wien.
Tel (+43) 1 523 4651. Fax. (+43) 1 523 46 51 50.
BirdLife partner in Austria and publishers of the journals *Vogelschutz in Österreich* and *Egretta*.

Birding in Austria (English and German)
www.bird.at
Birding website with information about birding trips, recent reports and SMS messaging service.

WWF Austria (English and German)
www.wwf.at
National branch of WWF with information about WWF Reserves near Vienna.

Club 300 Austria (English and German)
www.club300.at.tf
Austrian twitchers' website.

Auring (English and German)
www.auring.at
Local birding organisation located in the River March floodplain area. Information on recent observations, excursions, opening times of the ringing station etc.

Robb's Website (English and German)
http://robbsnet.com/robb.htm
Robbin Knapp's personal website with lots of useful information about birding in Austria.

Neusiedlersee National Park (German)
www.nationalpark-neusiedlersee.org
National park website with site descriptions, maps, access and bird checklist.

Donau-Auen National Park
(English and German) www.donauauen.at
Danube Floodplain National Park website with site description, maps, access and bird checklist.

Austrian Rarities Committee (English and German) www.birdlife.at/s2705.htm
AFK – Avifaunistische Kommission, c/o BirdLife Austria, Museumsplatz 1/10/8 A-1070 Wien.

Wiener Linien (English, German and Italian)
www.wienerlinien.co.at
Public transport for Vienna and district.

Verhehrsverbund Ost-Region – VOR (German)
www.vor.a
Web portal site for public transport in eastern Austria.

Post Bus (German) www.post.at
Regional and feeder bus service for provincial areas.

ÖBB (German and English)
www.oebb.at
Austrian railways.

Schneebergbahn (German and English)
www.schneebergbahn.at
Schneeberg Mountain Railway – The Salamander!

Books and publications

Atlas der Brutvögel Österreichs 1981–1985
M.Dvorak, A.Ranner, H.Berg (1993)
Österreichische Gesellschaft für Vogelkunde.

Finding Birds in Eastern Austria
Dave Gosney (1994) BirdGuides, ISBN 1898110107

Important Bird Areas in Österreich
Michael Dvorak and Eva Karner (1995)
BirdLife Austria

Collins Bird Guide
Lars Svensson *et al.* (1999) HarperCollins
ISBN 0002197286

Birds of Europe
Lars Jonsson (2006) Christopher Helm
ISBN 0713676000

Penduline Tit favours willows in riverine habitats in summer

Warsaw

Eastern Poland has become an increasingly popular destination for those in search of the eastern European forest and wetland species. Many birders will pass through Warsaw without realising that some of these species can be seen quite easily within or close to the city itself! The main physical feature of Warsaw, dividing the city east and west, is the River Vistula, whose watershed drains almost 50 per cent of Poland. This enormous volume of water periodically inundates large areas of the floodplain and creates the characteristic water meadows which are so important to sustain a unique flora and fauna. Substantial tracts of land on either side of the river are preserved as buffer zones, to absorb the seasonal floods, and these areas provide wonderful wildlife corridors right into the city.

The Kampinoski National Park to the west of the city is a vast forested area, unique in any European capital, which provides habitats for a number of the forest species which many birders travel great distances to see. There are also several other forested parks closer to the city centre which support most of the typical woodland species of eastern Europe.

Birding in Warsaw is very seasonal. The spring thaw produces a surge of melt water along the Vistula and its tributaries, creating flooded meadows along their banks which are important resting sites for migrating wildfowl. By late spring and early summer the forests and wetlands are alive with singing and displaying birds and this is the best time to visit Poland. Late summer, however, will be much quieter and the mosquitoes will have reached plague proportions. In autumn there is a build-up of migrating waterfowl and waders along the rivers, and numbers remain high until late winter, when areas of open water freeze-over. Winter is definitely the quietest season and the most uncomfortable time to be out birding due to the bitter cold, but you can still see an interesting selection of wildfowl and gulls wherever a patch of open water remains ice-free.

GETTING AROUND

Warsaw is a rather expansive city, so expect to cover quite a bit of distance if you hope to visit a number of sites. With a few exceptions most of the birding sites are located out of town and even those within the city limits will involve quite a bit of walking. Although rich in birdlife, the birds are distributed over large areas of habitat with no significant concentrations at convenient

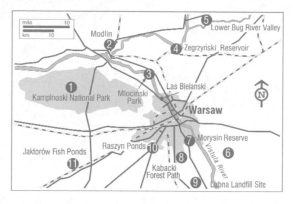

locations! There are no bird reserves as such, and no bird hides, so you have to tramp to find the birds rather than wait in one place for them to assemble in front of you.

The bus and tram network covers most of the city and all the sites within the city limits, and this is backed up by a metro system which serves the southern suburbs, including Kabacki Forest. The suburban train service covers some of the outlying sites, but expect the trains to be crowded at peak periods, and you will still need to walk a few kilometres when you get there. Most trains depart from Warsaw Central Station (Warszawa Centralna), but always check your travel plans as some depart from the other termini: Gdansk Station (Warszawa Gdanska), Warsaw West (Warszawa Zachodnia) or Warsaw East (Warszawa Wschodnia). Warsaw West is also the terminal for the (PKS) provincial bus service.

Although there are good cycling trails in the Kampinoski National Park and the Morysin Reserve, cycling on the open road is not recommended. Driving in Warsaw is reputed to be quite hazardous but is essential to visit some of the outlying sites and those that are very extensive. Traffic within the city can be very heavy but getting across town is made easy by two expressways running through the centre on either side of the Vistula. Make sure you are certain of the route to your destination, as once you have cleared the city centre there are no more bridges over the river for many kilometres upstream or downstream.

Binoculars are perfectly adequate for birding in the spring and summer, but a telescope would be desirable for late autumn viewing of wildfowl on the flat open floodplains. If visiting any of the wetland sites, you would be advised to invest in a pair of rubber boots. Even after the floods recede the ground remains extremely soft and wet and, unless you are very careful, you can easily sink up to your knees. Make sure that you obtain a map that covers the entire city and as far north as the Narew River and as far west as Kampinoski National Park.

Locations

1 Kampinoski National Park (7km)   (ZTM Bus Nos. 708/726)

Warsaw is unique among European capital cities in having a large national park on its doorstep. This is a huge area of Scots pine, alder and birch forests, surrounding low-lying bogs and wet meadows, which presents a daunting challenge to the visiting birder.

You could easily spend a week in this park and still only cover a fraction of the area, but it is a rich habitat for some of eastern Europe's most sought-after forest birds. As it is close to the city centre it is worth a visit at any time but particularly in spring and early summer. It lacks a significant stretch of water to attract waterfowl but the peat bogs and marshy areas flood in winter and spring, attracting migrant aquatic birds as well as providing breeding habitats for warblers and Corn Crakes. There are also other good wetland sites nearby on the banks of the Vistula River. As this is such a large area it is only possible to cover a small portion in a single visit and, in general, the best areas are the wet areas alongside any waterways, which support the old-growth trees. The drier sandy areas have been planted with a monoculture of pinewoods and are less interesting for birds.

If you have a car, a suggested route is to take **Route 7** northwest from the city to **Dziekanów Leśny** and enter the northeast corner of the park at **Sadowne**. The route southwest through **Sieraków** and **Truskaw** passes through some good habitat. Continue westwards to the village of **Roztoka** in the centre of the park, adjacent to which some of the best areas of old growth woodlands and wet meadows are located. You can then continue in a general southwest direction towards the southern perimeter, where you can exit the park at **Kampinos**: overall a distance of about 30km through the most productive areas of the park.

For those without a car, there is a network of walking and bicycle trails which traverse the park and provide a manageable way to access the forest habitat. There are bus services which serve some of the points between Kampinos and Sieraków. You can get the PKS provincial bus from the main bus station, PKS Zachodnia: this serves the southern perimeter of the park. Alternatively, you can get the ZTM Bus No. 708, which operates as far as Truskaw, or Bus No. 726 which serves Sieraków. Both operate from Pl. Wilsona in the northwest of the city centre.

KEY SPECIES

Resident
White-tailed Eagle, Northern Goshawk, Eurasian Sparrowhawk, Peregrine Falcon, Stock Dove, Grey-headed Woodpecker, Black Woodpecker, Middle Spotted Woodpecker, Lesser Spotted Woodpecker, Syrian Woodpecker,

Eurasian Treecreeper, Crested Tit, Willow Tit, Long-tailed Tit, Common Raven

Summer
Black Stork, European Honey-buzzard, Hobby, Corn Crake, Woodcock, Turtle Dove, Wryneck, Common Cuckoo, Tree Pipit, Grey

Wagtail, Blue-headed Wagtail, Common Nightingale, Thrush Nightingale, Mistle Thrush, Icterine Warbler, Barred Warbler, Wood Warbler, Collared Flycatcher, Red-breasted Flycatcher, Golden Oriole, Red-backed Shrike

2 Modlin Water Meadows (36km)

(Modlin Railway Station)

The seasonal flooding of the low-lying pastures along the banks of the rivers in central Poland keeps the vegetation low and has prevented intensive cultivation of the land. As water levels recede in late spring and summer, the short grass meadows provide optimum breeding habitat for waders and other ground-nesting species such as Corn Crakes, larks and pipits. Unfortunately this habitat is under threat throughout Europe and has almost disappeared in western Europe due to drainage and agricultural intensification. Some of the best examples of this habitat can be found along the north bank of the River Narew near Modlin. The area is bounded by the River Wkra, a tributary of the Narew, to the east and Route 630 to the west, and contains a mosaic of habitats including water meadows, marshes, reedbeds and small lakes.

From **Route 61**, take **Route 630** northwest to **Nowy Dwór Mazowiecki**. Cross the River Narew and turn east on **Route 623** which brings you to the village of **Stary Modlin**. There is a lake and large reedbed south of the village. Return to Route 623 and continue on towards **Pomiechówek**, where you can turn right (south) on a road just before the bridge over the Wkra, bringing you into an open area of meadows and flooded pastures. Stary Modlin is served by train from Warszawa Gdanska Station. However, it is best to cover this area by car, as apart from the large expanses to be covered, there are no bird hides or observation towers, making it difficult to approach the birds without alarming them.

KEY SPECIES

Resident	Crane, Water Rail, Spotted Crake,	Short-eared Owl
Greylag Goose, White-tailed Eagle,	Northern Lapwing, Snipe, Common	
Northern Goshawk, Common	Redshank, Black-tailed Godwit,	**Passage**
Buzzard, Gadwall, Goldeneye, Great	Common Tern, Hoopoe, Sky Lark,	Black Stork, Bean Goose, White-
Grey Shrike	Meadow Pipit, Blue-headed Wagtail,	fronted Goose, Eurasian Wigeon,
	Bluethroat	Northern Pintail, Golden Plover, Ruff,
Summer		Common Sandpiper, Common Snipe,
Little Bittern, Eurasian Bittern, White	**Winter**	Little Gull, Lesser Black-backed Gull,
Stork, Garganey, Northern Shoveler,	Whooper Swan, Bewick's Swan,	Black Tern
Marsh Harrier, Hobby, Common	Rough-legged Buzzard, Hen Harrier,	

3 Mlociński Park (10km)   (Tram Nos. 6/15/17/27)

The sewage outfall at Mlociny is one of the best locations on the Vistula River for gulls and in winter it provides an ice-free area of open water for wildfowl. The riparian woods along this stretch of the river also support an interesting selection of passerines and are favoured hunting grounds for raptors. The area is best accessed from the Mlociński riverside park, where there are a number of paths leading to the riverbank and south to the sewage outfall.

Take **Route 7 (Wybrzeze Gdynskie aka Wislostrada)** northwest from the city centre and take the first turn right (east) onto **Papirusov**, just after Mlociny. This road brings you to the entrance to Mlociński Park. If you are returning to the city centre, you could also check Las Bielański Park, which is just south of Mlociny and also has some interesting littoral habitat along the banks of the Vistula.

KEY SPECIES

Resident	Summer	Winter
Northern Goshawk, Gadwall,	Lesser Black-backed Gull, Common	Smew, White-tailed Eagle, Yellow-
Common Goldeneye, Middle	Tern, Little Tern, Common Kingfisher,	legged Gull, Caspian Gull, Common
Spotted Woodpecker, Lesser	Blue-headed Wagtail, Bluethroat,	Redpoll, Siskin, Brambling
Spotted Woodpecker	Marsh Warbler, Penduline Tit	

4 **Zegrzyński Reservoir** (25km)   *(ZTM Bus No. 705)*

The Zegrzyński Reservoir north of Warsaw is one of the best sites for winter birding, as it rarely freezes over completely since it is such a large body of water and because of the turbulence created by the confluence of the Rivers Bug and Narev. Consequently it draws in large numbers of wildfowl and gulls as other smaller waterbodies freeze over. In addition the reservoir is surrounded by woods, reedbeds and low-lying fields, which are prone to flooding in autumn, winter and early spring and attract geese and swans. However, this is a very large area and requires a telescope as the birds can be quite distant.

Take **Route 61** north from Warsaw and after about 25km the road crosses the reservoir at **Zegrze** bridge. Continue past the bridge for about 8km and turn east on **Route 624** for about 1km, crossing the River Narev and then turning south towards **Nowa Kania** and **Popowo Kościelne**. This brings you to the northern bank of the Bug River, which has the best wetland habitats, with wooded islands, creeks and reedbeds. Return to Route 61 and cross the bridge over the reservoir to the south shore. Turn east on Route 631 towards **Nieporet** and follow this road which tracks the south shore. The stretch of shoreline between the bridge and Nieporet is often the most productive.

It would be impossible to visit all these areas if you are relying on public transport but it is possible to cover the shore and flooded fields between Zegrze Bridge and Bialobrzegi from Nieporet, which is served by Bus No. 705 from the tram terminus at Żerań and by PKS provincial buses from Warszawa Zachodnia Bus Station.

KEY SPECIES

Resident
Greylag Goose, White-tailed Eagle, Northern Goshawk, Common Buzzard, Gadwall, Common Goldeneye, Greater Scaup, Goosander, Common Redpoll

Summer
Great Bittern, Garganey, Black Kite, Red Kite, Marsh Harrier, Hobby, Common Crane, Spotted Crake,

Common Tern, Common Kingfisher, Turtle Dove, Common Cuckoo, Hoopoe, Sky Lark, Meadow Pipit, Blue-headed Wagtail, Bluethroat, Bearded Tit, Penduline Tit

Winter
Black-throated Diver, Red-throated Diver, Whooper Swan, Bewick's Swan, Long-tailed Duck, Common Scoter, Velvet Scoter, Red-breasted Merganser,

Smew, Hen Harrier, Common Gull, Yellow-legged Gull, Caspian Gull, Shore Lark, Snow Bunting

Passage
Red-necked Grebe, Black-necked Grebe, Northern Pintail, Ruff, Common Redshank, Common Sandpiper, Snipe, Little Gull, Lesser Black-backed Gull, Black Tern, Redwing

5 **Lower Bug River Valley** (50km)   *(Wyszków Railway Station)*

The Bug is a tributary of the Narew and one of the rivers in the Vistula system. The lower reaches of the Bug between Wysków and the Zegrzyński Reservoir are an extensive area of backwaters, marshes and wet meadows which are bordered by large belts of mixed forest. The area is so good it is known locally as the 'Little' Biebrza Marshes. Although it is nowhere near as well known by visiting birders, its proximity to Warsaw makes it a very viable alternative to the 200km drive to Biebrza. The spring floods coincide with a large migration of ducks and waders along the river valleys and there are several sites where you can witness this spectacle. One of the best and most accessible places is at Wysków on the north bank of the river, where a short walk either side of the bridge provides good vantage points to scan the river and its banks. Another good spot is at Popowo Kościelne, which is about 20km west of Wysków, also on the north bank.

Take **Route 18** northeast from Warsaw and drive for about 50km to arrive at the bridge in **Wysków**. To get to **Popowo Kościelne** take Route 624 west from Wysków for about 20km. You can also travel by train and bus. There are trains to Wysków from Warszawa Centralna, although there is a more frequent service for Warszawa Wilenska Station on the east side of the city. The railway station in Wysków is less than ten minutes' walk north of the river bank. You can get a bus from Wysków to Popowo Kościelne: the bus station is alongside the railway station.

KEY SPECIES

Resident
Gadwall, Goosander, White-tailed Eagle, Northern Goshawk, Eurasian Sparrowhawk, Stock Dove, Black Woodpecker, Middle Spotted Woodpecker, Lesser Spotted Woodpecker, Syrian Woodpecker, Willow Tit, Long-tailed Tit, Hawfinch

Summer
Little Bittern, White Stork, Lesser Spotted Eagle, Marsh Harrier, Hobby, Corn Crake, Water Rail, Ringed

Plover, Little Ringed Plover, Common Tern, Little Tern, Yellow-legged Gull, European Roller, Turtle Dove, Wryneck, Common Cuckoo, Wood Lark, Tree Pipit, Thrush Nightingale, River Warbler, Marsh Warbler, Great Reed Warbler, Icterine Warbler, Barred Warbler, Firecrest, Collared Flycatcher, Red-breasted Flycatcher, Penduline Tit, Golden Oriole, Red-backed Shrike, Common Rosefinch, European Serin

Passage
Whooper Swan, Bewick's Swan, White-fronted Goose, Bean Goose, Black Stork, Garganey, Eurasian Wigeon, Pintail, Osprey, Oystercatcher, Northern Lapwing, Common Snipe, Common Redshank, Mediterranean Gull, Little Gull, Lesser Black-backed Gull, Caspian Tern, Black Tern, Redwing, Brambling

6 East Bank of Vistula River (10km)   *(Józefów Railway Station)*

The River Vistula is prone to flooding so large areas on both banks of the river are not built on, in order to act as a buffer zone absorbing the periodic inundations. These areas are excellent wildlife zones and one of the best sites is on the stretch of the river immediately south of Warsaw. Route 801 runs parallel to the east bank of the river and provides a convenient way to access the riverside. There are numerous small islands and gravel banks which are visible from the shore and provide safe havens for nesting gulls, waders and terns. The riparian woods and woodland edges are also excellent habitat for breeding and migrating passerines.

Take **Route 801** southeast from Warsaw for about 20km to **Zkrzypki**. At this point, the road runs closest to the river bank which is at a considerable height above the river level, providing a good observation point to scan the islands and sandbars. Continue south for about 6km to the ferry jetty at **Karczew**, which is a good vantage point to check upstream and downstream. It is less easy to cover this area on foot. The railway station at Józefów, served from Warszawa Wschodnia, is about 2.5km east of Route 801.

KEY SPECIES

Resident
Gadwall, Goosander, White-tailed Eagle, Northern Goshawk, Eurasian Sparrowhawk, Stock Dove, Black Woodpecker, Middle Spotted Woodpecker, Lesser Spotted Woodpecker, Syrian Woodpecker, Willow Tit, Long-tailed Tit, Hawfinch

Summer
Little Bittern, White Stork, Marsh Harrier, Hobby, Corn Crake, Water Rail, Ringed Plover, Little Ringed Plover, Common Tern, Little Tern, Yellow-legged Gull, Turtle Dove, Wryneck, Common Cuckoo, Tree Pipit, Common Nightingale, Thrush Nightingale, River Warbler, Marsh Warbler, Great Reed Warbler, Icterine Warbler, Barred Warbler, Lesser Whitethroat, Firecrest, Collared Flycatcher, Red-breasted Flycatcher, Penduline Tit, Golden Oriole, Red-backed Shrike, Common Rosefinch, European Serin

Passage
Black Stork, Garganey, Eurasian Wigeon, Pintail, Osprey, Oystercatcher, Northern Lapwing, Common Snipe, Common Redshank, Mediterranean Gull, Little Gull, Lesser Black-backed Gull, Caspian Tern, Black Tern, Redwing, Brambling

7 Morysin Reserve (13km)

(ZTM Bus 116/130/519/522)

The Morysin Reserve is in the south of Warsaw and set amidst a splendid area of parkland, lakes and riparian woodland extending for about 10km along the west bank of the Vistula. The Wilanówka and Jeziorka Rivers run through this area, fringed by small reedbeds and water meadows. There is abundant birdlife during late spring and early summer and a visit then will yield a wide selection of eastern European specialities. Several species of *Locustella* and *Acrocephalus* warblers occur here as well as both Common Nightingale and Thrush Nightingale, giving an excellent opportunity to distinguish between members of these challenging species-groups. There are many footpaths and trails through the park: the best area for birds is the path alongside the Wilanówka River.

The reserve is just north and east of **Wilanów Palace**, a popular tourist attraction. Take **Route 724** south from the city centre to Wilanów and turn east on **Vogla Street** to get to the palace.

KEY SPECIES

Resident
Northern Goshawk, Eurasian Sparrowhawk, Stock Dove, Black Woodpecker, Middle Spotted Woodpecker, Lesser Spotted Woodpecker, Syrian Woodpecker, Willow Tit, Long-tailed Tit, Hawfinch

Summer
Little Bittern, Marsh Harrier, Corn Crake, Water Rail, Turtle Dove, Common Kingfisher, Wryneck, Common Cuckoo, Sky Lark, Meadow Pipit, Tree Pipit, Common Nightingale, Thrush Nightingale, River Warbler, Marsh Warbler, Savi's Warbler, Great Reed Warbler, Icterine Warbler, Barred Warbler, Common Whitethroat, Firecrest, Collared Flycatcher, Red-breasted Flycatcher, Penduline Tit, Golden Oriole, Red-backed Shrike

8 Kabacki Forest (15km)

(Kabaty Metro Station)

If you are staying in the south of Warsaw and your schedule does not permit the time to visit Kampinoski National Park, then a visit to Kabacki Forest Park would be worth considering. This is about the best of Warsaw's 'small' parks, of which there are many, and it is a manageable place to explore on foot if you only have a few hours. It benefits from being a remnant of the much larger Mazowiecka Forest, which lies further south and east of Warsaw, and from being a mixed woodland with sufficient mature broadleaved trees to support a rich diversity of birdlife. The southern side is best, particularly the large meadow to the south of the Botanic Garden and the woodland edges along the perimeter of the meadow. It is popular with walkers and picnickers, particularly at weekends, so try to visit early in the morning or on weekdays.

KEY SPECIES

Resident
Northern Goshawk, Stock Dove, Black Woodpecker, Middle Spotted Woodpecker, Lesser Spotted Woodpecker, Syrian Woodpecker, Eurasian Treecreeper, Willow Tit,

Long-tailed Tit, Hawfinch

Summer
Hobby, Turtle Dove, Wryneck, Common Cuckoo, Tree Pipit, Grey Wagtail, Common Nightingale, Thrush

Nightingale, Mistle Thrush, Fieldfare, Icterine Warbler, Barred Warbler, Firecrest, Collared Flycatcher, Red-breasted Flycatcher, Golden Oriole, Red-backed Shrike, Common Rosefinch, Ortolan Bunting

9 Lubna Landfill Site (25km)

If you have the endurance to withstand the cold and the smell of a rubbish tip in winter, you can be rewarded with a fascinating selection of large gull species of every age and the potential of turning up a rarity. One of the best gull-watching sites in Poland, in winter, is the Lubna Landfill Site, southeast of the city. Access is controlled but a polite request to the staff will usually permit entry, and it is possible to watch from the perimeter. Up to 10,000 gulls can be seen including several maritime species.

Take **Route 723** southeast towards **Góra Kalwaria** and after about 25km, turn left (east) at the traffic lights after passing through the village of **Baniocha**. The large earth embankment is visible on the left-hand side of the road. There is a regular PKS bus service which operates from the Warszawa Zachodnia to Baniocha.

KEY SPECIES

Winter
Mediterranean Gull, Yellow-legged Gull, Caspian Gull, Lesser Black-backed Gull, Great Black-backed Gull, Glaucous Gull

10 Raszyn Ponds (13km)   (ZTM Bus No. 706)

The fish ponds at Raszyn are very close to Warsaw Airport in southwest Warsaw and would be well worth a look after picking up or returning a rental car. The ponds are dry and rather uninteresting during the winter but between April and October they support a fine selection of migrant and summer visitors. As with all such sites, water levels fluctuate considerably making some ponds more attractive than others at different times, and resulting in irregular breeding patterns for waterfowl. Wherever water levels are low, the exposed mud will attract small numbers of migrating waders. The ponds themselves have restricted access but you can easily scan the area from the approach roads and some of the roads which run between the ponds.

The ponds are located on either side of **Route 7** just south of Raszyn. Stopping on this road is inadvisable but there are several turn-offs on either side of the motorway from which you can view the ponds. This area is about 3km from the tram terminus at Okęcie, but there is a feeder bus service.

KEY SPECIES

Summer
Red-necked Grebe, Black-necked Grebe, Little Bittern, Eurasian Bittern, White Stork, Greylag Goose, Garganey, Northern Shoveler, Gadwall, Marsh Harrier, Hobby, Water Rail, Spotted Crake, Black-tailed Godwit, Common Tern, Little Tern,

Common Kingfisher, Blue-headed Wagtail, Bluethroat, Thrush Nightingale, Marsh Warbler, Great Reed Warbler, Penduline Tit, Golden Oriole

Passage
Eurasian Wigeon, Northern Pintail,

Rough-legged Buzzard, European Golden Plover, Ruff, Common Redshank, Common Sandpiper, Common Snipe, Little Gull, Yellow-legged Gull, Lesser Black-backed Gull, Black Tern

11 Jaktorów Fish Ponds (35km)   *(Jaktorów Railway Station)*

The fish ponds at Jaktorów are situated amidst some excellent habitat, including woodland and lowland wet meadows. The ponds themselves are reed-fringed for the most part with not much of a margin to attract waders. However, when flooded, the meadows to the north of the fish ponds are excellent for migrating waders. The mosaic of habitats created by the close proximity of woodland and wetlands makes this site one of the birding hotspots in the Warsaw area and it is worth visiting in summer as well as at migration times.

The ponds are just north of the town of Jaktorów. Take **Route 719** southwest from Warsaw and after about 30km you will come to Jaktorów. In the town centre, turn north on Pomorska Street and drive for about a kilometre. The fish ponds are on the left-hand side of the road. You can also take the train to Jaktorów from Warszawa Centralna or the bus from the tram terminus at Okęcie.

KEY SPECIES

Summer
Red-necked Grebe, Black-necked Grebe, Little Grebe, Eurasian Bittern, White Stork, Greylag Goose, Garganey, Northern Shoveler, Gadwall, Marsh Harrier, Hobby, Water Rail, Spotted Crake, Black-tailed Godwit, Common Tern, Common

Kingfisher, Blue-headed Wagtail, Bluethroat, Marsh Warbler, Great Reed Warbler, Penduline Tit, Golden Oriole

Passage
White-fronted Goose, Bean Goose, Eurasian Wigeon, Northern Pintail,

Hen Harrier, European Honey-buzzard, Rough-legged Buzzard, Northern Goshawk, European Golden Plover, Ruff, Common Redshank, Common Sandpiper, Common Snipe, Little Gull, Yellow-legged Gull, Lesser Black-backed Gull, Black Tern, Bearded Tit

Useful contacts and websites

Ogólnopolskie Towarzystwo Ochrony Ptaków (OTOP) (Polish) www.otop.org.pl
Polish Society for Ornithology, 02-026 Warszawa, ul. Raszyńska 32/44 lok.140. Warsaw branch of OTOP, BirdLife partner in Poland.

Komisja Faunistyczna (English and Polish)
www.biol.uni.wroc.pl/komfaun
Polish Rarities Committee, Sienkiewicza 21, Pl 50 – 335 Wroclaw.
Email: stawar@biol.uni.wroc.pl

BirdGuide.pl (English) www.birdguide.pl
Birding tour company website with information about sites and rarer species.

Birding in Poland (English)
www.birding.gt.pl
Jerzy Dyczkowski's personal website with information about birding sites near Warsaw.

KRASKA Bulletin (Polish) www.bocian.org.pl
Warsaw Natural History website with information about bird distribution around Warsaw.

Probirder www.probirder.com
Birding tour company operating in Poland.

Kampinoski National Park (English)
www.staff.amu.edu.pl/~zbzw/ph/pnp/kamp.htm
Park website with information about flora and fauna.

Zarząd Transportu Miejskiego (ZTM) (Polish)
www.ztm.waw.pl
Public transport service for Warsaw and district.

Polskie Koleje Państwowe S.A. (PKP)
(Polish and English) www.pkp.com.pl
Polish State Railways information.
Tel. (+48) 229436 or (+48) 429436.

Books and publications

Finding Birds in Poland
Dave Gosney (1993) Gostours
ISBN 0951792083

Where to Watch Birds in Eastern Europe
Gerard Gorman (1994) Hamlyn, ISBN 060057976X

Avifauna Polski (The Avifauna of Poland: Distribution, Numbers and Trends)
L. Tomialojc & T. Stawarczyk (2003)
PTPP Pro Natura, ISBN 839196261X

Collins Bird Guide
Lars Svensson *et al.* (1999) HarperCollins
ISBN 0002197286

Birds of Europe
Lars Jonsson (2006) Christopher Helm
ISBN 0713676000

TYPICAL SPECIES

Resident
Little Grebe, Great Crested Grebe, Grey Heron, Mute Swan, Mallard, Common Buzzard, Common Kestrel, Moorhen, Common Coot, Black-headed Gull, Wood Pigeon, Collared Dove, Great Spotted Woodpecker, Wren, Dunnock, Robin, Blackbird, Goldcrest, Marsh Tit, Coal Tit, Blue Tit, Great Tit, Eurasian Nuthatch, Short-toed Treecreeper, Eurasian Jay, Magpie, Carrion Crow, Common Starling, House Sparrow, Tree Sparrow, Chaffinch, Greenfinch, Goldfinch

Summer
Black Kite, Common Swift, Sand Martin, Barn Swallow, House Martin, White Wagtail, Black Redstart, Song Thrush, Reed Warbler, Garden Warbler, Blackcap, Common Chiffchaff, Firecrest, Spotted Flycatcher, European Serin

Winter
Great Cormorant, Tufted Duck, Common Pochard, Goosander, Yellow-legged Gull, Western Jackdaw, Siskin

Alpine Swift flying around Grossmünster Cathedral

Zurich

Although Switzerland is associated very strongly with the Alps and the dramatic Alpine scenery, Zurich is situated in the north of the country within the Mittelland Plateau and is actually quite distant from the high peaks. The main features of the area are several large lakes which are important wetlands for birds, although drainage and eutrophication have degraded the habitat somewhat. In a land-locked country such as Switzerland, they are focal points for migrating wildfowl, waders and even passerines, and consequently are well watched by local birders. The city is at the northern end of Lake Zurich. The lake is probably not as good for birds as some of the others nearby because the lakeshore is so built-up in the Zurich area that you need to travel quite a long distance to find suitable habitat. There are, however, other wetland sites on the shores of the neighbouring lakes which have good habitat and support interesting populations of birds. These lakes are significant for both breeding and wintering birds as, although winters can be quite cold, they do not ice-over for long periods.

Zurich is only 25km from the German border and less than 100km from Austria so it is quite feasible to visit either country in the course of a day's birding and there are good sites just over the borders. If you are prepared to travel further, a weekend trip to the Black Forest in southern Germany or south into the Bernese Alps would yield a number of species which are rare or absent from the Zurich area. However a good selection of Continental European species can be found within an hour's drive of the city.

GETTING AROUND

Zurich is a fairly small city but the urban area extends for a considerable distance along the shores of Lake Zurich, so it is necessary to travel some distance out of town to

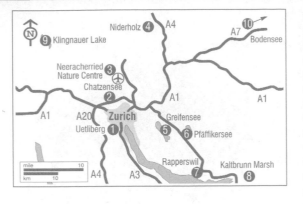

reach good habitats. Unless you have a car, getting to the sites is best achieved by train or S-Bahn, the suburban rail system. The Swiss railway system is, of course, legendary for its comprehensive network, its punctuality and its regularity. The claim of a train every hour to every town pretty much sums it up. Many of the sites are within walking distance of a railway station. Otherwise you can take a bicycle on the train, which is a really practical way to get around since the Zurich area is quite flat. You can even borrow bikes free at the main railway station in Zurich, on payment of a small returnable deposit. Nonetheless, as many sites are lakeside habitats and a telescope is recommended, riding a bike laden down with birding gear may not be your preferred mode of transport. However, almost all railway stations are served by the Post Bus system which will probably bring you close to the site. This is important as taxis in Switzerland are very expensive and not practical for birding purposes. If you are renting a car, make sure that you inform the rental company if you intend to drive into Germany or Austria as there will be a small surcharge to cover the breakdown recovery service.

Many birding sites are nature reserves of some kind, managed either by the local canton or by a conservation organisation such as Ala – Schweizerische Gesellschaft für Vogelkunde und Vogelschutz, the Swiss Bird Protection Society. There are no entrance fees and generally access is unrestricted except where there are breeding bird zones. These reserves are usually well provided with footpaths and hides so you will not need rubber boots. In fact, comfortable walking shoes are adequate for all the locations described. If you do, however, plan a trip into the Alps, then warm clothing and strong footwear are required even during the summer, as the weather at altitude can be very changeable. As most of the good sites are lakeshores, winter visits will be most productive if you have a telescope. Choose a city map which covers the whole of the Zurich Canton and preferably the neighbouring cantons as well.

Locations

1 **Uetliberg** (2km)   *(Uetliberg S-Bahn Station S10)*

The Uetliberg Mountain, almost 900m high, is just about 2km west of the city centre, and is a popular walking and hiking area. The best way to get there is to take the S-Bahn to the summit at Uetliberg and walk down the mountain. There are several hiking trails which you can follow back down through woodland, meadows and parkland, where it is possible to see all the typical woodland species of this area. Hiking trail maps are available at Uetliberg railway station.

A few stops on the S-Bahn before Uetliberg brings you to the Binz Nature Reserve. This is a small area of abandoned clay pits which have been rehabilitated in order to create wildlife habitats. Its main value for a visiting birder is its proximity to the city centre and if you have an hour or two to spare, it supports many of the species which are typical of woodland margins and aquatic habitats. The reserve is west of the city centre, not far from the slip-road to the **A3 Autobahn**. The entrance is at the intersection of **Gerenholzstrasse** and **Borrweg**.

KEY SPECIES

Resident	Summer	Winter
Little Grebe, Eurasian Sparrowhawk, Woodcock, Green Woodpecker, Lesser Spotted Woodpecker, Grey Wagtail, Mistle Thrush, Fieldfare, Willow Tit, Crested Tit, Reed Bunting	Black Kite, Hobby, Common Cuckoo, Alpine Swift, Turtle Dove, Tree Pipit, Common Redstart, Wood Warbler, Western Bonelli's Warbler, Willow Warbler, Pied Flycatcher	Gadwall, Eurasian Teal, Common Kingfisher, Long-tailed Tit, Redwing, Brambling, Hawfinch

2 Chatzensee (5km)

(Bus No. 32)

Chatzensee Nature Reserve is north of Zurich, about midway between the city centre and the airport. This small reserve consists of two lakes partly surrounded by mixed woodland. Both Eurasian and Short-toed Treecreepers breed in the woodland between the autobahn and the lakes whilst the stands of broadleaved woodland support breeding Golden Orioles. Small reedbeds, wetlands and water meadows bordering intensive agricultural land create a high diversity of habitats which can be seen when you follow the nature trail (Chatzensee Rundweg) around the reserve, a walk of about an hour and a half, without counting any birding stops. In autumn especially, there is a good chance of finding a rarity and, in winter, particularly at dawn, several Bitterns can be observed: the best place is from the trail east of the lakes.

From Zurich take the **A1 Autobahn** towards Bern, and exit east onto the **A20 Autobahn** towards the airport. After about ten minutes driving, exit north off the Autobahn at **Zuerich Affoltern**. Turn right and after 100m there is a large parking place situated near the Chatzensee. By public transport take a train to Zuerich Oerlikon, then change to Bus No. 62 (direction to Unteraffoltern). You will need to transfer at Glaubtenstrasse to Bus No. 32 which will bring you to the terminus at Holzerhurd. From there the nature trail is marked with yellow signs (Chatzensee).

KEY SPECIES

Resident	Winter	Passage
Little Grebe, Water Rail, Ruddy Shelduck (I), Red Kite, Lesser Spotted Woodpecker, Grey-headed Woodpecker, Eurasian Treecreeper, Yellowhammer, Reed Bunting	Grasshopper Warbler, Pied Flycatcher, Golden Oriole	Squacco Heron, White Stork, Black Stork, Red-crested Pochard, Garganey, European Honey-buzzard, Osprey, Northern Lapwing, Jack Snipe, Common Kingfisher, Red-backed Shrike, Great Grey Shrike, Wood Lark, Tawny Pipit, Northern Wheatear, Bluethroat, Whinchat, Common Stonechat

Resident: Little Grebe, Water Rail, Ruddy Shelduck (I), Red Kite, Lesser Spotted Woodpecker, Grey-headed Woodpecker, Eurasian Treecreeper, Yellowhammer, Reed Bunting

Summer: Black Kite, Hobby, Common Cuckoo, Alpine Swift, Blue-headed Wagtail, Marsh Warbler, Savi's Warbler, Grasshopper Warbler, Pied Flycatcher, Golden Oriole

Winter: Red-necked Grebe, Eurasian Bittern, Gadwall, Northern Shoveler, Eurasian Wigeon, Eurasian Teal, Hen Harrier, Peregrine Falcon, Common Snipe, Water Pipit, Linnet

Passage: Purple Heron, Great White Egret, Squacco Heron, White Stork, Black Stork, Red-crested Pochard, Garganey, European Honey-buzzard, Osprey, Northern Lapwing, Jack Snipe, Common Kingfisher, Red-backed Shrike, Great Grey Shrike, Wood Lark, Tawny Pipit, Northern Wheatear, Bluethroat, Whinchat, Common Stonechat

3 Neeracherried Nature Centre (17km)

(Oberglatt S-Bahn Station S5)
(Bus 510)

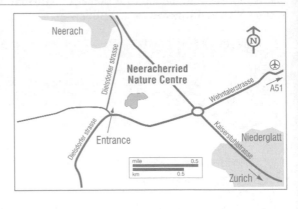

The SVS Nature Centre at Neerach is a small wetland reserve with reedbeds, ponds and water meadows. There were once much more extensive marshes in this area between the rivers Limmet and Glatt which have since been reclaimed, but the reserve is a protected zone where the surrounding meadows are grazed in such a way as to provide optimum breeding habitat for waders. Both Spotted and Little Crakes are believed to breed here and the provision of bird hides in strategic locations on the reserve offers some chance of seeing these secretive species. Apart from the crakes, there is a good selection of other freshwater species breeding at the reserve and small flocks of waders are regularly seen on migration. Unfortunately the reserve is only open at weekends from April to October and only occasionally during the winter. However it is close to Zurich Airport and, at the right time of year, a visit here would be a productive way to kill time if you are early for your flight. The hides also make it a worthwhile proposition even if you don't have a telescope.

Take the **A51 Autobahn** north from Zurich and drive about 8km past the airport exit. Exit west at **Bülach** in the direction of **Niederglatt**. At the roundabout on the outskirts of Niederglatt, continue straight on (west) for about half a kilometre, and the reserve is signposted on the right-hand side. If you are using public transport, take the S-Bahn to **Oberglatt** and from there, there is a bus service (**510**) to **Riedt bei Neerach**, which is just a five-minute walk from the entrance.

KEY SPECIES

Summer
Little Bittern, Eurasian Teal, Garganey, Black Kite, Red Kite, Spotted Crake, Little Crake, Water Rail, Little Ringed Plover, Northern Lapwing, Common Snipe, Common Tern, Common Kingfisher, Common Cuckoo, Wood Lark, Blue-headed Wagtail,

Grasshopper Warbler, Savi's Warbler, Great Reed Warbler, Golden Oriole, Red-backed Shrike, Reed Bunting

Passage
Little Egret, Great White Egret, Purple Heron, White Stork, Red Kite, European Honey-buzzard, Osprey,

Little Stint, Ruff, Spotted Redshank, Greenshank, Black-tailed Godwit, Eurasian Curlew, Green Sandpiper, Wood Sandpiper, Common Sandpiper, Whiskered Tern, Black Tern, Bluethroat, Penduline Tit

4 Niderholz (40km)

(Marthalen S-Bahn Station S33)

Most of the woodland around Zurich consists of spruce plantations and little remains of the original broadleaved forests which were cleared for agriculture. However, there are good tracts of oak forest along the south bank of the River Rhine which were conserved for charcoal and other forest uses. The forest area around Niderholz is north of

Zurich and is accessible by train and autobahn. It contains a lot of mature and storm-blown trees, and supports one sixth of Switzerland's population of Middle Spotted Woodpeckers. Five other species of woodpecker can be found here as well as other cavity-nesting passerines, including both Eurasian and Short-toed Treecreepers. The best

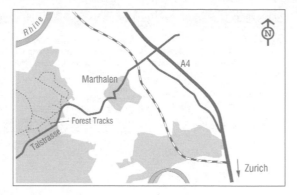

time to visit is early spring, when the woodpeckers are establishing their territories, and early summer, when the summer migrants have arrived.

Take the **A1** and then the **A4 Autobahn** north from Zurich and after about 38km exit west at **Marthalen**. In Marthalen, go southwest on **Talstrasse** towards **Ellikon** for about 1km. This road goes through the forest and there are tracks to either side.

KEY SPECIES

Resident		
Northern Goshawk, Eurasian Sparrowhawk, Common Pheasant, Woodcock, Stock Dove, Black Woodpecker, Green Woodpecker, Grey-headed Woodpecker, Middle Spotted	Woodpecker, Great Spotted Woodpecker, Lesser Spotted Woodpecker, Mistle Thrush, Fieldfare, Willow Tit, Long-tailed Tit, Crested Tit, Eurasian Treecreeper, Hawfinch	European Honey-buzzard, Red Kite, Hobby, Common Cuckoo, Turtle Dove, Tree Pipit, Common Redstart, Wood Warbler, Willow Warbler, Pied Flycatcher, Golden Oriole, Red-backed Shrike
	Summer	

5 Greifensee (15km)

(Schwerzenbach S-Bahn Station S5/S9/S14)
(Unster S-Bahn Station – Bus 840/842)

The Greifensee is a lake to the east of Zurich which is a nature protection area. Although much smaller than Lake Zurich, it supports much the same range of species. However, the area around the lake is mostly agricultural land with some woodland and is altogether less built-up than the shores of Lake Zurich. Also, there are reserves on the northern and southern shores, with footpaths, observation towers and a variety of habitats including marsh and reedbeds, making it better for birding. The lake supports the only Common Tern colony in the Zurich Canton and locally important breeding populations of other aquatic species. In winter it is good for ducks, grebes and, occasionally, Black-throated Divers.

Take the **A53 Autobahn** southeast from Zurich and exit west at the Greifensee exit. At Greifensee, scan the lake from the fishing jetty for gulls, ducks and grebes. Then take the **Schwerzenbacherstrasse** road north towards the **Schwerzenbach Bay** area, near the **River Glatt** outflow. There is an observation platform here and a footpath along the river bank and lakeshore. Returning to Greifensee, and continuing south on **Seestrasse**, leads to the **Riedikerried** Reserve on the south shore of the lake, where there is another observation tower and a tern nesting platform, as well as more footpaths by the lakeshore. The lake is about 7km from north to south but a bus service (840/842) connects the south shore of the lake with the train station at Unster. You can also take a rented bike on the train.

KEY SPECIES

Resident
Black-necked Grebe, White Stork (I), Red-crested Pochard, Red Kite, Water Rail, Black Woodpecker, Green Woodpecker, Lesser Spotted Woodpecker, Grey Wagtail, Dipper, Fieldfare, Mistle Thrush, Long-tailed Tit, Crested Tit, Tree Sparrow, Yellowhammer, Reed Bunting

Summer
Little Bittern, Garganey, Black Kite, Hobby, Common Quail, Northern Lapwing, Common Tern, Turtle Dove, Common Cuckoo, Grasshopper Warbler, Savi's Warbler, Marsh Warbler, Great Reed Warbler, Red-backed Shrike

Winter
Eurasian Bittern, Greylag Goose, Gadwall, Eurasian Teal, Eurasian Wigeon, Northern Shoveler, Common Goldeneye, Smew, Yellow-legged Gull, Common Gull, Common Kingfisher, Meadow Pipit, Water Pipit, Rook, Brambling

Passage
Great White Egret, Purple Heron, Night Heron, Marsh Harrier, Little Ringed Plover, Ruff, Dunlin, Little Stint, Spotted Redshank, Common Redshank, Greenshank, Eurasian Curlew, Common Snipe, Green Sandpiper, Wood Sandpiper, Common Sandpiper, Mediterranean Gull, Little Gull, Lesser Whitethroat, Penduline Tit

6 Pfäffikersee (17km)   (Seegräben-Aathal S-Bahn Station S14)

This is a small lake to the east of Zurich with a good selection of aquatic species in both summer and winter. Moreover, compared to the larger lakes, its small size and the network of footpaths around the shoreline make it easier for a birder on foot to cover the site. It supports good populations of both Reed and Marsh Warblers, so a visit here in early summer when the birds are singing will hone your skills in separating these two, often frustratingly similar, species. The southern shore of the lake, around the mouth of the River Aa, is probably the most interesting: with large reedbeds, wet meadows with breeding Lapwings, some woodland, and sheltered bays which are good for grebes and ducks in winter. However, there are a number of jetties around the other shores of the lake which are worth using as vantage points to scan the open waters.

Take the **A53 Autobahn** southeast from Zurich and continue southeast past the autobahn exit to **Seegräben**. The lakeshore is east of the town centre and there is a footpath leading along the southern shore to Strandbad. If you are using public transport, you could take the train to Seegräben and walk along the south and east shores of the lake towards Pfäffikon, from where you can take the train back to Zurich.

KEY SPECIES

Resident
Common Pheasant, Water Rail, Black Woodpecker, Green Woodpecker, Lesser Spotted Woodpecker, Dipper, Grey Wagtail, Mistle Thrush, Long-tailed Tit, Crested Tit, Yellowhammer, Reed Bunting

Summer
Little Bittern, Eurasian Teal, Garganey, Black Kite, Red Kite, Hobby, Spotted Crake, Little Ringed Plover, Northern Lapwing, Common Snipe, Common Tern, Common Cuckoo, Blue-headed Wagtail, Grasshopper Warbler, Savi's Warbler, Marsh Warbler, Great Reed Warbler, Golden Oriole, Red-backed Shrike

Winter
Black-necked Grebe, Red-necked Grebe, Eurasian Bittern, Gadwall, Eurasian Wigeon, Northern Shoveler, Yellow-legged Gull, Common Kingfisher, Great Grey Shrike

Passage
Great White Egret, Purple Heron, Night Heron, Ruff, Eurasian Curlew, Wood Sandpiper, Common Sandpiper, Mediterranean Gull, Black Tern, Penduline Tit

7 Rapperswil (30km)   (Rapperswil S-Bahn Station S5/S7/S16)

Rapperswil is located at the point in Lake Zurich where the shallow upper (northern) lake gives way to the deeper lower lake. There is a causeway across the lake here and the area offers easy access to the lakeshore and to several good birding spots in the vicinity of the town. A gull colony and a small tern colony are sited on offshore islets at Rapperswil. The Jona River inflow, just east of the town, is an area of marsh and riparian woodland. Overall, in terms of the variety of species and habitats, this area is probably the best part of the lakeshore for birds. Locally important numbers of wildfowl, including Eider, breed in this area and the lake itself acts as a conduit for many migrants in both directions.

Take **Route 17** south from Zurich along the east shore of Lake Zurich. The town of Rapperswil is about 30km south of the city on Route 17. First check the harbour for ducks, grebes and gulls. During the breeding season scan the offshore islets for breeding gulls and terns. At the railway station, go east for about 1.5km on **Bahnweg**, which runs alongside the railway line, to the bridge over the River Jona. There is a small marsh here which is interesting in spring and early summer for aquatic species. The small bay (Bucht von Wurmsbach) into which the Jona flows is good for grebes and diving ducks in winter.

KEY SPECIES

Resident
Little Grebe, Black-necked Grebe, Red-crested Pochard, Common Eider, Water Rail, Grey Wagtail, Long-tailed Tit, Dipper, Western Jackdaw, Bullfinch, Reed Bunting

Summer
Marsh Harrier, Hobby, Mediterranean Gull, Common Tern, Turtle Dove, Common Cuckoo, Common

Stonechat, Marsh Warbler, Great Reed Warbler

Winter
Great Bittern, Gadwall, Eurasian Teal, Eurasian Wigeon, Greater Scaup, Velvet Scoter, Hen Harrier, Common Snipe, Common Gull, Mediterranean Gull, Water Pipit

Passage
Great White Egret, Night Heron, Garganey, Pintail, Northern Shoveler, Red Kite, Northern Lapwing, Dunlin, Ruff, Common Redshank, Greenshank, Eurasian Curlew, Green Sandpiper, Wood Sandpiper, Little Gull, Black Tern, Pied Flycatcher, Penduline Tit

8 Kaltbrunn Marsh (45km)   *(Uznach Railway Station)*

Throughout western Europe, most of the water meadows associated with river flood-plains have been drained for agricultural use and few examples of this habitat now exist anywhere. However the marshes at Kaltbrunn (Kaltbrunner Riet), between Lake Zurich and Lake Wallen, have been conserved and the surrounding areas in the buffer zone are farmed sensitively for the benefit of wildlife. This is now an important breeding site for Black-headed Gulls and a feeding site for White Storks from the nearby colony at Uznach. It is also a stop-over point for small numbers of migrating waders. Even the best wetland sites can be useless for birding unless there is secure and discreet access. Many present an impenetrable wall of reedbeds and rank vegetation, behind which they are usually heaving with birds which you can't see! Fortunately, this site has a very well-positioned observation tower which overlooks the marsh, providing an excellent view of the reedbeds and open water. So although it is smaller than the large lakeshore sites, its facilities enable closer views of the birds, and if you don't have a telescope this might be a better option. It is very well watched and a surprising number of species have been found here.

Take the **A1** and then **A53 Autobahn** southeast from Zurich for about 40km. At this point the autobahn becomes the **A8**. After about 7km exit east at **Uznach**. The reserve entrance is about 1km south of the village on **Entenseestrasse**.

KEY SPECIES

Resident
Black-necked Grebe, Common Pheasant, Water Rail, Common Kingfisher, Grey Wagtail, Yellowhammer, Reed Bunting, Corn Bunting

Summer
Little Bittern, Spotted Crake, Common Tern, Turtle Dove, Wryneck, Common Cuckoo, Blue-headed Wagtail,

Common Nightingale, Great Reed Warbler, Golden Oriole, Red-backed Shrike

Winter
Little Grebe, Eurasian Bittern, Great White Egret, Greylag Goose, Common Shelduck, Gadwall, Pintail, Northern Shoveler, Eurasian Wigeon, Eurasian Teal, Hen Harrier, Peregrine Falcon, Great Grey Shrike

Passage
Little Egret, Purple Heron, Night Heron, White Stork, Garganey, Red Kite, European Honey-buzzard, Osprey, Common Crane, Northern Lapwing, Ruff, Common Redshank, Greenshank, Black-tailed Godwit, Eurasian Curlew, Green Sandpiper, Wood Sandpiper, Common Sandpiper, Black Tern, Bluethroat, Whinchat

9 Klingnauer Lake (35km)

(Koblenz S-Bahn Station S41)
(Döttingen Railway Station)

A barrage was constructed in the 19th century where the River Aare meets the Rhine, in order to control the flooding of the surrounding land. This led to the formation of a long artificial lake, and an area of marshland resulting from silted-up water channels. Some reclaimed land has been planted with forest and the whole area presents a diversity of habitats ranging from open waters to riparian woodland. This is now one of the best wetland sites in the north of Switzerland with over 270 species recorded. The lake and its surroundings provide habitats for breeding birds as well as winter and passage migrants, thus making a visit here worthwhile at any time of year. The only drawback to this site is the rather extensive area to be covered, stretching from Koblenz in the north to Böttstein in the south, a distance of about 7km. There is a foot-path along both banks of the lake which would be suitable for cycling or perhaps you can walk part of the way. Otherwise a car is necessary to cover all the likely areas.

Take the **A1 Autobahn** east from Zurich and exit north onto **Route 3** at **Wettingen**. Continue north on Route 3 and subsequently on **Route 5** to **Döttingen**. Turn west in Döttingen to the bridge across the Aare. There are paths going north on either side of the Aare which are worth walking. If using public transport, the stretch between Döttingen and Koblenz is one of the easiest to do quickly by train and walking. There is a railway station at Döttingen on the east bank of the Aare, and from here you could walk the east bank of the river to Koblenz, a distance of about 5km, from where you can get the S-Bahn back to Zurich: a round-trip of four to five hours from the main railway station in Zurich.

KEY SPECIES

Resident
Red-crested Pochard, Red Kite, Water Rail, Black Woodpecker, Green Woodpecker, Grey-headed Woodpecker, Lesser Spotted Woodpecker, Grey Wagtail, Dipper, Fieldfare, Mistle Thrush, Long-tailed Tit, Crested Tit, Tree Sparrow, Yellowhammer, Reed Bunting

Summer
Little Bittern, White Stork, Garganey, Black Kite, Hobby, Water Rail, Spotted Crake, Little Ringed Plover, Northern Lapwing, Common Tern,

Turtle Dove, Common Cuckoo, Grasshopper Warbler, Savi's Warbler, Marsh Warbler, Great Reed Warbler, Lesser Whitethroat, Red-backed Shrike

Winter
Black-necked Grebe, Eurasian Bittern, Greylag Goose, Gadwall, Eurasian Teal, Eurasian Wigeon, Northern Shoveler, Common Goldeneye, Smew, Common Gull, Common Kingfisher, Meadow Pipit, Water Pipit, Rook, Brambling

Passage
Great White Egret, Purple Heron, Night Heron, Marsh Harrier, Little Crake, Ruff, Dunlin, Little Stint, Spotted Redshank, Common Redshank, Greenshank, Eurasian Curlew, Common Snipe, Green Sandpiper, Wood Sandpiper, Common Sandpiper, Mediterranean Gull, Little Gull, Black Tern, Penduline Tit

Further Away

10 **Bodensee** (75km) *(Kreuzlingen Railway Station)*

Bodensee (Lake Constance) is a large lake on the River Rhine east of Zurich, where Austria, Germany and Switzerland meet. This is an immensely important wetland for breeding, wintering and migrating birds. Although it is a little distant from Zurich, the number and variety of birds to be seen here in a day would probably exceed that of all of the wetland sites

nearer to the city. If you have a car, then it is possible to visit sites in all three countries in a day and still get back to Zurich in the evening. This is less practical if you are relying on public transport, as it involves changing trains and making stops along the way. The stretch of shoreline between all the sites on the southern lakeshore is about 50km and, unless you are a champion cyclist, not really doable by bike.

Take the **A1** and then the **A7 Autobahns** west from Zurich to **Kreuzlingen**. At this point, you could cross the Rhine into **Konstanz** in Germany and turn immediately west onto the **B33** for about 1km. The entrance to the Wollmatinger Ried Reserve is on the left-hand (south) side of this road, near the Konstanz Sewage Ponds. This reserve supports a rich community of marsh and aquatic birds and is one of the best sites on the lake. It is an important nesting site for species such as Red-crested Pochards, Black-necked Grebes and Bearded Tits, which have a patchy breeding distribution in western Europe. There is a footpath through areas of reedbeds, wet meadows and small ponds with muddy margins to an observation tower, which provides an excellent vantage point from which to scan the lake and reedbeds. Migrant waders pass through in spring and summer, and in winter this part of the lake, known as the Untersee, holds up to 40,000 wildfowl.

In winter, the stretch of lakeshore between Kreuzlingen and the Austrian border is a well-known site for diving ducks and especially divers as the water is quite deep along here. Unfortunately, the road (Route 13) does not exactly run alongside the lakeshore, so you need to drive into the lakeside towns and scan from the boat marinas and harbours. The harbour at Kreuzlingen should be checked, but the most productive spots are at **Güttingen** and **Kesswil** about 10km further south. However, you should also check the **Luxburg Bay** further south at **Egnach** which has some mudflats which attract waders, when water levels are low: usually in spring.

If time permits, you could cross the Rhine into Austria in order to visit the Rhine Delta. This is a rather large area and at this point you are about 100km from Zurich. However this site could easily repay a whole weekend birding here. It is superbly positioned for birds migrating in autumn along the Rhine and heading for the Alpine passes. In poor weather conditions it acts as a holding zone for many passerines as well as waders and wildfowl. It is one of the more regular inland sites in western Europe for scarcer waders such as Marsh Sandpipers and Temminck's Stints and has even recorded

American waders. The area lies between the 'old' Rhine, which forms the Austrian border, and the newly constructed Rhine canal, a distance of some 8km of shoreline which would be difficult to cover on foot. One of the best zones is Fassbuch Bay, which is immediately west of the new Rhine canal inflow and is an area with sandbanks, reedbeds and marshland attracting gulls, terns and waders. Wetterwinkel Bay, which is east of the 'old' Rhine, is the best place in winter for diving ducks. There is a fairly extensive network of roads in the area so you can cruise the shoreline and inlets, but you must park and walk out onto the Rhine breakwater on foot.

KEY SPECIES

Resident
Black-necked Grebe, Red-crested Pochard, Red Kite, Water Rail, Bearded Tit, Tree Sparrow, Yellowhammer, Reed Bunting, Corn Bunting

Summer
Little Bittern, Garganey, Black Kite, Red Kite, Hobby, Common Quail, Spotted Crake, Little Ringed Plover, Northern Lapwing, Common Tern, Little Tern, Turtle Dove, Common Cuckoo, Common Stonechat, Whinchat, Grasshopper Warbler,

Savi's Warbler, Marsh Warbler, Great Reed Warbler, Red-backed Shrike, Golden Oriole

Winter
Black-throated Diver, Slavonian Grebe, Red-necked Grebe, Eurasian Bittern, Whooper Swan, Bewick's Swan, Gadwall, Eurasian Teal, Eurasian Wigeon, Northern Shoveler, Common Goldeneye, Greater Scaup, Velvet Scoter, Smew, Yellow-legged Gull, Common Gull, Common Kingfisher, Water Pipit

Passage
Great White Egret, Purple Heron, Night Heron, Marsh Harrier, Ruff, Dunlin, Little Stint, Spotted Redshank, Common Redshank, Greenshank, Eurasian Curlew, Black-tailed Godwit, Common Snipe, Green Sandpiper, Wood Sandpiper, Common Sandpiper, Mediterranean Gull, Little Gull, Black Tern, White-winged Black Tern, Caspian Tern, Hoopoe, Red-throated Pipit, Tawny Pipit, Northern Wheatear, Penduline Tit, Ortolan Bunting

Useful contacts and websites

Schweizer Vogelschutz SVS/BirdLife Schweiz (French and German) www.birdlife.ch
Wiedingstrasse 78, Postfach 8036 Zurich.
Tel: 044 457 7030. Email: svs@birdlife.ch
BirdLife partner in Switzerland and publishers of the journal *Ornis*.

Zürcher Vogelschutz (German)
www.zvs.ch
Local branch of SVS with information about activities in Zurich Canton.

Schweizerische Vogelwarte (French and German)
www.vogelwarte.ch
Swiss Ornithological Institute, CH - 6204 Sempach.
Tel. 041 462 9700. Fax. 041 462 9710.

Ala – Schweizerische Gesellschaft für Vogelkunde und Vogelschutz (German)
Sekretariat Werner Holliger, Breitestr. 22, CH-5015 Niedererlinsbach. www.ala-schweiz.ch
Local birding and conservation organisation, publishers of *Der Ornithologische Beobachter*.

Natur-und Vogelschutzverein Rapperswil-Jona (German) www.nvvrj.ch
Local birding and ringing group on Lake Geneva.

Arbeitsgruppe Naturschutz Greifensee (ASUG) (German) www.asug.ch
Local nature conservation group on Greifensee.

Greifensee Stiftung (German)
www.greifensee-stiftung.ch
Local nature conservation association for Greifensee.

Vereinigung Pro Pfäffikersee (German)
www.propfaeffikersee.ch
Local nature conservation association for Pfäffikersee.

CH Club 300 (German) www.chclub300.ch
Local birding group website with information and photos of recent rarities.

Swiss Birdline (French and German)
Tel. 021 61 61 222.

Swiss Rarities Committee (German)
www.vogelwarte.ch/sak_e_f.html
Schweizerische Avifaunistische Kommission,
c/o Schweizerische Vogelwarte, CH-6204 Sempach.

Zürcher Verkehrsverbund (English)
www.zvv.ch
Public transport for Zurich Canton.

Schweizerische Bundesbanen SBB (German and English) – Swiss Railways: www.sbb.ch

Post Bus (English) www.post.ch
Post Office Bus system – regional and feeder bus service.

Books and publications

Where to Watch Birds in Switzerland M.Sacchi *et al.* (1999) Christopher Helm, ISBN 0713651830

Brutvögel im Kanton Zürich Zürcher Vogelschutz.

Vögel in der Schweiz Burkhardt & Schmid, Schweizerische Vogelwarte

Schweizer Brutvogelatlas Schmid *et al.* Schweizerische Vogelwarte

Atlanta

Approaching Atlanta from the air, one is immediately struck by the extensive tracts of forest near to the airport, which only gives way to the urban areas on the final few seconds of the descent. This forest, mainly oak, pine and hickory, encroaches into the urban area, making Atlanta one of the most heavily wooded cities in America. This can be a tough habitat for birding due to the dense foliage and the dispersed nature of forest birds, but there are a number of sites around Atlanta which offer easy access and favourable viewing conditions.

Although Atlanta is an inland city, being about 250 miles from both the Atlantic or Gulf coasts, it nevertheless presents many excellent opportunities for the birder. The natural features of its hinterland include the Chattahoochee River, which flows through the northwest suburbs and acts as a conduit for migrating birds as well as providing some splendid riparian habitats. Just beyond the Chattahoochee River lie the foothills of the Appalachian Mountains, which can be reached within an hour from the city centre.

As the state of Georgia is situated just north of Florida, certain semi-tropical species occur. Yet it is just far north enough to experience cold spells in winter and some wintering wildfowl and sparrows which rarely venture as far south as Florida can be seen in the Atlanta area. Obviously migration times are best to see the greatest variety of species while midsummer is probably the quietest time due to the oppressive heat, the foliage growth and the fact that most breeding birds will have stopped singing.

Brown-headed Nuthatch and Carolina Chickadee are common visitors to bird-feeders in Atlanta

GETTING AROUND

For those visitors to Atlanta who rent a car at the airport, driving to most locations is actually quite easy because there is an orbital freeway (I-285) which skirts the city centre and airport to reach any point in the suburbs.

Public transportation in Atlanta is unified under the MARTA system, which operates the metro and the buses. Whereas this is perfectly adequate for the inner city birding locations, i.e. those

within the I-285 orbital freeway, those wishing to visit the
outer suburbs would be better
off with a car. The metro does
not extend much beyond the
I-285, and the MARTA bus
system tends to thin out in the
outer suburbs. There are local
transit companies which operate buses in the neighbouring
counties, but these are less
frequent and serve fewer
localities.

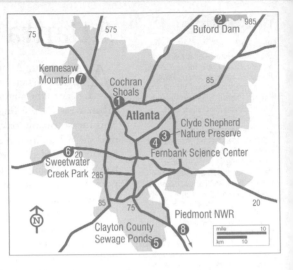

Birding in the Atlanta area
can be carried out with ease
and convenience as all of the
sites are very well paved. All
those mentioned in this guide
are public areas which are normally open from dawn to dusk, unless otherwise stated.
Access is free although there is a small charge for car-parking at some parks. As most of
the birding will be in woodland habitats a telescope is not really necessary unless you
need to scan the larger waterways.

The only need for caution occurs in the winter during the hunting season. Georgia
has some prime hunting areas and there is much hunting activity away from the public
parks. Shooting accidents can happen all too frequently in dense woodland. Pay careful
attention to any notices and it is probably best to avoid areas where gunfire is heard.

Maps, books and birding supplies are available in any of the large shopping malls that
are dotted along the I-285 and, of course, in the airport concourse.

Locations

1 Chattahoochee River – Cochran Shoals (12 miles) (CCT Bus #10B)

KEY SPECIES

Resident
Red-shouldered Hawk,
Belted Kingfisher, Red-
bellied Woodpecker, Pileated
Woodpecker, Brown-headed
Nuthatch, Brown Thrasher

Summer
Yellow-billed Cuckoo, Chuck-
will's-widow, Blue-gray
Gnatcatcher, Wood Thrush,
Gray Catbird, Yellow-
throated Vireo, Hooded
Warbler, Summer Tanager,
Indigo Bunting

Winter
Brown Creeper, Blue-headed
Vireo

Passage
Spotted Sandpiper,
Philadelphia Vireo, White-
eyed Vireo, Swainson's
Thrush, Tennessee Warbler,
Chestnut-sided Warbler,
Magnolia Warbler, Black-
throated Green Warbler,
Prairie Warbler, Bay-
breasted Warbler, Blackpoll
Warbler, Palm Warbler,
American Redstart,
Ovenbird, Scarlet Tanager,
Rose-breasted Grosbeak,
Baltimore Oriole

The Chattahoochee River National
Recreation Area is a linear park that
runs alongside the Chattahoochee
River for 48 miles. The section at
Cochran Shoals is easily accessed from
Atlanta and this provides some good
riparian habitat as well as broadleaved
woodland. The 'Hooch' is also a popular area with canoeists and the general
public, so can be quite crowded at
weekends and it is best visited early in
the mornings. Some trails at least are
prohibited to cyclists, so are safer for
walking. For birders, the main attractions are in spring and autumn, as there

is an appreciable passage of warblers.

The site is located on the west bank of the Chattahoochee and can be reached by taking **Exit 22** on **I-285** onto **Interstate North Parkway**. There is a Parking Lot just off the Interstate Parkway on the west bank of the river. The Cobb Community Transit Bus #10B operates a service from Five Points MARTA Station in central Atlanta to Interstate North Parkway, but only at peak hours on weekdays.

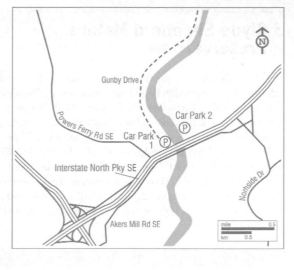

2 Buford Dam (50 miles)

Buford Dam is northeast of Atlanta. It was constructed on the Chattahoochee River in the 1950s and led to the creation of Lake Sidney Lanier. It is worth a visit in winter as the extensive lake hosts larger numbers of waterfowl than the smaller lakes near Atlanta, and the deep water in the vicinity of the dam is more attractive to species such as loons and grebes. It is also less than ten miles from the Mall of Georgia making it a convenient refuge for any birder whose spouse or partner wants to spend the day in the largest shopping mall in the southeastern USA. There are several good spots along the park shoreline from which to scan the lake.

Take **I-985** North to **Buford** and exit west at **Friendship Road (Exit 8)** which leads on to the **Buford Dam Road**. Buford Dam Park is on Buford Dam Road, about six miles from I-985. Gwinnett County Transit operates a bus service to Buford (Rte #50), but it involves an interchange with the MARTA Metro and its nearest set-down is over two miles from the lakeshore.

KEY SPECIES

Resident
Black Vulture, Wild Turkey, Killdeer, Belted Kingfisher, Red-bellied Woodpecker, Pileated Woodpecker, Brown-headed Nuthatch, Eastern Bluebird, Brown Thrasher, Chipping Sparrow, Eastern Meadowlark

Summer
Double-crested Cormorant, Green Heron, Osprey, Yellow-billed Cuckoo, Blue-gray Gnatcatcher, Wood Thrush,

Gray Catbird, Yellow-throated Vireo, Purple Martin, Hooded Warbler, Summer Tanager, Indigo Bunting, Orchard Oriole

Winter
Pied-billed Grebe, Common Loon, Eared Grebe, Horned Grebe, Gadwall, Green-winged Teal, Ring-necked Duck, Bufflehead, Lesser Scaup, Redhead, Hooded Merganser, Ruddy Duck, American Coot, Bonaparte's

Gull, Ring-billed Gull, Brown Creeper, Winter Wren

Passage
Blue-winged Teal, Common Goldeneye, Red-breasted Merganser, Swainson's Thrush, Blue-headed Vireo, Tennessee Warbler, Blackpoll Warbler, Palm Warbler, Ovenbird, Scarlet Tanager, Rose-breasted Grosbeak, Baltimore Oriole

3 Clyde Shepherd Nature Preserve (8 miles)

MARTA Bus #123 from Decatur Station

Peachtree Creek is a tributary of the Chattahoochee which flows through the northern suburbs of Atlanta. South Peachtree Creek Nature Preserve is a small wetland site located right behind the North Dekalb Shopping Mall. For such a small area, it contains a surprising variety of habitats and species. There is a marsh with an observation tower and walking trails through woods and some open areas. There are even beavers in the pond!

The site is about two miles north of Decatur MARTA Station, but Bus #123 will bring you to the North Dekalb Mall which is adjacent to the entrance on Wood Trail Lane. If driving, take **Hwy 78** northeast from the city centre to **Medlock Road** intersection. Turn left (north) onto Medlock Road and follow this road until it comes to a T-junction. Turn right at the T junction onto **Wood Trail Lane**: the preserve is at the end of this road.

KEY SPECIES

Resident
Pied-billed Grebe, Red-shouldered Hawk, Killdeer, Belted Kingfisher, Hairy Woodpecker, Red-bellied Woodpecker, Pileated Woodpecker, Brown-headed Nuthatch, Eastern Bluebird, Brown Thrasher, Chipping Sparrow

Summer
Green Heron, Yellow-billed Cuckoo,

Blue-gray Gnatcatcher, Wood Thrush, Gray Catbird, Yellow-throated Vireo, Purple Martin, Hooded Warbler, Louisiana Waterthrush, Summer Tanager, Indigo Bunting, Orchard Oriole

Winter
Ring-necked Duck, Hooded Merganser, Wilson's Snipe, Brown Creeper, House Wren, Winter Wren

Passage
Blue-winged Teal, Spotted Sandpiper, Sedge Wren, Marsh Wren, Swainson's Thrush, Blue-headed Vireo, Tennessee Warbler, Magnolia Warbler, Prairie Warbler, Blackpoll Warbler, Palm Warbler, Ovenbird, Scarlet Tanager, Baltimore Oriole

4 Fernbank Science Center (6 miles)

MARTA Bus #2 from Decatur Station

This is a museum and science centre set amidst 65 acres of broadleaved woodland which contains remnants of the old forests that once covered this part of the southeastern USA. This small area within the city supports a rich variety of flora and fauna: however, it doesn't contain any significant aquatic habitats, so this limits the range of birds that can be expected. The Science Center is a bird-banding station but also conducts bird walks every Saturday during the autumn and this is a great way to get familiar with the local birds and to meet other birders.

The center is located just east of the city centre on **Heaton Park Drive**. The best way to get there is to take **Ponce de Leon Avenue** eastwards from **Midtown**. About a mile past the **Druid Hills Golf Club**, take a left on **Artwood Road** and then the first right brings you onto Heaton Park Drive. Alternatively, take the metro to Decatur Station and from there the #2 bus will set you down at the turn-off for Artwood Road, a short walk from the center. Entrance to the woodland is free but it is only open to the public in the afternoons, except on Saturdays when it is open from 10.00.

KEY SPECIES

Resident
Red-bellied Woodpecker, Pileated Woodpecker, Brown-headed Nuthatch, Brown Thrasher, Common Grackle

Summer
Green Heron, Broad-winged Hawk, Yellow-billed Cuckoo, Blue-gray Gnatcatcher, Wood Thrush, Gray Catbird, Yellow-throated Vireo,

Louisiana Waterthrush, Hooded Warbler, Summer Tanager, Indigo Bunting

Winter	Swainson's Thrush, Tennessee	Blackpoll Warbler, Palm Warbler,
Brown Creeper, Winter Wren	Warbler, Chestnut-sided Warbler,	Ovenbird, Scarlet Tanager, Rose-
	Black-throated Green Warbler, Prairie	breasted Grosbeak, Baltimore Oriole
Passage	Warbler, Bay-breasted Warbler,	
Sharp-shinned Hawk, Veery,		

5 Clayton County Sewage Ponds (22 miles)

The E.L. Huie Land Application Facility and Newman Wetlands are two adjoining wetlands which include the sewage ponds. This is the best location in Atlanta for waders and surprisingly good for an inland site. Numbers are low compared with a coastal site, but there can be a good variety and occasionally a real rarity. The two areas are located in Clayton County about 20 miles south of Atlanta and are best visited by car. However, they are only about 20 minutes' drive south of Atlanta Airport, so

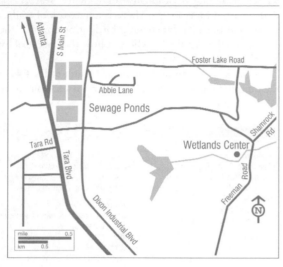

they are worth checking if you are collecting or dropping off a rental car. Birders are permitted to visit the sewage ponds and you can drive around the top of the dykes separating the ponds, using the car as a mobile hide. Newman Wetlands have been laid out to assist visitors with a boardwalk through the marsh. This area can be good at any time of the year, with the sewage ponds being particularly so during wader migration and in winter for waterfowl. Visiting hours are from 08.30 to 17.00 daily.

To get to the E.L. Huie Facility take **I-75** south to **Exit 235 (US Highway 19)**. Go south on Highway 19 for approximately eight miles and turn left onto **Freeman Road**. The sewage ponds are about a quarter of a mile from the junction on the left. Newman Wetlands Center is a further two miles along Freeman Road on the right.

C-Tran Transit System operate a bus service from the airport to Jonesboro, which is on Highway 19 about two miles north of the sewage ponds.

KEY SPECIES

Resident
Wild Turkey, Killdeer, Belted Kingfisher, Eastern Bluebird, Brown Thrasher, Chipping Sparrow, Eastern Meadowlark

Summer
Snowy Egret, Great Egret, Green Heron, Broad-winged Hawk, Osprey, Yellow-billed Cuckoo, Blue-gray Gnatcatcher, Wood Thrush, Gray Catbird, Purple Martin, Tree Swallow,

Prothonotary Warbler, Louisiana Waterthrush

Winter
Pied-billed Grebe, Gadwall, Northern Pintail, Northern Shoveler, Green-winged Teal, Ring-necked Duck, Redhead, Lesser Scaup, Hooded Merganser, American Coot, Wilson's Snipe, House Wren, Winter Wren

Passage
American Wigeon, Blue-winged Teal,

Red-breasted Merganser, Semipalmated Plover, Stilt Sandpiper, Least Sandpiper, Semipalmated Sandpiper, White-rumped Sandpiper, Pectoral Sandpiper, Greater Yellowlegs, Lesser Yellowlegs, Spotted Sandpiper, Solitary Sandpiper, Black Tern, Bank Swallow, Yellow Warbler, Palm Warbler, Scarlet Tanager, Rose-breasted Grosbeak, Bobolink

6 Sweetwater Creek State Park (16 miles)

This is a large state park (over 2,500 acres) just west of Atlanta. It has a quite extensive lake and system of waterways including Sweetwater Creek, so in winter it is good for waterfowl while the surrounding broadleaved woods host a thriving community of woodland species. The lake is deep enough to attract loons, grebes and diving ducks in winter, including occasionally some of the scarcer species such as Surf Scoters and Red-breasted Mergansers. There is a network of trails, all of which are worth exploring if time permits, but from a birding perspective the most productive are the Red Trail and Yellow Trail which both follow the course of Sweetwater Creek. A trail map is available at the park office.

The park is easy to find, being just south of **Exit 44** on **I-20,** and is signposted from the Interstate. It is, however, popular with the public and so is likely to be crowded at the weekends.

KEY SPECIES

Resident
Pied-billed Grebe, Black Vulture, Killdeer, Belted Kingfisher, Red-bellied Woodpecker, Pileated Woodpecker, Brown-headed Nuthatch, Eastern Bluebird, Brown Thrasher, Chipping Sparrow

Summer
Great Egret, Green Heron, Yellow-billed Cuckoo, Blue-gray Gnatcatcher, Wood Thrush, Gray Catbird, Yellow-throated Vireo, Purple Martin, Hooded Warbler, Louisiana Waterthrush, Summer Tanager, Indigo Bunting, Orchard Oriole

Winter
Common Loon, Gadwall, Ring-necked Duck, Lesser Scaup, Hooded Merganser, Ruddy Duck, American Coot, Ring-billed Gull, Brown Creeper, House Wren, Winter Wren

Passage
Blue-winged Teal, Surf Scoter, Red-breasted Merganser, Bonaparte's Gull, Swainson's Thrush, Blue-headed Vireo, Tennessee Warbler, Magnolia Warbler, Prairie Warbler, Bay-breasted Warbler, Blackpoll Warbler, Palm Warbler, Ovenbird, Scarlet Tanager, Baltimore Oriole

7 Kennesaw Mountain (20 miles)

Kennesaw Mountain National Battlefield Park commemorates a Civil War battle, but it is best known to Georgia birders as a warbler migration site. As is often the case with eastern US cities, certain greenbelt areas within the city or its suburbs can act as natural 'islands' attracting migrants. Just about every eastern American warbler and

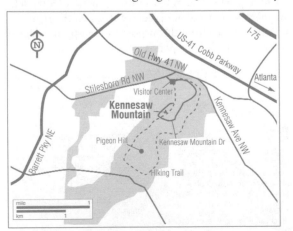

vireo has been recorded here and it is possible to see a dozen or more species on a good day in spring or autumn. Some, such as Blackpoll Warblers, are more abundant in spring than in autumn but for others, such as Tennessee Warblers, the opposite is true. As the mountain peak is almost 2,000 feet above sea-level it provides an excellent vantage point for those willing to make the hike of over a mile from the

Visitor Centre. At weekends there is a shuttle bus to take you to the top, but by taking the bus you will miss out on the birding opportunities along the way.

The park can be reached by taking **I-75** northwest to **Marietta**. Exit west onto **Barrett Parkway** at **Exit 269** and after two miles turn left onto **Old Highway 41**. Continue for another mile before turning right onto Stilesboro Road. The park entrance is a couple of hundred yards on the left.

KEY SPECIES

Resident Red-bellied Woodpecker, Hairy Woodpecker, Eastern Bluebird, Brown Thrasher **Summer** Yellow-billed Cuckoo, Chuck-will's-widow, Blue-gray Gnatcatcher, Wood Thrush, Gray Catbird, Yellow-throated Vireo, Hooded Warbler, Summer	Tanager, Scarlet Tanager, Indigo Bunting **Winter** Eastern Phoebe, Brown Creeper, House Wren, Winter Wren, Common Grackle **Passage** Swainson's Thrush, Blue-headed	Vireo, Philadelphia Vireo, Tennessee Warbler, Chestnut-sided Warbler, Magnolia Warbler, Black-throated Green Warbler, Prairie Warbler, Bay-breasted Warbler, Blackpoll Warbler, Palm Warbler, Cerulean Warbler, American Redstart, Ovenbird, Rose-breasted Grosbeak, Baltimore Oriole

Further Away

8 Piedmont National Wildlife Refuge (70 miles)

Although this site is rather distant from Atlanta, it is worth a mention as most birders visiting this part of Georgia include a trip to this site in the hope of seeing two of Piedmont's special birds: Red-cockaded Woodpecker and Bachman's Sparrow. At the right time of year (spring/early summer) there is a very high probability of seeing both species, as well as many other interesting species which are typical of the southeastern USA. Surprisingly, most of the forest cover in the area is secondary growth yet it still supports a healthy population of these specialist woodpeckers.

Take **I-75** southeast from Atlanta for about 50 miles and exit east at Forsyth onto **Juliette Road (Exit 186)**. Continue east on this road, past Juliette for about 18 miles and turn left into the Park Visitor Center, where you can ask for information on the most easily viewable woodpecker colony. Stay on the entrance road passing the Visitor Center to get to the car park beside Allison Lake. There are two trails here: Allison Lake Trail which leads down to the lakeshore and Red-cockaded Woodpecker Trail which leads to a colony of the aforementioned woodpeckers. The trees with nest holes are marked with white paint – it couldn't be easier!

KEY SPECIES

Resident Black Vulture, Red-shouldered Hawk, Wild Turkey, Northern Bobwhite, Belted Kingfisher, Red-bellied Woodpecker, Red-cockaded Woodpecker, Pileated Woodpecker, Eastern Phoebe, Brown-headed Nuthatch, Eastern Bluebird, Brown Thrasher, Bachman's Sparrow, Chipping Sparrow, Field Sparrow, Eastern Meadowlark	**Summer** Yellow-billed Cuckoo, Chuck-will's-widow, Blue-gray Gnatcatcher, Acadian Flycatcher, Wood Thrush, Gray Catbird, Yellow-throated Vireo, White-eyed Vireo, Prairie Warbler, Hooded Warbler, Kentucky Warbler, Louisiana Waterthrush, Yellow-breasted Chat, Summer Tanager, Blue Grosbeak, Indigo Bunting, **Winter** Brown Creeper, Savannah Sparrow, Brown-headed Cowbird	**Passage** Swainson's Thrush, Blue-headed Vireo, Blackpoll Warbler, Black-throated Green Warbler, American Redstart, Chestnut-sided Warbler, Magnolia Warbler, Black-throated Green Warbler, Prairie Warbler, Bay-breasted Warbler, Blackpoll Warbler, Palm Warbler, Cerulean Warbler, Ovenbird, Rose-breasted Grosbeak, Baltimore Oriole

Useful contacts and websites

Atlanta Audubon Society
www.atlantaaudubon.org
P.O. Box 29189, Atlanta, Georgia 30359.
Tel. 770 913 0511.

Georgia Ornithological Society
www.gos.org/index.html
P.O. Box 5825, Valdosta, GA 31603-5825.

Fernbank Science Center
www.fernbank.edu/
156 Heaton Park Drive, N.E., Atlanta,
Georgia 30307. Tel. 678 874 7102.

Kennesaw Mountain Birding
www.georgia-birding.com/KMT/index.htm

Clyde Shepherd Nature Preserve
www.cshepherdpreserve.org/

Chattahoochee River National Recreation Area
www.nps.gov/chat

Georgia Rare Bird Alert: Tel. 770 493 8862.

MARTA Public Transport in Atlanta
www.itsmarta.com

C-Tran Public Transport in Clayton County
www.web.co.clayton.ga.us/ctran

CCT Public Transport in Cobb County
www.cobbdot.org/cct.htm

Gwinnett County Transit
www.gctransit.com

Books and publications

Birds of Georgia John Parrish & Giff Beaton (2006) Lone Pine Publishing, ISBN 9768200057

Birding Georgia Giff Beaton (2000) Falcon Publishing, ISBN 1560447842

Birds of Kennesaw Mountain: An Annotated Checklist Giff Beaton (2004) Georgia Ornithological Society.

A Birder's Guide to Georgia Joel R. Hitt & Kenneth Turner Blackshaw. Georgia Ornithological Society, ISBN 9999000029

Annotated Checklist of Georgia Birds Giff Beaton et. al. (2003) Georgia Ornithological Society.

A Birder's Guide to Metropolitan Areas of North America Paul Lehman (2001) American Birding Association, ISBN 1878788159

A Field Guide to the Birds of Eastern and Central North America Roger Tory Peterson (2002) Houghton Mifflin, ISBN 0395740460

National Geographic Field Guide to the Birds of North America, 4th Edition National Geographic (2002), ISBN 0792268776

The North American Bird Guide David Sibley (2000) Pica Press, ISBN 1873403984

Field Guide to the Birds of Eastern North America David Sibley (2003) Christopher Helm ISBN 0713666579

National Geographic Guide to Bird Watching Sites: Eastern US Mel White (1999) National Geographic, ISBN 0792273745

• •

Boston

Any birder flying into Boston's Logan Airport by day will be immediately struck by the abundance of islands in Boston Harbour and the extensive creeks and inlets along the shoreline that suggest potential possibilities for seabirds, waders and waterfowl. The rich habitat combined with its strategic location on the Atlantic flyway make Boston a great city for a birder to visit. Massachusetts is one of the top destinations in the eastern USA for birding trips thanks to its diverse birdlife and a birding infrastructure that is enthusiastically maintained by a dedicated

Swallow, Barn Swallow, Yellow Warbler, Common Yellowthroat, Chipping Sparrow, Song Sparrow, Common Grackle, Red-winged Blackbird, Brown-headed Cowbird, Baltimore Oriole

Passage
Double-crested Cormorant, Wood Duck, Broad-winged Hawk, Killdeer, Northern Flicker, Eastern Kingbird, Red-eyed Vireo, Ruby-crowned Kinglet, Yellow-rumped Warbler, Blackpoll Warbler

corps of birders, whose activities through the years have given this small state one of the highest state lists in the country.

Boston itself is situated at the mouth of the Charles River which empties into Boston Harbour. The harbour is a broad, C-shaped bay with several promontories and many creeks and inlets, some of which have been dredged or otherwise developed for shipping. However, there are still a number of good sites for birds, particularly along the shore north of Boston. Inland, the low wooded hills to the west of the city contain a number of reservoirs which also have good habitat.

Birding in the Boston area tends to be seasonal. The winters are very cold and as a result many landbirds move south or to the coast. Inland areas can be very quiet in winter except where food is left out or where waterways are not frozen over. In winter, you are best advised to stick to the coastal areas as there are many possibilities for viewing waterfowl and gulls, even though this generally requires a telescope. Nonetheless, there is then always an opportunity to see the few passerines that remain in coastal parks and wetlands. The situation changes in summer when many migrant birds utilise the woodlands and inland wetlands. For those who do not wish to, or cannot, leave the city, there is always spring and autumn warbler migration. This annual phenomenon ensures that even city centre parks may support a bewildering variety of warblers that will astound visiting birders. For the visitor, spring migration is probably best, since the birds are in breeding plumage and the trees are not yet in full leaf, making viewing easier.

Wilson's Storm-Petrel is a summer visitor to Boston Harbour

GETTING AROUND

Boston is actually a rather small and compact city, but the towns on the outskirts have gradually coalesced into one continuous conurbation so that all of the area within the I-95/128 freeway has become part of Greater Boston. Happily for the birder, the coastal areas north of the city are well served by commuter rail service and the subway known as the 'T.' Also the bus system is quite straightforward and easy to use, and serves most of the western suburbs.

Driving is really only problematic in downtown Boston but, as a birder, you will not need to drive there since, with few exceptions, all birding sites are in the suburbs. Try to time your birding so that you are travelling away from the downtown area during the morning rush hour and towards the city in the evening.

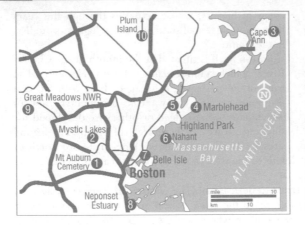

The best maps to use are the Rand McNally and Gousha Road Maps. For many of the coastal sites it may be advisable to wear boots or footwear that you won't mind getting wet and muddy. If you are travelling by public transport, the area around the South Station, a major transport hub, has lots of shopping facilities, if you need to stock up with supplies.

Locations

1 **Mount Auburn Cemetery and Fresh Pond** (5 miles)

Harvard T-Station

For some reason, Boston seems to be particularly well endowed with cemeteries. These are usually well wooded and undisturbed, which makes them good places for birds. Mount Auburn Cemetery is an extraordinary birding site just west of Harvard University in Cambridge. Apart from accommodating the graves of the poet Henry Longfellow and William Brewster, founder of the American Ornithologists' Union, it is also one of the better sites in North America to observe spring passerine migration. There is nothing particularly special about this spot at other times of the year, but from mid-April to the end of May it can be outstanding, and if a birder is in Boston at this time, it is a location not to be missed. Being a cemetery it is not crowded with children playing or noisy picnickers, so it makes for easy birding. Expect to run into lots of other birders who will be able to assist you with news and information, because it is very popular among local birders during spring migration.

Almost across the road from Mount Auburn is another migration hotspot, this one for aquatic species. Fresh Pond is a small reservoir set in a landscaped park, with a golf course on its western side and a few smaller ponds around the perimeter. In spring the wooded areas attract many of the same species as Mount Auburn, and in autumn it is well known for migrating wildfowl. In winter Fresh Pond freezes over and becomes less interesting.

Follow **Mt Auburn Street** west from **Harvard Square** for about a mile to reach the north entrance of Mt Auburn Cemetery. Fresh Pond is reached by going north for about 300 yards on Aberdeen Avenue, across from the north entrance of Mt Auburn. The entrance to Fresh Pond is about 300 yards further up on the right-hand (east) side of the park (Huron Avenue).

Even if you are on a tight schedule and restricted to the downtown Boston area, it is always worth checking the nearest public park during migration. Even the tiny Post Office Square Park, a highly urbanised and enclosed area in the centre of the Financial District, has turned up a surprising number and variety of migrants through the years.

Spring Passage – Mt Auburn
Yellow-bellied Sapsucker, Ruby-throated Hummingbird, Eastern Wood Pewee, Least Flycatcher, Eastern Phoebe, Blue-headed Vireo, Warbling Vireo, House Wren, Golden-crowned Kinglet, Veery, Gray-cheeked Thrush, Bicknell's Thrush, Swainson's Thrush, Hermit Thrush, Gray Catbird, Brown Thrasher, Nashville Warbler, Northern Parula,

Palm Warbler, Chestnut-sided Warbler, Magnolia Warbler, Blackburnian Warbler, Black-throated Blue Warbler, Black-throated Green Warbler, Black-and-white Warbler, American Redstart, Canada Warbler, Northern Waterthrush, Scarlet Tanager, Chipping Sparrow, Fox Sparrow, Lincoln's Sparrow, White-throated Sparrow, White-crowned Sparrow, Baltimore Oriole

Autumn Passage – Fresh Pond
Common Loon, Pied-billed Grebe, Green Heron, American Wigeon, Canvasback, Lesser Scaup, Greater Scaup, Ring-necked Duck, Bufflehead, Common Goldeneye, Hooded Merganser, Common Merganser, Ruddy Duck, American Coot

2 Mystic Lakes (8 miles)

West Medford Commuter Rail Station

The area around Medford in the northern suburbs of Boston contains a number of lakes and reservoirs which are open to the public. There are two lakes on the Mystic River that flows through this area. The wooded areas surrounding the lakes are good for passerines, while the lakes themselves are good for waterfowl unless they are frozen. The other reservoirs in the area – the Arlington Reservoir and the North and South Middlesex Reservoirs – are good in autumn for waders if the water levels are low enough.

There is a road around the southern and eastern shores of the Upper and Lower Mystic Lakes (Mystic Valley Parkway) but it is not possible to stop on the hard shoulder, and unfortunately there are only a few lay-bys. The best birding locations are on the eastern shore of the Upper Lake. Mount Pleasant Cemetery is at the southwest corner of the Lower Mystic Lake and is a good place to look for migrant passerines. The marshy woodland at the northwest corner of the cemetery is especially good.

Take the **I-93** freeway north from Boston and leave at **Exit 31** onto the **Mystic Valley Parkway**. This road follows the course of the Mystic River through the town of West Medford to the Lower Mystic Lake. If using public transport, the train station in West Medford is about half a mile from the lakeshore.

Resident
American Black Duck, Wild Turkey, Sharp-shinned Hawk, Red-bellied Woodpecker

Summer
Black-crowned Night Heron, Wood Duck, American Woodcock, Willow

Flycatcher, Eastern Kingbird, Marsh Wren, Warbling Vireo

Winter
Great Cormorant, Ring-necked Duck, Bufflehead, Common Merganser, Ruddy Duck

Passage
Osprey, Spotted Sandpiper, Belted Kingfisher, Swainson's Thrush, Wood Thrush, Magnolia Warbler, Palm Warbler, Black-and-white Warbler, Northern Waterthrush, Scarlet Tanager, Rose-breasted Grosbeak

3 Cape Ann Peninsula (40 miles)

Rockport Commuter Rail Station

This is basically the northern arm of Massachusetts Bay and provides a good vantage point for seawatching and for scanning for loons, grebes and sea ducks wherever there is access to the shore. The best seawatching conditions are during strong northeasterly winds in autumn, and the best seawatching points are at Halibut Point and Andrews Point in Rockport.

Take the **Callahan Tunnel** out of Boston and follow **Highway 107** as far as **Route 128**. Follow Route 128 all the way to **Rockport**. There is a commuter rail station at Rockport within walking distance of Rockport Harbour and Andrews Point. However, you need a car to access most of the other birding spots on the peninsula. The disadvantage of a car is that many of the prime coastal lookouts are actually on private property and there are only a few public areas where you can pull-off the road. On weekends in autumn and winter it can be virtually impossible to find a parking space.

KEY SPECIES

Spring
Red-throated Loon, Red-necked Grebe, Northern Gannet, Purple Sandpiper, Laughing Gull (uncommon), Common Tern, Black Guillemot

Autumn
Great Shearwater, Sooty Shearwater, Manx Shearwater, Northern Fulmar

(scarce), Leach's Storm-petrel (after storms), Northern Gannet, Harlequin Duck, phalaropes (after storms), Parasitic Jaeger, Pomarine Jaeger, Black-legged Kittiwake, Razorbill, Dovekie (after storms)

Winter
Red-throated Loon, Common Loon, Horned Grebe, Red-necked Grebe,

Great Cormorant, Northern Gannet, Harlequin Duck, Long-tailed Duck, Common Eider, King Eider (scarce), Surf Scoter, Black Scoter, White-winged Scoter, Red-breasted Merganser, Purple Sandpiper, Kumlien's Gull, Black-legged Kittiwake, Dovekie (scarce), Razorbill, Thick-billed Murre, Black Guillemot

---

4 Marblehead (16 miles)

Bus # 441 and # 442

The Marblehead Neck Peninsula is another site on the north shore of Massachusetts Bay. In particular there is Massachusetts Audubon's Marblehead Neck Sanctuary, located on Risley Road and nicely laid out with trails through wetland and woodland habitat that is excellent for migrant passerines in both spring and fall. There are both sandy and pebble beaches on the peninsula, and any vantage point is worth checking for gulls, sea ducks and waders in season.

Take the **Callahan Tunnel** northeast from downtown Boston onto **Highway 1A** and continue northeast for about 14 miles to **Marblehead**. In Marblehead, turn right (south) on **Pleasant Street** and then onto **Ocean Avenue**, which leads to the peninsula along a narrow isthmus.

There is no bus service out to the peninsula: the closest bus stop is on Pleasant Street, requiring a one mile walk. Take bus #441 or #442 from the Wonderland T-Station.

KEY SPECIES

Summer
Snowy Egret, Black-crowned Night Heron, Green Heron, Double-crested Cormorant, Osprey, American Oystercatcher (scarce), Killdeer, Willet, Spotted Sandpiper, Ruddy Turnstone, Laughing Gull, Common Tern, Red-eyed Vireo, Northern

Rough-winged Swallow, Carolina Wren, Gray Catbird, Brown Thrasher, Cedar Waxwing, Common Yellowthroat

Winter
Red-throated Loon, Common Loon, Horned Grebe, Red-necked Grebe,

Great Cormorant, Common Eider, Surf Scoter, White-winged Scoter, Common Goldeneye, Red-breasted Merganser, Bonaparte's Gull, Razorbill (occasional)

Passage
Blue-headed Vireo, Veery, Swainson's

Thrush, Gray-cheeked Thrush, Bicknell's Thrush, Northern Parula, Chestnut-sided Warbler, Magnolia Warbler, Blackburnian Warbler,	Black-throated Blue Warbler, Black-throated Green Warbler, Black-and-white Warbler, American Redstart, Canada Warbler, Northern	Waterthrush, Scarlet Tanager, White-throated Sparrow, Baltimore Oriole

5 Highland Park, Salem (13 miles)

Bus # 441 and # 442

Salem is northeast of Boston and an interesting tourist destination. However, there is also Highland Park, a nice public park near the town centre, which is readily accessible and often contains a very good variety of landbirds typical of New England. This is an area where you could encounter a Wild Turkey. Although historically indigenous to this area, Wild Turkeys have been introduced to replace extirpated populations and it is not always possible for a birder to know which birds are genuinely wild. However, as a visitor you may not be affected by such scruples so, unless the birds actually eat out of your hand, they can be counted!

Take the **Callahan Tunnel** northeast out of Boston onto **Highway 107** towards **Salem** and continue northeast for about 12 miles. About a mile before reaching Salem, turn right (south) onto **Wilson Street**. The entrance to Highland Park is about 400 yards on the right-hand side.

KEY SPECIES

Resident American Black Duck, Wild Turkey, Cooper's Hawk, American Kestrel, Hairy Woodpecker	**Winter** Bufflehead, Hooded Merganser, Red-bellied Woodpecker, American Tree Sparrow, Dark-eyed Junco
Summer Least Bittern (scarce), Black-crowned Night Heron, Wood Duck, American Woodcock, Spotted Sandpiper, Willow Flycatcher, Eastern Kingbird, Marsh Wren, Warbling Vireo, Swamp Sparrow, Baltimore Oriole	**Passage** Hermit Thrush, Swainson's Thrush, Wood Thrush, Palm Warbler, Black-and-white Warbler, Northern Waterthrush, Scarlet Tanager

6 Nahant (13 miles)

Lynn Commuter Rail Station

Along the north shore of Massachusetts Bay are a number of peninsulas that jut out into the Atlantic Ocean north of Boston and provide excellent vantage points from which to scan for ducks, grebes and seabirds. Nahant Peninsula is one such promontory, about ten miles northeast of the city. As well as being a good location to scan the northern portions of Boston Harbour, its orientation also makes it a migration stopover point for passerines and raptors in spring and autumn. The shoreline on either side of the causeway leading to Nahant is often good for waders.

Take the **Callahan Tunnel** northeast from downtown Boston onto **Route 1A** and continue northeast for about eight miles to **Lynn**. Turn south in Lynn onto **Nahant Road** that leads out onto the narrow Nahant Peninsula (one mile long). Check for migrants in the scrubby areas around the pond and golf course located just behind the beach at the base of the Nahant Peninsula.

KEY SPECIES

Resident Savannah Sparrow	crested Cormorant, Osprey, American Kestrel, American Oystercatcher (scarce), Killdeer, Spotted Sandpiper,	**Winter** Red-throated Loon, Common Loon, Horned Grebe, Red-necked Grebe,
Summer Great Egret, Snowy Egret, Black-crowned Night Heron, Double-	Laughing Gull, Common Tern, Bank Swallow, Eastern Towhee, Orchard Oriole	Great Cormorant, Brant, Greater Scaup, Common Eider, Harlequin Duck (occasional), Surf Scoter,

White-winged Scoter, Common Goldeneye, Bufflehead, Red-breasted Merganser, Dunlin, Sanderling, Bonaparte's Gull, Razorbill, Black Guillemot, Horned Lark, Common

Redpoll (irregular), Snow Bunting

Passage
Black-bellied Plover, Semipalmated Plover, Red Knot, Semipalmated

Sandpiper, Least Sandpiper, Short-billed Dowitcher, Greater Yellowlegs, Bonaparte's Gull, Little Gull (occasional), Belted Kingfisher, flycatchers, vireos, warblers, sparrows

7 Belle Isle Marsh (4 miles)

Suffolk Downs T-Station

This is an area of tidal wetland, with a T-station right beside it, that is only a few minutes from Logan Airport. It is one of the last remnants of saltmarsh remaining in the inner part of Boston Harbour and it provides an opportunity to visualise what the area must have been like before the city developed. The site is good for waders at high tide, since many birds use it as a roost, and the marsh is also frequented by wildfowl and herons. There are three areas to be checked: Rosie's Puddle, which is closest to the T-station; Belle Isle Park, where an observation tower and a boardwalk are located, and The Key at the end of Summer Street. The saltmarsh and the runways of nearby Logan Airport are also worth checking for Snowy Owls in winter.

Take the **Callahan Tunnel** towards **Logan Airport**. Pass the exit for the airport and take the next exit **(Bennington Street)** to the right. Stay on Bennington Street and turn right just before the Suffolk Downs T-Station. This is the area of Rosie's Puddle. Return to Bennington Street and continue past the T-Station. The entrance to Belle Isle Marsh is on the right-hand side of the road.

KEY SPECIES

Summer	Winter	Passage
Double-crested Cormorant, Great Egret, Snowy Egret, Black-crowned Night Heron, Osprey, American Kestrel, American Oystercatcher, Killdeer, Spotted Sandpiper, Laughing Gull, Common Tern, Least Tern, Marsh Wren, Saltmarsh Sharp-tailed Sparrow	Red-throated Loon, Common Loon, Horned Grebe, Great Cormorant, Bufflehead, Red-breasted Merganser, Rough-legged Hawk (uncommon), Northern Harrier, Dunlin, Sanderling, Bonaparte's Gull, Snowy Owl (uncommon)	Peregrine Falcon, Black-bellied Plover, Semipalmated Plover, Semipalmated Sandpiper, White-rumped Sandpiper, Least Sandpiper, Short-billed Dowitcher, Greater Yellowlegs, Lesser Yellowlegs

8 Neponset Estuary – Quincy (8 miles)

North Quincy T-Station

Quincy is on the south shore of Boston Harbour. The best birding area includes re-claimed saltmarsh, mudflats and a park bordering the Neponset River, which empties into Boston Harbour at Quincy. Some of the area is abandoned dockland that has been rehabilitated and transformed into habitat for birds, as well as for recreational use.

The best place to start is at Squantum Point Park where

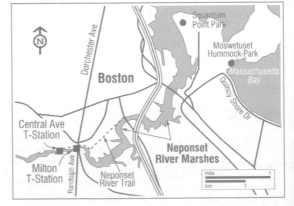

you can scan parts of Boston Harbour and its mudflats, then cross to the south shore of Quincy, where you can scan Quincy Bay with the benefit of some height gained at Moswetuset Hummock Park. After that walk upstream along the north bank of the Lower Neponset River Trail as far as the T-Station at Central Avenue.

KEY SPECIES

Summer	Winter	Passage
Great Egret, Snowy Egret, Black-crowned Night Heron, Green Heron, Glossy Ibis, Double-crested Cormorant, Osprey, American Oystercatcher, Killdeer, Willet, Spotted Sandpiper, Laughing Gull, Common Tern, Belted Kingfisher, Northern Rough-winged Swallow, Saltmarsh Sharp-tailed Sparrow, Savannah Sparrow	Red-throated Loon, Common Loon, Horned Grebe, Red-necked Grebe, Great Cormorant, Brant, American Black Duck, Common Eider, Surf Scoter, White-winged Scoter, Long-tailed Duck, Bufflehead, Common Goldeneye, Red-breasted Merganser, Sanderling, Bonaparte's Gull, Short-eared Owl (scarce), American Tree Sparrow, White-throated Sparrow	Black-bellied Plover, Semipalmated Plover, Ruddy Turnstone, Semipalmated Sandpiper, Least Sandpiper, Short-billed Dowitcher, Greater Yellowlegs, Lesser Yellowlegs

9 Great Meadows National Wildlife Refuge (18 miles)

Concord Commuter Rail Station

Great Meadows NWR to the west of Boston consists of about 3,000 acres of riparian forest, wetlands and grasslands which is distributed in several parcels along the Sudbury and Concord Rivers. It attracts large numbers of wildfowl during migration and the water levels are carefully managed to maintain the wetland in prime condition. The refuge also supports a good selection of typical New England landbirds. This is quite an extensive site and is probably best visited by car. However, the area just east of Concord on State Route 62, and which is served by commuter rail, is quite productive and has a viewing platform to scan the marsh. There are also several loop trails through the wetlands and alongside two shallow, freshwater pools. This area is best visited in summer and autumn: during midwinter there are few birds and the bitter cold can make for fairly miserable birding.

Take **Highway 2** (Cambridge Turnpike) northwest from Boston to **Concord**. In Concord centre go northeast on **Bedford Street** (SR 62) for about a mile and turn left at **Monsen Road**. This leads to the refuge entrance.

KEY SPECIES

Resident		Northern Shoveler, Northern Pintail,
Wild Turkey, Ruffed Grouse, Red-bellied Woodpecker, Fish Crow, Horned Lark, Cedar Waxwing, House Finch	Flycatcher, Eastern Kingbird, Tree Swallow, Bank Swallow, Marsh Wren, Warbling Vireo, Scarlet Tanager, Swamp Sparrow	Ring-necked Duck, American Coot, American Golden Plover (occasional), Semipalmated Sandpiper, Pectoral Sandpiper, Greater Yellowlegs,
Summer	**Winter**	Solitary Sandpiper, Wilson's Snipe, American Pipit, Wood Thrush, Palm Warbler, sparrows, Bobolink
Great Blue Heron, Black-crowned Night Heron, Turkey Vulture, Wood Duck, Spotted Sandpiper, Willow	Northern Shrike (irregular), American Tree Sparrow	
	Passage	
	American Black Duck, Gadwall,	

Further Away

10 Plum Island (40 miles)   *Newburyport Commuter Rail Station*

Plum Island and the adjacent Parker River National Wildlife Refuge comprise one of the best sites in the eastern USA and are less than an hour's drive north of Boston. The refuge contains upland woods, coastal thickets, mudflats, freshwater and saltwater wetlands, and an extensive sand-dune system. The refuge is well provided with strategically located boardwalks and observation platforms. The site is interesting at any time of the year, but especially during migration, and over 300 bird species have been seen here. It also sees some significant movements of raptors in spring, particularly during westerly winds.

Take **I-95** north for about 30 miles and exit east at **Exit 56** towards **Newburyport**. Continue east at Newburyport on **Plum Island Turnpike**, which leads to the refuge entrance after about three miles. There is also a commuter rail station in Newburyport. Although it involves a three-mile walk from the station to the refuge entrance, half the journey is along the southern shore of the **Merrimack River** estuary, an area offering many opportunities to scan for waterfowl, waders and gulls at any season of the year. Moreover, it is such a popular site that you might readily hitch a lift back to Newburyport from other passing birders. This is such a superb birding area that it is worth making the effort to get there.

KEY SPECIES

Resident
American Black Duck, Gadwall, Northern Harrier, Savannah Sparrow

Summer
Least Bittern, Great Egret, Snowy Egret, Green Heron, Black-crowned Night-Heron, Glossy Ibis, Wood Duck, Blue-winged Teal, Osprey, American Kestrel, Killdeer, Willet, Piping Plover, Spotted Sandpiper, Laughing Gull, Common Tern, Forster's Tern (scarce), Least Tern, Willow Flycatcher, Eastern Kingbird, Purple Martin, Marsh Wren, Gray Catbird, Eastern Towhee, Saltmarsh Sharp-tailed Sparrow, Field Sparrow

Winter
Red-throated Loon, Common Loon, Horned Grebe, Red-necked Grebe, Great Cormorant, Greater Scaup, Bufflehead, Common Eider, Long-tailed Duck, Surf Scoter, White-winged Scoter, Common Goldeneye, Barrow's Goldeneye, Red-breasted Merganser, Rough-legged Hawk, Northern Harrier, Purple Sandpiper, Dunlin, Sanderling, Kumlien's Gull, Bonaparte's Gull, Snowy Owl (irregular), Horned Lark, Northern Shrike, American Tree Sparrow, Snow Bunting, Lapland Bunting

Passage
Northern Pintail, Green-winged Teal, Northern Gannet, Sharp-shinned Hawk, Cooper's Hawk, American Kestrel, Merlin, Peregrine Falcon, Black-bellied Plover, Semipalmated Plover, Least Sandpiper, Baird's Sandpiper (scarce), White-rumped Sandpiper, Pectoral Sandpiper, Red Knot, Stilt Sandpiper, Short-billed Dowitcher, Long-billed Dowitcher, Greater Yellowlegs, Wilson's Phalarope, flycatchers, warblers

11 Boston Harbour Cruises   *Long Wharf – City Centre*

A number of whale-watching trips cruise Massachusetts Bay and south to Cape Cod Bay between April and October. Humpback, Fin and Minke Whales are all possible. These cruises can also be a good way to see seabirds, which are frequently attracted by the feeding activity of the whales. Boston Harbour is also peppered with islands, some of which support breeding colonies of seabirds, while others are roosts and feeding grounds for waders. Access is permitted to some of these islands: George's Island, Gallops Island and Peddocks Island being the most productive for visiting birders.

The departure point for both the harbour island cruises and the whale-watching trips is Long Wharf near the city centre. There are also a number of cross-harbour ferries which can be an alternative in winter, when there are no whale-watching trips, even though they don't stop for cetaceans or birds! The Boston to Hingham Harbour ferry runs year-round and offers the best chance of getting a sea view of the wintering sea-ducks around Hingham Bay. There is also a summer-only ferry to George's Island.

KEY SPECIES

Summer	Winter	Passage
Common Eider, Great Shearwater, Sooty Shearwater, Manx Shearwater, Wilson's Storm-petrel, Black-crowned Night-Heron, Snowy Egret, Double-crested Cormorant, Osprey, American Oystercatcher, Killdeer, Willet, Common Tern, Least Tern (shearwaters and petrels are possible only from whale-watching boats)	Red-throated Loon, Common Loon, Horned Grebe, Red-necked Grebe, Great Cormorant, Brant, Common Eider, Surf Scoter, White-winged Scoter, Common Goldeneye, Bufflehead, Red-breasted Merganser, Sanderling, Bonaparte's Gull, Razorbill (occasional), Black Guillemot	Northern Fulmar (occasional), jaegers, Black-legged Kittiwake, alcids (all of these are possible only from whale-watching boats)

Useful contacts and websites

Massachusetts Audubon Society
www.massaudubon.org
Massachusetts Audubon Society, 208 South Great Road, Lincoln, MA 01773. Tel. 781 259 9500.

Massbird.org
www.massbird.org
Portal site for bird clubs, activities and recent sightings in Massachusetts.

Brookline Bird Club
www.massbird.org/bbc

South Shore Bird Club
www.ssbirdclub.home.comcast.net/ssbc.html

Friends of Fresh Pond Reservation
www.friendsoffreshpond.org

Parker River National Wildlife Refuge
www.parkerriver.fws.gov

Pelagics - New England Seabirds
www.neseabirds.com/

Boston Harbour Cruises – Island ferries and whale-watching
www.bostonharborcruises.com

New England Aquarium – Whale-watching
www.neaq.org/visit/wwatch/index.html

Massachusetts Rare Bird Alert
Tel. 888 224 6444 Statewide.

Massachusetts Bay Transportation Authority
www.mbta.com

Tides for Massachusetts Bay
www.saltwatertides.com/dynamic.dir/massachusettssites.html#boston

Books and publications

Birds of Massachusetts (Natural History of New England Series) Richard R. Veit, Wayne R. Petersen (1993) Massachusetts Audubon Society ISBN 0932691110

A Birder's Guide to Eastern Massachusetts American Birding Association (1994) ISBN 1878788086

A Birder's Guide to Metropolitan Areas of North America Paul Lehman (2001) American Birding Association, ISBN 1878788159

A Field Guide to the Birds of Eastern and Central North America Roger Tory Peterson (2002) Houghton Mifflin, ISBN 0395740460

National Geographic Field Guide to the Birds of North America, 4th Edition National Geographic (2002), ISBN 0792268776

Birds of Boston Chris C. Fisher and Andy Bezener (1998) Lone Pine Publishing, ISBN 1551051826

The North American Bird Guide David Sibley (2000) Pica Press, ISBN1873403984

Field Guide to the Birds of Eastern North America David Sibley (2003) Christopher Helm ISBN 0713666579

National Geographic Guide to Bird Watching Sites: Eastern US Mel White (1999) National Geographic, ISBN 0792273745

A Peregrine Falcon perches high above the Chicago skyline

Chicago

Situated in northeastern Illinois on the western shores of Lake Michigan, Chicago is well placed for any birder wishing to experience the avifauna of the Great Lakes of North America. Although far removed from the oceans, its location between the Great Lakes and several tributaries of the Mississippi means that the State of Illinois lies in the spring migration path of birds moving up from the Gulf Coast, and the Chicago area is where many make first contact with Lake Michigan. Unfortunately, numerous birds perish as they strike the glass-fronted skyscrapers in the downtown area, so much so that the city has instituted a 'lights out' policy for high buildings during the migration period. In autumn, Lake Michigan acts like a funnel for birds moving south along its east or west shores and many pass through the Chicago area before moving onwards along the Illinois and Mississippi River systems. Northwesterly winds during these periods cause the birds to stop-over in areas of suitable habitat along the lakeshore and in pockets of woodland in the western suburbs of the city.

Although Chicago experiences a significant passage of passerines, it is just a little too far south to support significant numbers of breeding thrushes and wood-warblers, so midsummer can be a rather quiet period ornithologically. However, its location on the extreme eastern edge of the Great Plains, means that it is possible to see certain grassland species that are rare or absent further east. Much of the original prairie has been reclaimed for farming but pockets of habitat have been preserved.

There are good birding locations north, south and west of Chicago, and the opportunity to visit these sites is only limited by the distance you are prepared to travel. However, there are some very good sites within an hour's travelling time of the city and even in the city centre, the lakeshore parks are very interesting.

GETTING AROUND

As Chicago originally developed as a railway hub, its public transport system is very comprehensive. This is just as well, as it eliminates the need to use a car, particularly near the city centre. Driving in the city is complicated by heavy traffic, restrictions and trying to navigate the street system. Within the city you can use the Chicago Transit Authority

(CTA) buses and subway. The subway runs on elevated tracks over part of the system and is also referred to as the 'El'.

Out of town birding is facilitated by the suburban (METRA) rail system which can get you to the outskirts quickly, important in a city which is almost 30 miles from centre to outer suburbs. However, the bus system (PACE) in the suburbs is rather patchy and many routes operate only during rush hour and mostly to serve business parks and park-and-ride facilities. Particularly in West Chicago, the use of a car would be very beneficial and would enable you to visit several good sites within a ten-mile radius of Fermi Lab Campus, which acts as a kind of focal point for birding activity in DuPage County.

Most of the sites mentioned are public-access areas and are landscaped with trails and observation points, and have car-parking facilities. During spring thaw, however, some of the woodland trails get rather muddy, so boots are required. A telescope is necessary for scanning Lake Michigan or for wader-watching, but many of the smaller lakes and ponds can be covered adequately with binoculars. If you arrive without optics or fieldguide, there are many shopping areas both downtown and in the suburbs which are open late. The area near the University of Chicago has many bookstores that specialise in rare and obscure titles, so if you are interested in rare bird books you could combine birding in Jackson Park with a browse around the bookstores nearby.

The best motoring map to use is the AAA Road Map of Chicago and Vicinity or the Gousha Series Map of Metropolitan Chicago. If you are relying on public transport, make sure you also get a Transit System Map, obtainable at most bus and train stations.

Locations

1 **Chicago Lakeshore** (City Centre)

There is full public access along the entire Chicago shore of Lake Michigan, and along this stretch there are several public parks which provide sufficient cover to encourage migrant passerines to stop-over and to provide breeding sites for a number of species. The lake itself supports good numbers of wildfowl and gulls, depending on time of year, and the area is so well watched that it turns up a wide range of rarities, so that there is nearly always something interesting to see. Added to that, some of these sites are within walking distance of downtown and all are easily accessible by public transport. The process of creating more habitat is continuing as more publicly-owned land comes under the control of the Chicago Park District.

2 Grant Park (City Centre)

This is the closest lakeshore park to the city centre and, although one of the poorest from a habitat perspective, it still manages to attract a good selection of migrants. It is mostly manicured lawns, though there are many trees and some shrubs and these are worth checking for migrant warblers and flycatchers, particularly after a 'fall'. The main attraction of this park is that it is within walking distance of the city centre and the tourist areas on the waterfront from where the tourist cruises operate. So this is a good spot to check if you find yourself in the area with an hour or two to spare.

3 McCormick Bird Sanctuary and Meigs Field (City Centre)

27th St METRA Station

The sanctuary is a recently created, seven-acre prairie adjacent to McCormick Place Convention Center and the lakeshore. The prairie is fenced, but there is a viewing platform to the east and a 'water feature' (a small pond with flowing water) outside the fence. Although only recently planted, it looks like it will be a good migrant trap in the future.

Meigs Field is a disused airport (now called Northerly Island), one third of which has been converted into natural habitat, including wetlands, replanted grassland and the lakeshore beach. This is one of the most regular sites in Chicago for Snowy Owls, and it is also good for other winter visitors, such as sparrows and buntings.

Both sites are accessed from **Lakeshore Drive**: turn east on **18th Drive** to park in the McCormick Convention Center car park. The car park is north of the Convention Center, but the Bird Sanctuary is south of the center so this involves a walk of almost a mile (yes, it's that big!). A better way to get there is to take the METRA train to 27th Street and walk a couple of blocks south to 31st Street which crosses Lakeshore Drive and gets you to the birding areas quicker.

From Lakeshore Drive, turn east on **McFetridge Drive** to access the north end of Northerly Isle. The birding areas are at the south end of the island.

4 Lincoln Park (City Centre)

Wilson Avenue Red Line El Station
Addison Avenue Red Line El Station

Situated just north of the city centre or Loop, as it is known, Lincoln Park is a rather long, linear lakeside park, which is used by Chicagoans for a variety of leisure purposes but, at quiet times during the week and early mornings at the weekend, this can be a really good site for birds. The best locations for birds are in the northern half between Belmont Harbour and Montrose Harbour. The famous Magic Hedge is located on Montrose Point promontory. This is one of the best migrant traps in the Great Lakes region and in spring and autumn is the location for numerous sightings of rare passerines. Further south within the park, near the Chicago Academy of Sciences Nature Museum, is the North Pond, which provides habitat for aquatic birds. Any point along the shoreline provides a vantage point to scan the lake for gulls, terns and waterfowl. During strong onshore winds, Montrose Harbour can provide a sheltered area for loons, grebes and ducks.

5 Jackson Park (City Centre)   *57th St METRA Station*

This is one of several lakefront parks in the Chicago urban area, and is one of the city's premier birding sites, with records of some 300 species. It is easy walking distance from the University of Chicago campus. Its centre of attraction is Wooded Island, which is among the best urban parks in the area for migrant songbirds, particularly in spring. On an average day in the middle third of May, you can expect to encounter more than 20 species of wood-warblers as well as flycatchers, vireos, wrens, thrushes, orioles, etc. Scarcer sparrows are more likely to be in Bobolink Meadow, just east of Wooded Island. Species diversity is much lower in winter, when the lagoons are frozen.

On passage days, the lakefront promontory at 55th St can provide remarkably effortless and rewarding birding. This is also a good site for viewing waterfowl and gulls in winter. The beach and harbours at 63rd St can host a variety of gulls and waterfowl, and waders may stop at the beach on passage and should be looked for in the morning. Jackson Park is home to the oldest thriving colony of Monk Parakeets in the region: this species was admitted to the Illinois list of established species in 1999. There are weekly bird walks here every Saturday led by members of the local Audubon Chapter, and it is a good way to meet local birders.

KEY SPECIES (sites 1–5)

Resident Cooper's Hawk, Peregrine Falcon, Monk Parakeet (I), Hairy Woodpecker	**Winter** Canvasback, Lesser Scaup, Ring-necked Duck, Brown Creeper	Common Nighthawk, Yellow-bellied Sapsucker, Red-headed Woodpecker, Ruby-throated Hummingbird, Least Flycatcher, thrushes, vireos, warblers, sparrows
Summer Green Heron, Wood Duck, Spotted Sandpiper, Willow Flycatcher, Brown Thrasher, Blue-gray Gnatcatcher, Indigo Bunting, Baltimore Oriole	**Passage** Pied-billed Grebe, Greater Scaup, Semipalmated Plover, Semipalmated Sandpiper, Solitary Sandpiper, Spotted Sandpiper, Caspian Tern,	

6 Calumet Lake (17 miles)   *Hegewisch Railway Station*
93rd Street METRA Station

Calumet Lake, at the southern end of Lake Michigan, is a truly extraordinary place. This area is a mosaic of wetlands, open waters, riparian woodlands and restored prairie habitats which is completely hemmed in by steel mills, oil refineries and various industrial sites. That it was never totally reclaimed for development is thanks to the concerted efforts of conservationists and, today, attempts are being made to restore some of the disused industrial sites to wildlife habitats. Because of the dispersed nature of this site, it would be difficult to cover all areas without a car. However, the best areas – Lake Calumet, Powderhorn Prairie and Wolf Lake – can be visited in a day by using CTA Bus Route #30, which shuttles between Hegewisch Rail Station and 93rd St METRA Station. Waders are found on both Calumet Lake and Wolf Lake wherever there is mud exposed and this varies from year to year, depending on water levels.

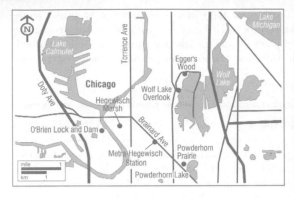

Take **I-94** south to **130th St**, exit east onto 130th St and turn right just before the bridge over the **Calumet River**. There is a car park at **O'Brien Lock & Dam** on the Calumet River, which is a good spot for gulls in winter. From here, return to 130th St but turn west and then north on **S. Doty Avenue** which runs parallel to I-94 but alongside the shore of Lake Calumet. There are several lay-bys on Doty Avenue giving views of the lakeshore.

Return to 130th St and continue east then south on **S. Brainard Avenue** to get to Powderhorn Lake and Prairie. Take **Avenue O** north from Brainard Avenue to get to **Wolf Lake Overlook** and **Egger's Grove**, which is the best area for Yellow-headed Blackbirds.

KEY SPECIES

Summer
Least Bittern, Great Egret, Yellow-crowned Night Heron, Green Heron, Wood Duck, Turkey Vulture, Sora, Killdeer, Spotted Sandpiper, American Woodcock, Blue-gray Gnatcatcher, Eastern Bluebird, Savannah Sparrow, Vesper Sparrow, Indigo Bunting, Bobolink, Eastern Meadowlark, Yellow-headed Blackbird

Winter
American Black Duck, Ring-necked Duck, Lesser Scaup, Northern Harrier, Rough-legged Hawk, Wilson's Snipe, Great Black-backed Gull, Thayer's Gull, Northern Shrike

Passage
Pied-billed Grebe, Canvasback, Ring-necked Duck, Bald Eagle, Osprey, Sandhill Crane, Semipalmated Plover,

Semipalmated Sandpiper, Least Sandpiper, Pectoral Sandpiper, Solitary Sandpiper, Short-billed Dowitcher, Greater Yellowlegs, Lesser Yellowlegs, Wilson's Phalarope, Caspian Tern, Forster's Tern, Yellow-bellied Sapsucker, American Pipit, thrushes, flycatchers, vireos, warblers

7 **Morton Arboretum** (25 miles)   *Pace Bus #829 from Lisle METRA Station*

This is a good year-round site, as it supports a diverse range of breeding birds, migrants and winter visitors. The arboretum has the only known breeding population of Yellow-throated Warblers in the Chicago area, and is important for other populations of passerines. It is most popular among Chicago birders as a migrant hotspot where over 30 species of warbler can be seen in spring, including some of the rarer species, as well as thrushes, flycatchers and vireos. Winter can also be equally good for raptors and for finches such as crossbills.

This is a rather large site but in the spring and autumn the best location is around Lake Marmo on the west side of the arboretum. Warblers frequent the streams near Parking 23 and other parking lots along the Main West Route. On the east side, Parking 7, Parking 15 and the Big Rock Visitors Station are also good locations. In winter the large pine collections on Frost Hill and Hemlock Hill should be checked for wintering owls and crossbills. In the summer, prairie grassland breeding birds are found in the grasslands in the northwest corner of the arboretum: park at Parking 22 then head north and west.

Hidden Lake, located just north of the arboretum on Route 53 is a good location for ducks and other waterfowl. There are two lakes fed by the DuPage River. Waders can occasionally be present when water levels are low and herons can be seen in small numbers in spring through autumn. Warblers can also be seen along the gravel trail that loops around the back lake. A couple of small paths go east along the river and can provide close views of aquatic birds. Grassland birds can also be seen in this area.

Take **I-88** west and exit north onto **Route 53** for about 300 yards. The arboretum lies on both sides of the road with the main entrance on the right-hand (east) side. There is a small entrance fee and you may drive around the arboretum.

There is a train service to Lisle just on the south side of I-88. The station is about a mile from the entrance and there is a rush-hour feeder bus service to Warrenville Road, about half a mile closer to the entrance.

KEY SPECIES

Resident Cooper's Hawk, Hairy Woodpecker, Great Horned Owl	Orchard Oriole, Scarlet Tanager, Veery, Savannah Sparrow, Field Sparrow	**Passage** Horned Grebe, Black Duck, Caspian Tern, Wilson's Snipe, Winter Wren, thrushes, flycatchers, vireos, warblers, sparrows
Summer Yellow-throated Warbler, Pine Warbler, Eastern Meadowlark, Bobolink, Brown Thrasher, Eastern Bluebird, Yellow-billed Cuckoo,	**Winter** Sharp-shinned Hawk, Red-breasted Nuthatch, Red Crossbill, White-winged Crossbill	

8 Springbrook Prairie (30 miles)

This is a preserved area of mixed habitat combining marsh, woodland and prairie grassland in southwest Chicago. It is a good site for breeding birds and for migrant species while, in winter, it is also good for raptors. Many birds can be seen at the marsh by the main parking lot on the west side of the reserve: particularly aquatic species but also several sparrow species in late autumn. In spring, warblers and other migrants can be seen in good numbers along the river and in the woods on the south end of the reserve. In summer, many grassland birds are present throughout the reserve including an occasional Dickcissel. A cross-country trek, away from the trails, is often required to view these species. However, be careful to avoid possible nesting areas and note that ticks can also be a problem in summer: tall rubber boots and/or insect repellent are recommended. Warblers and other migrants can be found along the river to the south of the parking lot and in the woods south of the river on both the east and west sides of the reserve. In late autumn and winter, Short-eared Owls can be seen hunting along Book Road at the west end of the reserve.

Take **I-55** west to the **Kingery Highway Interchange (Exit 274)**. Go north on Kingery Highway for about half a mile before turning west on **State Route 33**. Go west on SR33 for about 11 miles and turn south on **Plainfield-Naperville Road (CR 1)**. The main car park is on the right-hand (west) side of this road after about half a mile. This site is not so easy to get to by public transport. There is a bus service (#683) from Naperville Metra Station, about three miles away, but it is infrequent and operates only during rush hour.

KEY SPECIES

Summer Sedge Wren, Bobolink, Grasshopper Sparrow, Henslow's Sparrow, Savannah Sparrow, Field Sparrow, Eastern Meadowlark, Orchard Oriole	Northern Harrier, Northern Shrike, Short-eared Owl **Passage** Pied-billed Grebe, American Bittern, Least Bittern, Sora, Virginia Rail, Marsh Wren, Nelson's Sharp-tailed Sparrow, thrushes, flycatchers, vireos, warblers
Winter Rough-legged Hawk,	

9 Fermi Laboratory Campus (35 miles)

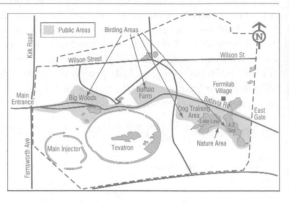

Of course, the Fermi Lab is famous among physicists for the many discoveries made at this site. It is, however, also well known to birders in the Chicago area because of the 6,800 acres of woodland, open grasslands and lakes which provide habitats for many bird species. Apart from a few restricted areas, access is permitted to the site to members of the general public who first obtain a pass at the Pine Street Entrance. Opening hours are from 08.00 to 20.00 and you may take a car into the grounds. This is a large site and a visitor on foot may want to consider taking two days to cover all of the suitable habitats. The best areas are the lakes in the southeast quadrant for waterfowl; the 'Big Woods' in the northwest quadrant for woodland species; the Buffalo Fields, which hold flocks of geese during migration and sparrows in the winter; and the open fields at the southern end of Eola Road for grassland species. Illinois is in the overlap zone for both Western and Eastern Meadowlarks, but you are most likely to see Eastern Meadowlarks at Fermi Lab.

To get there, take **I-88** west from Chicago, and exit north on **Farnsworth Avenue (State Route 77)**. Follow SR 77 for about three miles and the entrance to the campus is on the right. It is not so easy to get to the Fermi Lab Campus without a car and, moreover, as it is such a large site, it would be very tiring to have to walk around it. However, there are METRA stations at Geneva and at Aurora which are about five miles away so it would be possible to get a taxi from either station to the Education Center within Fermi Lab Campus and this would be convenient for the birding areas around the Big Woods and the Buffalo Field. Alternatively, there is a bus service (Pace Bus Route #676) which operates from Naperville METRA Station and, by a circuitous route, will bring you to Summer Lakes Park which is at the east entrance to Fermi Lab on Batavia Road. However, this service only operates on weekdays during rush hour.

KEY SPECIES

Resident
Cooper's Hawk, Hairy Woodpecker, Horned Lark, White-breasted Nuthatch

Summer
Great Egret, Green Heron, Wood Duck, Turkey Vulture, Virginia Rail, Sora, Killdeer, Spotted Sandpiper, American Woodcock, Yellow-billed Cuckoo, Ruby-throated Hummingbird, Willow Flycatcher, Bell's Vireo, Warbling Vireo, House Wren, Marsh Wren, Eastern Bluebird,

Yellow-breasted Chat, Scarlet Tanager, Henslow's Sparrow, Savannah Sparrow, Grasshopper Sparrow, Indigo Bunting, Dickcissel, Bobolink, Eastern Meadowlark, Orchard Oriole, Baltimore Oriole

Winter
Snow Goose, American Black Duck, Northern Pintail, Lesser Scaup, Northern Harrier, Sharp-shinned Hawk, Rough-legged Hawk, Wilson's Snipe, Northern Shrike

Passage
Pied-billed Grebe, Horned Grebe, Canvasback, Ring-necked Duck, Osprey, Sandhill Crane, Semipalmated Plover, Semipalmated Sandpiper, Least Sandpiper, Pectoral Sandpiper, Solitary Sandpiper, Short-billed Dowitcher, Greater Yellowlegs, Lesser Yellowlegs, Wilson's Phalarope, Caspian Tern, Yellow-bellied Sapsucker, thrushes, flycatchers, vireos, warblers

10 Pratt's Wayne Woods (35 miles)

Pace Bus #801 from Elgin METRA Station

Pratt's Wayne Woods is one of the largest forest reserves in DuPage County and because of the variety of habitats, one of the best for birds. There is a mixture of forest, grassland and marshes. As well as witnessing the spring and autumn warbler and fly-catcher migration, this site also supports an excellent selection of breeding species, particularly some grassland birds which are difficult to find elsewhere in the Chicago area. There are more than 12 miles of trails traversing all the main habitats. However, the wetland areas around the car parks and main lake are best if you are short of time.

Take **I-88** west from Chicago, and exit north on **State Route 59**. After about ten miles turn left (west) on **Army Trail Road** and then, after about two miles, turn right (north) onto **Powis Road**. The car park is on the left-hand side of this road. The park is about three miles from Elgin METRA Station, but there is a bus service from Elgin METRA Station to Charlestowne Mall, which is about a mile southwest from the park.

KEY SPECIES

Resident Cooper's Hawk, Hairy Woodpecker, White-breasted Nuthatch	Northern Rough-winged Swallow, Barn Swallow, House Wren, Willow Flycatcher, Eastern Bluebird, Field Sparrow, Indigo Bunting, Eastern Meadowlark, Baltimore Oriole	**Passage** Osprey, Solitary Sandpiper, Greater Yellowlegs, Yellow-bellied Sapsucker, Least Flycatcher, Acadian Flycatcher, Yellow-bellied Flycatcher, Yellow-throated Vireo, Veery, Hermit Thrush, Swainson's Thrush, Wilson's Warbler, Magnolia Warbler, Ovenbird, Winter Wren, Rose-breasted Grosbeak
Summer Great Egret, Green Heron, Wood Duck, Turkey Vulture, Killdeer, Spotted Sandpiper, Ruby-throated Hummingbird, Warbling Vireo,	**Winter** Sharp-shinned Hawk, Red-breasted Nuthatch	

11 O'Hare Airport Pools (17 miles)

Bus #332

If you travel much within the USA, the chances are that you will pass through O'Hare Airport at some stage, as it is probably the busiest hub airport in the world. If you are on a stop-over, even for just a few hours, this site offers an opportunity to see a fine selection of waders, close up and in a comparatively small area. It is best visited at migration times but also even in summer, when there is a chance of some stragglers or non-breeders.

From O'Hare Airport, turn south on **Mannheim Road (Route 12)**, and follow the airport perimeter to Irving Park Road. Turn right (west) on **Irving Park Road** until you come to the junction with Cargo Road. Turn right on **Cargo Road** and the pools are on either side of the road, adjacent to the large **Postal Depot**. There is a bus service to the Post Office (No. 332) from Rosemont Subway Station (one stop from the airport), but as it is just beyond the perimeter of the airport, it is much more convenient to take a short cab drive from the terminal.

KEY SPECIES

Passage Semipalmated Plover, Semipalmated Sandpiper, Least Sandpiper, Pectoral Sandpiper, Solitary	Sandpiper, Stilt Sandpiper, Short-billed Dowitcher, Greater Yellowlegs, Lesser Yellowlegs, Wilson's Phalarope

Further Away

12 Illinois Beach State Park (45 miles)

Waukegan METRA Station
Winthrop Harbour METRA Station

Located north of Chicago on the western shore of Lake Michigan and touching the border with Wisconsin, Illinois Beach Park is well worth the trip during autumn migration as it is one of the best sites in Illinois to observe raptor migration. It is also good at other times of the year because of its extensive size (over 4,000 acres) and variety of habitats, wetlands, woods and sand dunes. Its 6.5 miles of uninterrupted shoreline (apart from the nuclear power station right in the middle of it all!) gives some impression of what this area was like before the city spread out in all directions and creates a kind of wilderness effect which would be impossible to visualise in the more built-up sites on the southern lakeshore.

The best starting point is at the Park Headquarters, from where several trails fan out to the lakeshore and through marshy habitat and rather open woodlands. Take **I-94** north to **Zion** and exit east on **State Route 173**. After about seven miles, turn south on **Sheridan Road**. Go south for about two miles to **Wadsworth Avenue** where the entrance is signposted. About five miles further south from the park entrance on Sheridan Road, **Waukegan Harbour** is good for waterfowl and gulls in winter, and for waders in autumn. Alternatively, in autumn if the winds are westerly, go north on Sheridan Road to **Winthrop Harbour** from where, in these conditions, a steady southerly movement of buteos and accipiters may be seen. Winthrop Harbour is also an excellent site for gulls in winter. There are METRA stations at Waukegan, Zion and at Winthrop Harbour, adjacent to the park.

KEY SPECIES

Resident
Pied-billed Grebe, Cooper's Hawk, Hairy Woodpecker

Summer
Least Bittern, Great Egret, Green Heron, Wood Duck, Turkey Vulture, Virginia Rail, Sora, Killdeer, Spotted Sandpiper, American Woodcock, House Wren, Marsh Wren, Blue-gray Gnatcatcher, Eastern Bluebird, Veery, Ovenbird, Scarlet Tanager, Savannah Sparrow, Vesper Sparrow, Indigo

Bunting, Bobolink, Eastern Meadowlark, Baltimore Oriole

Winter
Horned Grebe, American Black Duck, Canvasback, Ring-necked Duck, Greater Scaup, Lesser Scaup, Ruddy Duck, Northern Harrier, Rough-legged Hawk, Great Black-backed Gull, Brown Creeper, Snow Bunting

Passage
White-winged Scoter, Bald Eagle,

Broad-winged Hawk, Red-shouldered Hawk, Osprey, Sharp-shinned Hawk, Peregrine Falcon, Merlin, Sandhill Crane, Semipalmated Plover, Semipalmated Sandpiper, Least Sandpiper, Dunlin, Pectoral Sandpiper, Ruddy Turnstone, Solitary Sandpiper, Stilt Sandpiper, Short-billed Dowitcher, Greater Yellowlegs, Lesser Yellowlegs, Wilson's Phalarope, Caspian Tern, Common Nighthawk, thrushes, flycatchers, vireos, warblers

13 Indiana Dunes (50 miles)

Ogden Dunes Station
Beverly Shores Station

The Indiana Dunes National Lakeshore Park consists of nearly 15 miles of sand dunes along the south shore of Lake Michigan behind which are wetlands, prairies and woodland. Although situated almost 50 miles south of Chicago in the State of Indiana, it is easy to get to by train and there are three stations within the park, enabling you to cover several stretches of the park if you do not have a car. This extensive area offers many different habitats and is worth visiting in any season. The south shore of Lake Michigan acts like a funnel for migrants and vagrants heading south and acts as a barrier for passerines heading north, which need to skirt the shoreline. For as long as the lake remains ice-free in winter large numbers of waterfowl can be seen and winter gull enthusiasts can search for Thayer's, Glaucous and Great Black-backed Gulls. There are also a few species to be found in this area which are rarer or more difficult to find in the neighbouring areas of Illinois: such as Red-headed Woodpecker, Carolina Wren, Tufted Titmouse and Cerulean Warbler.

The first stop is Ogden Dunes Station. There are two main sets of trails in this area. Closest to Ogden Dunes Station is the Inland Marsh Trail which is opposite the station on the south side of Highway 12. About 1.5 miles west of the station is the main park entrance on County Line Road, from where several trails lead to the shoreline dune system and Long Lake. The next stop is Dune Park Station from where several trails, north of the railway tracks, lead to the lakeshore. Good habitat can be encountered on the Cowles Bog Trail and the Dune Ridge Trail. The Beverly Shores Station is closest to the most easterly end of the Dune Ridge Trail and makes a convenient point to catch the train back to Chicago.

By car, you can reach the park by taking **I-90** south and exiting east onto **Route 12 at Gary (Exit 17)**.

KEY SPECIES

Resident
Pied-billed Grebe, Cooper's Hawk, American Coot, Red-headed Woodpecker, Hairy Woodpecker, Northern Flicker, Tufted Titmouse, Carolina Wren

Summer
Least Bittern, Great Egret, Green Heron, Wood Duck, Turkey Vulture, Virginia Rail, Sora, Killdeer, Spotted Sandpiper, American Woodcock, Acadian Flycatcher, Willow Flycatcher, Eastern Phoebe, House Wren, Marsh Wren, Blue-gray Gnatcatcher, Eastern Bluebird, Veery,

Prairie Warbler, Cerulean Warbler, Ovenbird, Scarlet Tanager, Field Sparrow, Savannah Sparrow, Vesper Sparrow, Bobolink, Eastern Meadowlark, Indigo Bunting, Baltimore Oriole

Winter
Horned Grebe, American Black Duck, Canvasback, Greater Scaup, Lesser Scaup, Ruddy Duck, Northern Harrier, Rough-legged Hawk, Wilson's Snipe, Thayer's Gull, Glaucous Gull, Great Black-backed Gull, Snow Bunting

Passage
White-winged Scoter, Bald Eagle, Red-shouldered Hawk, Osprey, Sharp-shinned Hawk, Sandhill Crane, Semipalmated Plover, Semipalmated Sandpiper, Least Sandpiper, Dunlin, Pectoral Sandpiper, Ruddy Turnstone, Solitary Sandpiper, Stilt Sandpiper, Short-billed Dowitcher, Greater Yellowlegs, Lesser Yellowlegs, Wilson's Phalarope, Bonaparte's Gull, Caspian Tern, Common Nighthawk, Yellow-bellied Sapsucker, thrushes, flycatchers, vireos, warblers

Useful contacts and websites

Chicago Audubon Society
www.chicagoaudubon.org
5801-C North Pulaski Road, Chicago, IL 60646-6057. Tel. 847 299 3505.

Chicago Ornithological Society
www.chicagobirder.org
10th Floor – #C980, 28 East Jackson Building, Chicago, IL 60604. Tel. 312 409 9678.

Illinois Ornithological Society
www.illinoisbirds.org
P.O. Box 931, Lake Forest, IL 60045.

DuPage Birding Club
www.dupagebirding.org/

Bird Conservation Network
www.bcnbirds.org
Coalition of bird conservation organisations in north-east Illinois and neighbouring areas.

City of Chicago Environmental Dept. –
information on birding in the city
www.cityofchicago.org/Environment/BirdMigration

Lake Calumet Wetlands – Sierra Club website:
www.sierraclub.org/chapters/il/calumet

Morton Arboretum
www.mortonarb.org

Indiana Audubon Society – site guide to
Indiana Dunes
www.indianaaudubon.org/Sites/IBA/IndianaDunes/
IndianaDunes.htm

Fermi National Accelerator Laboratory
Campus Wildlife Guide.
www.fnal.gov/pub/about/campus/
ecology/wildlife/index.html

Dupage County Forest Preserves
www.dupageforest.com

Illinois and Chicago Net-Birding
www.home.xnet.com/~ugeiser/Birds

Birding in DuPage
www.bcnbirds.org/dupage

The Gadwall Birding Page
www.gadwall.com/birding/aboutus/index.htm

ILbirds: Illinois birding mail-list
www.groups.yahoo.com/group/ILbirds

Chicago Rare Bird Alert Tel: 847 265 2118

DuPage Rare Bird Alert Tel: 630 406 8111

Chicago Transit Authority – public transport
www.transitchicago.com

METRA Rail Network – Chicago area
www.metrarail.com

Pace Bus Service – bus service within DuPage
County
www.pacebus.com

Books and publications

A Birder's Guide to the Chicago Region
Lynne Carpenter, Joel Greenberg & Kenn Kaufman
(2000) Northern Illinois University Press
ISBN 0875805825

Birds of Chicago Chris C. Fisher & David B.
Johnson (1997) Lone Pine Publishing
ISBN 1551051125

Chicago Area Birds Steven Mlodinow (1992)
Chicago Audubon Society, ISBN 0914091565

Birding Illinois Sheryl De Vore (2000)
Falcon Press, ISBN 1560446897

Birds of Illinois Field Guide Stan Tekiela (2000)
Adventure Publications, ISBN 1885061749

Birds of the Indiana Dunes Kenneth J Brock,
(Revised Edition 1997) Shirley Heinze
Environmental Fund, ISBN 0965935809

**The Chicago Region Birding Trail Guide: City
of Chicago** (2005)
www.egov.cityofchicago.org/webportal/COCWebPo
rtl/COC_ATTACH/birding_guide_finalsingle.pdf

**A Birder's Guide to Metropolitan Areas of
North America** Paul Lehman (2001) American
Birding Association, ISBN 1878788159

**A Field Guide to the Birds of Eastern and
Central North America** Roger Tory Peterson
(2002) Houghton Mifflin, ISBN 0395740460

**National Geographic Field Guide to the Birds
of North America, 4th Edition**
National Geographic (2002), ISBN 0792268776

The North American Bird Guide David Sibley
(2000) Pica Press, ISBN 1873403984

**Field Guide to the Birds of Eastern North
America** David Sibley (2003) Christopher Helm
ISBN 0713666579

**National Geographic Guide to Bird Watching
Sites: Eastern US** Mel White (1999)
National Geographic, ISBN 0792273745

Honolulu

Honolulu is the capital of O'ahu, the third largest and most developed of the Hawaiian Islands. If you had the chance to travel throughout the island chain you would not choose to spend much time on O'ahu as it has fewer endemics than most of the other islands, but for those with limited time there is still plenty to see, particularly in the two mountain ranges that dominate the skyline. The Ko'olau range, which runs along the eastern side of the island is particularly important, being one of the few places to see some of the endemic forest birds, including Apapane, I'iwi, O'ahu Elepaio and O'ahu Amakihi – the latter two being found nowhere else in Hawaii. As with all of the islands, the weather and rainfall are dictated by the trade winds and the windward coasts get more rain than the leeward sides. The mountains receive up to 400cm of rain per year whereas Waikiki gets just 60cm by comparison. The north coast of O'ahu is relatively quiet with fantastic beaches, marshes and scrub, while the south coast is dominated by Honolulu, Waikiki and the industrial developments at Pearl Harbor. The windward (east) coast has more beaches and several off-shore islands that are home to seabirds. By contrast, the leeward (west) coast is dry and provides little for birders, although Ka'ena Point State Park offers good sea-watching opportunities.

White Terns nest in Kapi'olani Park on Honolulu

GETTING AROUND

As the most populous of the Hawaiian Islands, O'ahu has a pretty good public transport service, so if you don't have the use of a car, you can still get around by bus. The main bus terminal is at Ala Moana shopping mall, just west of Waikiki. Most of the buses are fitted with bike racks so if you rent a bike you can take the bus over the mountain ranges and cycle to the birding sites. Taxis are also a practical option for sites close to Honolulu and on the south coast. If you do rent a car, you won't spend a lot of time driving, as few sites are more than an hour's drive from Honolulu and you will be able to visit several in a day.

Locations

1 Kapi'olani Park and Honolulu Zoo (2.5 miles)

These are popular tourist destinations. One of the major attractions in this park is the White Terns which you will see without any difficulty. They nest in banyan trees throughout the park between December and August and can be seen flying from here out to sea. The other species around the park and zoo are mainly introductions. Yellow-fronted Canaries are only found on one other island and are easy to see in the park. Red-whiskered Bulbuls are confined to O'ahu and can usually be seen easily on the zoo lawns. There is an extensive collection of exotic and native birds that is worth visiting. The zoo attracts Black-crowned Night Herons and Cattle Egrets which are used to humans so are easy to photograph. It is worth taking a look offshore as Great Frigatebirds are often seen and those who make the effort to seawatch might encounter shearwaters, Laysan Albatrosses, Brown and Red-footed Boobies and Pomarine Jaegers.

Access to Kapi'olani Park is free and unlimited. Honolulu Zoo is open from 09.00 to 16.30 daily, although it is closed on Christmas Day and New Year's Day. Entry fees apply. This area is very well served by several bus routes and taxis are plentiful.

> **KEY SPECIES**
>
> Great Frigatebird, Black-crowned Night Heron, White Tern, House Sparrow (I), Scaly-breasted Munia (I), Orange-cheeked Waxbill (I), Java Sparrow (I), Yellow-fronted Canary (I), House Finch, Northern Cardinal (I), Red-crested Cardinal (I)

2 Aiea Ridge Loop Trail (9 miles)   Bus #74

The Aiea Ridge Loop Trail is nearly four miles long. It begins in a Eucalyptus grove but soon enters modern Hawaiian forest with good numbers of Ohia and native trees, as well as introductions such as ironwood and Norfolk Pines. From the trail there are great views of Honolulu, Pearl Harbor and the Ko'olau Mountain Range. It is quite easy to see O'ahu Amakihi and Apapane, particularly just after the halfway point. O'ahu Elepaio have also been seen in this area in recent years. One species for which this trail is well known is the Mariana Swiftlet (Guam Swiftlet), which was introduced in the 1960s from Guam. The area where the Swiftlets nest and feed is out of bounds as it fringes the H3 Highway, but birds can occasionally be seen on the eastern side of the trail feeding over the valley and the highway. Sightings are usually from the end of the trail.

From Honolulu head west on **Highway 78** and take the Stadium/Aiea turn-off onto **Moanalua Road**. Turn right into **Aiea Heights Drive** at the traffic lights. Drive up the winding road through a residential area for 2.5 miles to **Keaiwa Heiau State Park**.

> **KEY SPECIES**
>
> Mariana Swiftlet (I), Red-billed Leiothrix (I), O'ahu Elepaio, Japanese Bush-Warbler (I), White-rumped Shama (I), O'ahu Amakihi, Apapane, Scaly-breasted Munia (I)

3 Kaneohe Marine Corps Base and Moku Manu Island (14 miles)

Bus #56/70

Visits to this site are currently (2006) not possible due to heightened security but details are given in case this situation changes. When access is allowed, close views can be obtained of a large colony of Red-footed Boobies in Ulupau Head crater. The birds are present all year but activity is at its greatest during the breeding season when the birds are frequent bombarded by Great Frigatebirds. Black and Brown Noddies nest on the cliffs below Ulupau Head and Laysan Albatrosses are also seen occasionally. Restrictions apply to using photographic and video equipment in the area. Offshore is the island of Moku Manu where breeding species include Sooty Terns, Brown Boobies, Brown Noddies, Grey-backed Terns, Christmas Shearwaters and a few Masked Boobies. Distant views can be obtained.

At the entrance to the Marine Corps Base there are the Nu'upia Ponds, the eastern end of which can be viewed without entering the base. The best views are from the Kane'ohe Bay Drive/H-3 on-off ramp and from the path to the H-3 gate. These ponds have nesting Black-necked Stilts and Common Moorhens, and are good for migrants during spring and autumn.

When visits are allowed you must apply at least 30 days in advance for permission to visit. Applications for a permit should be made to: Commanding General/PAO, Marine Corps Base Hawaii, Box 63002, MCBH Kaneohe, HI 96863-3002, USA.

Tel. 001 808 257 8839. Fax. 001 808 257 2511.

KEY SPECIES

Christmas Shearwater, Laysan Albatross, Great Frigatebird, Red-footed Booby, Brown Booby, Masked Booby, Black-necked Stilt, Common Moorhen, Sooty Tern, Brown Noddy, Black Noddy, Grey-backed Tern, wildfowl, gulls and terns, Orange-cheeked Waxbill (I).

4 Kuli'ou'ou Valley Trail (9 miles)

Bus #22/58/80/80A

This trail is the best site to observe the endangered O'ahu Elepaio. The birds are most vocal in February and March and can be hard to find at other times. If you are patient, you should be able to spot them working their way through the thick vegetation hunting insects. The O'ahu Elepaio is an endangered subspecies of Elepaio endemic to O'ahu. Many authorities consider it to be a separate species and it is likely that it will be split in the future.

Kuli'ou'ou Valley Trail is in east Honolulu about 4.5 miles east of Kahala, just west of Hawaii Kai. To reach the trailhead, get on **Kalanianaole Highway, Route 72**, and head east from Kahala or west from Hawaii Kai. Turn north, towards the mountains, on **Kuliouou Road**. About 500 yards north of Route 72, Kuliouou Road bends sharply to the left, then back to the right. Continue about 500 yards further north, almost to the end of Kuliouou Road, then turn right on **Kala'au Place**. Continue north about 200 yards to the end of Kala'au Place and park along the road, but not in the no parking zone at the very end. The trail starts right at the end of Kala'au Place and follows a streambed, which is usually dry. The trail runs about a mile up the valley through thick, mainly non-native, forest. About 200 yards up the trail,

KEY SPECIES

O'ahu Elepaio, Red-billed Leiothrix (I), White-rumped Shama (I), Scaly-breasted Munia (I), House Finch (I)

Kuli'ou'ou Ridge Trail branches off to the right: however you should remain on the valley trail if you want to see O'ahu Elepaio.

5 Manana Islet and Makapu'u Point (14 miles)

Bus #22/58

Manana Islet is a seabird sanctuary which can only be visited by special arrangement. However using a telescope it is possible to obtain distant views from Makapu'u Point. The most obvious species is Sooty Tern and large numbers fly around the island. Red-footed and Brown Boobies, Great Frigatebirds and Red-tailed Tropicbirds are fairly easily observed. In addition Masked Boobies, plus Black and Brown Noddies can be seen. Dusk is a good time to see Wedge-tailed and Christmas Shearwaters, which nest on nearby islands and are then preparing to return to their nesting burrows. The nearby Sealife Park has a colony of Red-footed Boobies which originally consisted of rehabilitated birds. They are sometimes joined by Brown and Masked Boobies. Occasionally the Hawaii Audubon Society undertakes trips to the islet and it may be worth enquiring when their next trip is planned.

The point is reached via **Highway 72** from the eastern end of Waikiki, or **Highway 61** and then **Highway 72** from downtown Honolulu.

KEY SPECIES

Wedge-tailed Shearwater, Christmas Shearwater, Red-footed Booby, Brown Booby, Masked Booby, Great	Frigatebird, Red-tailed Tropicbird, Sooty Tern, Black Noddy, Brown Noddy

6 Ka'ena Point State Park (38 miles)

Bus #40/52/83

The state park is an undeveloped area of 853 acres of coastal strip on both the north and west sides of Ka'ena Point. The main reason for visiting this site is the presence from December to July of Laysan Albatrosses which occasionally breed. Over 20 have been present in recent years. Ka'ena Point Trail, which is about two miles long, starts at Yokohama Bay (at the end of the paved road) on the south side and follows the old railroad, or starts at the end of the Farrington Highway on the north side. The trail has no shade and can become very hot.

The park is on the northwest corner of O'ahu and is reached on the north shore along the **Farrington Highway (Route 930)** from Waialua or Haleiwa and **Highway 803** from Wahiawa (Bus #52; 83). Continue past Dillingham Airfield which is on the south side of Farrington Highway until the road ends. From the south the park is reached via **Highway 93**, also called Farrington Highway (Bus #40) from Honolulu via Kapolei on the southwest side of O'ahu.

KEY SPECIES

Laysan Albatross, Great Frigatebird, Red-footed Booby, Ruddy Turnstone, Wandering Tattler,	Sanderling, Northern Mockingbird (I), African Silverbill (I)

7 James Campbell National Wildlife Refuge (35 miles)

Bus #55

Situated near the north tip of O'ahu, this is one of Hawaii's premier wetland sites and represents the best chance of seeing the rare Bristle-thighed Curlew which is present all year in small numbers, with up to 20 in winter. There are two refuge units, Kii and Punamano. There is no public access to Punamano and only limited access to Kii from 1 August to 15 February. Guided tours are offered on Thursdays and Saturdays. The main purpose of the refuge is to provide habitat for four endangered birds: the endemic Black-necked Stilt, Hawaiian Coot, Common Moorhen and Hawaiian Duck (although these are now Hawaiian Duck/Mallard hybrids).

The refuge is on the north side of **Kamehameha Highway** on the northeast tip of O'ahu. The Kii unit of James Campbell NWR is located off Kamehameha Highway between the town of Kahuku and the Turtle Bay resort at Kuilima. If the refuge is closed birds may be observable from a cemetery on a rise that overlooks the pastures and dunes between Kahuku Golf Course and the marsh. To reach the cemetery, follow the road at the north end of the parking area on foot (not by car) until it ends at a fence adjacent to the cemetery. The nearby aquafarm is also a good place to look for migrants.

KEY SPECIES

Resident
Black-crowned Night Heron, Hawaiian Duck, Common Pheasant (I), Hawaiian Coot, Common Moorhen, Black-necked Stilt, Bristle-thighed Curlew (easiest in winter), Short-eared Owl, Northern Mockingbird (I), Japanese Bush Warbler (I), Scaly-breasted Munia (I), Java Sparrow (I), House Finch (I), Northern Cardinal (I), Red-crested Cardinal (I)

Winter
Cackling Goose, American Wigeon, Eurasian Wigeon, Green-winged Teal, Northern Shoveler, Blue-winged Teal, Garganey, Northern Pintail, Ring-necked Duck, Lesser Scaup, Greater Scaup, Tufted Duck, Bufflehead, Semipalmated Plover, Ruff, Pectoral Sandpiper, Sharp-tailed Sandpiper, Sanderling, Wandering Tattler, Lesser Yellowlegs, Long-billed Dowitcher, Common Snipe, Ruddy Turnstone

Useful contacts and websites

Birding Hawaii
www.birdinghawaii.co.uk
A very informative site about birding in Hawaii.

Hawaii Audubon Society
www.hawaiiaudubon.com
850 Richards Street, Suite 505, Honolulu, HI 96813-4709. Tel. 808 528 1432. Email: hiaudsoc@pixi.com

Bird Checklists of the United States – Hawaii
www.npwrc.usgs.gov/resource/othrdata/chekbird/r1/15.htm
Useful checklist to help sort out which species are indigenous and which are introduced.

James Campbell NWR
www.fws.gov/pacificislands/wnwr/ojamesnwr.html

Oahu Nature Tours
www.oahunaturetours.com/custombirding.html
Local bird-tour company which organises custom birding tours including pelagic trips from Oahu.

Hawaii Tourist Board
www.visit.hawaii.org

Oahu Transit Services Inc.
www.thebus.org
Bus services throughout Oahu.

Books and publications

A Field Guide to the Birds of Hawai'i and the Tropical Pacific H. Douglas Pratt, Phillip L. Bruner & Delwyn G. Berrett (1987) Princeton University Press, ISBN 0691023999

Enjoying Birds and other Wildlife in Hawai'i H. Douglas Pratt (2002) Mutual Publishing ISBN 0935180001

The Birdwatchers Guide to Hawai'i Rick Soehren (1996) University of Hawai'i Press ISBN 0824816838

Hawai'i's Birds Hawaii Audubon Society (1995) ISBN 1889708003

Hawai'i's Beautiful Birds H. Douglas Pratt (2001) Mutual Publishing, ISBN 1566471206

Hawai'i Les Beletsky (2005) Interlink Publishing ISBN 1566566134

Los Angeles

The greater Los Angeles area constitutes one of the largest urban areas in the USA and, for a birder, would not immediately raise hopes of finding many birds. However, there is a long coastline extending from Santa Monica in the north to Newport Beach in the south, and access from the northern and eastern suburbs to the San Gabriel Mountains. This whole area is actively patrolled by an enthusiastic and committed army of local birders, who have amassed a list of almost 500 species for the LA area.

Without human settlement and irrigation, the area occupied by Los Angeles would be semi-desert and the indigenous avifauna reflects this biotype. However, in order to experience the true desert habitat, it is necessary to drive 50 miles or more over the San Gabriel Mountains, where the avifauna of southern California has more in common with that of Arizona and Mexico than that of northern California. There are, however, public parks and reservoirs surrounding the city which hold key pockets of habitat for land-birds, and irrigation has created other possibilities, resulting in Los Angeles having such a varied and interesting avifauna.

Offshore, the mixing of the cold California Current from the north with warm ocean currents from the south has created rich feeding grounds for marine life and for seabirds, and pelagic trips off this stretch of coast are among the most productive in the world.

The whole of Los Angeles County comprises the core urban area considered for the purposes of this book. However there are adjacent wetland areas in Orange County which are well worth visiting as they are particularly good for birds and would provide a welcome diversion for any birder who has been dragooned into a vacation at nearby Disneyland. These sites are conveniently located on the Pacific Coast Highway which is reasonably well served by public transport. European birders may marvel at just how tame Nearctic waders can be on the American west coast.

GETTING AROUND

In many ways, Los Angeles truly is a city of superlatives but one exception is its public

Phainopepla is a summer visitor to Hollywood

transport. As a low-density sprawling city, it is not particularly suited to public transport and consequently the system is very fragmented and infrequent. Only a selection of bus routes and the metro rail radiate out from Downtown LA and most other routes connect points within the suburbs, so that getting to many destinations involves a combination

of buses or the metro. Apart from the coastal strip, it is not practical for the birder to rely on public transport to get to the inland sites. Although the freeway system can be very bewildering, and the traffic is relentless and unforgiving, most visitors to Los Angeles will have to summon up the courage to rent a car.

Outside public areas, birding can be very frustrating as much of the land is in private hands and trespassing is strictly forbidden, so it is not possible simply to pull off the road to stroll into a likely looking spot. For this reason, it is best to stick to the known public areas and avoid stopping on roads that have no hard shoulder.

Equip yourself with a good map or motor atlas which covers all of LA County and the adjacent areas of the neighbouring counties. Light footwear and clothing are generally adequate for southern California's benign climate.

Locations

1 Debs Regional County Park (5 miles) *MTA Bus #81*

This compact city park is only about three miles northwest of Dodger Stadium in Downtown LA. You could drive there along the Pasadena Freeway, although it is well served by public transport: both by metro and bus. The northern end of the park contains the best mix of habitats, from rolling grassy slopes to mixed woodland. Unlike Sepulveda Park, Debs Park does not have much habitat for aquatic birds but it nevertheless supports a significant selection of landbirds at all times of the year and can be particularly productive during spring migration. The local Audubon Society Chapter lead bird walks in the park on Fridays and Saturdays: details are posted at the Audubon Center in the park.

KEY SPECIES

Resident
White-throated Swift, Nuttall's Woodpecker, Bewick's Wren

Summer
Allen's Hummingbird, Black-headed Grosbeak, Brewer's Blackbird, Bullock's Oriole

Winter
Say's Phoebe, Ruby-crowned Kinglet,

Loggerhead Shrike, Townsend's Warbler, American Goldfinch

Passage
Vaux's Swift, Rufous Hummingbird, Pacific-slope Flycatcher, Barn Swallow, Warbling Vireo, Black-throated Gray Warbler, Wilson's Warbler, Lazuli Bunting, Western Tanager

Much closer to Downtown LA is Echo Lake Park which, although smaller than Debs Park, has a good-sized lake which attracts waterfowl. This park is just a few blocks west of the Dodger Stadium off Sunset Boulevard. To get there from downtown, take the #2, #4, or #304 bus to Echo Park Avenue/Sunset Boulevard and walk one block south.

2 Mount Wilson (27 miles)

Mt Wilson is one of the closest peaks of the San Gabriel Mountains to Downtown LA, and serves a useful purpose for checking the smog level: if Mt Wilson is visible from Downtown LA, smog levels are low; if it is not visible, levels are high! It is also accessible by road right up to the peak, where the famous Mt Wilson Observatory is located as well as a large communications centre for TV and radio stations. At 5,700ft, the mountains offers a habitat for a number of summer visitors which would normally be seen only as passage migrants at lower levels, as well as for more specialised montane species. The road up to the mountain from Pasadena is known as the Angeles Crest Highway and offers the best opportunity to see some of the higher elevation species of southern California within an hour's drive of the city centre. Unfortunately, there is no practical way to bird this area by public transport. To get to the **Angeles Crest Highway**, exit north off **I-210 (Foothill Freeway)**.

KEY SPECIES

Resident
Golden Eagle, Red-shouldered Hawk, Mountain Quail, Acorn Woodpecker, Nuttall's Woodpecker, White-throated Swift, Common Raven, Steller's Jay, Mountain Chickadee, Canyon Wren, Bewick's Wren, Lark Sparrow, Lesser Goldfinch

Summer
Olive-sided Flycatcher, Dusky Flycatcher, Western Tanager, Chipping Sparrow, Black-headed Grosbeak, Lazuli Bunting, Lawrence's Goldfinch

Winter
Williamson's Sapsucker, Ruby-crowned Kinglet, Golden-crowned Kinglet, Townsend's Solitaire, Hermit Warbler, Townsend's Warbler

Passage
Pacific-slope Flycatcher, Black-throated Gray Warbler, Wilson's Warbler

3 San Joaquin Wildlife Sanctuary (45 miles)   *OCTA Bus #79*

This is an area of riparian vegetation and scrub habitat surrounding the freshwater reservoirs for the city of Irvine. It has been managed over the years to enhance its attractiveness for birds and, despite the fact that the surrounding area has become more built-up, it continues to attract large numbers of birds into a relatively small area. It is quite close to Newport Bay.

Going north on **Jamboree Road** turn right on **Michelson** and follow the sign to **Irvine Water District plant**. A car park and the Audubon Society buildings are at the entrance. You can also take the OCTA Bus #79 from the Newport Beach Transit Center, which serves both the Upper Newport Bay Reserve Area and Jamboree Road.

KEY SPECIES

Resident
American White Pelican, Green Heron, Snowy Egret, Black-shouldered Kite, California Quail, American Avocet, Greater Roadrunner, Common Ground Dove, Marsh Wren, Great-tailed Grackle

Summer
Least Tern, Black Skimmer, Allen's Hummingbird, Tree Swallow, Barn Swallow, Wilson's Warbler, Yellow-breasted Chat, Bullock's Oriole

Winter
American Wigeon, Cinnamon Teal, Green-winged Teal, Bufflehead, Marbled Godwit, Western Sandpiper, Least Sandpiper, Long-billed Dowitcher, Wilson's Snipe, Ruby-crowned Kinglet, American Pipit

4 Bolsa Chica Ecological Reserve (35 miles) OCTA Bus #1

Bolsa Chica is a fantastic wetland site in Orange County which is surrounded by 'nodding donkey' oilwell pumps and just across the highway from Huntington Beach, which is usually thronged with sunbathers and surfers. Birding is excellent all year round.

In summer it supports a large tern colony including Elegant, Royal and Least Terns. Waterfowl are abundant in winter and a huge number and variety of waders occur on passage. It is on the Pacific Coast Highway, which is quite well served by bus. The Orange County Transit Authority Bus Route #1 connects Bolsa Chica and the other coastal locations with Long Beach. Long Beach can be reached from all points in LA by the Metro BlueLine.

KEY SPECIES

Resident
Clark's Grebe, Brown Pelican, Snowy Egret, Black-shouldered Kite, Snowy Plover, American Avocet, Marsh Wren, Savannah Sparrow, Tricolored Blackbird, Western Meadowlark,

Summer
Least Tern, Elegant Tern, Royal Tern, Caspian Tern, Common Tern, Black Skimmer

Winter
American White Pelican, Blue-winged Teal, Cinnamon Teal, Green-winged Teal, Greater Scaup, Bufflehead, Black-bellied Plover, Semipalmated Plover, Willet, Least Sandpiper, Western Sandpiper, Dunlin, Marbled Godwit, Sanderling, Whimbrel, Long-billed Curlew, Short-billed Dowitcher, Long-billed Dowitcher, Heermann's Gull

Passage
Red-necked Phalarope, Wilson's Phalarope, Lesser Yellowlegs, Spotted Sandpiper, Mew Gull, rare gulls, terns and waders

5 Newport Bay Ecological Reserve (45 miles) OCTA Bus #1

Just about ten minutes further south from Bolsa Chica is Newport Bay, another excellent wetland site. It is also served by bus from Downtown LA and is close to the San Joaquin Wildlife Sanctuary, which makes it a worthwhile area to spend a day's birding. However, unlike Bolsa Chica, the birds tend to be quite distant and a telescope is needed to get good

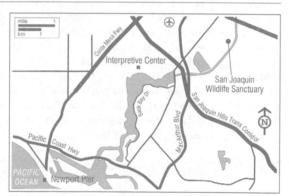

views. It is one of the best sites near Los Angeles for two Californian specialties: the California Gnatcatcher and the levipes race of Clapper Rail. The pier at Newport Beach is also a good spot for seawatching during onshore winds.

From the **Pacific Coast Highway**, the best way to approach the reserve is from the northeast corner on **Jamboree Road**, taking a left turn onto **Back Bay Drive**.

KEY SPECIES

Resident
Clark's Grebe, Black-crowned Night Heron, Snowy Egret, Clapper Rail,

Black-shouldered Kite, American Avocet, Burrowing Owl, Cactus Wren, California Gnatcatcher, Savannah

Sparrow, Tricolored Blackbird

Summer	Teal, Greater Scaup, Bufflehead,	Seabird Passage
Least Tern, Elegant Tern, Royal Tern, Caspian Tern, Common Tern, Black Skimmer	Black-bellied Plover, Semipalmated Plover, Willet, Least Sandpiper, Western Sandpiper, Dunlin, Marbled Godwit, Sanderling, Whimbrel, Long-billed Curlew, Short-billed Dowitcher, Long-billed Dowitcher, Heermann's Gull	Pacific Loon, Common Loon, Sooty Shearwater, Black-vented Shearwater, Brant, Surf Scoter, Red Phalarope, Pomarine Jaeger, Parasitic Jaeger, Common Murre, Cassin's Auklet, Rhinoceros Auklet
Winter		
American White Pelican, Blue-winged Teal, Cinnamon Teal, Green-winged		

6 **Palos Verdes Peninsula** (29 miles)   *MTA Bus #225 and #226*

This area near Long Beach is definitely the migration hotspot for Los Angeles. It has probably hosted more rarities than anywhere else within the LA area. It is also a site for California Gnatcatchers and is an excellent seawatching point. However, the area is over ten square miles in extent and much of the land is private property, with many exclusive and extremely wealthy communities. There

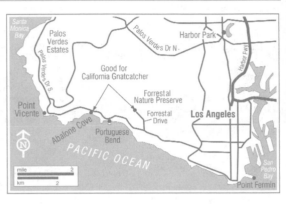

is a bus service, which serves the western side of the peninsula, but a car is needed to have any chance of exploring the area in its entirety.

Seawatching takes place from Point Vicente, which is the southwesterly tip of the peninsula and on the 225 bus route. Onshore winds in spring and autumn are obviously the best conditions, but even in summer it is a good vantage point to see local seabirds, and loons, grebes and gulls can be seen in winter. About two miles further east from Point Vicente is Abalone Creek, a good location for Californian Gnatcatchers and for other passerines at migration time. The golf courses in the area are good for pipits, meadowlarks and sparrows, and aquatic birds in the water hazards! One other location that is also worth visiting, is the Harbor Park near the intersection of the Harbor Freeway and the Pacific Coast Highway. This is a very well-watched migrant 'trap' and good for passerines at all times of the year.

KEY SPECIES

Resident
Black Oystercatcher, Forster's Tern, Cactus Wren, Marsh Wren, California Gnatcatcher, Blue-gray Gnatcatcher, Loggerhead Shrike, Lesser Goldfinch

Summer
Least Tern, Caspian Tern, Common Tern, Common Murre, Hooded Oriole, Tricolored Blackbird

Winter
Black Turnstone, Surfbird, Heermann's Gull, American Pipit, Western Meadowlark, Rufous-crowned Sparrow

Passage
Pacific Loon, Common Loon, Sooty Shearwater, Black-vented Shearwater, Sabine's Gull, jaegers, Cassin's

Auklet, Rhinoceros Auklet, Lazuli Bunting, Black-headed Grosbeak, Blue Grosbeak, Western Tanager, flycatchers, warblers, passerine rarities

7 Marina del Rey (14 miles; 5 miles from LAX) *MTA Bus #33*

What was once the sea-outlet for Ballona Creek is now a built-up harbour area for expensive yachts but is worth mentioning, as it is the closest birding site to the airport. It is at the coastal end of Venice Boulevard (take Bus #33), and a good location for first-time visitors to the American west coast to familiarise themselves with the typical gulls and waders. Pelagic trips sometimes operate from here.

KEY SPECIES

Resident
Red-breasted Merganser, Black Oystercatcher, Forster's Tern

Summer
Heermann's Gull, Least Tern, Elegant Tern, Caspian Tern

Winter
Wandering Tattler, Black-bellied Plover, Surfbird, Willet, Least Sandpiper, Western Sandpiper, Long-billed Dowitcher, Short-billed Dowitcher, Black Turnstone, Ruddy Turnstone, Glaucous-winged Gull, Bonaparte's Gull

8 Malibu Creek State Park (25 miles) *MTA Bus #434*

Malibu Creek is a coastal lagoon north of LA which is a good site for waders, gulls and terns. The surrounding hinterland is a state park consisting of a sizeable parcel of the Santa Monica Mountains. The two areas provide a good combination of habitats, between coastal lagoons and chaparral-covered hillsides, where several Californian near-endemics can be found.

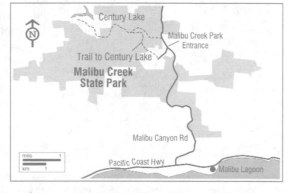

Take the **Pacific Coast Highway** north from LA to **Malibu**. Malibu state beach is on the left-hand (south) side of the road. Just after you cross a bridge over Malibu Creek, there is a car park on the left. From the car park there is a trail down to the beach and lagoon. The waders may be on the lagoon mudflats if the tide is in: if not, check the beach. Obviously, early mornings on weekdays are best to avoid the crowds.

To get to Malibu Creek State Park, continue west on the Pacific Coast Highway for another mile and turn north onto **Malibu Canyon Road**. Continue north for about seven miles: the main entrance to the park is on the left-hand side of the road. From here, there are several trails into the mountains through live oak and chaparral, the highest of which reach an altitude of over 2,000ft. The main trail, which heads west from the car park and ascends to Century Lake, passes through some good habitat. If you don't have a car, there is a bus service along Pacific Coast Highway, serving Malibu Beach, but no service up into the hills.

KEY SPECIES

Resident
Malibu Canyon: Black-shouldered Kite, Red-shouldered Hawk,

California Quail, Nuttall's Woodpecker, Acorn Woodpecker, Cassin's Kingbird, Cactus Wren,

Bewick's Wren, Common Raven, Lark Sparrow, Savannah Sparrow, Tricolored Blackbird. Lesser Goldfinch

Malibu Creek: Brown Pelican, Snowy Egret, Black-necked Stilt, Killdeer, Forster's Tern, Marsh Wren **Summer** **Malibu Creek:** Elegant Tern, Royal Tern, Caspian Tern, Common Tern	**Winter** **Malibu Creek:** Green-winged Teal, Lesser Scaup, Sora, Black-bellied Plover, Semipalmated Plover, Willet, Least Sandpiper, Western Sandpiper, Dunlin, Marbled Godwit, Sanderling,	Whimbrel, Short-billed Dowitcher, Long-billed Dowitcher, Spotted Sandpiper, Heermann's Gull, Bonaparte's Gull, Glaucous-winged Gull

9 Sepulveda Basin Wildlife Area (16 miles)   *MTA Bus #236*

The hills surrounding the San Fernando Valley in the north of the city are laced with man-made reservoirs. Not all are open to the public, but a few are accessible and Sepulveda Basin is one of the closest to the city centre (if Los Angeles can be considered to have a centre). The area has been landscaped for the public, with trails and viewpoints at the water's edge offering very easy access. To make it even easier, the San Fernando Valley Audubon Society organise regular bird walks in the park and this also presents an opportunity to meet local birders and to get further information. Details are available on their website.

It is located just northwest of the **intersection** of **I-101** and **I-405** and is served by Metro Transit Authority Bus #236 from Universal City Metro Station on the Red Line.

KEY SPECIES

Resident Acorn Woodpecker, Nuttall's Woodpecker **Summer** Black-chinned Hummingbird, Black-headed Grosbeak, Blue Grosbeak, Bullock's Oriole **Winter** Cinnamon Teal, Northern Shoveler, Say's Phoebe,	Blue-gray Gnatcatcher, Townsend's Warbler **Passage** Vaux's Swift, Rufous Hummingbird, Allen's Hummingbird, Pacific-slope Flycatcher, Tree Swallow, Barn Swallow, Western Kingbird, Warbling Vireo, Black-throated Gray Warbler, Wilson's Warbler, Western Tanager

Further Away

10 San Pedro Pelagic Trips (24 miles)

Some of the best pelagic birding anywhere in the world is to be experienced off the coast of southern California. Some trips depart from San Pedro, at the end of I-110 near Long Beach, and others from Marina del Rey (see page 393). Trips are operated right through the year and have the added bonus of encountering whales and other marine mammals. Trips are organised for both inshore waters (up to 8 hours duration) and offshore (8 to 18+ hours) as different birds occur in each oceanic zone. Schedules and trip lists are posted online and bookings can also be made online. Useful websites are given on page 395. The vessels are very well appointed and crewed by experienced pelagic trip leaders. Although the sea journeys can be both long and expensive, they offer the chance to see some of the most enigmatic species in the world.

KEY SPECIES

Summer/Autumn Buller's Shearwater, Flesh-footed Shearwater, Pink-footed Shearwater, Sooty Shearwater, Black Storm-petrel, Ashy Storm-petrel, Leach's Storm-petrel, South Polar Skua, Sabine's Gull, jaegers	**Winter** Black-vented Shearwater, Black-legged Kittiwake, Xantus's Murrelet, Rhinoceros Auklet, Cassin's Auklet

Useful contacts and websites

Southern California Audubon Chapters
www.socalaudubon.org/socal/
6042 Monte Vista St, Los Angeles, CA 90042.
Tel. 323 254 0252.

San Fernando Valley Audubon Society
www.socalaudubon.org/sfvas/
Audubon Society Chapter centred on Hollywood and
the San Fernando Valley.

Los Angeles Audubon Society
www.laaudubon.org
7377 Santa Monica Blvd, West Hollywood, CA
90046-6694. Tel. 323 876 0202.
LA County Audubon website, with details of field
trips, bird sightings and pelagics.

Bolsa Chica Conservancy
www.bolsachica.org
Local conservation group's website with
information, maps and checklist for Bolsa Chica.

Newport Bay Naturalists & Friends
www.newportbay.org
Local conservation group's website with reports,
maps and general information about Newport Bay.

California Pelagics
www.surfbirds.com/Pelagic/wcoast.html
Information on pelagic trips available off the
Californian coast.

Mitch Heindel's Website
www.angelfire.com/ca5/pelagics
Packed with information about pelagics and many
other aspects of birding in southern California.

Metro Transit Authority
www.mta.net
Public transport within LA County

Orange County Transit Authority
www.octa.net
Public transport within Orange County

Tidal Schedule for San Pedro Channel
www.saltwatertides.com/dynamic.dir/
californiasites.html#pedro

Books and publications

**Birds of Los Angeles: Including Santa
Barbara, Ventura and Orange Counties**
Chris C. Fisher & Herbert Clarke
Lone Pine Publishing (1997) ISBN 1551051044

A Birder's Guide To Southern California
Brad Schram (1998) American Birding Association
ISBN 1878788175

**A Birder's Guide to Metropolitan Areas of
North America** Paul Lehman (2001) American
Birding Association, ISBN 1878788159

Peterson Field Guides: Western Birds
Roger Tory Peterson (1998) Houghton Mifflin
ISBN 061813218X

**National Geographic Field Guide to the Birds
of North America, 4th Edition**
National Geographic (2002) ISBN 0792268776

The North American Bird Guide David Sibley
(2000) Pica Press, ISBN 1873403984

**Field Guide to the Birds of Western North
America** David Sibley (2003) Christopher Helm
ISBN 0713666587

**National Geographic Guide to Bird Watching
Sites: Western US** Mel White (1999)
National Geographic, ISBN 0792274504

• •

Mexico City

The largest city in the world is also, at 2,250m, among the world's highest capitals. It occupies a flat basin, surrounded by volcanic peaks and ridges, where several large lakes once extended over the valley. The growth of a city this size is not without an environmental impact: drainage and depletion of aquifers have reduced the valley's great lakes to dry salt flats in many places, and the city's topography results in smog getting trapped between the mountains at high altitude and falling as acid rain on the mountain forests. Despite this, Mexico City is close to some wonderful birding areas in the surrounding mountains and in some of the restored wetlands. In addition,

A Curve-billed Thrasher at the Teotihuacan Pyramids

the magnificent sites of the Pre-Colombian civilisations nearby themselves support some interesting birds, so you can enjoy both birds and culture at the same time.

The altitude gives Mexico City a cooler climate than that experienced on the adjacent coasts and although the city lies just within the Neotropical biogeographical region it is also near the southern breeding limit for many North American species, while also being well within the wintering zone for a significant number of Nearctic warblers and other passerines. North American visitors will find a distinctly western bias to the wintering species but during migration there is a more diverse mix. In addition to the North American birds, the mountains surrounding the city also support a distinct community of Central American highland species with representatives of several bird families which otherwise occur in greater abundance and diversity further south.

GETTING AROUND

Finding your way around the largest city in the world is not without its challenges. Traffic congestion, smog and the sheer scale of the city make overground travel a wearisome, frustrating and very time-consuming exercise for even the most intrepid urban birder. Luckily the underground metro system offers a cheap and efficient means of getting around the city. Although it does not serve all of the birding sites there is a comprehensive system of feeder buses which connect the metro stations with the

outlying areas, including the national park entrances, and you can get to many of the sites by public transport alone, with some supplementary walking.

If you are driving, the key routes with which to get familiar are Avenida Insurgentes, which runs along a north/south axis connecting with major highways at both ends, and the Anillo Periférico (peripheral ringroad) which is actually a semicircle around the southern outskirts of the city. The latter more or less connects the airport in the east and Chapultepec Park in the west, with interchanges with all of the major highways going south and west. You will need a city map which includes all of the Distrito Federal as well as the adjoining parts of the neighbouring states of México and Morelos.

A number of the sites mentioned in this guide are national parks or monuments of some kind, which means that they don't open until after 08.00. While early morning is the best time for birding, often the last hour before closing time can be good as many people will have left and few are entering, making for undisturbed birding in pleasant temperatures.

Although well within the tropics, its altitude gives Mexico City a relatively cool climate, particularly on winter mornings when there can be a substantial ground frost. The second effect of the altitude is the greater physical effort required to climb the mountain trails. Some of the mountain sites reach an elevation of over 3,500m which will certainly tax the stamina of any lowlander, although unlikely to result in altitude sickness if ascended slowly (which, at these heights, is the only way to ascend!).

Locations

1 Chapultepec Park (City Centre)

Chapultepec Metro Station Line1
Constituyentes Metro Station Line 7

Chapultepec is a large park located just west of the city centre and bisected by the Anillo Periférico. Like all city parks it is well used by the public and just a little too well maintained to provide habitats for skulking species and those requiring more specialist niches. However, a surprising variety of resident birds can be found here, as well as common migrant passerines, which are less particular in their requirements. There are a number of lakes and waterways to add to the species diversity. The west end of the park tends to be the least crowded and is also the most wooded so you should concentrate your efforts here. There are substantial stands of introduced trees, such as eucalyptus, which are less suitable for the birds, but the park supports some native woodland and efforts are being made to replant with native species.

The park is easy to find, being at the west end of Paseo de le Reforma and served by a number of stops on the metro.

KEY SPECIES

Resident	Winter
Black-crowned Night Heron, American Coot, Canyon Wren, Canyon Towhee, Song Sparrow, Black-backed Oriole	Red-tailed Hawk, American Kestrel, Cedar Waxwing (irregular), Black-and-white Warbler, Bullock's Oriole

2 Xochimilco Lake Ecological Park (22km)

Xochimilco Railway Station

Xochimilco Lake was one of the largest of the lakes which existed in the Mexican Basin during Pre-Colombian times. As the city grew and these lakes were drained a system of interconnecting canals for drainage, irrigation and transportation was constructed. The lake and canals at Xochimilco are the best-preserved examples of this system of waterways and the area now attracts a wide diversity of aquatic birds.

The best way to see the waterways and the birds is to hire a boat, which is a bit touristy but if you ask the boatman he will bring you to the best areas for birds. In addition there are several trails which lead alongside the canals and through cultivated plots and plantations, all of which are good for birds. Local birders organise bird walks here on Saturdays at 10.00, departing from the boat landing stage.

Xochimilco is on the south side of the city. Take the **Anillo Periférico** south to Xochimilco and continue east past the exit for **Avenida División del Norte**. The park entrance is a further 3km on the south side of the Periférico. There is a light rail service from Tasqueña Metro Station to Xochimilco town centre but this is about 5km from the Ecological Park and would require a taxi ride.

KEY SPECIES

Resident
Pied-billed Grebe, Eared Grebe, Black-crowned Night Heron, Great White Egret, Green Heron, Cinnamon Teal (rare breeder), Mallard (Mexican Duck), Ruddy Duck, American Coot, Common Moorhen, Killdeer, American Avocet, Black-necked Stilt, Mourning Dove, Vermilion Flycatcher, Cassin's Kingbird, Loggerhead Shrike, Sedge Wren, Marsh Wren,

Yellow Warbler, Common Yellowthroat, Song Sparrow, Red-winged Blackbird, Lesser Goldfinch

Winter
White-faced Ibis, American White Pelican, Snowy Egret, Great Blue Heron, Tricolored Heron, American Wigeon, Northern Shoveler, Northern Pintail, Green-winged Teal, Blue-winged Teal, Turkey Vulture, Red-

tailed Hawk, Northern Harrier, American Kestrel, Sora, Least Sandpiper, Spotted Sandpiper, Lesser Yellowlegs, Long-billed Dowitcher, Belted Kingfisher, Black Phoebe, American Pipit, Orange-crowned Warbler, Nashville Warbler, Yellow-rumped (Audubon's) Warbler

Passage
Baird's Sandpiper, Wilson's Phalarope

3 Desierto de los Leones National Park (23km)

The mountains to the west of Mexico City rise to an altitude of over 4,000m and despite deforestation there are still areas of high quality deciduous and pine forest across large stretches of hillside. The Desierto de los Leones National Park is a protected area of pine and oak forest on the eastern slopes of these mountains, spanning an elevation of between 2,600 and 3,700m, with a stunted pine and juniper zone at the highest altitudes. The park is criss-crossed with many hiking and mountain-biking trails, providing access to each altitudinal zone. As it is quite close to the city, it is very popular with the public at weekends, but also large enough to provide secluded areas which are undisturbed and good for birding. Most visitors tend to congregate around the old monastery, but by continuing past there on the mountain road to the summit at El Pantano, you will pass through good birding habitat. A second area worth looking at is the Cruz Blanca recreational area on the west side of the park.

Take **Highway 15** west towards **Toluca** and after about 20km exit south at the sign for Desierto de los Leones National Park, at the first toll-booth. This road leads directly into the park and continues uphill to the monastery for about 3km, providing birding opportunities along the way. The trail to Cruz Blanca begins just before the entrance to the park on the right-hand side and ascends the mountain for about 1.5km. If you don't have a car, take the Metro Line 7 to Barranca del Muerto, from where there is a bus service to the monastery, but only at weekends.

KEY SPECIES

Resident
Whip-poor-will, Green Violet-ear (August to January), Blue-throated Hummingbird, Magnificent

Hummingbird, Broad-tailed Hummingbird, Hairy Woodpecker, Strickland's Woodpecker, Northern Flicker, Cordilleran Flycatcher,

Steller's Jay, Mexican Jay, Common Raven, Mexican Chickadee, White-breasted Nuthatch, Pygmy Nuthatch, Brown Creeper, Gray-barred Wren,

Golden-crowned Kinglet, Western
Bluebird, Brown-backed Solitaire,
Russet Nightingale-Thrush, Ruddy-
capped Nightingale-thrush, Hutton's
Vireo, Crescent-chested Warbler, Red
Warbler, Slate-throated Redstart,

Golden-browed Warbler, Olive
Warbler, Black-headed Grosbeak,
Rufous-capped Brush Finch, Green-
striped Brush Finch, Spotted Towhee,
Striped Sparrow, Red Crossbill, Pine
Siskin

Winter
Plumbeous Vireo, Warbling Vireo,
Orange-crowned Warbler, Nashville
Warbler, Yellow-rumped (Audubon's)
Warbler, Townsend's Warbler, Hermit
Warbler, Black-and-white Warbler

4 Bosque de Tlalpan (18km)   *Universidad Metro Station Line 3*

The Bosque de Tlalpan is a national park on the lower slopes of Ajusco, an extinct volcano to the south of the city. It is mainly deciduous woodland which still supports a good variety of forest species in a comparatively small area, although it is degraded to some extent. The upper slopes lie within the pine-oak forest zone, the typical biome for this part of highland Mexico.

The park is just south of the **Anillo Periférico** at the intersection of the Anillo and **Avenida Insurgentes**. Exit south onto Avenida Insurgentes and take the first turn right (**Calle Santa Teresa**) which leads to the main park entrance. There is a bus service from the Universidad Metro Station to Bosque de Tlalpan Park.

KEY SPECIES

Resident	Winter
Broad-billed Hummingbird, Ladder-backed Woodpecker, Greater Pewee, Vermilion Flycatcher, Cassin's Kingbird, Canyon Wren, Hooded Yellowthroat, Blue Grosbeak, Rufous-crowned Sparrow	Sharp-shinned Hawk, Dusky Flycatcher, Buff-breasted Flycatcher, Orange-crowned Warbler, Nashville Warbler, Yellow-rumped (Audubon's) Warbler, Black-throated Gray Warbler, Townsend's Warbler, Hermit Warbler, MacGillivray's Warbler, Summer Tanager, Western Tanager, Bullock's Oriole

5 UNAM Botanic Gardens (18km)   *Universidad Metro Station Line 3*

The UNAM Botanic Gardens are on the north side of the Anillo Periférico quite near the Bosque de Tlalpan, and both sites can be visited in a day. The gardens have been developed on a bed of lava rock known as Pedregal and, apart from the geological interest, the stunted oak forest and other vegetation is also of botanical value. The variety of flowering trees and shrubs and the constant irrigation during the dry season attract many birds, particularly hummingbirds and warblers, and this is quite a reliable spot for the endemic Hooded Yellowthroat.

From the Anillo Periférico, go north on **Avenida Insurgentes** through the university campus. The Botanic Gardens (**Jardin Botanico**) are signposted on the left-hand (west) side of Avenida Insurgentes, about 1.5km north of the intersection with the Anillo Periférico. There is a bus service from the Universidad Metro Station to the UNAM University Campus, but this would still require a 2km walk, or you could get a taxi to the Botanic Gardens.

KEY SPECIES

Resident	Winter
Broad-billed Hummingbird, Ladder-backed Woodpecker, Greater Pewee, Vermilion Flycatcher, Cassin's Kingbird, Rock Wren, Canyon Wren, Hooded Yellowthroat, Blue Grosbeak, Rufous-crowned Sparrow, Black-chinned Sparrow	Sharp-shinned Hawk, Dusky Flycatcher, Buff-breasted Flycatcher, Orange-crowned Warbler, Yellow-rumped (Audubon's) Warbler, Black-throated Gray Warbler, Townsend's Warbler, Hermit Warbler, Nashville Warbler, MacGillivray's Warbler, Summer Tanager, Western Tanager, Bullock's Oriole

Further Away

6 La Cima and Coajomulco (60km)

Highway 95 to Acapulco on the Pacific Coast cuts through a pass in the mountains between Cerro Ajusco and Cerro Chichinautzin on the southern flank of the city. Just on the border between the Distrito Federal and the neighbouring state of Morelos, there is a well-known stakeout on the roadside at La Cima for the Sierra Madre Sparrow, a rare and localised endemic. The high altitude bunch grassland at this elevation is a critical habitat for this species, but there are also many other birds to be seen in the general area, particularly in the pine forests and forest edges, which make a visit here worthwhile. At these heights the nights get very cold so that bird activity in the pine forests peaks just after dawn. An early morning visit, therefore, is recommended. The grassland birds are active through the day.

Take **Highway 95** (not the Toll Road) south from Mexico City towards the town of **Tres Marias** (Tres Cumbres) and after about 25km on the highway you will pass under a railway bridge. One kilometre beyond the bridge, there is a track to the right (west) marked **La Cima** which cuts through an area of scattered pines and grassland. The Sierra Madre Sparrows, along with other grassland species, can be seen in this area. Return to Highway 95 and go south to Tres Marias. At this point you are now descending the southern slope of the mountains. At Tres Marias, continue south, on Highway 95, towards **Coajomulco**. This road passes through some forested hillsides for about 5km, and is good for pine woodland species. The best location is 5km after Tres Marias, on the south side of the road opposite the turn-off for Coajomulco.

KEY SPECIES

Resident
Green Violet-ear (August to January), Blue-throated Hummingbird, Magnificent Hummingbird, Broad-tailed Hummingbird, Mountain Trogon, Ladder-backed Woodpecker, Hairy Woodpecker, Strickland's Woodpecker, Northern Flicker, White-striped Woodcreeper, Greater Pewee, Cordilleran Flycatcher (mainly summer), Buff-breasted Flycatcher, Rose-throated Becard, Violet-green Swallow, Steller's Jay, Mexican Jay, Mexican Chickadee, White-breasted Nuthatch, Pygmy Nuthatch, Brown Creeper, Gray-barred Wren, Golden-crowned Kinglet, Western Bluebird, Brown-backed Solitaire, Blue Mockingbird, Ocellated Thrasher, Hutton's Vireo, Chestnut-sided Shrike-Vireo, Crescent-chested Warbler, Red Warbler, Rufous-capped Warbler, Slate-throated Redstart, Olive Warbler, Blue-hooded Euphonia, Hepatic Tanager, Rufous-capped Brush Finch, Green-striped Brush Finch, Spotted Towhee, Yellow-eyed Junco, Red Crossbill, Pine Siskin, Black-headed Siskin, Hooded Grosbeak

Winter
Hammond's Flycatcher, Dusky Flycatcher, Plumbeous Vireo, Warbling Vireo, Orange-crowned Warbler, Colima Warbler, Nashville Warbler, Yellow-rumped (Audubon's) Warbler, Townsend's Warbler, Hermit Warbler, Black-and-white Warbler

Resident (La Cima)
Horned Lark, American Pipit, Sedge Wren, Blue Mockingbird (forest), Striped Sparrow, Sierra Madre Sparrow, Eastern Meadowlark, Brewer's Blackbird

7 Almoloya Marshes (56km)

The Almoloya Marshes were formed by the River Lerma and lie west of the mountains which circle the Mexico City Basin. The flow of the river has become much reduced in recent years and the marshes have dried out to form a much smaller area than previously. They are still, however, a very important wetland and provide a crucial habitat for aquatic birds in what is generally a very arid area. This site is of major importance for the Black-polled Yellowthroat, a rare and localised Mexican endemic, but it is also a

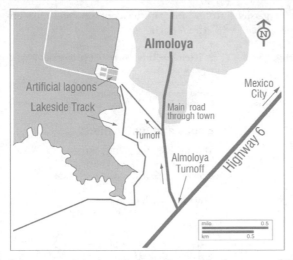

superb place for herons, wildfowl and waders. Also check any flocks of Red-winged Blackbirds for the local endemic form known as the 'Bicolored Blackbird'.

Take the **Toluca Highway (15)** west from Mexico City to **La Marqueza** and turn south on **Highway 6** towards **Tianguistenco**. Stay on Highway 6, go past the turn off for Tianguistenco and after about 5km you will come to the turn-off for Almoloya. Ignore this turn-off but turn right (north) at the next crossroads and proceed towards the town of Almoloya. Just before entering Almoloya turn left (west) onto a tarmac road which runs down to the lakeshore, where there is a set of artificial lagoons and a causeway. South from here, there is a track alongside the marshes, with reedbeds and the lake on the right-hand (west) side of the road and cultivation plots on the left-hand side. You can drive along this stretch of road for a few kilometres around to the south shore of the lake.

KEY SPECIES

Resident
Pied-billed Grebe, Green Heron, Cattle Egret, White-faced Ibis, Mallard (Mexican Duck), Ruddy Duck, American Coot, Common Moorhen, King Rail, Virginia Rail, Black-necked Stilt, Killdeer, Loggerhead Shrike, Vermilion Flycatcher, Marsh Wren, Common Yellowthroat, Black-polled Yellowthroat, Red-winged Blackbird,

Tri-colored Blackbird, Song Sparrow, Savannah Sparrow

Winter
Snowy Egret, Great Egret, Black-crowned Night Heron, Great Blue Heron, American Wigeon, Northern Shoveler, Northern Pintail, Blue-winged Teal, Cinnamon Teal, Ruddy Duck, Northern Harrier, Sora,

American Avocet, Least Sandpiper, Spotted Sandpiper, Wilson's Snipe, Lesser Yellowlegs, Long-billed Dowitcher, Belted Kingfisher, Striped Sparrow, Chipping Sparrow, Yellow-headed Blackbird

Passage
Baird's Sandpiper, Wilson's Phalarope

8 Teotihuacán (45km)

It is most likely that your trip to Mexico City will include a visit to the pyramids at Teotihuacán. Although not particularly noted for its birdlife, the site does offer an opportunity to see some of the species which are more typical of the dry plateau to the northeast of the city. This habitat supports a lower diversity of species than the mountain forests, and the birds are more thinly distributed, but in the early morning before the day heats up the birds tend to be more vocal and more visible. So be sure to bring your binoculars if sightseeing in the area.

KEY SPECIES

Resident
Burrowing Owl, Ladder-backed Woodpecker, Vermillion Flycatcher, Cassin's Kingbird, Cactus Wren, Rock Wren, Canyon

Wren, Phainopepla, Loggerhead Shrike, Blue Grosbeak, Canyon Towhee, Black-chinned Sparrow, Lesser Goldfinch

The site is very easy to get to, as there are many tour operators and a regular bus service, operating from Indios Verdes Metro Station (Line 3) which will bring you to the park gates.

As the area is very dry, check the gardens around the park headquarters and museum for sprinklers or flowering shrubs which will attract birds.

Useful contacts and websites

CIPAMEX (Spanish)
www.iztacala.unam.mx/cipamex
Consejo Internacional para la Preservación de las
Aves, A.C Tel. (+55) 5272 9689
Apartado Postal 77-297, Lomas de Sotelo, C.P.
11201, México, D.F.
BirdLife International partner in Mexico.
Email: m.cervantes@conservation.org

Sociedad Audubon de Mexico (English)
www.audubonmex.org
APDO Postal 834, San Miguel de Allende,
Guanajuato, Mexico 37700 Tel. (+52) 415 154 8470

Neotropical Bird Club (English)
www.neotropicalbirdclub.org
c/o The Lodge, Sandy, Bedfordshire, SG19 2DL,
United Kingdom

HUITZIL (English and Spanish)
Journal of Mexican Ornithology
www.huitzil.net

Servicios de Transportes Eléctricos del DF
(Spanish)
www.ste.df.gob.mx
Portal site for public transport system in
Mexico City.

Books and publications

The Birds of Mexico City Richard G. Wilson &
Hector Ceballos-Lascurain (1993) BBC Printing and
Graphics, Ontario, Canada, ISBN 092164101X

Where to Watch Birds in Mexico
Steve Howell (1999) Christopher Helm
ISBN 0713650877

**A Guide to the Birds of Mexico and
Northern Central America**
Steve N.G. Howell, Sophie Webb (1995) Oxford
University Press, ISBN 0198540124

**A Field Guide to the Birds of Mexico and
Adjacent Areas: Belize, Guatemala, and El
Salvador** Ernest Preston Edwards (1998)
University of Texas Press, ISBN 0292720920

• •

Resident
Great Blue Heron, Great White Egret, Cattle Egret, Green Heron, Black-crowned Night Heron, White Ibis, Turkey Vulture, Common Moorhen, Killdeer, Laughing Gull, Mourning Dove, White-crowned Pigeon, Northern Flicker, Red-bellied Woodpecker, Great Crested Flycatcher, Blue Jay, Fish Crow, Carolina Wren, White-eyed Vireo, Common Yellowthroat, European Starling, Northern Cardinal, Eastern Towhee, House Sparrow, Red-winged Blackbird, Boat-tailed Grackle

Summer
Chimney Swift, Gray Kingbird, Purple Martin, Black-whiskered Vireo

Winter
Pied-billed Grebe, American Coot, Ring-billed Gull, Ruby-throated Hummingbird, Belted Kingfisher, Yellow-bellied Sapsucker, Tree Swallow, Eastern Phoebe, House Wren, Hermit Thrush, Cedar Waxwing, Yellow-rumped Warbler, Black-and-white Warbler, Palm Warbler, American Goldfinch

Miami

Unlike most of the cities in this book, Miami is frequently a birding destination in its own right. The whole of South Florida is a birders' wonderland and Miami is often a staging post for a birding trip lasting several days or weeks. Most of those who are on a birding trip will, of course, leave Miami almost immediately and head for the hotspots elsewhere in the state. However, there is much to see for those who are restricted to Miami and have only a short time available.

Situated near the southern tip of the Florida Peninsula, Miami enjoys an avifauna that is unique in the USA. As well as the standard eastern North American species, it supports an infusion of tropical species which reside here at the northern extremity of their ranges. The best time to visit is mid-winter when numbers and variety are at a maximum. Many species which are summer visitors further north are resident or are winter visitors in south Florida. The climate is at its coolest and driest at this time of the year so birding

Passage
Barn Swallow, Eastern Wood Pewee, Eastern Kingbird, Red-eyed Vireo, Blackpoll Warbler, Yellow Warbler, Rose-breasted Grosbeak, Indigo Bunting, Baltimore Oriole, Summer Tanager

Boat-tailed Grackle is a typical resident of the golf course in Miami

activity is much more pleasant. Also, in addition to the winter visitors which may be present, a steady passage of migrants passes through on their northerly or southerly movements. Summer is very hot and can also be very wet, and while there are plenty of birds to be seen, the heat, humidity and brutal onslaughts from biting insects can limit the amount of time that you may wish to spend in the field.

South Florida also hosts a bewildering array of introduced, mainly tropical, species which have become naturalised. Its tropical climate and extensively cultivated exotic flora in suburban gardens, golf courses and hotel grounds means that these species have all the requirements of their natural habitat with none of the predators and they are thriving. For most North American birders these birds are 'tickable' and whether you want to do this is largely a matter of personal preference. This account does not dwell too strongly on the best sites for these species, other than making a passing mention, but the reference texts on page 408 may prove useful if you want to pursue them.

GETTING AROUND

Public transport in and around Miami is very smooth and efficient, and this is generally the case for the coastal strip north and south of the city. In addition to the bus routes, there is the Metrorail which, although it doesn't serve any birding sites, is a quick way to get to the outer suburbs to catch one of the buses which connect the Metrorail stations with the

sites in south Miami. However, public transport alone is inadequate for reaching the best birding areas, in particular the Everglades National Park, which is about 40 miles southwest of Miami and is not served by public transport. In this case, it is worth renting a car. For some reason, Florida is one of the least expensive states in the USA for car rental, and the extra mobility this gives you makes it well worthwhile. Moreover, the sheer extent of the Everglades Park means that without transport it would be impossible to cover more than one or two good spots on foot. The car also acts as a hide, carries your birding gear and water containers, and offers some shade when the sun is at its most intense.

The best map to use is probably the AAA Road Map of Miami and Fort Lauderdale. However, if you are planning to travel more extensively in Florida, it would be worth investing in the Florida Atlas and Gazetteer by Delorme. Most of the birding sites are, in fact, county, state or national parks and there is a small entrance fee (usually $5 or less). This fee is usually valid for the whole day, so you can leave and return in the late afternoon or evening to take advantage of cooler temperatures. At most of the bigger parks, you can obtain books, trail maps and bird checklists from the visitor centres. Although many visitors may be a little disconcerted by the abundance of alligators in many areas, and their apparent fearlessness, humans do not form any part of their menu. Venomous snakes are probably a greater threat but unlikely to cause a problem since there is rarely any reason to stray from the well-walked trails. Apart from the heat and the mosquitoes in summer, and the odd hurricane in autumn, birding in Florida is a wonderful experience.

Locations

1 Enchanted Forest Park (9 miles)   Bus #3

If you are based in north Miami, the Enchanted Forest and adjoining Arch Creek Park offer some good birding opportunities without you having to travel all the way to the sites in south Miami. As well as the parkland, the Arch Creek flows between the parks and both have several trails through areas of indigenous woodland. There is a nature centre within Arch Creek Park and there are occasional guided nature tours given by park officials.

KEY SPECIES

Resident
Yellow-crowned Night Heron, Red-shouldered Hawk, Brown Thrasher, Prairie Warbler, Northern Mockingbird

Winter
Orange-crowned Warbler, Northern Parula, Yellow-throated Warbler, Ovenbird, Northern Waterthrush

Passage
Black-throated Blue Warbler, Black-throated Green Warbler, Magnolia Warbler, American Redstart, Painted Bunting

Take the **Biscayne Boulevard (Route 1)** northwards to North Miami Beach and turn left (west) on NE 135th St. The park entrance is about 450 yards on the right-hand side.

2 Key Biscayne (10 miles)   Bus Route B

This is a low-lying island off the coast just south of Miami and joined to the mainland by Rickenbacker Causeway. The island forms the northern boundary of Biscayne National Park, a marine park with coral reefs, small islands and mangroves which is very popular with scuba divers. On Key Biscayne you will find woodland, wetlands and mangroves as well as shoreline habitats. There are two areas on the island worth checking: Crandon County Park on the north end and Bill Baggs State Park at the southern tip. Crandon County Park is best for herons and waterfowl: make sure you check every 'white' heron as white phases of the Reddish Egret and both white and intermediate (Würdemann's Heron) morphs of the Great Blue Heron are prevalent in south Florida. The beaches are good for gulls, terns and waders and Bill Baggs Park is good for passerines and raptors at migration times.

KEY SPECIES

Resident
Double-crested Cormorant, Brown Pelican, Anhinga, Little Blue Heron, Tricolored Heron, Reddish Egret, Great Blue Heron, Red-shouldered Hawk, Limpkin, Willet, Black Skimmer, Mangrove Cuckoo, Pine Warbler, Prairie Warbler

Summer
Magnificent Frigatebird, Swallow-tailed Kite, Wilson's Plover, Least

Tern, Yellow-billed Cuckoo, Common Nighthawk, Chuck-will's-widow

Winter
Horned Grebe, American White Pelican, Northern Gannet, Roseate Spoonbill, Blue-winged Teal, Ring-necked Duck, Lesser Scaup, American Kestrel, Semipalmated Plover, Greater Yellowlegs, Lesser Yellowlegs, Willet, Marbled Godwit, Least Sandpiper, Western Sandpiper,

Sandwich Tern, Caspian Tern, Royal Tern, Forster's Tern, Yellow-throated Warbler, Ovenbird, Northern Waterthrush, Savannah Sparrow

Passage
Black-necked Stilt, Black-bellied Plover, Pectoral Sandpiper, Short-billed Dowitcher, Painted Bunting, Bobolink

---

3 Doug Barnes Park (8 miles)   *Bus #40 from Douglas Metrorail Station*

Although not on the coast, this rather unassuming city park is one of the best sites in Miami for passerine migration in spring and autumn. There is some aquatic habitat for herons and waterfowl but it is primarily known as a migrant trap for warblers. You can also see a number of introduced exotic species here, most notably Spot-breasted Orioles. It is only about 15 minutes drive from Miami Airport so it is very convenient if you have a couple of hours to spare before your flight. You can get there by bus and Metrorail, but a taxi ride from the airport would be quickest and not too expensive. The park is on the aptly-named Bird Road, one of the main east-west thoroughfares in south Miami.

KEY SPECIES

Resident
Anhinga, Little Blue Heron, Tricolored Heron, Red-shouldered Hawk, Pileated Woodpecker, Pine Warbler, Prairie Warbler, Spot-breasted Oriole (I)

Summer
Yellow-billed Cuckoo, Common

Nighthawk

Winter
Orange-crowned Warbler, Northern Parula, Yellow-throated Warbler, Ovenbird, Northern Waterthrush, Gray Catbird

Passage
Swainson's Thrush, Yellow-throated Vireo, Black-throated Blue Warbler, Magnolia Warbler, Cape May Warbler, Kentucky Warbler, American Redstart, Painted Bunting, Bobolink, Orchard Oriole

---

4 Matheson Hammock (10 miles)   *Bus #65 from Douglas Metrorail Station*

Matheson Hammock County Park is one of a series of public areas on the coast south of Miami which provide a mixture of habitats from woodland to shoreline, as well as a vantage point to scan Biscayne Bay. This park is a good spot for passerine migration in spring and autumn, and its shoreline is good for waders. It is also one of the city parks which hosts several colonies of exotic species such as Hill Mynas and Yellow-chevroned Parakeets.

Matheson Hammock is on the **Old Cutler Road** about a mile south of **Coral Gables**.

Further Away

5 **Everglades National Park** (40 miles)

This world-famous birding site provides all the thrilling spectacles that make up the Florida birding experience. It is a huge area of swamps, open water and raised areas of woodland known as hammocks, and a birder could spend a week or more at this site. Not only is it a fantastic habitat for aquatic birds, in winter it is probably the best site in North America for wintering wood-warblers and other insectivorous passerines. For those on a tighter schedule, the best birding locations are the trails around the Royal Palm Visitor Center, Snake Bight Trail and Long Pine Key Campground. The trails are specially constructed with boardwalks through the wetter areas so walking shoes are all that are needed. The park can be very crowded during vacation times but most visitors seem to stick to the campgrounds so it is possible to walk the trails comparatively undisturbed. Some of the trails are guided by park personnel and these can be well worthwhile.

There are basically three focal points for birders within the park. There is a cluster of sites around the main park entrance at **Homestead**. This is reached by taking the Florida Turnpike southwest to its terminus at Florida City. From there the **Ingraham Highway (Route 9336)** is signposted and this takes you to the Royal Palm Visitor Center. Fanning out from here are several trails including the Anhinga Trail, and the Gumbo-Limbo Trail. Try to get to these sites as early as possible, as they tend to get very crowded with tourists later in the day. **Long Pine Key Road** is off the main park road, two miles past the turnoff to Royal Palm.

The second area worth a visit is around the 'town' of **Flamingo** at the southern tip of the park. It is about 30 miles from the Royal Palm Visitor Center and there is accommodation available here, which probably needs to be booked in advance during the winter season. The Snake Bight Trail and Eco Pond are in this area which also happens to be one of the most regular sites in the USA for Greater Flamingos that occasionally stray from the Caribbean. The Snake Bight Trail is 3.2 miles round trip and one of the best trails for warblers, but it is also infested with mosquitoes and you will need very strong repellent or consider wearing protective clothing, including gloves! The trailhead is on the main park road, five miles before reaching Flamingo. The boardwalk from the Snake Bight Trail leads to a viewing platform where you can scan the mudflats for waders. It is best to arrive here as the tide is

coming in. There are also ranger-led boat tours from Flamingo which can bring you into wetland habitats which would be inaccessible on foot: enquire at the Royal Palm Visitor Center about these. A shorter alternative would be the half-mile boardwalk at **Mahogany Hammock**. The turnoff for this is also on the main park road, 20 miles from the park entrance or 17 miles from Flamingo.

One other area that is worth checking, if time permits, is the **Shark Valley Entrance** at the northern extremity of the park. There is a visitors' tram which takes you into the park and this is a great vantage point from which to view the birds and other wildlife. This area is the best location for Snail Kites and Limpkins. Shark Valley is 18 miles west of Miami on **US Route 41**, also known as the **Tamiami Trail**.

There is no public transport service to the Everglades National Park although several tour companies operate day trips from Miami which enable you to explore the area around the Royal Palm Visitor Center. However to get the most out of your day you really need your own transport and daily car rentals are not much more than the cost of a tour bus ticket.

KEY SPECIES

Resident
Double-crested Cormorant, Anhinga, Little Blue Heron, Tricolored Heron, Black-crowned Night Heron, Wood Stork, Glossy Ibis, Mottled Duck, Black Vulture, Snail Kite, Bald Eagle, Osprey, Red-shouldered Hawk, Short-tailed Hawk, Northern Bobwhite, Clapper Rail, King Rail, Purple Gallinule, Limpkin, Black Skimmer, Mangrove Cuckoo, Pileated Woodpecker, Pine Warbler, Prairie Warbler, Seaside Sparrow, Eastern Meadowlark

Summer
Swallow-tailed Kite, Black-necked Stilt, Wilson's Plover, Least Tern, Yellow-billed Cuckoo, Common Nighthawk, Chuck-will's-widow

Winter
American White Pelican, Roseate Spoonbill, Blue-winged Teal, Ring-necked Duck, Lesser Scaup, Sora, American Avocet, Greater Yellowlegs, Lesser Yellowlegs, Willet, Marbled Godwit, Least Sandpiper, Western Sandpiper, Caspian Tern, Royal Tern,

Forster's Tern, Orange-crowned Warbler, Northern Parula, Yellow-throated Warbler, Ovenbird, Northern Waterthrush, Savannah Sparrow, Painted Bunting

Passage
Black-bellied Plover, Pectoral Sandpiper, Short-billed Dowitcher, Black-throated Blue Warbler, Bobolink

6 Lucky Hammock (40 miles)

If you are visiting the Everglades and have some extra time to spare on the way back, then it is worth checking out the so-called 'Lucky Hammock', so-called by local birders because of its impressive track record of attracting rarities. This is an area of restored hammock woodland just outside the Everglades National Park and surrounded by agricultural land. The area is more correctly known as Frog Pond Water Management Area and there are a number of waterways and swampy patches which are good for aquatic species. However, it is the variety of migrant passerines and vagrants for which the Lucky Hammock is known to be 'Lucky'.

Take the **Ingraham Highway (Route 9336)** towards the Everglades National Park, but about half a mile before the park entrance, turn south on **SW 232 Ave (Aerojet Road)**. The Lucky Hammock is about 200 yards on the right-hand (west) side of the road. If you stay on Aerojet Road for about the next three miles you will come to the entrance of the **Southern Glades Water Management District**, an extensive area of sawgrass marshland with a small population of the endangered Cape Sable Seaside Sparrow. You cannot drive past the entrance as the gate is locked, but you may walk along the road beyond the locked gate.

KEY SPECIES

Resident
Little Blue Heron, Tricolored Heron, Glossy Ibis, Red-shouldered Hawk, White-crowned Pigeon, Pileated Woodpecker, Pine Warbler, Prairie Warbler, Seaside Sparrow

Summer
Yellow-billed Cuckoo, Common Nighthawk, Whip-poor-will, Blue Grosbeak

Winter
Lesser Nighthawk, Yellow-throated Warbler, Ovenbird, Northern Waterthrush, Savannah Sparrow, Painted Bunting

Passage
Cliff Swallow, Least Flycatcher, Alder Flycatcher, Black-throated Blue Warbler, Black-throated Green Warbler, Magnolia Warbler, Cape May Warbler, Worm-eating Warbler, Orchard Oriole

Useful contacts and websites

Tropical Audubon Society
www.tropicalaudubon.org/
Tropical Audubon Society, 5530 Sunset Drive, Miami, Florida 33143.
General information about birding in South Florida including the TAS Birdboard with recent sightings.

Everglades National Park www.nps.gov/ever

Bill Baggs Cape Florida State Park – Key Biscayne
www.floridastateparks.org/capeflorida

Florida Breeding Bird Atlas
www.wildflorida.org/bba

Florida Ornithological Society Records Committee www.fosbirds.org

South Florida Birding Connection
www.geocities.com/TheTropics/2380/

Miami-Dade Transit
www.miamidade.gov/transit
Public transport in Miami – Dade County.
Tel. 305 770 3131.

Tidal Information www.freetides.com

Books and publications

A Birder's Guide to Florida
Bill Pranty (1996) American Birding Assocation
ISBN 1878788043

A Birdwatching Guide to Florida
Derek Moore (1997) Arlequin Press
ISBN 1900159554

National Audubon Society Regional Field Guide to Florida National Audubon Society (1998) Knopf, ISBN 067944677X

Florida's Birds: A Field Guide and Reference
Herbert W. Kale & David S. Maehr (2005) Pineapple Press, ISBN 1561643351

Birds of Florida Field Guide Stan Tekiela (2005) Adventure Publications, ISBN 159193107X

A Birder's Guide to Metropolitan Areas of North America Paul Lehman (2001) American Birding Association, ISBN 1878788159

A Field Guide to the Birds of Eastern and Central North America Roger Tory Peterson (2002) Houghton Mifflin, ISBN 0395740460

Field Guide to the Birds of Eastern North America David Sibley (2003) Christopher Helm ISBN 0713666579

National Geographic Field Guide to the Birds of North America, 4th Edition National Geographic (2002) ISBN 0792268776

National Geographic Guide to Bird Watching Sites: Eastern US Mel White (1999) National Geographic, ISBN 0792273745

The North American Bird Guide David Sibley (2000) Pica Press, ISBN 1873403984

●●●●●●●●●●●●●●●●●●●●●●●●●●●●●●●●●●

TYPICAL SPECIES

Resident
Double-crested Cormorant, Canada Goose, Red-tailed Hawk, Ring-billed Gull, American Herring Gull, Mourning Dove, Northern Flicker, Downy Woodpecker, Blue Jay, American Crow, Black-capped Chickadee, White-breasted Nuthatch, American Robin, Cedar Waxwing, European Starling, Northern Cardinal, Song Sparrow, Common Grackle, Red-winged Blackbird, House Finch, House Sparrow

Summer
Chimney Swift, Ruby-throated Hummingbird, Tree Swallow, Gray Catbird, Yellow Warbler, Baltimore Oriole

Winter
Red-breasted Nuthatch, Ruby-crowned Kinglet, White-throated Sparrow, Dark-eyed Junco

A Red-tailed Hawk flies past the Statue of Liberty

New York

If the city of New York did not exist, the area around New York Bay and the mouth of the Hudson River would be a wonderland for birds and wildlife in general. The combination of salt and freshwater areas together with the dense woodlands would have provided habitats for a huge variety of species. However, the city is now at the hub of a huge metropolitan area that has just about swallowed up the entire original environment. Despite this, there are still pockets of habitat which are extremely productive as birding areas, often in the most unlikely locations.

Situated midway along the Atlantic seaboard of North America, New York is ideally situated to observe the migration of many species in spring and autumn. Sufficient habitat has been preserved on the coast to permit huge numbers of wildfowl and waders to stop off on their migration and for some to winter in the area. The Hudson River provides a perfect conduit for migrating waterfowl, raptors and passerines, and it is possible to see just about every migratory species typical of eastern North America in New York, so much so, that the New York metropolitan area has a checklist of over 400, one of the highest for any comparable area in the country and extraordinary when you consider that this is one of the largest cities in the world.

GETTING AROUND

New York City could not function without an extremely efficient public transport system. The buses and subway are comparatively cheap and comprehensively serve all the birding locations in the city and suburbs. In fact, due to traffic congestion and the sheer density of roads, freeways, flyovers and exits, the visitor would be ill-advised to drive anywhere in the New York area, unless accompanied by a local driver. Interchange between subway lines is very

easy and you can reach most birding locations by subway alone. Where a bus is required, you need to familiarise yourself with the prefix code of the bus route number which signifies which borough it serves. It is generally a New York Transit Authority bus, the largest bus line, although there are separate companies serving some outlying areas, such as Green Bus Lines serving The Rockaways.

Birding in the New York area is generally very easy, thanks to the well-laid out parks and public areas. However, be prepared for extreme weather conditions, from baking heat with high humidity in the summer to unendurable cold with biting winds in the winter. Comfortable walking shoes are adequate for all the locations described, even the saltmarshes, as there is no need to walk onto the marshy areas. As is typical throughout the USA, most county and state parks charge a small entrance fee. Birding activity may be divided between passerine watching in the spring and autumn in the city parks and wader and wildfowl watching on the coastal marshes in autumn and winter. The former requires little more than a pair of binoculars, whereas the latter needs a telescope in order to do full justice to these superb locations. Despite what you might read about New York, the city parks are quite safe during the day: wandering around the park with binoculars is not regarded as an unusual pursuit and, in fact, is often a great way to make contact with other birders. Make sure you obtain a city map which covers all the five boroughs of New York, to use in conjunction with this guide.

Locations

1 **Central Park** (City Centre)

72nd St Subway Station Lines 'B' and 'C'
103rd St Subway Station Lines 'B' and 'C'

This is an urban birding area that is almost too good to be true. Situated right in the heart of midtown Manhattan and sandwiched between 8th and 5th Avenue, this legendary park acts like a green oasis in the middle of a concrete desert. It possesses a rich diversity of resident species but additionally attracts many migrants and occasionally a real rarity. Central Park is also a focal point for local and visiting birders during spring and autumn, and you will most likely meet fellow birders during your visit.

Like all public parks, Central Park is best visited early in the morning when there is least disturbance. The best birding locations are in the central area around The Lake. In particular, the Ramble area, where bird feeders are kept well stocked, is one of the best locations for migrants and for wintering passerines. At the northern end of the park, the Harlem Meer and the Jackie Onassis Reservoir are good for waterfowl, and the Ravine connecting Harlem Meer with The Pool has good riparian habitat.

An early morning visit will produce all the typical common species of eastern North America, winter or summer. However, an absolutely bewildering variety of passerines can be found at migration times if a fall of migrants occurs, rivalling the best coastal migration hotspots for both numbers and species diversity. In optimum conditions it would not be unusual to record over 20 different species of warbler.

KEY SPECIES

Resident	Winter	Warbler, Yellow-rumped Warbler,
American Kestrel, Red-bellied Woodpecker, Tufted Titmouse	Bufflehead, Hooded Merganser, American Coot, Yellow-bellied Sapsucker, Fox Sparrow	Black-throated Blue Warbler, Blackburnian Warbler, Blackpoll Warbler, American Redstart, Common Yellowthroat, Ovenbird
Summer		
Great Egret, Green Heron, Black-crowned Night Heron, Eastern Kingbird, Wood Thrush	**Passage** Hermit Thrush, Veery, Red-eyed Vireo, Warbling Vireo, Magnolia	

2 Van Cortlandt Park (11 miles)   *Woodlawn Subway Station Line '4'*

The park is at the northern edge of the Bronx City Borough and has very extensive woodland as well as a freshwater marsh. Rather like Central Park, it has the effect of being an 'island' of rich habitat amongst a sea of concrete. Therefore, as well as providing a habitat for breeding species, it also acts like a magnet for migrating passerines. The grassland areas at Vault Hill are managed to provide optimum habitat for Eastern Bluebirds, which have declined greatly due to nest-site competition with European Starlings. The best area for migrants is the Croton Woods at the northern perimeter and this area is best accessed from the **Woodlawn Subway Station** on **East 233rd St**.

KEY SPECIES

Resident	Sandpiper, Laughing Gull, Eastern Phoebe, Eastern Kingbird, Eastern Bluebird	Passage
Pied-billed Grebe, American Black Duck, Wood Duck, American Coot, Red-bellied Woodpecker, Hairy Woodpecker		Red-eyed Vireo, Blue-headed Vireo, Black-throated Blue Warbler, Yellow-rumped Warbler, Palm Warbler, Canada Warbler, Blackpoll Warbler, Magnolia Warbler, Common Yellowthroat, Ovenbird, Veery, American Goldfinch
	Winter	
Summer	Northern Shoveler, Ring-necked Duck, Hooded Merganser, Ruddy Duck, Hermit Thrush, American Tree Sparrow, Fox Sparrow	
Great Egret, Green Heron, Black-crowned Night Heron, Spotted		

3 Pelham Bay Park (14 miles)   *Pelham Bay Subway Station Line '6'*

This park is north of the city in The Bronx borough and is quite easy to get to from La Guardia Airport. It is the largest park in New York and comprises a variety of habitats, which makes it an excellent birding location at any time of the year. In view of its size, birders are best advised to confine their activity to the northern end which is the most productive for birds. However, this involves a three-mile walk from the subway station to the best birding areas but, at least during the summer, there is a bus service between the subway station and the nearby beach, by which you can return. In addition to deciduous woodland, there is extensive saltmarsh and an excellent view of Long Island Sound to permit scanning for waterfowl in winter, for which a telescope is advisable. Like many New York parks, it experiences falls of passerine migrants in spring and autumn, but additionally it is also a well-known raptor watchpoint in autumn. It is also quite a good site for owls, such as Great Horned Owls: although you are unlikely to come across one by chance, their roosts are often well known to local birders who may be able to help you.

KEY SPECIES

Resident
Great Blue Heron, Black-crowned Night Heron, Gadwall, Sharp-shinned Hawk, American Coot, Great Horned Owl, Marsh Wren, Northern Mockingbird, Sharp-tailed Sparrow.

Summer
Yellow-crowned Night Heron, Green Heron, Snowy Egret, Great White Egret, Osprey, American Oystercatcher, American Woodcock,

Willet, Spotted Sandpiper, Laughing Gull, Common Tern, Forster's Tern, Least Tern, Black Skimmer, Red-eyed Vireo, Wood Thrush, Chestnut-sided Warbler, Brown Thrasher, Eastern Towhee.

Winter
Horned Grebe, Brant, American Wigeon, Green-winged Teal, Northern Pintail, Greater Scaup, Lesser Scaup, Ring-necked Duck, Bufflehead,

Ruddy Duck, Northern Harrier, Song Sparrow

Passage
Bald Eagle, Broad-winged Hawk, Red-shouldered Hawk, Sharp-shinned Hawk, American Kestrel, Black-bellied Plover, Semipalmated Plover, Red Knot, Short-billed Dowitcher, Solitary Sandpiper, Greater Yellowlegs, Semipalmated Sandpiper, flycatchers, warblers

4 Prospect Park (7 miles) *Parkside Subway Station Line 'D'*

Prospect Park in Brooklyn is smaller than Central Park. However, it is a larger site if the adjoining locations, which are also good for birds, are included and it also has some wetland habitat. It is also nearer to JFK Airport and more convenient for anyone who is staying in the eastern suburbs and does not want to go into the city centre. The deciduous woodland supports all of the typical eastern US passerines and experiences large falls of migrants in both spring and autumn. If the park is particularly crowded with people, try the adjoining Greenwood Cemetery which is quieter than Prospect Park and makes for more pleasant birding as well as hosting the largest Monk Parakeet colony in New York.

The best areas in the park are Prospect Lake, the Lullwater area, Pagoda Swamp and the Vale of Cashmere. There are winter bird-feeding stations at the Tennis House and near the Terrace Bridge over the Lullwater. The park is well served by subway stations and bus routes: the **Parkside Station** is adjacent to Prospect Lake.

KEY SPECIES

Resident
American Black Duck, Wood Duck, American Kestrel, Red-bellied Woodpecker, Hairy Woodpecker, Monk Parakeet (I)

Summer
Great Egret, Green Heron, Black-crowned Night Heron, Laughing Gull,

Eastern Phoebe, Eastern Kingbird, Wood Thrush

Winter
Pied-billed Grebe, Northern Shoveler, Ring-necked Duck, Hooded Merganser, Ruddy Duck, American Coot, Fish Crow, American Tree Sparrow, Fox Sparrow, Swamp Sparrow, American Goldfinch

Passage
Spotted Sandpiper, Hermit Thrush, Veery, Red-eyed Vireo, Blue-headed Vireo, Magnolia Warbler, Yellow-rumped Warbler, Black-throated Blue Warbler, Palm Warbler, Canada Warbler, Blackpoll Warbler, Common Yellowthroat, Ovenbird

5 Brooklyn Marine Park (11 miles)

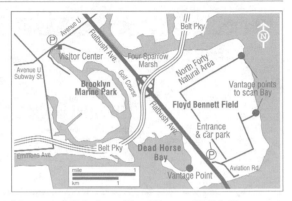

This is a large area on the northwest shore of Jamaica Bay and alongside a disused airfield called Floyd Bennett Field. The Marine Park consists of saltmarsh and mudflats, while the airfield is one of the few areas of extensive non-manicured grassland in the city and is good for pipits, sparrows and meadowlarks. The area is best explored by car as it is quite extensive and the car is very useful as a mobile hide on Floyd Bennett Field. However it is also feasible to cover most of the areas on foot as the **Q35 Green Line bus** serves the stretch of Flatbush Avenue between the two sites. The inlets in the Marine Park are quite narrow, so ducks and waders can be viewed at low tide from the paths.

Additionally, it is worth checking the aptly-named Four Sparrow Marsh just north of the Belt Parkway on Flatbush Avenue. Song, Savannah, Swamp and Sharp-tailed Sparrows all breed in this small area.

KEY SPECIES

Resident
Great Blue Heron, Gadwall, Sharp-shinned Hawk, Peregrine Falcon, Ring-billed Gull, Carolina Wren, Marsh Wren, Northern Mockingbird, Sharp-tailed Sparrow, Swamp Sparrow, Savannah Sparrow

Summer
Snowy Egret, Osprey, American Oystercatcher, American Woodcock,

Clapper Rail, Laughing Gull, Common Tern, Forster's Tern, Least Tern, Black Skimmer, White-eyed Vireo, American Redstart, Brown Thrasher, Eastern Towhee

Winter
Red-throated Loon, Common Loon, Horned Grebe, Great Cormorant, Brant, American Wigeon, Green-winged Teal, Greater Scaup,

Bufflehead, Surf Scoter, Red-breasted Merganser, Hooded Merganser, Northern Harrier, Horned Lark

Passage
American Kestrel, Cooper's Hawk, Wilson's Snipe, Willet, Upland Sandpiper, Spotted Sandpiper, Solitary Sandpiper, Gull-billed Tern, American Pipit, warblers, Eastern Meadowlark, Bobolink, American Goldfinch

6 Jamaica Bay (13 miles)

Jamaica Bay is probably the best-known birding location in the New York area, and not without reason. It consists of over 9,000 acres of saltmarsh, mudflats and woodland, providing habitats for thousands of birds, particularly during migration. It is also very easy to get to by public transport from New York and it is very conveniently located for anyone going to or from JFK Airport.

The Jamaica Bay Wildlife Refuge is well provided with trails, walkways and hides at all the strategic locations, so that rubber boots are not absolutely necessary. The refuge is open all year round with access from dawn to dusk. Like all wetland sites, a telescope is needed for best results. To get to the best birding locations from the Broad Channel Subway Station, walk about half a mile north along Cross Bay Boulevard: West Pond will be on the left-hand side and East Pond on the right. When the water levels are lowered in the East Pond in the autumn to expose the mud, it becomes a magnet for waders. In these conditions, the shore can be very muddy and boots are then required. The Gardens area further north is the best location for passerines during migration.

KEY SPECIES

Resident
Gadwall, Peregrine Falcon, Ring-billed Gull, Northern Mockingbird

Summer
Great Blue Heron, Glossy Ibis, Black-crowned Night Heron, Yellow-crowned Night Heron, Green Heron, Snowy Egret, Osprey, Clapper Rail, American Oystercatcher, Willet, Laughing Gull, Common Tern, Forster's Tern, Least Tern, Black

Skimmer, Willow Flycatcher, White-eyed Vireo, Brown Thrasher, Eastern Towhee, Boat-tailed Grackle

Winter
Horned Grebe, Snow Goose, Brant, American Wigeon, Green-winged Teal, Northern Pintail, Greater Scaup, Lesser Scaup, Ring-necked Duck, Bufflehead, Ruddy Duck, Northern Harrier, Sanderling, Great Black-backed Gull

Passage
Sharp-shinned Hawk, Black-bellied Plover, Semipalmated Plover, Semipalmated Sandpiper, Least Sandpiper, Red Knot, Hudsonian Godwit, Short-billed Dowitcher, Greater Yellowlegs, Lesser Yellowlegs, raptors, flycatchers, warblers

---

7 Rockaway Peninsula (16 miles)

Q35 Green Line Bus from Flatbush Subway Station

The narrow peninsula that extends along the southwest tip of Long Island acts as a breakwater for Jamaica Bay and is often the first landfall for migrating birds arriving from the south. It is also an excellent vantage point from which to scan Lower New York Bay for divers, grebes, ducks and seabirds. For this reason it is probably best in autumn, winter and spring.

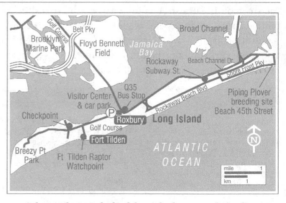

In summer, the beaches are very popular and crowded, although there is a breeding site for the rare Piping Plover and for terns.

There are three distinct sites in this location which are all worth a look, particularly in winter and during migration: Jacob Riis Park, which has good open areas and scrub habitat around the golf course and park buildings; Fort Tilden, which is more densely wooded but also has a number of open areas around the sports fields and is a superb raptor watchpoint; and the extreme tip of the peninsula at Breezy Point, an area of sand dunes and coastal shrubs. Simply strolling around these areas, checking likely-looking areas of habitat will turn up birds. The area is best covered by car as it is approximately three miles from the bus stop at Fort Tilden to Breezy Point. However, it is such an interesting area that it will repay a day-long visit. Much of the area is part of the Gateway National Recreation Area, so you will need to pay a small fee for a parking permit. Breezy Point is a gated residential area, with no access off the main road, but you can park at the tip and walk along the trails through the dunes.

KEY SPECIES

Resident
American Black Duck, Great Black-backed Gull, Carolina Wren, Northern

Mockingbird, Savannah Sparrow, Sharp-tailed Sparrow

Summer
Great Blue Heron, Snowy Egret, Osprey, American Oystercatcher,

Willet, Piping Plover, Laughing Gull, Common Tern, Forster's Tern, Least Tern, Black Skimmer, House Wren, Marsh Wren, Brown Thrasher, Eastern Towhee

Winter
Red-throated Loon, Common Loon,

Horned Grebe, Great Cormorant, Brant, Greater Scaup, Bufflehead, Surf Scoter, Red-breasted Merganser, Hooded Merganser, Northern Harrier, Purple Sandpiper, Dunlin, Sanderling, Bonaparte's Gull, Horned Lark, Fox Sparrow, Swamp Sparrow, Snow Bunting

Passage
Black-bellied Plover, Semipalmated Plover, Red Knot, Short-billed Dowitcher, Greater Yellowlegs, Ruddy Turnstone, Caspian Tern, raptors, flycatchers, warblers

8 Great Kills Park – Staten Island (18 miles)

S78 Bus from Ferry Terminal or 1X Bus from Manhattan

For a birder who is staying on Staten Island or an adjacent area of New Jersey, Great Kills Park on the east shore of Staten Island offers some excellent shoreline birding. Although no substitute for Jamaica Bay, it does provide an alternative to the rather time-consuming trip across the Lower New York Bay to get to the birding areas in Brooklyn. The Great Kills Park consists of a wide, south-facing inlet with access out to the promontory, which gives views of the bay as well as offering a variety of habitats including sand-dunes, saltmarsh, coastal scrub and some substantial tracts of woodland.

Access to the park is from **Hyland Boulevard**, the main road running along the east side of Staten Island. There is good woodland and reedbed habitat along the main drive into the park from Hyland Boulevard to the car park at the park headquarters. You can walk to the tip of the promontory (Crooke's Point) and back to the car park, taking in several areas. The outer beach is good for gulls, terns and waders; the woods around the car park at Crooke's Point are good for migrants and the inner bay has sheltered areas which are good in winter for ducks and grebes. The jetties at the yacht marina in the inner bay provide useful vantage points to scan for waterfowl.

KEY SPECIES

Resident
Great Blue Heron, Gadwall, Sharp-shinned Hawk, Great Black-backed Gull, Carolina Wren, Northern Mockingbird, Sharp-tailed Sparrow

Summer
Snowy Egret, Osprey, American Oystercatcher, Willet, Laughing Gull, Common Tern, Forster's Tern, Least Tern, Black Skimmer, American

Redstart, Brown Thrasher, Eastern Towhee

Winter
Red-throated Loon, Common Loon, Horned Grebe, Great Cormorant, Brant, Greater Scaup, Bufflehead, Surf Scoter, Red-breasted Merganser, Hooded Merganser, Northern Harrier, Purple Sandpiper, Dunlin, Sanderling, Bonaparte's Gull, Belted Kingfisher,

Horned Lark, Savannah Sparrow, Snow Bunting

Passage
Black-bellied Plover, Semipalmated Plover, Red Knot, Short-billed Dowitcher, Greater Yellowlegs, Ruddy Turnstone, Caspian Tern, raptors, flycatchers, warblers

Useful contacts and websites

New York City Audubon Society
www.nycas.org/
New York City Audubon Society, 71 West 23rd Street, Suite 1529, New York, NY 10010
Tel. 212 691 7483

Brooklyn Bird Club
www.brooklynbirdclub.org
609 Greenwood Ave, Brooklyn, NY 11218
Tel. 718 875 1151

Hudson River Audubon Society
www.hras.org
P.O. BOX 616, Yonkers, New York 10703

Prospect Park Audubon Center
www.prospectparkaudubon.org/
Tel. 718 2873400

Queens County Bird Club
www.geocities.com/herbirdroth

Fire Island Raptor Enumerators
www.pipeline.com/~merlin/firehw.htm

Jamaica Bay Wildlife Refuge
www.nps.gov/gate/homepage/jb-bp.htm

Friends of Van Cortlandt Park
www.vancortlandt.org

Central Park Conservancy
www.centralparknyc.org

Federation of New York State Bird Clubs
www.fnysbc.org

New York City Bird Report
Online Bird Report updated daily
www.nycbirdreport.com

New York City Rare Bird Alert
Tel. 212 979 3070

National Parks Service – Gateway National
Recreation Area www.nps.gov/gate

New York City Transit – buses and subways
www.mta.nyc.ny.us/nyct

Books and publications

**The New York City Audubon Society Guide to
Finding Birds in the Metropolitan Area**
Marcia T. Fowle & Paul Kerlinger (2001) Cornell
University Press, ISBN 0801485657

**New York's 50 Best Places to Go Birding in
and Around the Big Apple** John Thaxton (1998)
City and Company, ISBN 1885492669

Birds of New York Field Guide
Stan Tekiela (2000) Adventure Publications
ISBN 1885061757.

The Atlas of Breeding Birds in New York State
Robert F. Andrle & Janet R. Carroll (1998) Cornell
University Press, ISBN 0801416914

**Birds of New York City: Western Long Island
and Northeastern New Jersey** Chris C. Fisher
& Andy Bezener (1998) Lone Pine Publishing
ISBN 1551051745

**A Birder's Guide to Metropolitan Areas of
North America** Paul Lehman (2001) American
Birding Association, ISBN 1878788159

**A Field Guide to the Birds of Eastern and
Central North America** Roger Tory Peterson
(2002) Houghton Mifflin, ISBN 0395740460

**National Geographic Field Guide to the Birds
of North America, 4th Edition**
National Geographic (2002) ISBN 0792268776

The North American Bird Guide David Sibley
(2000) Pica Press, ISBN 1873403984

**The Sibley Field Guide to Birds of Eastern
North America** David Sibley (2003) Christopher
Helm, ISBN 0713666579

**National Geographic Guide to Bird Watching
Sites: Eastern US** Mel White (1999) National
Geographic, ISBN 0792273745

• •

San Francisco and San José

As the cities of San Francisco and San José form a conurbation adjacent to the southern shore of San Francisco Bay, better known as Silicon Valley, there is some merit in considering the whole urban area for the purposes of a birders' guide. Moreover many people fly into San Francisco when making a business trip to Silicon Valley. This area covers a number of different habitats, including rocky seashore, saltmarsh, lowland scrub and coastal forest, each with its own distinct population of birds.

The city of San Francisco is at the northwestern tip of a peninsula that runs northwest to southeast, with San José at the southeastern end. It is bounded by San

Brown Pelicans flying past the Golden Gate Gridge

Francisco Bay on the northeast and the Pacific Ocean to the southwest. The spine of this peninsula is formed by the Santa Cruz Mountains with the Diablo Mountains further to the east. San José is in the heart of Santa Clara Valley which lies between the two mountain ranges. San Francisco has quite a mild, rather moist, climate, whereas San José is much hotter and drier.

In such a densely populated area, it is not surprising that much habitat has been lost and in many cases the birds have been forced into ever-decreasing pockets of suitable territory. Nevertheless, like most west coast cities in the USA, environmental concerns figure highly among the citizenry and what has been preserved is well maintained and supports a wide variety of birds. This area of coastal California offers some of the most spectacular scenery in the USA and is famous for its seascapes and stands of ancient Redwoods. So even if the birds aren't cooperating on any particular day the sightseeing more than compensates. You may be tempted to make the trip a little further north to Marin County to visit the well-known bird observatory at Point Reyes. However it is difficult to get to without a car, as public transport in the area is rather poor. Moreover all the typical species of the area can be seen much more easily at sites within and close to San Francisco itself. So, if you are without a car and your time is limited you would be better birding closer to the city unless, of course, you have a compelling reason to visit Point Reyes – e.g. a mega-rarity!

GETTING AROUND

The city of San Francisco has a very efficient public transport system (MUNI) for getting around the city locations, and this is a much better option than driving within the city. For birding south of San Francisco in San Mateo County, the SamTrans Bus Service offers some possibilities to get to the key birding areas. Within Silicon Valley, the Valley Transportation Authority (VTA) operates bus and light rail services, although it is most likely that anyone visiting the area will rent a car as it is the most practical means of travelling from the airport and within the valley. Most people commute by car to Silicon Valley so

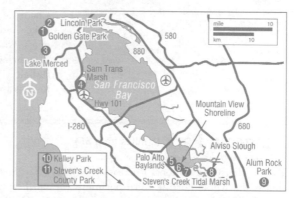

expect to encounter major traffic jams on the freeways at peak times and try to avoid travelling during rush hours. The roads through the foothills of the Diablo and Santa Cruz Mountains are rather twisting and narrow, and it is often difficult to find a safe place to 'pull-off' the road if you spot something interesting, therefore extreme caution should be exercised. Also, much of the land is privately owned, thus restricting activity to the public areas.

Apart from the odd earthquake, birding in the San Francisco area is generally very easy with very few impediments from the weather, although it is susceptible to fog on the coast. So generally a light rain jacket is sufficient. Comfortable walking shoes are adequate for all the locations, even the saltmarshes, as there is no need to walk onto the marshy areas. As is typical throughout the USA, most county and state parks charge a small entrance fee. You will probably need two maps to cover the area adequately: one for San Francisco and one for San José.

Locations

1 Golden Gate Park (3 miles from SF)   MUNI Bus #5

Bounded on one side by the Pacific Ocean and on the east by Stanyan St, Golden Gate is a fairly large but safe city park that contains a good selection of the passerines of the San Francisco Bay area, and even a herd of Bison! The eastern end has most of the tourist attractions, so birders are best advised to confine their attention to the less crowded western end. An early morning visit will produce all of the typical common species of the Pacific Coast of North America and occasionally turns up rarities at migration time. The numerous watercourses within the park are good for gulls and other waterfowl.

There are several entrances to the park along **Fulton Street**, which is served by Bus Route #5. Take the 25th Ave entrance which leads onto Cross Over Drive. East of Cross Over Drive is Stow Lake, which is good in winter for gulls and waterfowl, and the Strybing Arboretum, which is worth checking for passerines during migration.

KEY SPECIES	
Resident Great Egret, Red-shouldered Hawk, Pygmy Nuthatch, Brown Creeper	Bufflehead, Ruddy Duck, Thayer's Gull, Glaucous-winged Gull, Red-breasted Sapsucker, Varied Thrush
Summer Violet-green Swallow	**Passage** Flycatchers, vireos, warblers, Western Tanager
Winter Green-winged Teal,	

From there concentrate on the area west of Cross Over Drive, particularly the bushes and woodland around the Chain of Lakes (particularly Middle Lake and North Lake) which hold many woodland species and migrants in spring and autumn, and the Buffalo Paddock which is good for finches and blackbirds.

2 Lincoln Park (5 miles from SF)   MUNI Bus #18/38

The Golden Gate National Recreation Area is a large coastal public park that covers almost the whole northwestern perimeter of San Francisco, extending from Golden Gate Park round the tip of the peninsula to the Golden Gate Bridge, and provides excellent access to the shoreline. This is a huge area which includes an old military base (the Presidio) as well as the coastal strip, and is worthy of full exploration. However, if time is limited, the most productive areas are in and around Lincoln Park.

If you are starting early in the morning it is best to commence at Sutro Heights Park so that the sun is behind you when looking offshore. A telescope is recommended for viewing birds offshore and on Seal Rocks, which are just off the coast at Sutro Heights. Waders, ducks and auks can all be seen from here. From there follow El Camino del Mar trail (on foot only – no cars) through Lincoln Park which offers spectacular views over the ocean and check the trees and bushes for passerines: this area is a real migrant trap during migration. The two hotspots which are most frequented by birders are the East Wash and West Wash, two north-facing wooded gullies, at the east and west ends respectively of the Lincoln Park Golf Course, which have turned up many rarities. Many birds of prey migrate across the narrowest part of San Francisco Bay during the autumn and this area offers an opportunity to see them. All of the western buteos and accipiters are seen in good numbers with the added possibility of a rarity. A better site for raptor migration is the aptly-named Hawk Hill, across the bay in Marin County.

Take **Geary Boulevard** west and fork right at **Point Lobos Avenue** where the street becomes one-way. This will bring you to the entrance to Sutro Heights Park. You can walk from there along El Camino del Mar to get to the East Wash and West Wash, or else you can drive back east along Geary Boulevard to **34th Avenue**: turn left and drive north to **Legion of Honor Drive**. There is a large car park at Legion of Honor Museum and the East Wash lies just north of this car park. If you don't have a car, the cross-town bus route (No. 18) shuttles between Sutro Heights Park and the Legion of Honor Museum.

KEY SPECIES

Resident
Double-crested Cormorant, Brandt's Cormorant, Black Oystercatcher, Pygmy Nuthatch, Brown Creeper, Hutton's Vireo, Common Raven

Summer
Sooty Shearwater, Brown Pelican, Heermann's Gull, Pigeon Guillemot,

Common Murre, Allen's Hummingbird, Olive-sided Flycatcher, Orange-crowned Warbler, Hooded Oriole

Winter
Red-throated Loon, Clark's Grebe, Western Grebe, Bufflehead, Common Goldeneye, Black

Turnstone, Willet, Marbled Godwit, Surfbird, Sanderling, Ring-billed Gull, Glaucous-winged Gull, Golden-crowned Kinglet

Passage
Wandering Tattler, Parasitic Jaeger, Elegant Tern, flycatchers, warblers, raptors

3 Lake Merced (8 miles from SF)

Zoo Metro Station Line 'L'
MUNI Bus #18/29/88

Lake Merced actually consists of two fairly deep lakes which are good for grebes and diving ducks in winter. They are surrounded by a variety of woodland and more open scrub habitats. The lakes are fringed with reeds and riparian undergrowth and so have no habitat for waders in normal conditions, but the ocean beaches lie just on the other side of Highway 35, so there is easy access to the sea to scan for seabirds. There is a paved path around both lakes with trails off the main path into the woods and around the perimeter of the adjacent golf courses. The best areas for access to the lakes are at the north end of the North Lake (Sunset Boulevard) and both car parks at the south end of the South Lake (one on the east shore, one on the west).

Take **I-280** south and exit west onto **Daly Boulevard**. After about half a mile turn north onto **Lake Merced Boulevard**. Drive north for about half a mile to a lay-by and car park on the left-hand side of the road. This leads to the south shore of South Lake. Returning to Lake Merced Boulevard, continue north to the junction with **Sunset Boulevard**, where there is a car park and several trails down to the north shore of North Lake. A further location worth checking is **Fort Funston** which is reached by taking **Highway 35** along the west side of the lake for about a mile south of the North Lake and pulling in on the right-hand side of the road. The scrubby areas around the fort are good for migrant passerines, whilst the high dunes provide a good seawatch point.

KEY SPECIES

| **Resident**
Great Egret, Green Heron, Black-crowned Night Heron, Clapper Rail, Virginia Rail, Sora, Common Moorhen, Belted Kingfisher, Marsh Wren, Bewick's Wren, Pygmy Nuthatch, Brown Creeper, Common Yellowthroat, Purple Finch, American Goldfinch, Pine Siskin | **Summer**
Heerman's Gull, Forster's Tern, Caspian Tern, Northern Rough-winged Swallow, Cliff Swallow, Violet-green Swallow, Pacific-slope Flycatcher, Wilson's Warbler

Winter
Red-throated Loon, Pacific Loon, | Common Loon, Western Grebe, Clark's Grebe, Eared Grebe, Northern Shoveler, Blue-winged Teal, Cinnamon Teal, Green-winged Teal, Canvasback, Ruddy Duck, Red-breasted Merganser, Thayer's Gull, Glaucous-winged Gull |

4 SamTrans Marsh (12 miles from SF)   *SamTrans Bus #292/397*

Although there are better and larger wetland sites at the southern end of San Francisco Bay, this site's significance lies in the fact that it is adjacent to San Francisco Airport and can easily be visited if you have just a couple of hours to spare. Also, its small size means that most of the birds can be viewed just with binoculars. The habitat consists of saltmarsh, reedbed and some exposed mud at low tide, making it good for ducks, waders, gulls and a few passerines. Winter tides bring out the Clapper Rails as well as the occasional Soras and Virginia Rails. Tricolored Blackbirds nest and can usually be found.

The site is just east of **Bayshore Freeway (101)** at the northern perimeter of SFO Airport. If exiting from the airport terminal, turn north on 101 and then take the next right onto **San Bruno Avenue** which leads to the **United Air Lines Maintenance Hangar**. There is a T-junction at the end of this road: turn left and follow the road alongside the perimeter fence to the north side of the hangar. The marsh is on the left-hand side of the road (**N. Access Road**). Pull in to the turn-off on the left-hand side where there is a car park near the entrance to the **SamTrans Bus depot**. From here, there is a footpath alongside the marsh which also encircles the SamTrans Peninsula, providing access to the bay foreshore.

KEY SPECIES

| **Resident**
Black-crowned Night Heron, Clapper Rail, Virginia Rail, Sora, Common Moorhen, Marsh Wren, Common Yellowthroat, Savannah Sparrow, Tricolored Blackbird

Summer
American Avocet, Black-necked Stilt, | Least Tern, Forster's Tern, Caspian Tern

Winter
Western Grebe, Gadwall, Northern Shoveler, Northern Pintail, Blue-winged Teal, Green-winged Teal, Black-bellied Plover, Willet, Least Sandpiper, Western Sandpiper, | Marbled Godwit, Common Snipe, Herring Gull, Thayer's Gull, Ring-billed Gull

Passage
Osprey, Semipalmated Plover, Dunlin, Whimbrel, Greater Yellowlegs |

San Francisco Bay

The inner section of San Francisco Bay is characterised by an extensive area of tidal saltmarshes. The marshes are criss-crossed by meandering tidal creeks or sloughs from the streams that drain into the bay from the surrounding countryside. Large areas of shallow water have been impounded to create saltpans and recreational waterways. There are also conservation areas as well as floodwater control and sewage treatment facilities, all of which combine to produce a mosaic of aquatic habitats, both freshwater and saltwater. Large numbers of wildfowl and waders migrate through and winter here, and even in

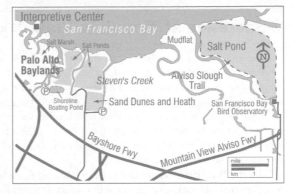

summer, there is always something interesting to be seen. This is a very large area, but access is fragmented by areas that are off-limits to the public. However, there are a number of access points which can be visited in the course of a day.

5 Palo Alto Baylands (32 miles from SF)

Of all of the locations along the southern shore of San Francisco Bay, Palo Alto Baylands is about the best. It is also closest to San Francisco Airport, being less than 30 minutes' drive on the Bayshore Freeway from the airport terminal. It is a popular birding spot and you are likely to meet local birders who will be able to provide you with information about the general area. The habitat combines saltmarsh, reedbed, mudflats and a freshwater pond, in a fairly compact area. At the Interpretive Center, a boardwalk provides access out to the saltmarsh, so that rubber boots are not required. Although many of the waders in the tidal creeks are quite approachable, a telescope would be very beneficial.

The freshwater pond on the left of the entrance supports the typical freshwater ducks but more importantly provides an opportunity to study the various plumages of a variety of gulls at close quarters, plus the added opportunity to pick out one of the rarer species. The gulls frequent the nearby refuse dump and use the freshwater pond to bathe and to take advantage of visitors feeding the ducks. A muddy tidal creek adjacent to the car park supports a good number of waders. The birds seem to be accustomed to the cars and permit quite a close approach. All the typical West Coast waders are possible. At high tide, the saltmarsh boardwalk is one of the best places to see Clapper Rails, and the viewing platform at the end of the boardwalk enables you to scan the surrounding mudflats and areas of open water for ducks and grebes. Picking out the odd Clark's Grebe among the large numbers of Western Grebes is a typical Californian birding challenge! The tracks leading off to the west and to the east of the Interpretive Center are good for sparrows and other passerines.

Take the **Bayshore Freeway (101)** southeast from San Francisco and exit north at **Embarcadero Road**. Follow Embarcadero Road for about a mile to reach the Interpretive Center. The **Palo Alto Shuttle** service, which is free, connects with the train station in **Palo Alto** weekday mornings and evenings, as there are a number of offices and hi-tech facilities in the area.

6 Shoreline at Mountain View (35 miles from SF)   VTR Bus #40

The newly created Shoreline Park at Mountain View offers, uniquely in this area, a large freshwater lagoon which is deep enough to attract grebes and diving ducks. There are also several areas of saltmarsh and tidal estuary with breeding birds as well as winter and passage migrants. Access is easy from Bayshore Freeway and there is a network of jogging trails and footpaths which offer views of most of the best areas. The one exception is the salt evaporation ponds on the northern perimeter which are private property, but which could be scanned with a telescope if the birds are feeding or roosting near the embankment.

Take the **Bayshore Freeway (101)** southeast from San Francisco and exit north at **San Antonio Road**. Continue north on San Antonio Road to **Terminal Boulevard** where there is a parking area and where the park entrance is located. Directly east of the park entrance is the Shoreline Boating Pond, a freshwater lagoon which supports loons, grebes and diving ducks in winter and is one of the regular sites in the bay area for Barrow's Goldeneyes. Going north from the park entrance provides access to the embankment that separates the saltpans, and also to a walkway along the Charleston Slough, which is one of the best areas for waders.

7 Steven's Creek Tidal Marsh (37 miles from SF)   VTR Bus #40

Steven's Creek is one of the small rivers which flow though Santa Clara Valley and empty into San Francisco Bay. For much of its course inland, there is a public walking trail along its banks, and where it enters the bay, the trail runs long the top of the tidal embankment. This provides an excellent vantage point to scan the marshes and mudflats formed at the estuary. This is also a good spot for wintering ducks and migrating waders. The open area of dunes and grassland to the west of Steven's Creek, which is part of Shoreline Park, is good for passerines and also a small population of Burrowing Owls.

Take the **Bayshore Freeway (101)** southeast from San Francisco and exit north at Shoreline Boulevard. Continue north on **Shoreline Boulevard** for about a mile to Crittenden Lane and turn right. You can park at the end of **Crittenden Lane** where you have access to the Steven's Creek Trail. A bridge over Steven's Creek at this spot allows you to scan up and down the river banks.

8 Alviso Slough (10 miles from San José)   VTR Bus #58

Alviso Slough is at the extreme southeastern end of San Francisco Bay and is the furthest access point to the bay shoreline from San Francisco, although closest to San José. There are large areas of saltpans which can be good for ducks. A ten-mile loop trail from Alviso follows the embankments around the saltpans. As well as ducks, there can be large gull roosts on the saltpans as many feed at the nearby sewage treatment facility.

Take the **Bayshore Freeway (101)** southeast from San Francisco and exit east onto **Highway 237**. Continue east to **Alviso** and exit north at Alviso onto **Gold Street**. Go north on Gold Street as far as you can (about half a mile) and the road swings round to the left and ends at a disused yacht marina. Check the muddy channels in the harbour for waders. At the north end of the marina is the trailhead for the Alviso

Slough Trail. Before setting out on the trail, it is worth checking with the San Francisco Bay Bird Observatory (SFBBO) which is just before the marina, to see 'if there is anything about'.

KEY SPECIES (sites 5–8)

Resident
Black-crowned Night Heron, Clapper Rail, Virginia Rail, Sora, Common Moorhen, Burrowing Owl, Marsh Wren, Western Meadowlark, Common Yellowthroat, Savannah Sparrow

Summer
American Avocet, Black-necked Stilt, Least Tern, Forster's Tern, Caspian Tern

Winter
Red-throated Loon, Eared Grebe, Western Grebe, Clark's Grebe, American White Pelican, Gadwall, Northern Shoveler, Northern Pintail, Blue-winged Teal, Cinnamon Teal, Green-winged Teal, Lesser Scaup, Barrow's Goldeneye, Redhead, Ruddy Duck, White-tailed Kite, Northern Harrier, Peregrine Falcon, Black-bellied Plover, Willet,

Least Sandpiper, Western Sandpiper, Dunlin, Marbled Godwit, Common Snipe, Herring Gull, Thayer's Gull, Ring-billed Gull, American Pipit, Swamp Sparrow

Passage
Osprey, Semipalmated Plover, Dunlin, Whimbrel, Greater Yellowlegs

9 Alum Rock Park (8 miles from San José)   VTR Bus #64

Alum Rock Park is in the foothills of the Diablo Mountain Range and offers a somewhat different range of birds from that of the Santa Cruz Mountains with species typical of a drier habitat. In summer, the area is very hot and very dry, and adequate drinking water should be carried even over short distances. Some of the trails are quite steep and on a hot day the effort required to scale them should not be underestimated. From the car park, parallel trails either side of the creek run east/west, and separate trails lead up to the rim of the canyon. The trails along the creek are the most heavily forested and are best for woodland species. Feeders at any of the park buildings should be checked for hummingbirds. A climb to the rim via the North Rim Trail offers a chance to find species more adapted to open country and grassland, including Yellow-billed Magpies, and gives fantastic views from Eagle Rock over the whole of Silicon Valley.

Alum Rock Park is just east of San José and is signposted from **I-680**. It can be reached by following **Alum Rock Avenue** east from the city centre: however, this involves numerous traffic-light junctions. A bus service (No. 64) operates along Alum Rock Avenue from the Alum Rock Light Rail Station and will leave you within a mile of the park entrance, but be prepared for an uphill walk!

KEY SPECIES

Resident
Golden Eagle, California Quail, Nuttall's Woodpecker, Acorn Woodpecker, Brown Creeper, White-breasted Nuthatch, Wrentit, Canyon Wren, Yellow-billed Magpie, Steller's

Jay, California Thrasher, Rufous-crowned Sparrow, Lark Sparrow, Lawrence's Goldfinch

Summer
Northern Rough-winged Swallow,

Bullock's Oriole, Black-headed Grosbeak

Winter
American Dipper, Red-breasted Sapsucker, Say's Phoebe

10 Kelley Park (San José City Centre)

For anyone who is 'marooned' in San José without the use of a car, a practical option is Kelley Park just south of I-280 which bisects the downtown section of San José and is near the San José State University Campus. It is an urban park but with good cover for birds along the banks of Coyote Creek and, provided you visit early in the morning, it is not too crowded. Although the habitat is rather artificial, it is possible to see a selection of the typical birds of the central Californian Valley.

If you feel like walking, the park adjoins the Coyote Creek Park, a long (about 18 miles) linear park which runs along Coyote Creek as far as Anderson Reservoir.

<table>
<tr><td colspan="2">KEY SPECIES</td></tr>
<tr><td>Resident
Northern Flicker, Nuttall's Woodpecker, Wrentit</td></tr>
</table>

11 Steven's Creek County Park
(13 miles from San José)

VTR Bus #23

This is one of several county parks in the foothills of the Santa Cruz Mountains and is a good location to sample the avifauna of this range. The trails through the wooded areas can be steep but are generally easy to negotiate. If you don't want to walk the entire length of the main trail, you can drive along Steven's Canyon Road which runs parallel to the trail and there are lay-bys and picnic areas which enable you to access the better areas, including a lookout over the Steven's Creek Reservoir. However you will see more birds by carefully walking the trails and it is probably more satisfying. As the vegetation is so dry, it is generally possible to listen for the rustle of dry leaves to give away the presence of ground-feeding passerines such as towhees, thrashers and sparrows. Spring and early summer are the best times to visit.

The park is just east of San José on Steven's Canyon Road, and is signposted from the **I-280** freeway (**Foothill Boulevard Exit**). There is a car park and visitors' centre, and the park is open from 08.00. From the visitors' centre, the main trail heads south following the course of Steven's Creek and passing the reservoir. If you don't have a car, there is a bus service (No. 23) to Foothill Boulevard from Valley Fair Mall but this will leave you about two miles from the park entrance.

KEY SPECIES

Resident	**Summer**
California Quail, Band-tailed Pigeon, Nuttall's Woodpecker, Acorn Woodpecker, Hairy Woodpecker, Wrentit, Spotted Towhee, California Thrasher	Spotted Sandpiper, White-throated Swift, Vaux's Swift, Pacific-slope Flycatcher, Solitary Vireo, Hutton's Vireo, Warbling Vireo, MacGillivray's Warbler, Wilson's Warbler, Western Tanager

12 San Francisco Bay Pelagic Trips

The coast of California is revered among seabird fanatics for its superb pelagic seabirds. Pelagic trips are operated from right along the coast, although for some strange reason only a few operate from San Francisco itself. However, in addition to seabird pelagics, whale-watching trips to the Farallon Islands, 27 miles offshore also offer opportunities to see seabirds. These islands contain seal and seabird colonies and can be visited by permit only, but the whale-watching trips sail around the islands, offering the chance to view seabirds. Late summer to late autumn is considered the best season, but there are possibilities even in midwinter. Some of the trips originate from the **Yacht**

KEY SPECIES

Summer	
Black-footed Albatross, Buller's Shearwater, Pink-footed Shearwater, Sooty Shearwater, Black Storm-petrel, Ashy Storm-petrel, Common Murre, Pigeon Guillemot, Tufted Puffin,	Rhinoceros Auklet, Cassin's Auklet, South Polar Skua, Sabine's Gull
	Winter
	Loons, grebes, Pelagic Cormorant, Brandt's Cormorant, Black Scoter, Surf Scoter, gulls

Marina on Marina Green, about a mile east of the Golden Gate Bridge. Others operate from **Sausalito**, just north of San Francisco and served by Golden Gate Transit Bus and also by Golden Gate Ferry from Market St Ferry Terminal.

Useful contacts and websites

Golden Gate Audubon Society
www.goldengateaudubon.org
2530 San Pablo Ave, Suite G, Berkeley, CA 94702-2047. Tel: 510 843 2222. Fax. 510 843 5351.

Golden Gate National Recreation Area
www.nps.gov/goga
Information, access points and maps.

Golden Gate Raptor Observatory
www.ggro.org
Golden Gate Raptor Observatory, Building 201, Fort Mason, San Francisco, CA 94123. Tel. 415 331 0730.

San Francisco Bay Bird Observatory
www.sfbbo.org

Shearwater Journeys – pelagic trips:
www.shearwaterjourneys.com

SF Bay Whale Watching – pelagics to Farallon Islands: www.sfbaywhalewatching.com

Oceanic Society – pelagic trips:
www.oceanic-society.org

Santa Clara Valley Audubon Society
www.scvas.org
415 Cambridge, Suite21, Palo Alto, CA 94306.

Marin County Audubon Society
www.marinaudubon.org
Marin Audubon Society, Box 599, Mill Valley, California 94942-0599.

San Francisco Field Ornithologists
www.sffo.org

California Birding
www.fog.ccsf.cc.ca.us/~jmorlan/rare.htm
Packed with information about birding in California.

Northern California Rare Bird Alert
Tel. 415 681 7422.

San Francisco Municipal Railway (MUNI)
www.sfmuni.com

SamTrans – bus services in San Mateo County and to San Francisco Airport:
www.samtrans.com

Valley Transportation Authority (VTA)
www.vta.org
Public transport in Santa Clara County.

Palo Alto Baylands Shuttle Service
www.city.palo-alto.ca.us/shuttle

Tides for San Francisco Bay
www.saltwatertides.com/dynamic.dir/californiasites.html#francisco

Books and publications

Birds of San Francisco and the Bay Area
Chris C. Fisher & Joesph Morlan (1996) Lone Pine Publishing, ISBN 1551050803

Birding Northern California John Kemper (1999) Falcon Press, ISBN 1560448326

Birder's Guide to Northern California
L. and J. Westrich (1991) Gulf Publishing ISBN 0872010635

Birding at the Bottom of the Bay B. Wyatt, A. Stoye and C. Harris (1990) Santa Clara Valley Audubon Society.

San Francisco Peninsula Birdwatching
C. Richer (ed) (1996) Sequoia Audubon Society. ISBN 0961430117

A Birder's Guide to Metropolitan Areas of North America Paul Lehman (2001) American Birding Association, ISBN 1878788159

Peterson Field Guides: Western Birds
Roger Tory Peterson (1998) Houghton Mifflin ISBN 061813218X

National Geographic Field Guide to the Birds of North America, 4th Edition
National Geographic (2002) ISBN 0792268776

The North American Bird Guide David Sibley (2000) Pica Press, ISBN 1873403984

Field Guide to the Birds of Western North America David Sibley (2003) Christopher Helm ISBN 0713666587

National Geographic Guide to Bird Watching Sites: Western US Mel White (1999) National Geographic, ISBN 0792274504

Seattle

With Puget Sound and Lake Washington on either side, and its well-wooded parks and suburbs, Seattle has a great feel to it as a birding location and any birder with half a day or more to spare there will not be disappointed. The city and its environs offer a chance to see a good selection of the typical avifauna of the Pacific Northwest region.

Located between the Pacific Ocean and the Cascade Mountains, Seattle experiences the typical mild but wet weather of the Pacific Northwest. Indeed the Olympic Peninsula to the west of the city is one of the wettest locations in North America. The Cascade Mountain Range, running along a north/south axis, acts as a great natural divide for the state of Washington. To the east of the Cascade Range the climate is much drier and more continental in character. For this reason, the birds of Washington State are separated into two groups: those typical of the dry region, east of the Cascades, and those which are more typical of west of the Cascades. There are many eastern species which would be very rare or even unknown in the west and vice versa.

Options for birding depend very much on the time of year. Winter months offer the best possibilities for wildfowl, the summer months offer more in the way of passerines and migration times are good for waders, wildfowl and passerines. The main challenge for the short-term visitor is to decide which sites to visit from the abundant selection of good birding locations. Inevitably time and transport facilities will determine what is realistic.

Any visitor to Seattle will be captivated by the views of Mount Rainier (when weather permits!) and may wish to combine some sightseeing or skiing with birding. However, even in summer, mountain birding requires a full kit of weatherproof clothing and strong footwear and therefore may not be practical. Moreover, with the exception of high-altitude species such as Ptarmigan, found above 5,000ft, most montane birds can be seen closer to the city. Likewise a boat trip to the San Juan Islands to combine whale-watching with birdwatching may initially seem attrac-

A Great Blue Heron patrols the shallows in sight of Seattle's Space Needle

TYPICAL SPECIES

Resident
Double-crested Cormorant, Pelagic Cormorant, Great Blue Heron, Canada Goose, Mallard, Red-tailed Hawk, American Coot, Killdeer, Glaucous-winged Gull, Northern Flicker, Downy Woodpecker, Anna's Hummingbird, Steller's Jay, American Crow, Black-capped Chickadee, Chestnut-backed Chickadee, Bushtit, Bewick's Wren, Winter Wren, Red-breasted Nuthatch, American Robin, Golden-crowned Kinglet, European Starling, Spotted Towhee, Song Sparrow, White-crowned Sparrow, Red-winged Blackbird, House Sparrow, American Goldfinch, House Finch, Dark-eyed Junco

Summer
Rufous Hummingbird, Tree Swallow, Violet-green Swallow, Barn Swallow, Warbling Vireo, Orange-crowned Warbler, Yellow Warbler, Yellow-rumped Warbler, Common Yellowthroat, Wilson's Warbler, Cedar Waxwing, Savannah Sparrow

Winter
Horned Grebe, Western Grebe, American Wigeon, Greater Scaup, Lesser Scaup, Canvasback, Bufflehead, Barrow's Goldeneye, Common Goldeneye, White-winged Scoter, Surf Scoter, Mew Gull, Ring-billed Gull, Bonaparte's Gull, Ruby-crowned Kinglet, Golden-crowned Sparrow

tive, but the departure port for the ferries is at Anacortes, over 50 miles north of Seattle and served by only one bus per day, so it may not represent the best use of valuable time. For the purposes of this guide, the sites covered are those which are relatively close to the city and (mostly) accessible by public transport.

GETTING AROUND

Seattle and its airport (Sea-Tac Airport) are well served by both freeways and public transport and some birding locations are easily accessible to a visitor, with or without a car. In addition, the ferry services which criss-cross Puget Sound, offer excellent opportunities for birding and provide access to some good habitats on the islands in the Sound and further afield on

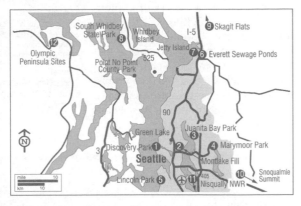

the Olympic Peninsula. However, many bus routes operate at reduced frequency outside peak periods and some operate on weekdays only and others at weekends, so very careful study of timetables is required. The relevant websites are listed on page 435. For those whose business may take them to the outer suburbs of Redmond or Bellevue, the public transport options are not so abundant and it is probably more likely that they will have a car at their disposal.

In addition to a good city map, you will also need a road map which covers the neighbouring counties, as well as the opposite shore of the Olympic Peninsula. No special footwear other than comfortable walking shoes is necessary to visit all the lowland areas mentioned in this guide. Hiking boots and warm clothing are a necessity for any trips into the mountains and, wherever you are, always be prepared for wet weather!

Locations

1 Discovery Park (5 miles)   *Metro Transit Bus #19, 24, 33*

Just a few miles north of the city centre on the shores of Puget Sound, this city park is in an excellent location to scan the waters of the bay and acts as a migrant trap for birds moving north or south. Part of its attraction is the variety of habitats it supports, from coniferous and deciduous woodland to open park and seashore habitats. For the new arrival in Seattle, it presents an undemanding opportunity to acquaint yourself with the avifauna of the Pacific Northwest. A walk from the South Parking Lot along the South Beach Trail to the West Point Lighthouse will offer woodland, open grassland and some stretches of beach and deeper water. These habitats support all the typical birds of the region. One note of caution: it is a fair bet that all of the crows will be American Crows – since even local birders are reluctant to identify a Northwestern Crow, the casual visitor has no chance! It is believed that most crows on the central and southern Washington coasts are probably intergrades between the two species/races. Pure Northwestern Crows are found further north, on the Strait of Juan de Fuca and in Canada.

From Downtown, go west on Elliott Avenue, and continue north as Elliott changes to 15th Avenue NW. Follow the signs for **Emerson Avenue** and cross the Ship Canal, where Emerson Avenue changes to **W. Emerson Place**. Follow Emerson to **Gilman Avenue W.**, turn right and continue on Gilman as it turns into **W. Government Way**, which will take you to the park entrance. Maps and bird lists for the park are available at the Visitor Center, open daily from 08.30 to 17.00. From here, walk the 2.8-mile Loop Trail, from which other trails go to the beach, or drive to the North Parking Lot, and walk the beach or Loop trails from there. The best seabird viewing is from the area north of the West Point Lighthouse.

KEY SPECIES

Resident Rhinoceros Auklet, Pigeon Guillemot, Band-tailed Pigeon, Pileated Woodpecker, Hutton's Vireo **Summer** Caspian Tern, Willow Flycatcher, Pacific-slope Flycatcher	**Winter** Red-breasted Merganser, Varied Thrush, Fox Sparrow, Pine Siskin **Passage** California Gull, Common Tern, waders, hirundines, flycatchers, warblers, seaducks

2 Montlake Fill (5 miles)

Metro Transit Bus #25/65

Montlake Fill (Union Bay Natural Area) is a disused landfill site northeast of the city centre on the shores of Lake Washington and near the University of Washington Campus. When landfill operations ceased in the 1960s the site was redeveloped into an amenity area, planted with trees and shrubs, and wetland habitats were created. This site and the adjacent Union Bay shoreline offer a good cross-section of species typical of the Seattle area.

Take **I-5** north and exit east onto **NE 45th Street**. Pass the University Campus and turn south onto **Mary Gates Memorial Drive**, which leads to the **Center for Urban Horticulture** on **NE 41st Street**. The entrance to Montlake Fill is through the grounds of the Horticultural Center.

Flocks of gulls in winter can be quite confusing as the Glaucous-winged and Western Gulls hybridise producing some odd-looking offspring. The best place to study this phenomenon at close quarters is at Green Lake, which is just two miles northeast of Montlake Fill on the western side of I-5.

To get to **Green Lake**, go back to NE 45th Street and turn left. Continue across I-5 and turn right on **Stone Way N**. Stone Way turns to **Greenlake Way N** and then E Green Lake Way. Follow this around the lake, stopping along the way to walk down to the water. The gulls may be studied here at close quarters, but the lake itself may only interest gull enthusiasts as there is little else there.

KEY SPECIES

Resident	**Winter**	**Passage**
Pied-billed Grebe, Gadwall, Northern Shoveler, Belted Kingfisher **Summer** Green Heron, Cinnamon Teal, Spotted Sandpiper, Vaux's Swift, Violet-green Swallow, Cliff Swallow, Marsh Wren	Green-winged Teal, Pintail, Ring-necked Duck, Ruddy Duck, Hooded Merganser, California Gull, Thayer's Gull, Glaucous-winged Gull, Western Gull	Western Grebe, Sora, Western Sandpiper, Least Sandpiper, Long-billed Dowitcher, American Pipit

3 Juanita Bay Park (13 miles)   *Metro Transit Bus #255*

Located on the northeast shore of Lake Washington, this prime wetland site is well worth a visit at any time of the year, although numbers of waterfowl are naturally at their highest in winter. This is a popular birding spot for many Seattle birders and offers the opportunity to meet local enthusiasts and to exchange information. The site is very well maintained with walkways and viewing platforms so there is no danger of getting your feet wet! Osprey nesting platforms have been constructed that are visible from the walkway. The park is on Market Street and is served by public transport.

Take **I-405** north and exit west on **116th Avenue (Exit 20A)**. Go west on 116 for about a mile and turn left (south) onto **98th Avenue**. The park entrance and a parking lot are about 200 yards along on the right-hand side.

KEY SPECIES

Resident Pied-billed Grebe, Green Heron, Wood Duck, Virginia Rail, Marsh Wren **Summer** Osprey, Spotted Sandpiper, Vaux's Swift, Rufous Hummingbird, Willow	Flycatcher, Black-headed Grosbeak, Bullock's Oriole **Winter** Green-winged Teal, Ring-necked Duck, Hooded Merganser, Bald Eagle, Wilson's Snipe, California Gull, Herring Gull, Belted Kingfisher, Fox Sparrow, Pine Siskin	**Passage** Western Sandpiper, Greater Yellowlegs, Warbling Vireo, Orange-crowned Warbler, Wilson's Warbler, Black-throated Gray Warbler, Townsend's Warbler

4 Marymoor Park (15 miles)   *Metro Transit Bus #256 from Seattle*
Metro Transit Bus #249 from Bellevue

This is a rather large public park located beside Redmond Town Center at the northern end of Lake Sammamish. It lacks the saltwater habitat of the Seattle shoreline parks but has a much richer variety of landbirds. A nature trail runs alongside the waterway and offers good viewing opportunities of the lake from a specially-constructed platform. Like most public parks it is very crowded at weekends and also has a large dog-walking area where dogs are allowed 'off-leash'. However, the park is large enough to provide enough secluded areas where birds are undisturbed, particularly on the eastern side. The best spots are around the Clise Mansion and on the loop trail along the river, which begins in the off-leash dog area.

Travelling east from Seattle on **520**, take the **West Lake Sammamish Parkway NE** exit and turn left into the park.

KEY SPECIES

Resident Pied-billed Grebe, Double-crested Cormorant, Gadwall, Virginia Rail, Belted Kingfisher **Summer** Green Heron, Wood Duck, Osprey, Spotted Sandpiper, Band-tailed	Pigeon, Western Wood Pewee, Willow Flycatcher, Vaux's Swift, Black Swift, Cliff Swallow, Red-eyed Vireo, Swainson's Thrush, Marsh Wren, Black-headed Grosbeak **Winter** Northern Shoveler, Green-winged	Teal, Ring-necked Duck, Hooded Merganser, Common Merganser, Bald Eagle, Fox Sparrow, Lincoln's Sparrow **Passage** Warblers, Western Tanager

5 Lincoln Park (6 miles)

Metro Transit Bus #54
Sound Transit #570 from SeaTac Airport

This is another of Seattle's shoreside parks on Puget Sound. It is further south of the city centre, making it a good location for anyone on their way to SeaTac Airport. It is also right beside the Fauntleroy Ferry Terminal and a trip on one of the ferries to the western shore of Puget Sound is a good way to see some of the aquatic birds typical of the south Puget Sound: such as loons, grebes, diving-ducks, gulls and terns.

The park is immediately north of the **Fauntleroy Ferry Terminal**, which is sign-posted from the **509 Freeway**.

KEY SPECIES	
Resident Band-tailed Pigeon	Red-breasted Merganser, Common Murre, Rhinoceros Auklet
Summer Caspian Tern, Pigeon Guillemot, Purple Martin	**Passage** California Gull, seaducks, waders, flycatchers, hirundines, warblers
Winter Common Loon, Black Scoter, Harlequin Duck,	

6 Everett Sewage Ponds (27 miles)

The city of Everett is just 25 miles north of Seattle, well within the 'commuting' radius of the city and served by several bus routes and the Sounder Train. Just east of the town centre, the Snohomish River divides into several channels where it enters Puget Sound, forming several marshy islands, including Spencer Island. This island and the adjacent sewage ponds are a surprisingly rich wetland area. To enter the ponds, you must request a permit at the Site Office.

The **I-5 Freeway** through Everett runs right past the ponds, but exiting the freeway and navigating back to the ponds requires some care, as there is no exit adjacent to the site. The easiest approach is to exit west on **Marine Drive** at **Exit 195**, cross the Snohomish River Bridge and loop back south on **Ross Avenue**. There is no bus to Spencer Island but it should be possible to take a cab from **Everett Bus Terminal**.

KEY SPECIES	
Resident Pied-billed Grebe, American Bittern, Gadwall, Virginia Rail, Belted Kingfisher, Marsh Wren	**Winter** Northern Shoveler, Green-winged Teal, Ring-necked Duck, Bald Eagle, Wilson's Snipe, gulls
Summer Osprey, Spotted Sandpiper, Black Swift, Black-headed Grosbeak	**Passage** Cinnamon Teal, waders, hirundines, warblers

7 Jetty Island – Everett (28 miles)

Everett Transit Bus #23

This is a man-made island, originally created from spoil dredged out of the mouth of the Snohomish River, but it has now been consolidated and developed into a public amenity. The island is reached by free ferry from **10th Street** in downtown **Everett**. However, this service is only available in summer when it runs daily every half-hour. At any other time, the island and the intertidal mudflats can be scanned from the shore. There are also Harbour Cruises operating from 10th Street which offer the opportunity to view the marine birdlife on this stretch of Puget Sound.

KEY SPECIES

Resident
Belted Kingfisher, Marsh Wren, Savannah Sparrow

Summer
Osprey, Spotted Sandpiper, Caspian

Tern, Arctic Tern, Pigeon Guillemot, Purple Martin

Winter
Green-winged Teal, Bald Eagle, Dunlin

Passage
Semipalmated Plover, Least Sandpiper, Western Sandpiper, Sanderling, Short-billed Dowitcher, Long-billed Dowitcher, Greater Yellowlegs, California Gull

8 Whidbey Island (30 miles)

Whidbey is the largest island in Puget Sound and one of the nearest to Seattle. It is a rather elongated narrow island, about 40 miles long, with good birding spots all along the way. It is really only practical to explore this island by car and there is a **car ferry terminal** at **Mukilteo**, about 25 miles north of Seattle, near Everett. Some of the wetland locations require rubber

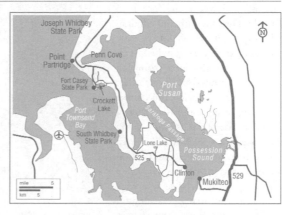

boots and preferably a telescope. There is a tremendous variety of habitats on the island, both saltwater and freshwater, as well as forested areas and open fields. Due to its location at the northern end of Puget Sound, it is much better than the areas around Seattle for some marine species such as Pelagic Cormorants, Black Oystercatchers, gulls and alcids. To describe all of the sites on this brilliant birding location would require a small booklet in its own right, and this guide can only give a brief overview, but consult the Whidbey Island Audubon Society's website for more information.

From where the car ferry docks at Clinton on the southeast tip of the island, the closest birding locations are: Lone Lake for freshwater ducks and riparian passerines; South Whidbey Island State Park, an old-growth forest habitat for woodpeckers and a good variety of passerines; Crockett Lake for waders and ducks; and Fort Casey State Park for alcids, gulls, Black Oystercatchers and also cetaceans.

If you have the time and the transport to go further north, then it would be worth checking Penn Cove (east side) and the shoreline and lagoons at its western end: the best spot on the island for seaducks and for rocky shore birds. Nearby Point Partridge (west side) and the shoreline between there and Joseph Whidbey State Park offer rocky shoreline habitat for Black Oystercatchers and also vistas over the oceanic waters of the Strait of Juan de Fuca.

KEY SPECIES

Resident
Bald Eagle, Virginia Rail, California Quail, Black Oystercatcher, Pigeon Guillemot, Belted Kingfisher, Marsh Wren, Pileated Woodpecker

Summer
Osprey, Spotted Sandpiper, Caspian Tern, Cliff Swallow, Swainson's Thrush, Savannah Sparrow

Winter
Red-throated Loon, Common Loon, Green-winged Teal, Harlequin Duck, Dunlin, Sanderling, Black Turnstone, Surfbird, Black-bellied Plover,

| Common Murre, Marbled Murrelet, Rhinoceros Auklet | **Passage** Semipalmated Plover, Ruddy Turnstone, Least Sandpiper, Western Sandpiper, Short-billed Dowitcher, | Long-billed Dowitcher, Greater Yellowlegs, California Gull, flycatchers, warblers |

Further Away

9 Skagit Flats (65 miles)

Skagit Flats is a low-lying area of the Skagit River Delta, north of Seattle. This is a truly outstanding birding location but it is only practical to access it with a car, owing to its distance from the city and the need to cover a flat, extensive area during the course of a day, often using the car as a hide. In winter it is one of the best locations in North America for watching swans, geese and, in particular, raptors. As many as 13 raptor species can be seen in winter and for many Washington birders seeing five falcon species in a day: Peregrine, Merlin, American Kestrel, Gyr Falcon and Prairie Falcon, is an attainable goal, especially if combined with a trip to the Samish Flats further north. In summer, the wet areas and meadows host good populations of aquatic and riparian species.

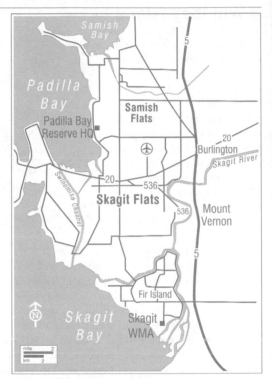

Exit west off **I–5** Interstate Highway at **Interchange 221** and travel west on **Fir Island Road**. The fields on either side of the road support thousands of Snow Geese as well as Tundra and Trumpeter Swans. Keep a lookout on fence posts and overhead wires for perched raptors. The tidal mudflats are good for ducks and waders. Turn left at the signpost for the **Skagit Wildlife Area**: the area beyond the car park is good for landbirds. Just north of Skagit Flats is the equally good Padilla Bay which supports thousands of Brant geese in winter.

KEY SPECIES

Resident
Wood Duck, Virginia Rail, Marsh Wren

Summer
Cinnamon Teal, Blue-winged Teal, Sora, Osprey, Caspian Tern, Brown-headed Cowbird

Winter
Tundra Swan, Trumpeter Swan, Snow Goose, Brant, Bald Eagle, Northern Harrier, Red-tailed Hawk, Rough-legged Hawk, American Kestrel, Merlin, Peregrine Falcon, Gyr Falcon, Prairie Falcon, Snowy Owl, Short-eared Owl, Black-bellied Plover,

Dunlin, Greater Yellowlegs, Northern Shrike, American Pipit, Western Meadowlark, ducks, gulls, sparrows

Passage
Western Sandpiper, Long-billed Dowitcher, Short-billed Dowitcher, Whimbrel

10 Snoqualmie Summit Ski Area (50 miles)

This is one of the most popular ski resorts in the Seattle area and the nearest to the city, being just an hour's drive away. It is also served by Greyhound Bus and by numerous Ski Tour buses which operate from the city centre. The ski lifts operate from a base level of approximately 3,000ft to over 5,000ft. Both offer possibilities for birding, winter or summer, and an opportunity to see some of the montane species typical of the Cascade Mountain Range. During the winter, the few species that remain on the mountain will be at lower elevations and the areas around the ski lifts would be the most productive. In summer, the cross-country ski runs provide a network of trails that traverse the mountains up to the tree line. Be prepared for a reasonably strenuous hike, with an ascent of over 1,500ft. Wear strong footwear, and carry adequate fluids and warm clothing as the weather can be very changeable even in summer.

Snoqualmie Pass is located about 50 miles east of Seattle on the **I-90 Interstate**.

KEY SPECIES

Resident	Summer
Red-tailed Hawk, Blue Grouse, Hairy Woodpecker, Pileated Woodpecker, Common Raven, American Dipper, Red Crossbill, Pine Siskin	Red-breasted Sapsucker, Olive-sided Flycatcher, Western Wood Pewee, Pacific-slope Flycatcher, Willow Flycatcher, Swainson's Thrush, McGillivray's Warbler, Townsend's Warbler

11 Nisqually National Wildlife Refuge (50 miles)

The Nisqually River enters the south end of Puget Sound, flanked on both sides by an expanse of saltmarshes and mudflats. A sea dyke protects an area of freshwater marshland interspersed with lagoons. There is also some mixed woodland on the drier patches. The main area of interest lies between the river and McAllister Creek. A five-mile loop trail (Brown Farm Dike Trail) skirts the perimeter of this area and provides access to all the main habitats, although it is closed during hunting season. There is an observation tower on this trail, and there are several shorter trails within the refuge area, with boardwalk access through marshy patches.

This site is very conveniently located just off the **I-5 Interstate** between Olympia and Tacoma. The Visitor Center is sign-posted at **Exit 114**. Also worth checking is the Nisqually Reach Nature Center at Luhr Beach, at the mouth of McAllister Creek. The jetty there provides a good vantage point to scan the adjacent shoreline and mudflats. From Exit 114, go west on **Martin Way** for about a mile and then turn north onto **Meridian Road** for about 2.5 miles before turning right (east) onto **46th Avenue**. Take the first left (**D'Milluhr Drive**) and follow it to the parking lot and jetty.

KEY SPECIES

Resident	Summer	Winter
Pied-billed Grebe, American Bittern, Gadwall, Ruddy Duck, Bald Eagle, Northern Harrier, Virginia Rail, Pigeon Guillemot, Belted Kingfisher, Band-tailed Pigeon, Red-breasted Sapsucker, Pileated Woodpecker, Marsh Wren	Wood Duck, Cinnamon Teal, Spotted Sandpiper, Caspian Tern, Cliff Swallow, Northern Rough-winged Swallow, Western Wood Pewee, Willow Flycatcher, Pacific-slope Flycatcher, Swainson's Thrush, Black-throated Gray Warbler, MacGillivray's Warbler, Savannah Sparrow, Black-headed Grosbeak	Red-throated Loon, Common Loon, Brant, Green-winged Teal, Northern Shoveler, Northern Pintail, Ring-necked Duck, Common Merganser, Red-breasted Merganser, Hooded Merganser, Rough-legged Hawk, Sharp-shinned Hawk, Dunlin, Sanderling, Wilson's Snipe, Common Murre, Marbled Murrelet, Rhinoceros Auklet

| **Passage** Greater White-fronted Goose, Black-bellied Plover, Semipalmated Plover, | Least Sandpiper, Western Sandpiper, Short-billed Dowitcher, Long-billed Dowitcher, Greater Yellowlegs, | California Gull |

12 Olympic Peninsula sites (50 miles)

If your visit to Seattle involves a trip across Puget Sound to the Olympic Peninsula, there are several sites along the east and north shores of the peninsula which offer good birding opportunities, particularly for the more pelagic species, which are less likely to occur in the Seattle area. A typical route would be to take the ferry from Edmonds (15 miles north of Seattle) to Kingston on the Kitsap Peninsula. During migration it would be worth driving north from Kingston to Point No Point County Park, which provides a vantage point to observe loons, ducks and alcids. Tidal surges in this area produce turbulence which attracts large flocks of feeding birds, particularly gulls and terns.

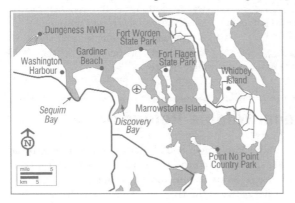

Continue west, driving over the Hood Canal Bridge onto the Olympic Peninsula, and turn north towards Port Townsend. Two of the best sites in the Port Townsend area with shoreline access are Fort Flagler State Park at the northern tip of Marrowstone Island, and Fort Worden State Park within the town.

The north coast of the Olympic Peninsula has some excellent wetland sites and sheltered bays and these can be encountered by driving west from Port Townsend to Discovery Bay: check Gardiner Beach at the northwest corner of the bay. Continue west to Sequim Bay, which has extensive tidal flats which are good for waders, ducks and gulls. The mudflats at the south end of the bay and the northwest corner (Washington Harbour) support the largest numbers of birds. The Yacht Marina on the west shore of the bay is a good vantage point to scan for deeper water species such as loons, diving ducks and alcids.

Further west from Sequim Bay is Dungeness National Wildlife Refuge, a narrow sandy spit of land which encloses tidal mudflats and is an important wintering area for Brant geese and other wildfowl.

KEY SPECIES

Resident
Brandt's Cormorant, Bald Eagle, Black Oystercatcher, Pigeon Guillemot, Belted Kingfisher

Summer
Wood Duck, Osprey, Caspian Tern, Tufted Puffin

Winter
Red-throated Loon, Pacific Loon, Common Loon, Red-necked Grebe,

Brant, Green-winged Teal, Northern Shoveler, Northern Pintail, Ring-necked Duck, Redhead, Harlequin Duck, Oldsquaw, Black Scoter, Hooded Merganser, Black-bellied Plover, Dunlin, Sanderling, Black Turnstone, Surfbird, Black-bellied Plover, Common Murre, Ancient Murrelet, Marbled Murrelet, Rhinoceros Auklet

Passage
Turkey Vulture, Semipalmated Plover, Ruddy Turnstone, Least Sandpiper, Western Sandpiper, Short-billed Dowitcher, Long-billed Dowitcher, Greater Yellowlegs, Whimbrel, Parasitic Jaeger, California Gull, Heerman's Gull, Common Tern, flycatchers, warblers

Useful contacts and websites

Washington Ornithological Society
www.wos.org/

Seattle Audubon Society
www.seattleaudubon.org

BirdWeb
www.birdweb.org/birdweb
Seattle Audubon's online guide to the birds of
Washington.

Whidbey Island Audubon Society
www.whidbeyaudubon.org

Skagit Audubon Society
www.fidalgo.net/~audubon

East Lake Washington Audubon Society
www.elwas.org

Friends of Marymoor Park
Useful information website
www.marymoor.org

Kitsap Audubon Society
www.kitsapaudubon.org

Black Hills Audubon Society
www.blackhillsaudubon.com

Olympic Peninsula Audubon Society
www.olympicpeninsulaaudubon.org

Washington State Ferries
www.wsdot.wa.gov/ferries

King County Transit – public transport within
Seattle area: www.transit.metrokc.gov

Sound Transit – public transport beyond Seattle
area: www.soundtransit.org

Everett Transit – public transport within Everett:
www.everettwa.org/transit

Greyhound Bus Company – bus service to
Snoqualmie Summit: www.greyhound.com

Books and publications

Birds of Seattle and Puget Sound Chris C.
Fisher (1996) Lone Pine Publishing
ISBN 1551050781

**Birding in Seattle and King County: Site
Guide and Annotated Lists** Eugene S. Hunn,
(1982) Seattle Audubon Society, ISBN 0914516051

A Birder's Guide to Washington Hal Opperman
(2003) American Birding Association
ISBN 1878788205

Birding Washington Rob and Natalie McNair-
Huff (2005) Falcon Press, ISBN 076272577X

**A Birder's Guide to Metropolitan Areas of
North America** Paul Lehman (2001) American
Birding Association, ISBN 1878788159

Peterson Field Guides: Western Birds Roger
Tory Peterson (1998) Houghton Mifflin
ISBN 061813218X

**National Geographic Field Guide to the Birds
of North America, 4th Edition**
National Geographic (2002), ISBN 0792268776

The North American Bird Guide David Sibley
(2000) Pica Press, ISBN 1873403984

**Field Guide to the Birds of Western North
America** David Sibley (2003) Christopher Helm
ISBN 0713666587

**National Geographic Guide to Bird Watching
Sites: Western US** Mel White (1999)
National Geographic, ISBN 0792274504

●●●

TYPICAL SPECIES

Resident
Mute Swan, Canada Goose, Mallard,
American Black Duck, Red-tailed Hawk,
Ring-billed Gull, American Herring Gull,
Great Black-backed Gull, Mourning
Dove, Downy Woodpecker, Hairy
Woodpecker, Blue Jay, American Crow,
American Robin, Black-capped
Chickadee, White-breasted Nuthatch,
European Starling, Song Sparrow,
Northern Cardinal, House Sparrow,
American Goldfinch, House Finch

Summer
Great Blue Heron, Turkey Vulture,
Killdeer, Spotted Sandpiper, Northern
Flicker, Chimney Swift, Ruby-throated

Toronto

Many birders will be familiar with Toronto as the arrival point in Canada for a trip to the migration hotspots of Point Pelee and Long Point on the north shore of Lake Erie. While both of these superb sites are just a few hours' drive away, for all but the super-rarities, Toronto has much to offer a visiting birder on a tight schedule wishing to experience the avifauna of the Great Lakes region. Situated on the north shore of Lake Ontario, the waterfront is just a short walk from the city centre. The city is very well landscaped with parks offering habitats for land-

Hummingbird, Eastern Wood Pewee, Great Crested Flycatcher, Purple Martin, Tree Swallow, Bank Swallow, Winter Wren, Gray Catbird, Brown Thrasher, Red-eyed Vireo, American Redstart, Yellow Warbler, Chipping Sparrow, Red-winged Blackbird, Common Grackle, Brown-headed Cowbird, Baltimore Oriole

Winter
Gadwall, Lesser Scaup, Greater Scaup, Long-tailed Duck, Bufflehead, Goldeneye, Dark-eyed Junco

Passage
Common Loon, Broad-winged Hawk, Sharp-shinned Hawk, flycatchers, vireos, thrushes, warblers

Upland Sandpipers favour the short turf of Toronto Airport

birds and an extensive shoreline from which to scan the lake. Although Lake Ontario is the smallest of the Great Lakes, it is still so vast as to create the impression of Toronto being a maritime city. Indeed, it is possible to 'seawatch' from headlands during spring and autumn to observe a continuous stream of ducks, grebes and divers migrating along the shore. Over much of the province of Ontario human population density decreases substantially the further north you move from the shores of the Great Lakes. Therefore the greatest environmental pressures have tended to be on those wetland sites along the littoral areas of Lake Ontario. Nevertheless, those sites that have been given protection support a diverse range of species.

Like many temperate zone cities, the best time to be in Toronto is during migration, when birds are on the move and many non-resident species can be observed on passage. Toronto, like other cities in eastern North America, experiences falls of passerines during spring and autumn which can produce a bewildering variety of warblers, flycatchers and other migrants even in city centre parks. Summer is also quite good as Toronto lies within the overlap of the breeding ranges of the northern breeding thrushes, warblers and sparrows, as well as the more southerly breeding species. Unfortunately, summer brings with it swarms of the most relentless and persistent biting insects which can penetrate light clothing, so be prepared for an onslaught in wooded and wetland areas. Winter, on the other hand, can be very quiet. The extremely cold weather and frozen waterways do not encourage many birds to winter in the area, and the few resident species can be quite depleted in numbers as a proportion of the population move south. However, there are often southerly movements of northern birds such as Snowy Owls, Snow Buntings and northern finches as some compensation.

GETTING AROUND

In Toronto, like many Canadian cities, great emphasis is placed on providing a comprehensive public transport system. Within the city boundaries, the system is operated by the Toronto Transit Commission. Immediately to the west, the buses are operated by Mississauga Transit. Additionally there is the commuter rail network, the GO-Train, which can be used to reach the outer suburbs quickly. The buses and trams, or streetcars as they are called, are the principal method of getting around and many routes operate from hubs at the suburban rail stations. There is also a subway system which although limited in size serves a few of the locations which interest birders.

A car enables much more flexibility. Also, if birding in winter, the car can act as a draught-proof heated room if you need to warm yourself after a spell outdoors. Driving in Toronto is not really problematic except when trying to park in the city centre. This would be necessary if you were planning to visit the Toronto Islands or visit the lakeshore parks, and in such circumstances it is best to use public transport. If driving, try to time your birding so that you are travelling away from the city centre during morning rush hour and towards the city in the evening.

The best maps to use are MapArt Road Maps. For many of the sites it is advisable to wear boots or at least footwear that you won't mind getting wet and muddy, particularly after the spring thaw when much of the ground is very soft.

Locations

1 Toronto Islands (3km)

There are several low-lying islands just offshore from Toronto collectively known as the Toronto Islands. They are popular recreation areas for city dwellers and can be accessed by ferries from the south end of Bay Street in the city centre. The best for birders are Ward's Island (east end) and Algonquin Island (west end) which are joined by a bridge and, in summer, served by an inter-island tram system. There is a ferry to Hanlan's Point on the main Toronto Island, and a worthwhile trip would be to take the ferry to Hanlan's Point and walk the trail (or use the tram) to Ward's Island ferry terminal, birding along the way, particularly at the lagoons and inlets along the north shores of Toronto Island.

KEY SPECIES

Summer
Double-crested Cormorant, Black-crowned Night Heron, Green Heron, Wood Duck, Osprey, Virginia Rail, Common Tern, Caspian Tern, Belted Kingfisher, House Wren, Marsh Wren, Savannah Sparrow, Indigo Bunting, Eastern Meadowlark

Winter
Redhead, Oldsquaw, Hooded

Merganser, Common Merganser, Red-breasted Merganser, Rough-legged Hawk, Glaucous Gull, Iceland Gull, Northern Shrike, American Tree Sparrow, Snow Bunting

Passage
Red-throated Loon, Red-necked Grebe, Horned Grebe, Tundra Swan, Black-bellied Plover, Semipalmated Plover, Ruddy Turnstone, Red Knot,

Semipalmated Sandpiper, Least Sandpiper, Short-billed Dowitcher, Greater Yellowlegs, Upland Sandpiper, Wilson's Phalarope, Little Gull, Bonaparte's Gull, Black Tern, Horned Lark, American Pipit, flycatchers, vireos, warblers

2 Tommy Thompson Park (7km)

Tommy Thompson Park (also known as Leslie Street Spit) is one of Toronto's lakeshore public parks and is on a headland which juts out into Lake Ontario. The park is on a

5km long, narrow, man-made peninsula, whose tip provides a superb vantage point to scan the lake. Luckily, a free shuttle bus will take you about two-thirds of the way to the tip. A number of lagoons or embayments have been constructed on each side of the spit and are managed to encourage wildlife. The muddy margins of these lagoons are good for waders during spring and autumn, and the open waters hold significant numbers of breeding and wintering wildfowl. The wooded areas act as migrant traps for passerines as well as holding the largest heronry in Toronto. There are even Coyotes roaming the park!

Go east from the city centre on **Lakeshore Boulevard** and turn south at **Leslie Street**. The park entrance is about 1km from the junction with Lakeshore Boulevard.

KEY SPECIES

Summer	Winter	
Double-crested Cormorant, Black-crowned Night Heron, Green Heron, Wood Duck, Blue-winged Teal, Green-winged Teal, Osprey, American Kestrel, Virginia Rail, American Woodcock, Ring-billed Gull, Common Tern, Caspian Tern, Belted Kingfisher, Marsh Wren, Savannah Sparrow, Indigo Bunting, Eastern Meadowlark	Redhead, Hooded Merganser, Common Merganser, Red-breasted Merganser, Rough-legged Hawk, Glaucous Gull, Snowy Owl, Northern Shrike, American Tree Sparrow, Snow Bunting **Passage** Red-necked Grebe, Horned Grebe,	Black-bellied Plover, Semipalmated Plover, Red Knot, Semipalmated Sandpiper, Least Sandpiper, Short-billed Dowitcher, Greater Yellowlegs, Lesser Yellowlegs, Wilson's Phalarope, Little Gull, Bonaparte's Gull, Black Tern, American Pipit, flycatchers, vireos, warblers

3 Guildwood Park (18km)

TTC Bus #116 from Kennedy Subway Station

This is one of a string of parks and public recreation areas along the shores of Lake Ontario, east of Toronto, in an area known as the Scarborough Bluffs. These are low cliffs (some 60m high) along the shore, some of which are indented by steep overgrown ravines. Because of the terrain, they were not built on as the city grew so that much of the natural vegetation is still intact. Guildwood Park contains a good mix of deciduous and coniferous woodland as well as a shrub understorey. There are good observation points within the park from which to view the lake.

Take **Highway 2** east from Toronto for about 16 km. After passing **Guildwood GO-Train Station** turn south on **Galloway Road** to bring you to the park entrance. It is approximately 2km from Guildwood GO-Train Station and, if travelling on foot, an easier alternative is to take a bus from Kennedy Subway Station, which passes by the park entrance.

KEY SPECIES

Resident	Winter	
Pileated Woodpecker, Brown Creeper, Carolina Wren, Northern Mockingbird, Cedar Waxwing **Summer** Double-crested Cormorant, Black-crowned Night Heron, Wood Duck, American Kestrel, Virginia Rail, Common Tern, Belted Kingfisher, Willow Flycatcher, Eastern Phoebe, Eastern Kingbird, Winter Wren, House Wren, Brown Thrasher	Hooded Merganser, Common Merganser, Red-breasted Merganser, Rough-legged Hawk, Red-breasted Nuthatch, American Tree Sparrow, Snow Bunting **Passage** Red-necked Grebe, Horned Grebe, Tundra Swan, American Wigeon, Ruddy Duck, Northern Harrier, Yellow-bellied Flycatcher, Least Flycatcher, Warbling Vireo, Golden-	crowned Kinglet, Ruby-crowned Kinglet, Blue-gray Gnatcatcher, Veery, Wood Thrush, Hermit Thrush, Tennessee Warbler, Blackburnian Warbler, Magnolia Warbler, Yellow-rumped Warbler, Wilson's Warbler, Scarlet Tanager

4 Thickson's Woods (45km)
Oshawa GO-Train Station

This is one of the last remnants of old-growth pine forest left on the shores of Lake Ontario, which has now been preserved. Apart from the tall stands of white pines, there is a diverse mix of deciduous and coniferous woodland where fallen trees are allowed to decay. There are also open clearings which are actively managed to provide habitat for grassland species, and wetlands areas around the Corbett Creek Marsh. This site is at its best during migration as the visible passage along the lakeshore presents an ever-changing variety of birds.

Go east on **Highway 401** for about 40km and exit south on **Thickson Road (Exit 412)**. Continue past **Wentworth Street** and there is a trail through the woods on the left-hand side of Thickson Road. This trail traverses all of the key habitat zones, including mixed woodland, open meadow and marsh. If using public transport, the site is about a 3km walk from Oshawa GO-Train Station.

KEY SPECIES

Resident
Brown Creeper, Northern Mockingbird, Cedar Waxwing

Summer
Double-crested Cormorant, Black-crowned Night Heron, Wood Duck, American Kestrel, Virginia Rail, Belted Kingfisher, Eastern Phoebe, Eastern Kingbird, House Wren, Wood Thrush, Northern Waterthrush, Ovenbird, Common Yellowthroat

Winter
Hooded Merganser, Common Merganser, Rough-legged Hawk, Red-breasted Nuthatch, American Tree Sparrow

Passage
Red-necked Grebe, Horned Grebe, American Wigeon, Yellow-bellied Flycatcher, Least Flycatcher, Philadelphia Vireo, Warbling Vireo, Golden-crowned Kinglet, Ruby-crowned Kinglet, Blue-gray Gnatcatcher, Gray-cheeked Thrush, Swainson's Thrush, Hermit Thrush, Tennessee Warbler, Northern Parula, Black-throated Blue Warbler, Blackburnian Warbler, Chestnut-sided Warbler, Magnolia Warbler, Yellow-rumped Warbler, Blackpoll Warbler, Palm Warbler, Mourning Warbler, Canada Warbler, Wilson's Warbler, Scarlet Tanager, Rose-breasted Grosbeak, Bobolink

5 Second Marsh (55km)
Oshawa GO-Train Station

Second Marsh is one of the best restored wetland areas in Ontario and the site of the first successful breeding attempt by Little Gulls in North America, back in 1962. The Second Marsh and the nearby McLaughlin Bay Wildlife Reserve provide a continuous wetland and wet woodland habitat along the lakeshore. There is a network of trails through the marshes

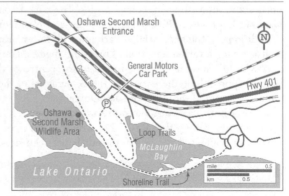

which are mapped and signposted over a distance of several kilometres, as well as two viewing platforms. This is still the best site in North America for Little Gulls, and up to 100 birds can be seen on spring mornings.

If you are driving, take the **#419** Exit off **Highway 401** and go south on **Farewell Street**, and turn left at Col. Sam Drive. The entrance to Second Marsh is 500m on the right-hand side. There is no public transport directly to the marsh, but it should be possible to get a taxi from the Oshawa GO-Train Station and return from the General Motors HQ building near the marsh.

KEY SPECIES

Summer
Double-crested Cormorant, American Bittern, Least Bittern, Black-crowned Night Heron, Green Heron, Wood Duck, Osprey, Virginia Rail, Little Gull, Common Tern, Caspian Tern, Belted Kingfisher, Eastern Phoebe, Eastern Kingbird, Marsh Wren, Savannah Sparrow, Swamp Sparrow

Winter
Redhead, Hooded Merganser,

Common Merganser, Red-breasted Merganser, Rough-legged Hawk, Pileated Woodpecker, Northern Shrike, Snow Bunting, Evening Grosbeak

Passage
Red-throated Loon, Red-necked Grebe, Horned Grebe, Tundra Swan, Trumpeter Swan, American Wigeon, Lesser Scaup, Ruddy Duck, Dunlin, Pectoral Sandpiper, Semipalmated

Sandpiper, Greater Yellowlegs, Lesser Yellowlegs, Upland Sandpiper, Bonaparte's Gull, Horned Lark, Yellow-bellied Flycatcher, Least Flycatcher, Ruby-crowned Kinglet, Veery, Wood Thrush, Tennessee Warbler, Nashville Warbler, Magnolia Warbler, Yellow-rumped Warbler, Blackburnian Warbler

6 Humber Bay Park (11km)

High Park Subway Station
Mimico GO-Train Station

Humber Bay Park is just west of the city centre, where the Humber River empties into Lake Ontario. It is one of many parks located along the lakeshore in Toronto and, being so close to the city centre and so easy to access, it tends to be very popular with the general public and can be very crowded. Nevertheless, it is excellent for migrants and also sup-

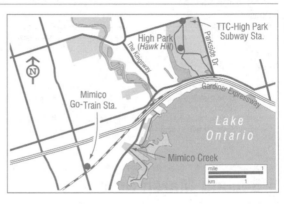

ports some interesting breeding species. The park is on the waterfront, south of the Gardiner Expressway. It is divided into East and West Humber Parks by Mimico Creek.

Additionally, the nearby High Park is a well-known raptor watchpoint in autumn, as well as having a varied selection of breeding species. The best point for raptor watching is Hawk Hill, an area of high ground in the centre of the park. The birds are usually spotted approaching the park over the skyscrapers of the city centre, and then turning west to follow the shoreline of Lake Ontario.

To cover both areas would require an 8km round trip, but there are plenty of birding opportunities along the way. The streetcar service (TTC #501or #508) along Lakeshore Boulevard allows you to shuttle between sites.

KEY SPECIES

Summer
Double-crested Cormorant, Black-crowned Night Heron, Green Heron, Wood Duck, Osprey, Virginia Rail, Common Tern, Caspian Tern, Belted Kingfisher, Eastern Phoebe, Eastern Kingbird, Marsh Wren, Savannah Sparrow, Swamp Sparrow

Winter
Redhead, Hooded Merganser, Common Merganser, Red-breasted Merganser, Rough-legged Hawk, Northern Shrike, Snow Bunting

Passage
Red-throated Loon, Red-necked Grebe, Horned Grebe, Tundra Swan, Trumpeter Swan, American Wigeon,

Lesser Scaup, Ruddy Duck, Ruddy Turnstone, Red Knot, Semipalmated Sandpiper, Bonaparte's Gull, Horned Lark, Yellow-bellied Flycatcher, Least Flycatcher, Ruby-crowned Kinglet, Veery, Wood Thrush, Tennessee Warbler, Nashville Warbler, Magnolia Warbler, Yellow-rumped Warbler, Canada Warbler

7 Credit Valley Conservation Area
(36km from City Centre; 15km from airport)

Mississauga Transit Bus #42 and #82

The Credit River Valley runs through the centre of Mississauga, west of Toronto, and is only about a 15-minute drive from the airport, so this is worth a visit if you are going to or coming from the latter. The Credit Valley Conservation Area is a large, discontinuous, linear park along either side of the Credit River which drains into Lake Ontario. The site also includes some of the watershed areas. The closest sections to the airport are where the valley is bisected by Highway 401. North of the highway is the CVC Administration Centre, with trails leading north and west along the river from the car park. South of the highway is Credit Meadows Park, also with a network of trails.

Take **Highway 401** west from Toronto (the highway that follows the southern perimeter of the airport) and exit north onto **Mavis Road (Exit 340)**. Go north on Mavis Road and after about 1.5km turn left onto **Crawford Mill Avenue**. At the T-junction turn right into **Gooderham Estate Boulevard** and then take the first left onto **Old Derry Road**. Go west on Old Derry Road, until after you cross the **Credit River bridge** and the park entrance is on the right-hand side. If you return to the Old Derry Road, follow it as it turns sharply south and continue south for about 2km, passing over Highway 401; you will come to the entrance of Credit Meadows Park on the left-hand side of the road.

Bus #42 runs from Westwood Mall just north of the airport (and would be a short cab ride from the terminal building) to the CVC Administration Centre.

KEY SPECIES

Resident
Brown Creeper, Carolina Wren, Northern Mockingbird, Cedar Waxwing

Summer
Black-crowned Night Heron, Wood Duck, American Kestrel, Virginia Rail, American Woodcock, Belted Kingfisher, Eastern Phoebe, Wood Thrush

Winter
Hooded Merganser, Common Merganser, Rough-legged Hawk, Red-breasted Nuthatch, American Tree Sparrow

Passage
Yellow-bellied Flycatcher, Least Flycatcher, Warbling Vireo, Golden-crowned Kinglet, Ruby-crowned Kinglet, Blue-gray Gnatcatcher, Gray-cheeked Thrush, Swainson's Thrush, Hermit Thrush, Tennessee Warbler, Northern Parula, Black-throated Blue Warbler, Black-throated Green Warbler, Blackburnian Warbler, Chestnut-sided Warbler, Magnolia Warbler, Yellow-rumped Warbler, Blackpoll Warbler, Palm Warbler, Mourning Warbler, Canada Warbler, Wilson's Warbler, Scarlet Tanager, Rose-breasted Grosbeak, Bobolink

Further Away

8 Niagara Falls (125km)

Winter can be a very quiet time for birds in the Toronto area and anyone who is already familiar with the winter avifauna may find the birding a little dull. Niagara Falls is a prime tourist attraction that is about a 90-minute drive from Toronto or about two hours by train or Greyhound Bus. You may find that a trip to the falls is part of your itinerary or you may contemplate making the trip anyway. In winter, gull fanatics will need absolutely no prompting. Niagara Falls is a quite exceptional site for gulls: up to 14 species can be seen in a day in winter, and a total of 19 have been recorded. As well as having large gull breeding colonies, the Great Lakes are an important wintering area for gulls from the Arctic which migrate along the St Lawrence River. The turbulence created by the falls ensures that the water doesn't ice over and gulls converge on the feeding opportunities. The best time of year is from mid-November to the end of December when numbers can reach up to 100,000, and many vagrant gulls are attracted. The train and bus stations are situated close to Whirlpool Bridge and some of the best gull action can be observed between Whirlpool and Horseshoe Falls. Other vantage points are the lookout at the Adam Beck Generating Station and the river bank in the village of Queenston.

KEY SPECIES

Summer	Winter	Passage
Double-crested Cormorant, Black-crowned Night Heron, Great Egret, Wood Duck, Blue-winged Teal, Green-winged Teal, Osprey, American Kestrel, Common Tern, Belted Kingfisher, Marsh Wren, Savannah Sparrow, Swamp Sparrow	American Wigeon, Redhead, Canvasback, Hooded Merganser, Common Merganser, Red-breasted Merganser, Rough-legged Hawk, Peregrine Falcon, Purple Sandpiper, Iceland Gull, Thayer's Gull, Glaucous Gull, Northern Shrike, Snow Bunting	Red-necked Grebe, Horned Grebe, Black-bellied Plover, Ruddy Turnstone, Hudsonian Whimbrel, Wilson's Phalarope, Little Gull, Bonaparte's Gull, Sabine's Gull, Black-legged Kittiwake, California Gull, Caspian Tern, Black Tern

Useful contacts and websites

Ontario Field Ornithologists
www.ofo.ca
Contains link to Ontbirds (OFO listserve).

Toronto Ornithological Club
www.torontobirding.ca

Toronto Field Naturalists
www.sources.com/tfn/

South Peel Naturalists Club
www.spnc.ca
Covers Mississauga and areas west of Toronto.

Pickering Naturalists
www.pickeringnaturalists.org
Local naturalist group with field trips in east Toronto.

Durham Region Field Naturalists
www.drfn.ca
Covers Oshawa and areas east of Toronto.
Tel. 905 576 2738.

Second Marsh Wildlife Area
www.secondmarsh.com

McLaughlin Bay Wildlife Reserve
www.mclaughlinbay.org

Toronto Islands Park
www.toronto.ca/parks/island
Park information and ferry details.

Greater Toronto Raptor Watch
www.torontobirding.ca/~gtrw

Thickson's Woods Land Trust
www.thicksonswoods.com

Toronto and Southern Ontario Birding Message Board
www.outdoorontario.net/birds/

Toronto Transit Commission
www.city.toronto.on.ca/ttc
Buses, trams, trains and subways. Tel. 416 393-INFO

Greyhound Bus Company
www.greyhound.com
Public transport to Niagara Falls.

Mississauga Transit
www.mississauga.ca/portal/residents/publictransit
Bus service for areas west of the airport.

Books and publications

A Birdfinding Guide to the Toronto Region
Clive Goodwin (1988) The Goodwins, Toronto
ISBN 0969366906

A Bird-finding Guide to Ontario
Clive Goodwin (1995) University of Toronto Press
ISBN 0802069045

Birds of Ontario Andy Bezener (2005) Lone Pine Publishing, ISBN 1551052369

The ROM Field Guide to Birds of Ontario
Janice M. Hughes (2001) McClelland & Stewart
ISBN 0771076509

Birds of Toronto and Vicinity
Gerald McKeating (1990) Lone Pine Publishing
ISBN 0919433634

Greater Toronto Area Bird Checklist and Reporting Guidelines G. Coady & R. Smith (2000) Toronto Ornithologists Club

A Birder's Guide to Metropolitan Areas of North America Paul Lehman (2001) American Birding Association, ISBN 1878788159

A Field Guide to the Birds of Eastern and Central North America Roger Tory Peterson (2002) Houghton Mifflin, ISBN 0395740460

National Geographic Field Guide to the Birds of North America, 4th Edition
National Geographic (2002) ISBN 0792268776

Field Guide to the Birds of Eastern North America David Sibley (2003) Christopher Helm ISBN 0713666579

The North American Bird Guide David Sibley (2000) Pica Press, ISBN 1873403984

National Geographic Guide to Bird Watching Sites: Eastern US Mel White (1999) National Geographic, ISBN 0792273745

• •

TYPICAL SPECIES

Resident
Great Blue Heron, Canada Goose, Turkey Vulture, Red-tailed Hawk, Red-shouldered Hawk, American Kestrel, Killdeer, Ring-billed Gull, Mourning Dove, Northern Flicker, Downy Woodpecker, Hairy Woodpecker, Red-bellied Woodpecker, Pileated Woodpecker, Cedar Waxwing, Blue Jay, American Crow, Fish Crow, Carolina Chickadee, Tufted Titmouse, White-breasted Nuthatch, Carolina Wren, American Robin, Northern Mockingbird, European Starling, Eastern Towhee, Song Sparrow, Northern Cardinal, Red-winged Blackbird, Common Grackle, Brown-headed Cowbird, House Sparrow, House Finch, American Goldfinch

Summer
Broad-winged Hawk, Osprey, Laughing Gull, Chimney Swift, Common Nighthawk, Yellow-billed Cuckoo, Ruby-throated Hummingbird, Eastern Wood-pewee, Acadian Flycatcher, Great-crested Flycatcher, Eastern Phoebe, Eastern Kingbird, Purple Martin, Northern Rough-winged Swallow, House Wren, Blue-gray Gnatcatcher, Gray Catbird, Brown Thrasher, Red-eyed Vireo, White-eyed Vireo, Veery, Wood Thrush, Yellow Warbler, American Redstart, Black-and-white Warbler, Northern Parula, Common Yellowthroat, Ovenbird, Baltimore Oriole, Orchard Oriole, Scarlet Tanager, Summer Tanager, Chipping Sparrow, Indigo Bunting

Winter
Herring Gull, Great Black-backed Gull, Yellow-bellied Sapsucker, Winter Wren, Golden-crowned Kinglet, Ruby-crowned Kinglet, Hermit Thrush, Yellow-rumped Warbler, Brown Creeper, White-throated Sparrow, Dark-eyed Junco

Passage
Sharp-shinned Hawk, Blue-headed Vireo, Swainson's Thrush, Chestnut-sided Warbler, Magnolia Warbler, Black-throated Blue Warbler, Black-throated Green Warbler, Canada Warbler, Blackpoll Warbler, Palm Warbler, Rose-breasted Grosbeak, Bobolink

Washington DC

Most cities try to interpose some parks and woodland amidst the buildings in order to provide some green lungs for the conurbations. Washington DC, on the other hand, is rather like one large park with a few big buildings scattered around it. Taken together with the Potomac and Anacostia Rivers running through its centre, and a number of other waterways, it all adds up to some really good birding sites within a short distance of each other. There is a wide variety of species to be seen at any time of the year and, because it lies a little further south than the other cities of the northeastern USA, several species are resident all year round which would be summer migrants further north.

Flanking the Potomac River just upstream from Chesapeake Bay, Washington DC is situated along the Atlantic Flyway and experiences a wave of migration twice a year. Despite its distance from the open sea, the river is tidal at Washington which means that exposed mudflats exist on each bank at low tide, which can be good for waders at migration time. The wide open vistas of the parks and monuments mean that, unlike many North American city centres, you have a great view of the skies and can occasionally spot raptors such as Turkey Vultures and Ospreys on migration. The city also experiences Nighthawk flights during spring and late summer, usually in the evenings, when migrating flocks descend to lower altitudes due to weather conditions and you can be surrounded by a flock of feeding Common Nighthawks!

It is just a little too far inland to make a trip to the Atlantic coast practical for the short-term visitor. Hardcore birders may be tempted to do so, however, as the migration hotspot of Cape May N.J. can be reached by ferry from Lewes on the coast of Delaware. Also, and without resorting to ferries, the Delaware and Maryland coastlines and Kiptopeke

(Virginia) on the southern tip of the Delmarva Peninsula can rival Cape May in autumn, while some of the best pelagic trips in the North Atlantic operate out of Virginia Beach. However, this guide is restricted to areas within 60 minutes' travel time of the city centre.

GETTING AROUND

The metro subway system is very comprehensive so with a single day ticket or weekend ticket you can visit all the central birding sites as well as the tourist attractions. The bus sys-

A Common Nighthawk flies across the White House lawn

tem can be used to reach locations a little further outside the city. The buses are operated by Metrobus, the same company that operates the metro, and many of the bus routes operate from suburban metro stations offering a feeder service. Check the schedules as some of these buses only operate during peak periods or in one direction only. Most of the traffic during rush hours will be towards the centre of the city so, if staying out of town, try to avoid these times when travelling to Rock Creek Park, East Potomac Park or Theodore Roosevelt Island.

Driving through Washington DC is surprisingly easy as the system of freeways whisks you past the city streets very efficiently. However, it is very easy to miss the exit you are looking for and then often very difficult to navigate your way back to your destination. So try to stay in the slow lane wherever possible and keep an eye on the road signs. Coincidentally, many of the birding sites lie along the George Washington Memorial Parkway, which runs parallel to the Potomac River on the Virginia side.

The best maps are the Rand McNally and Gousha Road Maps. At many of the birding sites, you can obtain trail maps from the visitor centre. If you are visiting Washington DC in late spring or summer, you will definitely need insect repellent to ward off the brutes, otherwise birding near woodland or wetlands is almost impossible. There are plenty of drugstores that have products best suited to local conditions although I have yet to come across a product that offers 100 per cent protection and doesn't dissolve the rubber on your binoculars! Most of the sites are paved with gravel or timber walkways so you are unlikely to need rubber boots. However for most of the sites along the Potomac River you need a telescope to scan distant flocks of wildfowl and waders.

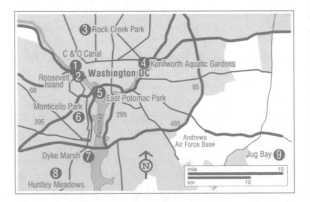

Locations

1 Chesapeake & Ohio Canal (4 miles) — Foggy Bottom Metro Station

The C & O Canal runs parallel to the Potomac, northwestward from the centre of the city for a distance of about 180 miles. Over much of this distance there is a linear park that provides riparian and woodland habitat for birds. The relevant stretch for city-bound birders is from the end of the canal at Georgetown upriver to Great Falls, a distance of about 13 miles. The river acts as a conduit for migrants and this, together with the insect life associated with the proximity of water, means that spring and autumn can produce falls of migrant flycatchers, warblers and other insectivorous passerines. In winter, the open stretches of the river support small numbers of ducks.

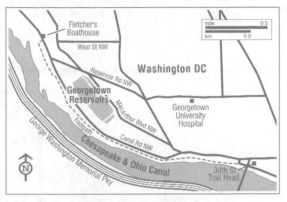

There is only one access point to the canal towpath from the city centre and this is at **34th Street**, just on the north side of Key Bridge. The canal towpath runs along the south bank of the canal and there are very few bridges over the canal along its length which will bring you back to the city side, so plan how far you want to walk before returning. There is a second bridge over the canal at Fletcher's Boathouse about three miles upstream from Key Bridge and if you wish to cross back to the city side here, you could also visit Georgetown Reservoir on McArthur Boulevard, which supports a wider variety of wintering wildfowl than the canal and Potomac River.

KEY SPECIES

Resident
Black Vulture, Bald Eagle, Belted Kingfisher

Summer
Double-crested Cormorant, Black-crowned Night Heron, Green Heron, Wood Duck, Spotted Sandpiper, Barn

Swallow, Warbling Vireo, Prothonotary Warbler, Yellow-throated Warbler, Louisiana Waterthrush

Winter
Ring-necked Duck, Bufflehead, Common Merganser

Passage
Caspian Tern, Tennessee Warbler

2 Theodore Roosevelt Island (3 miles) — Rosslyn Metro Station

This is a wooded island in the centre of the Potomac River which can be reached from the west shore by a pedestrian bridge. It is part of the George Washington Memorial Parkway (GWMP), an almost continuous riverside park which extends along the west bank of the Potomac River through the city of Washington DC and beyond. There is a footpath around the perimeter of the island and a small marsh at the southern end. At low tide you can scan the mudflats from the southern tip for gulls and any waders that might be present. From Rosslyn Metro Station, walk east on 19th Street to the footbridge over the motorway which brings you to the GWMP and the west bank of the Potomac.

KEY SPECIES

Resident
American Black Duck, Black Vulture, Bald Eagle, Belted Kingfisher

Summer
Double-crested Cormorant, Black-crowned Night Heron, Green Heron,

Great Egret, Wood Duck, Forster's Tern, Caspian Tern

Winter
Ring-necked Duck, Bufflehead, Common Merganser, Swamp Sparrow

Passage
Spotted Sandpiper, Bonaparte's Gull, Gray-cheeked Thrush, Tennessee Warbler, Northern Waterthrush, Marsh Wren

3 Rock Creek Park (5 miles)

Van Ness Metro Station

Rock Creek is a long narrow park which runs more or less North/South along the central axis of Washington DC. Although it does not have much aquatic habitat, this is probably the best site in the city to observe the phenomenon of warbler migration during both spring and autumn. All of the eastern North American warblers have been recorded here, and many are regular in large numbers. Bear in mind that despite their bright colours many American wood-warblers occupy the top canopy

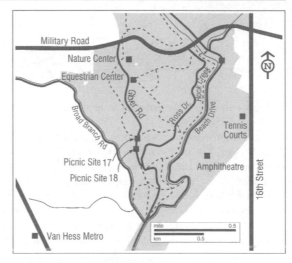

of tall trees and can be difficult to pick out: almost impossible once the trees are in full leaf. Most American birders can detect warblers and other passerines by call and very often this is how the birds are found. The best areas for birding are bounded by **Military Road**, **Glover Road** and **Ross Drive**, especially around the Nature Center and the Equestrian Center. For winter passerines, there are feeders at the Nature Center on Glover Road, while further south on Glover Road, Picnic Sites 17 and 18 are the best areas for migrants. The nearest metro station is at the intersection of Van Ness St and Connecticut Avenue, which is about half a mile west of the park.

KEY SPECIES

Summer
Acadian Flycatcher, Louisiana Waterthrush

Winter
Red-breasted Nuthatch, Hermit Thrush

Passage
Least Flycatcher, Gray-cheeked Thrush, Swainson's Thrush, Blue-winged Warbler, Tennessee Warbler, Bay-breasted Warbler, Blackburnian Warbler, Worm-eating Warbler,

Hooded Warbler, Northern Waterthrush, Field Sparrow

4 Kenilworth Aquatic Gardens (4.5 miles)   *Deanwood Metro Station*

The Kenilworth Gardens is an area of marshes on the east bank of the Anacostia River which were created originally to grow water lilies but have now been developed into a garden for displaying many kinds of aquatic vegetation. Apart from the system of ponds and dykes, which provides habitats for herons and other aquatic birds, the gardens provide access to saltmarsh on the banks of the river where waders can be observed. There is a system of trails through the gardens and out onto the marshes, leading to a boardwalk which extends to the banks of the Anacostia River. You can obtain a map from the Nature Center in the gardens. The wooded areas of the park support passerine migration in spring and autumn, as well as some resident woodland species, and the more unkempt areas are good for sparrows.

Go northeast on **Kenilworth Avenue (Route 295)** and at the exit for **Eastern Ave**, take the right-hand slip road to the overpass, where you can make a U-turn and then go south on the opposite slip road to **Quarles Street**. Turn right on Quarles Street, which leads to the entrance. Getting to the gardens on foot is a little problematic as the nearest stop on the Metro (**Deanwood**) leaves you on the 'wrong' side of 295. You need to walk up to Eastern Ave to get to the overpass which leads to Quarles Street and the entrance.

KEY SPECIES

Resident
American Black Duck, Belted Kingfisher

Summer
Green Heron, Great Egret, Wood Duck, Spotted Sandpiper, Forster's

Tern, Marsh Wren, Warbling Vireo, Prothonotary Warbler, Yellow-breasted Chat, Blue Grosbeak, Grasshopper Sparrow

Winter
American Wigeon, Green-winged Teal, Northern Pintail, Hooded Merganser,

Common Merganser, Swamp Sparrow

Passage
American Bittern, Greater Yellowlegs, Lesser Yellowlegs, Wilson's Snipe, Solitary Sandpiper, Tennessee Warbler, Field Sparrow

5 East Potomac Park (City Centre)   *Smithsonian Metro Station*

This location is almost in the dead centre of Washington DC and adjacent to the main tourist sites. It is probably the best place in Washington DC for gulls and regularly produces rarities such as Lesser Black-backed, Glaucous and Iceland Gulls in late winter/early spring. The park is on an island which extends from the tidal basin on the Washington Waterfront southeastwards to the confluence of the Potomac and Anacostia Rivers. The southeasterly tip, Hains Point, provides an unimpeded vantage point to scan the lower Potomac and Anacostia Rivers, the Washington Channel and the associated mudflats, although there is little to see here in summer. The river is very broad at this point and views may be distant even with a telescope, but good views of the gulls can be obtained in the tidal basin area where many roost. Also worth checking is the golf course, which can be good for waders after heavy rain.

KEY SPECIES

Resident
American Black Duck, Belted Kingfisher

Summer
Double-crested Cormorant, Forster's Tern, Caspian Tern, Barn Swallow,

Chipping Sparrow

Winter
Horned Grebe, Bald Eagle, Merlin, Ring-necked Duck, Lesser Scaup, Bufflehead, Common Goldeneye, Common Merganser, Red-breasted

Merganser, Lesser Black-backed Gull, Glaucous Gull, Iceland Gull

Passage
Common Loon, Pied-billed Grebe, Spotted Sandpiper, Bonaparte's Gull, Field Sparrow

6 **Monticello Park** (6 miles)

This is a small suburban woodlot which seems to act like a magnet for migrating warblers and other passerines, particularly in spring. As Washington parks go, it is a pretty unassuming place but it attracts a huge variety of species every year, and if you are visiting in spring or autumn, then this site is worth a visit. It is also very small, so can be 'done' in a few hours and at migration times there are usually a few local birders present who will be able to advise on recent sightings. The small

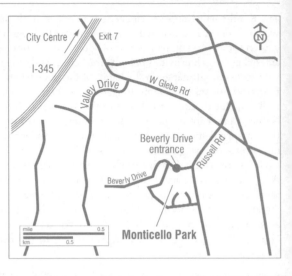

size and topography of the site mean that you can often obtain better views of the birds than in the bigger, more densely-wooded areas. Outside the migration seasons, it is probably not worth spending your time here.

Take the **I-395** south to **Glebe Road (Exit 7)**. Exit east on Glebe Road and continue east to **Russell Road**. Turn right (south) onto **Russell Road**, and after about a quarter of a mile, turn right onto **Beverly Drive**. The park entrance is on the left-hand side of Beverly Drive. This site would be less than a ten-minute taxi ride from Ronald Reagan National Airport, so could be easily visited on a stop-over.

KEY SPECIES

Passage
Philadelphia Vireo, Yellow-throated Vireo, Gray-cheeked Thrush, Blue-winged Warbler, Nashville Warbler, Tennessee Warbler, Bay-breasted Warbler, Blackburnian Warbler, Cape May Warbler, Mourning Warbler, Hooded Warbler, Wilson's Warbler, Northern Waterthrush, Louisiana Waterthrush, Blue Grosbeak

7 **Dyke Marsh** (10 miles)

Dyke Marsh is situated about ten miles south of Washington DC city centre on the west bank of the Potomac. This is a freshwater tidal marsh which is periodically subjected to inundation by saltwater from the estuary. Although it is quite small, a surprising number and variety of birds can be found within the marsh and on the river estuary. Local birders conduct bird walks at 08.00 every Sunday morning, so this is a good way to meet birders from the area. The marsh is just east of the George Washington Memorial Parkway beside the Belle Haven Marina.

Take the **GWMP** south passing under the I-495 Beltway and after crossing the bridge (**Stone Bridge**) over **Hunting Creek**, take the first turn east for **Belle Haven Marina**. Just before the marina, the first turn on the right (south) brings you into Dyke Marsh/Picnic Area. If, instead of turning right after the turn-off from the GWMP, you immediately turn left (north), you can drive to the most northerly car park. From here, you can walk north along a track to the Stone Bridge over Hunting Creek. The mudflats on either side of the bridge are good for waders.

KEY SPECIES

Resident
American Black Duck, Bald Eagle, Belted Kingfisher, Swamp Sparrow

Summer
Double-crested Cormorant, Great Egret, Snowy Egret, Black-crowned Night Heron, Green Heron, Wood Duck, Forster's Tern, Caspian Tern, Tree Swallow, Barn Swallow, Warbling

Vireo, Marsh Wren, House Wren, Yellow Warbler, Prothonotary Warbler

Winter
Common Loon, Gadwall, American Wigeon, Redhead, Ring-necked Duck, Lesser Scaup, Bufflehead, Common Goldeneye, Hooded Merganser, Common Merganser, Red-breasted Merganser, Ruddy Duck, American Coot

Passage
Northern Pintail, Green-winged Teal, Blue-winged Teal, Semipalmated Sandpiper, Least Sandpiper, Pectoral Sandpiper, Short-billed Dowitcher, Solitary Sandpiper, Spotted Sandpiper, Greater Yellowlegs, Lesser Yellowlegs, Field Sparrow

8 Huntley Meadows (11 miles)   *Bus #108 from Huntington Metro Station*

This area is a short distance south of the city centre but its wet grassland habitat and freshwater marsh support a number of species, such as King Rails and Eastern Bluebirds, which are not so regular in the more central areas of the city. Although it is quite large (well over 1000 acres), there is a boardwalk, observation tower and trail system which enable you to access the best areas with comparative ease.

Take the **George Washington Memorial Parkway** south to **Alexandria** where it connects with the **Richmond Highway (Route 1)**. After driving about 3.5 miles exit west off the Richmond Highway into Lockheed Boulevard. The park entrance is at the junction of **Lockheed Boulevard** with **Harrison Lane**. Metro Bus #108 departs from Huntington Metro Station and serves the neighbourhood just west of Harrison Lane. The nearest stop to the park is on Rolling Stone Way.

KEY SPECIES

Resident
American Black Duck, Bald Eagle, King Rail, American Woodcock, Belted Kingfisher, Eastern Bluebird

Summer
Black-crowned Night Heron, Yellow-crowned Night Heron, Little Blue Heron, Green Heron, Great Egret,

Snowy Egret, Wood Duck, Spotted Sandpiper, Forster's Tern, Caspian Tern, Barn Swallow, Tree Swallow, Prothonotary Warbler, Field Sparrow

Winter
Hooded Merganser, American Coot, American Tree Sparrow, Swamp Sparrow

Passage
Pied-billed Grebe, American Bittern, Green-winged Teal, Blue-winged Teal, Solitary Sandpiper, Wilson's Snipe, Greater Yellowlegs, Swainson's Thrush

9 Jug Bay (20 miles)

Jug Bay is an extensive wetland about 20 miles east of the city centre, situated in Maryland, on the east bank of a sweeping bend on the Patuxent River. It is in a less built-up area than the wetlands in the immediate vicinity of the city, so the habitats are less constricted and there is an opportunity to see species which might be difficult or impossible to see from the city sites. You can also see River Otters and Beavers. The drawback is that you really need a car to get here as there is no public transport. However, it is worth a visit because of the size of the area and the variety of habitats, which include freshwater marshes, deciduous forest and cultivated land.

Take **Pennsylvania Avenue (State Highway 4)** east as far as **Bristol**, then exit west on Wrighton Road. **Wrighton Road** terminates after about 2.5 miles at Jug Bay Wetlands Sanctuary.

KEY SPECIES

Resident
Wood Duck, Black Vulture, Bald Eagle, Virginia Rail, American Woodcock, Belted Kingfisher, Marsh Wren, Eastern Bluebird

Summer
Least Bittern, Green Heron, Great Egret, Snowy Egret, Little Blue Heron, Black-crowned Night Heron, Wood Duck, Forster's Tern, Caspian Tern,

Barn Swallow, Tree Swallow, Yellow-throated Vireo, Prothonotary Warbler, Prairie Warbler, Hooded Warbler, Summer Tanager, Field Sparrow

Winter
Gadwall, Northern Shoveler, Ruddy Duck, Common Merganser, Wilson's Snipe, Red-breasted Nuthatch, American Tree Sparrow, Swamp Sparrow

Passage
Pied-billed Grebe, Glossy Ibis, Tundra Swan, Green-winged Teal, Blue-winged Teal, Sora, American Coot, Greater Yellowlegs, Lesser Yellowlegs, Solitary Sandpiper, Spotted Sandpiper, Semipalmated Sandpiper, Least Sandpiper, Dunlin, Short-billed Dowitcher, Swainson's Thrush, Tennessee Warbler, Pine Warbler, Northern Waterthrush

Useful contacts and websites

Audubon Society of the District of Columbia
www.dcaudubon.org
P.O. Box 15726, Washington, DC 20003
Tel. 202 547 2355.

Fairfax Audubon Society
www.fairfaxaudubon.org
4022 Hummer Road, Annandale, VA 22003
Tel. 703 256 6895.

Maryland Ornithological Society
www.mdbirds.org
Cylburn Mansion, 4915 Greenspring Avenue,
Baltimore, MD 21209 Voicemail: 800 823 0050

Northern Virginia Bird Club
www.nvabc.org
P.O. Box 9291, McLean, VA 22102

Virginia Society of Ornithology
www.ecoventures-travel.com/vso
520 Rainbow Forest Drive, Lynchburg, VA 24502

Friends of Jug Bay Wetlands Sanctuary
Information about this site: www.jugbay.org

Friends of Dyke Marsh
Information about this site: www.fodm.org

Friends of Huntley Meadows Park
Information about this site:
www.friendsofhuntleymeadows.org

Kenilworth Park and Aquatic Gardens
www.nps.gov/nace/keaq

Rock Creek Park
www.nps.gov/rocr

Chesapeake and Ohio Canal
www.nps.gov/choh

Theodore Roosevelt Island Park
www.nps.gov/this

Washington DC Rare Bird Alert
Tel. 301 652 1088

MDOsprey: Discussion of Maryland birds
Subscription: listserv@home.ease.lsoft.com

VA-Bird: Discussion of Virginia birds
www.freelists.org/list/va-bird

Metro Bus and Rail
www.wmata.com/timetables/bus_timetables.cfm

Books and publications

Finding Birds in the National Capital Area
Claudia Wilds (1992) Smithsonian Institution Press
ISBN 156098175X

Atlas of the Breeding Birds of Maryland and the District of Columbia Chandler S. Robbins and Eirik A. T. Blom (Editors) (1996) University of Pittsburgh Press, ISBN 0822939231

A Birder's Guide to Virginia David W. Johnston (1997) American Birding Association
ISBN 1878788124

Birds of Virginia Field Guide Stan Tekiela (2002) Adventure Publications, ISBN 1885061366

A Birder's Guide to Metropolitan Areas of North America Paul Lehman (2001) American Birding Association, ISBN 1878788159

A Field Guide to the Birds of Eastern and Central North America Roger Tory Peterson (2002) Houghton Mifflin, ISBN 0395740460

Field Guide to the Birds of North America
National Geographic (2002) ISBN 0792268776

The North American Bird Guide David Sibley (2000) Pica Press, ISBN 1873403984

Field Guide to the Birds of Eastern North America David Sibley (2003) Christopher Helm
ISBN 0713666579

National Geographic Guide to Bird Watching Sites: Eastern US Mel White (1999) National Geographic, ISBN 0792273745

Rufous Hornero at its characteristic nest in Buenos Aires

Buenos Aires

Buenos Aires is at the mouth of one of the great river systems of South America. The confluence of the Rio Paraná and Rio Uruguay has resulted in a broad estuary – the Rio de la Plata, where the huge volume of water can overwhelm tidal action and produce surges which inundate the river banks. Originally the banks from the Paraná Delta to Buenos Aires were a huge wetland, teeming with wildlife. The growth of the city has gradually reclaimed much of the wetlands but, where isolated pockets remain, it is still possible to see many of the birds which were once more widespread. The city also lies on the edge of the humid pampas with many of the species typical of this habitat present where suitable habitat remains.

For the visitor from the northern hemisphere, the birds of Buenos Aires present a fantastic variety of unfamiliar bird families, as well as some of the more familiar ones. In addition to the large number of resident species, the Rio de la Plata area is visited in summer by breeding birds which winter in the tropics, as well as non-breeding migrants from the northern hemispheric winter. During the southern winter it also receives a small number of migrants from further south.

All the good birding sites are on the coast, from the city north to the Paraná Delta, and can be visited comfortably in the course of a day trip. For those prepared to travel further, there are some excellent sites on the Atlantic coast which offer more variety in terms of seabirds and migrant waders. Inland from the city, to the south and west lie the open expanses of the pampas, which support a unique community of birds and are particularly good after heavy rains swell their lakes and wetlands. However, to visit these areas would require a round trip of over 500km, beyond the scope of this book.

GETTING AROUND

Taxis, which are reasonably cheap and plentiful, are the best way for getting around if you are planning to go no further than the city centre sites. For longer distances, you could also consider using a 'remise', a private mini-cab with driver that can be hired for the day at a reasonable cost. There are train services which operate along the coast north of the city as far as the Paraná Delta and these will get you to the sites north of

Buenos Aires. All these trains depart from Retiro, one of the main railway stations in the city centre. Provincial buses serving Otamendi operate from Once Bus Station (Plaza Miserere), on the Subte (subway) line.

Driving through Buenos Aires is a very daunting challenge for a visitor. Due to the numerous one-way streets and lack of clear signage in many areas, visiting drivers attempting to cross this enormous city using secondary roads will almost certainly end up lost. To find your way out of the city, the key route northwards is Avenida del Libertador, which will take you from the city centre to the birding sites north of the city, and to the Pan-American Highway, the main road to the Paraná Delta.

Once you have got clear of Buenos Aires, getting around is quite easy. Except in midsummer, the climate is equable and the nature reserves are well appointed with wardening, nature trails and boardwalks where necessary. Some, not all, nature reserves close for one day a week (typically Mondays) or have restricted opening times. Also, this stretch of low-lying coast is prone to flooding during onshore gales, so you are advised to phone ahead to make sure the reserve is open when you plan your visit: contact details are provided.

Locations

1 **Parks in Central Buenos Aires** (City Centre)

Costanera Sur (2) is a much better option for city centre birding than any of the parks. However if you happen to be restricted to Buenos Aires on a Monday, the only day that Costanera is closed, then you might consider one of the city parks. There are a number of large urban parks in Buenos Aires and if you are staying in the city centre, it is quite likely that you are within walking distance of one. The Bosques de Palermo (Parque 3 de Febrero) is one of the best in terms of variety of habitat, as it includes the Japanese Garden (Jardín Japonés), the Rose Garden (Rosedal) and a number of small lakes. It is, however, popular with the public and gets very crowded, particularly at weekends. It is north of the city centre in the Palermo neighbourhood, between Avenida del Libertador and Avenida Figueroa Alcorta. The Botanical Gardens (Jardín Botánico), also in Palermo at Plaza

TYPICAL SPECIES

Passage
White-eyed Parakeet (I), Ñanday Parakeet, Canary-winged Parakeet (I)

Italia (on metro line D), are a good spot for the more common urban species, as well as several small populations of escaped parrots.

2 Costanera Sur Reserve (City Centre)   *Bus Route No. 2*

This 360 hectare coastal lagoon, just east of the centre, is the site of an abandoned land reclamation project. Known to locals simply as the Reserva Ecologica, Costanera Sur is one of the best birding sites on the Rio de la Plata. It is popular with joggers and cyclists and can get rather crowded on weekend afternoons, particularly on sunny days. In a fairly small area there is a variety of marshland habitats as well as an undisturbed area of pampas grassland. There is a network of trails through the reserve and a path along the shoreline, which provides an opportunity to scan the open waters of the River Plate estuary. Three large freshwater lagoons are surrounded by rushbeds, riparian woodland and marsh. For an urban site, it has attracted a very large number of species (over 290) and is definitely a 'must-see' for any birder visiting Buenos Aires. On a typical spring/summer day, you could expect to find 75 or more species in a morning. The reserve is open daily, except Mondays, from 08.00 until sunset. It may also be closed after heavy rainstorms.

The main entrance is at the intersection of **Avenida Brasil** and **Avenida Tristan Achával Rodríguez**, east of the city centre, beyond the trendy restaurant district and former docks known as 'Puerto Madero'. There is a second entrance at the north end of the reserve which is even closer to the city centre and to Retiro Train Station: this is at the intersection of **Avenida Viamonte** and **Avenida Carlos Noel**, just beyond the Buquebus ferry terminal (from which ferries leave for Uruguay). The north entrance provides access to more extensive woodland, which often yields birds not found at the more popular southern end of the reserve. Both entrances are well served by taxis and the No. 2 Bus Route (but only those with a sign in the window indicating 'Ciudad Deportiva').

KEY SPECIES

Resident
Whistling Heron, Striated Heron, Fulvous Whistling Duck, Red Shoveler, Cinnamon Teal, Black-headed Duck, Snail Kite, Common Moorhen, Spot-flanked Gallinule, American Painted-snipe (occasional if water levels low), Grey-headed Gull, Kelp Gull, White-tipped Dove, Ñanday Parakeet, Sooty-fronted Spinetail,
Chicli Spinetail, Sulphur-bearded Spinetail, Freckle-breasted Thornbird, Many-colored Rush Tyrant, Long-tailed Reed Finch, Yellow-billed Cardinal, Grassland Yellow Finch, Epaulet Oriole, Solitary Cacique, White-browed Blackbird, Scarlet-headed Blackbird

Summer
Blue-billed Black Tyrant, Black-
backed Water Tyrant, Yellow-browed Tyrant, White-winged Becard, Rusty-collared Seedeater

Winter
Silvery Grebe, Cinnamon Teal, Bar-winged Cinclodes, White-banded Mockingbird, Sharp-billed Canastero, Hudson's Canastero, White-lined Tanager

3 Ribera Norte Reserve (20km)   *Barrancas Railway Station*

The Ribera Norte Reserve is a coastal marsh on the shores of the Rio de la Plata, north of Buenos Aires. This is a rather small reserve but contains several different habitats and a good selection of aquatic birds. There are some exposed mudflats on the shoreline at low tide. Although it supports fewer species than Costanera Sur, it attracts fewer non-birders and so is less crowded. It is also easier to see some of the shyer reedbed species, such as Rufous-sided Crakes.

Take **Avenida del Libertador** north from the city towards **San Isidro**. After about 15km, where the road becomes a tree-lined boulevard, you will reach the suburb of **Acassuso**. Turn right (east) here on **Calle Peru**. Go to the end of Calle Peru and turn left at **Barrancas Railway Station**, then take the next turn right. Turn left at the end of this street (Calle Almafuerte), and the reserve entrance is about 150m on the right-hand side, just over a steep embankment. Inside the reserve, there is a loop trail which leads through all of the main habitats (although parts are frequently closed due to flooding).

KEY SPECIES

Resident
Whistling Heron, Brazilian Teal, Rufous-sided Crake, Grey-necked Wood Rail, White-tipped Dove, White-throated Hummingbird, Gilded Sapphire, Checkered Woodpecker, Narrow-billed Woodcreeper, Mottle-cheeked Tyrannulet, Yellow-browed Tyrant, Curve-billed Reedhaunter, Tropical Parula, Red-rumped Warbling Finch, Scarlet-headed Blackbird

Summer
Dark-billed Cuckoo, White-winged Becard, Black-backed Water Tyrant, Bran-colored Flycatcher, Southern Beardless Tyrannulet, Masked Yellowthroat

Winter
White-cheeked Pintail

---

4 Vicente Lopez Ecological Reserve (15km)

Vicente Lopez Railway Station

This public park, which includes a small, artificial, freshwater lagoon, attracts a good selection of aquatic species. It is just north of the city on the shores of Rio de la Plata. There is extensive growth of willow and other riparian trees, as well as a dense understorey providing good cover for the more skulking passerines. Trails along the shore of the lagoon provide good access and this site is well worth a visit if you are staying on the north side of Buenos Aires. Local birders are usually on-hand on Saturdays, making this a good location to make contact and maybe get some guidance if you are new to birding in the Neotropics.

Take **Avenida del Libertador** north from the city to the suburb of **Vicente Lopez**, and turn right (east) onto **Calle Paraná** and continue to the end of this street. The reserve entrance is at the end of the street.

KEY SPECIES

Resident
Rufous-sided Crake, Grey-necked Wood Rail, White-tipped Dove, Checkered Woodpecker, Blue-and-yellow Tanager, Epaulet Oriole, Screaming Cowbird

Summer
White-winged Becard, Black-backed Water Tyrant, Southern Beardless Tyrannulet, Yellow-browed Tyrant, Bran-colored Flycatcher, Masked Yellowthroat

Further Away

---

5 Otamendi Reserve (70km)

Otamendi Railway Station

The delta of the River Paraná is an extensive area of marshes, floodplains, riparian forest and open waterways. Some parts have been reclaimed for agriculture whilst other areas were once farmed and are now abandoned. Overall, this has created a mosaic of different wetland habitats with abundant birdlife. Many typical delta species can be found in the Otamendi Nature Reserve which provides convenient access to some of the key habitats. Unlike the reserves closer to the city, Otamendi is not surrounded by urban developments and is contiguous with its natural hinterland of the Paraná Delta, supporting a richer diversity of species, and is also part of the natural migration pathway of the River Paraná itself.

The reserve lies between the Rivers Paraná and Luján, and within the reserve there are two trails with observation platforms enabling you to scan the reedbeds and stretches of open water. From the visitors' centre you can walk the El Talar loop trail. This is about 1km long, leading through woodland and open uncultivated areas to a viewing platform on a bluff, overlooking the wetlands. There is also a second path, the Laguna Grande trail, accessible only by those visitors accompanied by a licensed guide or park warden. The Laguna Grande trail begins at the bottom of the hill below the El Talar viewing platform and leads out across the reedbeds on boardwalks to a viewing platform in the lagoon.

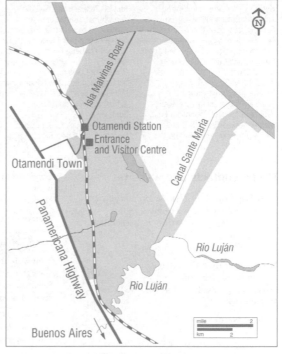

Species such as Stripe-backed Bitterns, American Painted-snipes and Crested Doraditos are only likely to be spotted along this trail, so it is well worth the minimal expense of hiring a warden to take you there. If you have a car, you can drive the Islas Malvinas Road, which is about 6km long, through an area of floodplain to the banks of the Paraná de las Palmas, one of the branches of the River Paraná. However, you might prefer to walk this road as it provides the best birding in the reserve. All three areas are worth checking as each different habitat supports a different community of birds.

Take the **Pan-American Highway** north from Buenos Aires and at the **Pilar/ Campana** intersection where the motorway suddenly splits, take the right-hand lanes signposted **Campana – Zarate** (the sign only appears at the last moment!). After about 68km from the centre of Buenos Aires turn off into the town of **Ing. Rómulo Otamendi**. Drive through and the reserve headquarters is on the right just past the end of town. Continue along the main road which doubles back, dropping down past the train station, and drive across the tracks to reach the lower and more interesting Paraná Deltaic marsh habitats. The **Islas Malvinas Road** begins northeast of the train station and heads east. The Straight-billed Reedhaunter (the highly sought-after skulker for which Otamendi is best known) is known to occur in the roadside reedbeds 1.1km beyond the railway tracks.

There is a train service from Buenos Aires to Otamendi, but it is not recommended. A better option is the bus service from Once Bus Station (Plaza Miserere) to Campana, which stops at the turn-off for Otamendi.

KEY SPECIES

Resident	Bittern, Maguari Stork, Roseate	Brazilian Teal, Red Shoveler, Black-
Whistling Heron, Stripe-backed	Spoonbill, Southern Screamer,	headed Duck, White-tailed Kite,

Harris's Hawk, American Kestrel, Aplomado Falcon, Dusky-legged Guan, Rufous-sided Crake, Common Moorhen, Spot-flanked Gallinule, American Painted-snipe, White-tipped Dove, Striped Cuckoo, Striped Owl, Gilded Sapphire, Checkered Woodpecker, Straight-billed Reedhaunter, Stripe-crowned Spinetail, Yellow-chinned Spinetail, Variable Antshrike, Correndera Pipit, Mottle-cheeked Tyrannulet, Tawny-crowned Pygmy Tyrant, Rufous-browed Peppershrike, Masked Yellowthroat, Golden-crowned Warbler, Tropical Parula, Diademed Tanager, Long-tailed Reed Finch, Red-rumped Warbling Finch, Grassland Yellow Finch, Ultramarine Grosbeak, Screaming Cowbird, Epaulet Oriole, Solitary Cacique, Brown-and-yellow Marshbird, Scarlet-headed Blackbird, White-browed Blackbird

Summer
Ash-colored Cuckoo, Dark-billed Cuckoo, Curve-billed Reedhaunter, Sooty-fronted Spinetail, Chicli Spinetail, Sulphur-bearded Spinetail, Freckle-breasted Thornbird, Southern Beardless Tyrannulet, Crested Doradito, Warbling Doradito, Bran-colored Flycatcher, Yellow-browed Tyrant, Blue-billed Black Tyrant, White-winged Becard, Brown-chested Martin, Blue-and-white Swallow, Cliff Swallow, Red-eyed Vireo, Glaucous-blue Grosbeak

Useful contacts and websites

Aves Argentinas – Asociación Ornitológica del Plata (Spanish) www.avesargentinas.org.ar
25 de Mayo 749 2°6 (C1002ABO), Buenos Aires, Argentina. Tel./Fax. (+54) 11 4312 1015/2284/8958. BirdLife partner for Argentina.

Neotropical Bird Club (English)
www.neotropicalbirdclub.org
c/o The Lodge, Sandy, Bedfordshire, SG19 2DL, United Kingdom.

Seriema Nature Tours (English and Spanish)
www.seriematours.com
25 de Mayo 758 10°G (C1002ABO), Buenos Aires, Argentina. Tel./Fax. (+54) 11 4312 0928/6345. The only Buenos Aires-based company of professional bird guides. A free detailed map and annotated bird checklist for Costanera Sur (pdf) can be downloaded from their website in the 'What's New' section.

Birding Sites in Argentina (English and Spanish)
www.geocities.com/fotosaves/index_english.html

Administración de Parques Nacionales (Spanish) www.parquesnacionales.gov.ar
National parks of Argentina website with maps and information about each national park.

Parques Nacionales de Buenos Aires (Spanish)
www.todoargentina.net/Geografia/Parques/buenosaires/

Vicente Lopez Ecological Reserve (English and Spanish)
www.lincoln.edu.ar/reserve/index.html

Ribera Norte Nature Reserve (English)
www.geocities.com/riberan
Calle La Ribera al 400, Acassuso, B.A. Tel (54) 011 4747 6179. Email: ariberanorte@hotmail.com

Otamendi Nature Reserve (Spanish)
www.parquesnacionales.gov.ar/03_ap/28_otamendi_RNE/28_otamendi_RNE.htm
Rawson 1.080, (2.804), Campana, B.A. Tel: 034 8944 7505. Email: otamendi@apn.gov.ar

Servicios Ferroviarios en Argentina (Spanish)
www.ar.inter.net/transportes.htm
Main train services operating out of Buenos Aires.

Metrovías S.A. (English and Spanish)
www.metrovias.com.ar
Metro system in Buenos Aires.

Trenes de Buenos Aires S.A. (Spanish)
www.tbanet.com.ar
Suburban trains serving Otamendi and other areas north of Buenos Aires.

Tren de la Costa (Spanish)
www.trendelacosta.com.ar
Suburban trains serving the coast north of Buenos Aires.

Los Colectivos (Spanish)
www.loscolectivos.com.ar/lineas.htm
Bus services in Buenos Aires area.

Books and publications

Birds of Argentina & Uruguay (Gold Edition) T. Narosky & D. Yzurieta (2003) Vazquez Mazzini Editores, ISBN 987913205X

Birds of Southern South America and Antarctica Martin de la Peña & Maurice Rumboll (1997) HarperCollins, ISBN 0002200775

Annotated Checklist of the Birds of Argentina Juan Mazar Barnett & Mark Pearman (2001) Lynx Edicions, ISBN 8487334326

Where to Watch Birds in South America Nigel Wheatley (1994) Christopher Helm ISBN 0713639091

Las Aves de la Provincia de Buenos Aires: Distribucion y Estatus T. Narosky & A. de Giacomo (1993) Asociación Ornitológica del Plata, Buenos Aires

Lista De Campo Para Las Aves Argentinas R Straneck & G Carrizo (1999) Literature of Latin America, ISBN 9509725110

Caracas

As the capital city of one of the most bird-rich countries in the world, Caracas is a gateway for many bird tour groups and others intending to travel into the interior of the country. This is a tropical birding paradise: the luxuriant growth of foliage spilling off the roofs and balconies of apartment blocks and office buildings is testament to how quickly the jungle would reclaim the cityscape if left to its own devices. Caracas lies at an altitude of almost 1,000m, occupying a rectangular valley in the coastal mountain range (Cordillera de la Costa). The valley floor is completely built-over apart from a number of city parks, and the city is gradually encroaching up the sides of the valley. Although only 10km from the Caribbean Sea, the valley is separated from the coast by El Ávila ridge, 2,765m high.

The city's proximity to these elevations is the key to the rich diversity of species on its doorstep. At sea-level, the natural vegetational zone is dry xerophytic thorn scrub which changes to dry deciduous forest higher up before becoming cloud forest and ultimately sub-páramo vegetation at the highest elevations, as a consequence of the variation in average temperature and rainfall. All these zones support distinct bird communities. Luckily, and despite the growth of the city, there are still many intact areas of habitat within easy reach of the short-term visitor.

Compared to other stretches of coastline in Venezuela, the coastal strip just north of Caracas is rather rocky with no shallow estuaries or mangroves and little evidence of the large seabird colonies which exist on the offshore islands. In fact, there are few good wetland habitats in the vicinity of Caracas, which means that most birding involves carefully working the forest trails. During the dry (and cooler) season, which extends from November to April, and which also corresponds to the boreal winter, local bird populations are augmented by North American migrants. However many forest species can be quite silent, at least for part of this time of year, which can make finding birds in the forest a very difficult and challenging undertaking. For this reason many seasoned Neotropical birders actually prefer the wet season, when courtship and breeding activity is more advanced. There are also some austral winter visitors to Venezuela during the wet season but they are thinly dispersed and there are few opportunities to observe them.

GETTING AROUND

For the visitor, Caracas is a large, complex and potentially very confusing metropolis. For city centre

Ornate Hawk-Eagle can be seen close to the city centre of Caracas

birding and for access to El Ávila National Park, you are best advised to use the metro and/or city taxis, which are cheap and plentiful. Out of town, bus transport is available throughout the Caracas hinterland, and Venezuela possesses a very comprehensive and practical system of informal shared taxis or *por puestos* which operate in even

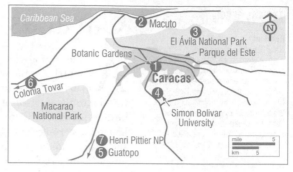

the remotest areas. Most buses and por puestos which serve the birding sites operate from the main Bus Terminal near La Bandera Metro Station, just south of the city centre, and also from the Aeroejecutivos Bus Terminal, near the Altamira Metro Station.

A car is a definite advantage for birding outside Caracas, although renting a car is quite expensive. However, the overall cost is not as ruinous as you might first imagine since petrol is very cheap. The main thing to remember when navigating the city is that the only practical way out of the valley is via the main Autopistas: Route 2 (Autopista La Guaira) to the coast; Route 9 (Autopista Petare) to the east; and Route 1 (Autopista del Valle) to the south and west. If you attempt to climb out of the valley using any of the minor mountain roads, you will most likely end up in a dead end or, worse still, get lost in another, completely unfamiliar, part of the city. For this and other practical reasons, a short-term visiting birder may prefer to travel with a local bird guide who has his own transport.

One of the disadvantages of birding by car in remote areas is that you will frequently need to park on the roadside while you enter the forested areas, leaving the car and its contents vulnerable to theft. The national parks usually have supervised car parks at the park headquarters with trails fanning out from the car park, but if you must stop in a remote location you should seek out a roadside snack bar (of which there are many in Venezuela) near the trailhead, where the proprietor will usually be happy to watch your car in return for your custom.

Birding in these forests is both thrilling and frustrating. They can appear to be completely birdless, in the absence of birds calling, and require very careful and silent progress along the trails watching for any movement in the foliage and listening for the rustle of leaves on the forest floor. In this respect, the lone birder is probably at an advantage over a tour group although two observers is probably the optimum number. Pay close scrutiny to any flowering or fruiting trees and also observe, from a discreet distance, any swarms or columns of army ants which cross the trail, as all will attract a steady stream of feeding birds.

As some of the sites are quite distant from Caracas, you will also need a map of the neighbouring states of Aragua and Miranda, in addition to a map of the Distrito Federal.

Locations

1 **Parks in Central Caracas**
Parque del Este Metro Station
Plaza Venezuela Metro Station

For the first-time visitor to the Neotropics, the city's many parks and suburban gardens will produce a rich assortment of interesting birds. Two parks, in particular, are worth mentioning: the Parque del Este and the Botanic Gardens. The Parque del Este is a large park to the east of the city centre, with good tree cover as well as ponds and waterways which support some aquatic species. The park has been planted with examples of xerophytic shrubs as well as trees typical of the humid forests, offering habitats for a cross-section of species. If you are commencing a birding trip, the park provides a gentle introduction to the birdlife of South America. The park is, of course, very popular with city-dwellers but, unlike many city parks around the world, it opens very early in the morning, thus permitting a dawn visit when the birding is at its best.

The Botanic Gardens (**Jardín Botánico**) and the adjacent **Parque los Caobos** are right in the centre of the city, close to many city hotels and offices. It is worth visiting early in the morning, although the opening hours are restricted to 08.30–16.30. The dominant forest cover is deciduous woodland, typical of the drier slopes, while the constant irrigation and profusion of flowering shrubs and trees make it a good site for parakeets, hummingbirds and euphonias.

KEY SPECIES

Resident
Scarlet Ibis, Chestnut-fronted Macaw, Orange-winged Parrot, Wattled Jacana, Common Moorhen, Lilac-tailed Parrotlet, Copper-rumped Hummingbird, Scaled Piculet, Crested Spinetail, Mouse-colored Tyrannulet, Stripe-backed Wren, Fulvous-headed Tanager, Rosy Thrush Tanager, Thick-billed Euphonia, Trinidad Euphonia, Black-faced Grassquit, Yellow-hooded Blackbird

2 Macuto (17km)

Macuto is a coastal town which is close to the airport at Maiquetia and many visitors who are catching an early morning flight stay at the numerous hotels here. As it is on the coast it offers an opportunity to see some of the seabirds which are typical of the Caribbean Sea, although this stretch of coastline is rather built-up with few natural habitats. It also offers an opportunity to explore the coastal slopes of the Cordillera de la Costa, as the entrance to El Ávila National Park is on the south side of the town. The road through the national park from Macuto to Galipán, which is passable only by 4WD, is worth checking. The temperatures on the coast are significantly higher than in Caracas and the coastal slopes of the Cordillera support xerophytic vegetation at their lower levels. This is characterised by cacti and drought-tolerant bushes which support a unique community of birds, several of which are endemic to Venezuela.

KEY SPECIES

Resident
Brown Pelican, Brown Booby, Neotropic Cormorant, Magnificent Frigatebird, Laughing Gull, Royal Tern, Scaled Dove, White-eared Parakeet, Red-eared Parakeet, Yellow-headed Parrot, Pale-bellied Hermit, Buffy Hummingbird, Russet-throated Puffbird, Scaled Piculet, Short-tailed Ant Thrush, White-fringed Antwren, Tawny-crowned Pygmy Tyrant, Stripe-backed Wren, Trinidad Euphonia, Glaucous Tanager, Black-faced Grassquit

3 El Ávila National Park (2km)   *Metro bus #314*

El Ávila (2,250m) is one of the peaks of the Cordillera de la Costa which separates Caracas from the coast, and lies at the centre of a large national park which extends from the northern edge of the city to the coast and for about 45km along the mountain range. Despite its proximity to the city, it is largely ignored by bird-tour groups in a hurry to get to the interior of the country. However, for the short-term visitor with no transport of their own, this is probably the best site within the urban area and will

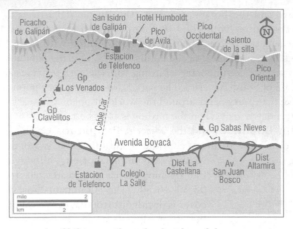

astonish and delight the birder who is more familiar with the birdlife of the temperate zones. Although most of the forest cover is secondary growth, the area within El Ávila National Park encompasses the full range of tropical vegetational zones, from the dry xerophytic scrub on the coast to cloud forest and sub-páramo at the summit. In general, the coastal slopes are more humid than the inland slopes. There is a good network of hiking trails on both sides of the mountain, and the old road from La Guaira to Caracas passes through the park.

Avenida Boyacá which runs west to east along the foothills of the coastal mountains on the northern edge of the city is the key route for accessing El Ávila National Park. From the city side of the mountain there are trails from Boyacá Avenue which ascend the south slope of the mountain. In particular, you can take the cable-car from Mariperez on Boyacá Avenue to the summit (Pico Ávila). At the summit of the mountain there is an old hotel building, The Humboldt Hotel, and there is a trail west from here along the ridge of the mountain which passes through cloud forest and cleared areas of high altitude grassland towards the neighbouring peak of Galipán. At this point you can descend the mountain through cloud forest and deciduous forest zones to the guard post at Los Venados. Continue along this trail and you can exit the park through the Clavelitos Guard Post, which has car-parking, and is located about 1km west of the cable-car station at Mariperez.

An alternative route is to take **La Castellana** exit off Boyacá Avenue, park at the **Sabas Nieves Guard Post** and ascend the mountain towards the **Silla** (Saddle). This is a rather steep climb but leads through some good cloud forest to an open area of sub-páramo habitat. You can get to the Mariperez cable-car station by taxi or the Metrobus #314 from Plaza Venezuela Metro Station.

KEY SPECIES

Resident
Ornate Hawk Eagle, Band-tailed Guan, Venezuelan Wood Quail, Red-eared Parakeet, Lilac-tailed Parrotlet, Striped Cuckoo, White-tailed Nightjar, Rufous Nightjar, Green-tailed Emerald, Rufous-shafted Woodstar,

Violet-chested Hummingbird, White-tipped Quetzal, Groove-billed Toucanet, Black-throated Spinetail, Crested Spinetail, Guttulated Foliage-gleaner, Rufous-lored Tyrannulet, Schwartz's Ant-thrush, Plain-backed Antpitta, Rusty-breasted Antpitta,

Venezuelan Tapaculo, Mountain Elaenia, Handsome Fruiteater, Black-hooded Thrush, Pale-breasted Thrush, Inca Jay, Fulvous-headed Tanager, Trinidad Euphonia, Glaucous Tanager, Blue-black Grassquit, Ochre-breasted Brush Finch, Troupial

4 Simon Bolivar University (8km)

The Simon Bolivar University Campus is on the south side of the city and has extensive surrounding forest cover. The campus itself is in a valley separated from the rest of the city by a ridge of the Cordillera, which is forested. From the campus, there are a number of trails by which you can access the cloud forest at the top of the ridge, and a rich diversity of species can be seen in the area. However, even for a short visit, you can see much by birding along the forest edge habitats, which can often be more productive than the forest interior. The best habitat is on the south side of the campus in the vicinity of the swimming pool and sportsfields. The campus is not served by metro, but the university has a bus service operating from El Silencio in the centre of the city which serves the campus.

KEY SPECIES

Resident
Band-tailed Guan, Venezuelan Wood Quail, Red-eared Parakeet, Lilac-tailed Parrotlet, Striped Cuckoo, Green-tailed Emerald, Rufous-shafted Woodstar, Violet-chested Hummingbird, White-tipped Quetzal, Groove-billed Toucanet, Black-throated Spinetail, Crested Spinetail, Guttulated Foliage-gleaner, Rufous-lored Tyrannulet, Schwartz's Ant-thrush, Plain-backed Antpitta, Rusty-breasted Antpitta, Caracas Tapaculo, Mountain Elaenia, Handsome Fruiteater, Black-hooded Thrush, Pale-breasted Thrush, Fulvous-headed Tanager, Trinidad Euphonia, Glaucous Tanager, Blue-black Grassquit, Ochre-breasted Brush Finch, Troupial

Further Away

5 Guatopo National Park (80km)

The Serranía del Interior mountain range lies south of Caracas creating a barrier between the valley in which the city lies and the interior of the country. These mountains are subjected to heavy rainfall and are densely forested, and the area has national park status in order to protect the watershed for the coastal river systems and those of the interior. As such it is subject to less pressure through fragmentation and destruction of the forest edges that is characteristic of the forested areas nearer the city. A number of shyer species such as the Harpy Eagle, which require extensive undisturbed habitat for breeding territories, still inhabit this area. There are some hiking trails through the park but for casual visitors entry is generally restricted to the area near the park headquarters and adjacent to the main road through the park.

Take **Autopista 1** south towards **Charallave** and exit east onto **Route 11** to **Santa Teresa del Tuy**. From Santa Teresa continue east on **Route 4** until it joins **Route 12** at **Alpes del Tuy**, and turn south on **Route 12** towards **Altagracia de Orituco**. After about 20km the entrance to the **Agua Blanca** campground is on the left-hand (east) side of the road. You may park here and walk the hiking trail to **Santa Crusita** (about two hours each way). There are spring-fed lakes at both Agua Blanca and at Santa Crusita. Returning to Route 12 you can continue south towards Altagracia for about 10km and there is a reservoir (**Embalse de Guanapitos**) on the right-hand (west) side of the road with a loop trail around the shore.

KEY SPECIES

Resident
King Vulture, Swallow-tailed Kite, Plumbeous Kite, Solitary Eagle, Black Hawk-Eagle, Ornate Hawk-Eagle, White Hawk, Band-tailed Guan, Crested Guan, Helmeted Curassow, Gray-necked Wood Rail, Venezuelan Wood Quail, White-eared Parakeet, Military Macaw, Lilac-tailed Parrotlet, White-tipped Dove, Sooty-capped Hermit, Rufous-breasted Hermit, Violet-headed Hummingbird, Rufous-tailed Jacamar, Groove-billed Toucanet, Black-mandibled Toucan, Lineated Woodpecker, Rufous-winged Antwren, Schwartz's Ant-thrush, Masked Tityra, Wire-tailed Manakin, Lance-tailed Manakin, Brown-capped Tyrannulet, Venezuelan Tyrannulet, Mountain Elaenia, Forest Elaenia, Golden-fronted Greenlet, Red Siskin, Yellow Oriole

6 **Colonia Tovar** (50km)

A community of German immigrants founded an alpine village in the Cordillera de la Costa to the west of Caracas at an altitude of 1,800m. The village has developed into a resort complete with alpine chalets and German restaurants. There, however, the similarity with the Alps ends. The mountains in this area span the subtropical and temperate altitudinal zones, and support a full selection of forest birds typical of each zone, despite the fragmentation of the forest cover. There are several widely spaced trails through the mountains and you could easily spend several days in the area in order to cover all of them adequately. The best strategy for the short-term visitor is to drive beyond the resort and stop and explore some of the better-known trailheads.

Take the **Carretera al Junquito** (**Route 4**), west from Caracas to Colonia Tovar. This is a branch off the **Autopista Francisco Fajardo**, the main thoroughfare through the southwestern suburbs of the city. Proceed through Colonia Tovar on the road towards **La Victoria** and about 4km past Tovar turn left (east) towards **Capachal**. This road descends through some good habitat. As there is little traffic on this road, you can pull over and walk along the roadside, looking and listening. Return to the **Tovar/La Victoria** road and after a further 4km south, turn right (west) at the sign for **Buenos Aires** (there is a bus shelter at the turn-off). This road leads to the Maya River valley (Cortada de Maya) and it is also worth walking for about a kilometre from the main road.

On returning to Colonia Tovar take the main road back to Caracas and as you pass the main gate for the resort, turn left (north) towards **El Limón**. After about 7km there is a turn-off to the right (east) towards **Chichiriviche**. The roadside forest in the vicinity of this turn-off is also worth checking.

Finally, the main road between Tovar and El Junquito passes along the northern perimeter of the **Macarao National Park** and although this is a rather busy road, it is worth stopping at suitable lay-bys to check the roadside habitat.

This is a very popular tourist site and it is well served by tour companies as well as by por puestos. It is even possible to get a taxi from Caracas as the resort is very well known. However it gets extremely busy at weekends.

KEY SPECIES

Resident
Black Hawk-Eagle, White-collared Swift, Bronzy Inca, Long-tailed Sylph, Green Violetear, Sparkling Violetear, Green-tailed Emerald, Rufous-shafted Woodstar, Groove-billed Toucanet, Strong-billed Woodcreeper, Montane Woodcreeper, Black-throated Spinetail, Rufous Spinetail, Streaked Tuftedcheek, Montane Foliage-gleaner, White-throated Tyrannulet, Chestnut-crowned Antpitta, Caracas Tapaculo, White-throated Flycatcher, Mountain Elaenia, Blue-and-white Swallow, Green-and-black Fruiteater, Gray-breasted Wood Wren, Glossy-black Thrush, Yellow-legged Thrush, Brown-capped Vireo, Slate-throated Redstart, Three-striped Warbler, Black-crested Warbler, Bluish Flowerpiercer, Blue-winged Mountain Tanager, Fawn-breasted Tanager, Trinidad Euphonia, Oleaginous Hemispingus, Blue-capped Tanager, Beryl-spangled Tanager, Blue-black Grassquit, Ochre-breasted Brush-finch, Rufous-collared Sparrow, Grayish Saltator, Yellow Oriole

7 **Henri Pittier National Park** (110km)

The Henri Pittier National Park is one of the world's birding hotspots, and although it is quite distant from Caracas (about two hours' drive), to pass up a chance to visit this site would be to miss a really exceptional birding opportunity. It is certainly possible to visit the park in a day trip from Caracas, but an overnight visit will permit a dawn forest walk which will yield a superb haul of forest species. The national park incorporates a large tract of cloud forest along both the northern and southern slopes of the Cordillera de la Costa, as well as coastal zone scrub along the Caribbean Coast. The range of elevations and habitats supports a bewildering range of species including ten Venezuelan endemics. The mountain passes through the park are key migration routes for many Nearctic migrants, and the bird-ringing station is one of the main centres of Neotropical bird study.

The park is between the city of **Maracay** and the coast, and there are two main access roads from Maracay. Both are well worth birding, but if your time is limited, prioritise the **Maracay/Ocumare Road**. Going north from Maracay, the **Rancho Grande Ecological Station** is located about 20km on the right-hand (east) side of the road. The rooftop of this building has a veranda at treetop height with feeding stations for the birds. Visiting birders may go up on the roof and be treated to superb views of many upper-canopy species as well as tanagers and hummingbirds attracted to the feeding stations. There are several forest trails at the rear of the building compound which ascend into the cloud forest and give access to the highest elevations along the mountain ridges. The most productive trail is the Pico Guacamayo which consistently delivers the longest list of birds seen en route. However access to the RGES and Pico Guacamayo requires a permit which takes several days to obtain, so unless you have pre-arranged your visit with the assistance of a local bird guide, this is not really an option for a visit at short notice. It is hoped that the situation may change in the future.

About 200m north from Rancho Grande is the **Portachuelo Pass**, the highest point on the road and location of the bird-ringing station, operated by the Venezuela Audubon Society. During migration, a fantastic variety of New World warblers, vireos and flycatchers (as well as insects) may be seen here. The road descends through wetter forest towards the coastal scrub zone. At the lower elevations, the more open, dry forest supports a different range of species from the higher slopes. You may continue on the main road through Ocumare to **Cata** on the coast. The dry coastal scrub supports yet another range of species, although fewer in number than the forests. Unfortunately, there is no good wetland habitat along the coast which might attract aquatic species, but you can scan the sea offshore for boobies, terns and frigatebirds.

Should you have gone due north from Maracay on the **Maracay/Choroni Road**, you will cross the mountains at a higher elevation and will actually be driving through the cloud forest. The Choroni Road pass at **Cumbre** is a good place to stop and check the roadside forest, particularly at dawn if you can make it up there in time. The road winds down to the coast at Choroni, and just north of Choroni is the **Museo Cadafe**, where another trail leads through dry deciduous forest.

KEY SPECIES

Resident
Fasciated Tiger Heron, Solitary Eagle, Black-and-white Hawk-Eagle, Black Hawk-Eagle, Ornate Hawk-Eagle, White Hawk, Band-tailed Guan, Helmeted Curassow, Venezuelan Wood Quail, Chestnut-fronted Macaw, Orange-chinned Parakeet, Scarlet-fronted Parakeet, Red-eared Parakeet, Red-billed Parrot, White-collared Swift, Gray-rumped Swift, White-tipped Swift, Fork-tailed Palm Swift, Black-throated Mango, White-vented Plumeleteer, Violet-chested

Hummingbird, Bronzy Inca, Long-tailed Sylph, Collared Trogon, White-tipped Quetzal, Green Kingfisher, Rufous-tailed Jacamar, Groove-billed Toucanet, Scaled Piculet, Crested Spinetail, Montane Foliage-gleaner, Guttulated Foliage-gleaner, Scallop-breasted Antpitta, Plain Antvireo, Black-crested Antshrike, Black-backed Antshrike, Black-faced Ant-thrush, Lance-tailed Manakin, Common Tody Flycatcher, Black Phoebe, Smoke-colored Pewee, Tropical Pewee, Venezuelan

Flycatcher, White-fronted Tyrannulet, Cinnamon Flycatcher, Yellow-breasted Flycatcher, Blue-and-white Swallow, Handsome Fruiteater, Buff-breasted Wren, Gray-breasted Wood Wren, Scaly-breasted Wren, Black-hooded Thrush, Pale-breasted Thrush, Bare-eyed Thrush, Rufous-browed Peppershrike, Brown-capped Vireo, Tropical Parula, Slate-throated Redstart, Golden-crowned Warbler, Three-striped Warbler, White-eared Conebill, Fulvous-headed Tanager, Silver-beaked Tanager, Blue-winged

Mountain Tanager, Thick-billed Euphonia, Orange-bellied Euphonia, Golden Tanager, Bay-headed Tanager, Swallow Tanager, Blue-black Grassquit, Yellow-bellied Seedeater, Ochre-breasted Brush Finch,

Rufous-collared Sparrow, Grayish Saltator, Buff-throated Saltator, Shiny Cowbird, Russet-backed Oropendola

Oct–Mar
Tennessee Warbler, Cerulean Warbler,

Blackburnian Warbler, Blackpoll Warbler, Black-and-white Warbler, American Redstart, Northern Waterthrush, Summer Tanager

Useful contacts and websites

Sociedad Conservacionista Audubon de Venezuela (SCAV)
www.audubonvenezuela.org
Apartado 80.450, Caracas 1080-A, Venezuela.
Tel. (+58) 212 9922812/ 9923268.
BirdLife International partner in Venezuela.
Email: audubon@cantv.net

Neotropical Bird Club (English)
www.neotropicalbirdclub.org
c/o The Lodge, Sandy, Bedfordshire, SG19 2DL, United Kingdom.

Birds and Birding in Venezuela (English)
www.birdvenezuela.com
Email: sharpebirder@gmail.com or rodsha@telcel.net.ve

Birding Venezuela (English and Spanish)
www.birdingvenezuela.com
Bird guides in the Caracas area and information about bird tours to the interior of Venezuela.
Tel. (+58) 212 2665766/2667467/2662445.
Email: birdingvenezuela@cantv.net

Instituto Nacional de Parques INPARQUES (Spanish)
www.inparques.gov.ve
Venezuelan National Parks website.

BioParques (English and Spanish)
www.bioparques.org

Information about Venezuelan National Parks.
Tel. (+58) 212 7314016. Fax. (+58) 212 7317189.
Email: bioparques@bioparques.org

Venezuela Voyage (English)
www.venezuelavoyage.com
Web portal site with very helpful ecotourist information.

The Venezuela EcoPortal (English)
www.ecoportal8.tripod.com/birdwatch.htm

C.A. Metro de Caracas (Spanish)
www.metrodecaracas.com.ve
Metro and bus service in Caracas.

Books and publications

Birds of Venezuela: photographs, sounds and distributions Peter Boesman (1999) Bird Songs Inernational, ISBN 9075838034

Birds of Venezuela Steven Hilty (2003) Christopher Helm, ISBN 0713664185

Where to Watch Birds in South America Nigel Wheatley (1994) Christopher Helm ISBN 0713639091

Birding in Venezuela Mary Lou Goodwin (2003) Lynx Edicions, ISBN 8487334482

Birds of Northern South America: An Identification Guide Vols 1&2 Robin Restall et.al. (2006) Christopher Helm, ISBN 0713660260

Resident
Pied-billed Grebe, Neotropic Cormorant, Peruvian Pelican, Snowy Egret, Great Egret, Cattle Egret, White-cheeked Pintail, Black Vulture, American Kestrel, Common Moorhen, Killdeer, Grey-headed Gull, Grey Gull, Inca Tern, Pacific Dove, Croaking Ground Dove, Eared Dove, Lesser Nighthawk, Groove-billed Ani, Amazilia Hummingbird, Peruvian Sheartail, Blue-and-white Swallow, Yellowish Pipit, Vermillion Flycatcher, Southern Beardless Tyrannulet, Long-tailed Mockingbird, Bananaquit, Blue-gray Tanager, Shiny Cowbird, House Sparrow, Rufous-collared Sparrow

Lima

Lima lies close to several endemic bird areas. Both onshore and offshore this region is teeming with many rare, unique and highly specialised birds, and Peru has become one of the ultimate destinations for birders, drawn by the immense diversity of species. In many ways, the area around Lima offers just a taster of the birding opportunities that exist in the interior of the country. The key to these riches is the variety of different habitats which are juxtaposed together in a unique combination of latitude, climate and altitude.

The climate of the coastal area where Lima is situated is determined by the cold Humboldt Oceanic Current, which forms the richest marine area in the world. The cool air that drifts onshore carries little moisture creating desert conditions which dominate the coast and adjacent mountains. Most precipitation here is in the form of drizzly fog that rolls in off the ocean and rain falls only in El Niño years. The climate is generally cool and damp, especially in the 'winter' months of June to October, though it can be uncomfortably hot from December to February.

Inca Tern is one of the most beautiful of all terns

The Andes are less than 100km to the east of the city and rise to over 4,000 metres within a two-hour drive. Several altitudinal zones, each with its own unique community of birds, are accessible from the city. The dry coastal plain is a desert which has been irrigated in places to facilitate agriculture, but some areas of higher elevation can trap enough moisture from the mist and fog to provide habitats, known as the *lomas*, for a range of plant and animal specialists. The few wetland areas are frequented by resident aquatic species and winter migrants from North America as well as some altitudinal migrants from the Andes.

Whether your visit to Lima is a short business or pleasure trip, or the start of a lengthy birding expedition to the interior of Peru, there is plenty to see in the area including some species which would be difficult to find anywhere else. If you have extra time on your hands you could consider a two- or three-day trip along the Central Highway up into the High Andes or further south on the Pan-American Highway to the coastal habitats at Paracas.

GETTING AROUND

As almost all of the good birding sites are quite some distance out of town, for all practical purposes you will need a car in order to get around. This unfortunately is going to plunge you into the traffic chaos of Lima which will challenge the most intrepid of urban birders. The key route for all the sites is the Pan-American Highway (Norte: heading north and Sur: heading south), which bypasses the city on its eastern outskirts, where there is also an interchange with the Central Highway: the main access route to the Andes. You should attempt to travel from the city centre to the outskirts during non-peak periods, to avoid the worst of the traffic jams. This means starting either early (before 07.00) or mid-morning. This is one reason why many visiting birders hire a bird-tour company to do the driving. The incremental cost of hiring a guided tour, when compared with self-drive car rental is more than offset by the frustration of trying to find your own way around, and a guide will bring you to the birds far more efficiently than you can on your own. Another reason is that outside Lima the roads are rather poor, especially up in the mountains and require some pretty advanced driving skills to negotiate the steep, narrow and winding mountain roads. Moreover, most car rental companies don't permit their cars to be driven on unsurfaced roads, so this rules out the Santa Eulalia Valley. If, in spite of all this, you still want to do your own driving up in the mountains, make sure you have a car with a good-sized engine which can pull up the steep inclines and with fairly generous ground clearance: it doesn't have to be a four wheel drive, but of course if you get stuck in the middle of a raging mountain torrent, 4WD definitely has some advantages!

If you are restricting yourself to the city and its immediate environs, then taxis are the best way to get around, as they are cheap and plentiful, and you will always be able to flag one down for your return journey, but make sure you have the address of your accommodation and, preferably, a map to show the taxi-driver. If you have time to spare, the buses which ply up and down the Pan-American Highway are also a good way to get to any of the sites north or south of Lima. However, the lack of published timetables and schedules means that you cannot be sure when you will get a return bus, so buses are not recommended for birders on a really tight schedule.

Despite its tropical location, Lima is rather cool for most of the year. It can even be cold from June to October, particularly if there is low cloud and mist, so you will require some warm clothing especially early in the morning. However, bear in mind that this area is a desert, so make sure you carry sufficient drinking water and be aware of the strong solar radiation, even on cloudy days, and use sunscreen and a hat. All of these factors become even more extreme if you are birding at a high elevation. Additionally, it is generally accepted that 4,000 metres is the maximum altitude to climb to on your first day. If you are planning to go higher, stay overnight lower down in order to acclimatise before making your ascent the next day. In and around Lima there are some safety issues so prudence is advisable: don't flaunt your optics in public places and get a known taxi-driver from the hotel where you are staying to wait for you when you travel to local sites.

Locations

1 Parque el Olivar (3km)

The urban parks in central Lima are quite small and with no naturally wooded hinterland, so bird populations are restricted. However, the irrigation produces an abundance of vegetation which creates an oasis of habitat in what is naturally a very arid area. The cover attracts several forest edge species and, of course, the year-round provision of blooming flowers and shrubs attracts both nectar feeders and insectivorous species. The Parque el Olivar in San Isidro is one of the best parks. Recently, small signs (in Spanish) have been put up depicting the birds in the park. These will also help visiting birders on what may be their first day of birding in Peru.

The park is opposite the Camino Real Shopping Mall in **San Isidro** (some three blocks) which is on the south side of the city centre. Just ask someone to point you in the right direction from Camino Real.

2 Pantanos de Villa Reserve (18km)

This coastal lagoon is just south of the city and is one of the few freshwater wetlands in what is basically a desert area. However, the water has a tendency to be brackish at times as a result of depletion of the underground springs and saltwater encroachment. There is a nice mix of different habitats: an abundant growth of reedbeds and vegetation, as well as exposed mud and open water, and this is an important staging point for migrating waders and for wintering ducks. The reserve is landscaped with nature trails and bird hides, making it a very convenient and accessible site near the city. The beach behind the marsh is also well worth a look.

Take the **Avenida Panamericana Sur** south to the intersection with **Avenida Huaylas**. Turn right on Hauylas and go north for about 1km and turn left (west) onto **Avenida Lavalle**, which is marked by a row of palm trees. The park entrance and headquarters building is on the left side of this road. If travelling alone, you should check in with the park headquarters as it is advisable for a staff member to accompany you for safety reasons.

3 Ventanilla (15km)

Set in the midst of a dry sandy desert, the coastal marshes at Playa Ventanilla provide a habitat for a range of aquatic species as well as feeding and bathing opportunities for passerines. There are a number of shallow ponds among the reedbeds which can dry out to some extent and the muddy margins are good for waders. This site is often the best in the Lima area for both numbers and variety of waders, particularly at migration time.

Take the **Avenida Panamericana Norte** north for about 10km, and turn left (west) towards Playa Ventanilla. The marsh is between the turn-off and the coast on the right-hand side of the road. Along the continuation of the road towards the beach there are mudflats that sometimes hold water and can be very good for waders. It is conveniently close to the airport so this site is a good option if you have a few hours to kill before your flight.

KEY SPECIES

Resident
White-tufted Grebe, Striated Heron, Green Heron, Cinnamon Teal, Andean Duck, Turkey Vulture, Slate-colored Coot, Plumbeous Rail, Peruvian Thick-knee, Black-necked Stilt, Band-tailed Gull, Common Miner, Wren-like Rushbird, Vermilion Flycatcher,

Many-colored Rush Tyrant, Chestnut-throated Seedeater, Parrot-billed Seedeater.

Oct–Mar
Osprey, Black-bellied Plover, Semipalmated Plover, Least Sandpiper, Western Sandpiper,

Semipalmated Sandpiper, Baird's Sandpiper, Pectoral Sandpiper, Sanderling, Ruddy Turnstone, Short-billed Dowitcher, Spotted Sandpiper, Lesser Yellowlegs, Greater Yellowlegs, Wilson's Phalarope, Franklin's Gull, Black Skimmer

Further Away

4 **Lomas de Lachay Reserve** (100km)

Although this reserve is a little distant from Lima, with a car it is easily 'done' in a day trip from the city. Also, if your flight out of Lima departs in the evening, it is practical to do a day's birding in Lomas de Lachay and then get to the airport, which is also north of Lima. This site is at a point where the foothills of the Andes penetrate into the Peruvian desert and, because of their eleva-

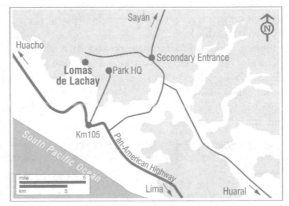

tion, trap the moisture in the fog rolling in from the sea, producing a green oasis in the middle of the desert. There is a mosaic of different habitats, from gallery forest to open grassland, which provides niches for a wide variety of species. It can be tough going during the humid season (June to November) because of the dense fog and drizzle, and visibility can be poor even at other times of the year.

Take the **Pan-American Highway** north for about 100km and when you come to the **Km 105** marker turn right (east) to get to the reserve entrance, which is about 3km from the highway. The road into the reserve passes through some good habitat which is worth checking. Once in the reserve there are several trails from the reserve headquarters. A second entrance to the reserve can be reached by turning off to the right (east) about 1km before the main entrance and taking the road towards Sayán. The entrance is on the left-hand side of the road after about 6km, leading into a valley with extensive areas of cactus which is a good location for the endemic Cactus Canastero: you need to check first with the reserve headquarters whether this area is open to the public.

If you have some extra time you may want to add a visit to Paraiso Lagoons (Km 135), where breeding of the very rare Peruvian Tern was confirmed (Oct–Nov) in 2003 and 2004. It is also a great place for many waders and usually has a large number of Chilean Flamingos. Look out for Tawny-throated Dotterels near the entrance.

KEY SPECIES

Resident		
Black-chested Buzzard-Eagle, Variable Hawk, Harris's Hawk, Peregrine Falcon, Tawny-throated Dotterel, Peruvian Thick-knee, Least Seedsnipe, Bare-faced Ground Dove, White-tipped Dove, Mountain Parakeet, Andean Swift, Oasis	Hummingbird, Peruvian Sheartail, Coastal Miner, Grayish Miner, Thick-billed Miner, Cactus Canastero, Southern Beardless Tyrannulet, Vermillion Flycatcher, Blue-and-white Swallow, Chiguanco Thrush, Yellowish Pipit, Peruvian Meadowlark, Cinereous Conebill, Blue-black	Grassquit, Chestnut-throated Seedeater, Band-tailed Seedeater, Grassland Yellow Finch, Raimondi's Yellow Finch (seasonal), Band-tailed Sierra Finch, Collared Warbling Finch, Hooded Siskin

5 Santa Eulalia Valley (75km)

The Andes rise quite steeply from the plains east of Lima to altitudes of over 4,000m. The landscape transforms from dry desert to greener valleys with low bushes and a few hardy eucalyptus trees. Some cultivation takes place on terraced hillsides and near villages. Despite the fact that the overall area looks quite barren, these valleys support a rich avifauna including several endemics. The Santa Eulalia Valley ascends to over 3,000m and is often the starting point for birding trips which take in the famous sites around Marcapomacocha at 4,000m. Such a trip is probably too ambitious for the lone birder with limited time, but the road up to Santa Eulalia from the Central Highway provides plenty of birding opportunities and is certainly feasible in a day trip from Lima.

Go east on the **Central Highway** as far as **Chosica**, pass the town and turn left (north) at the exit to Santa Eulalia. After about 5km the road ascends into the mountains and the vegetation changes from dry cactus and scattered shrubs to more extensive ground-cover with cultivated plots and some isolated trees. From Chosica up to **San Pedro de Casta** is about 30km and an ascent of about 3,000m. This is a long and tortuous drive on a narrow mountain road, so despite the obvious temptation, do not try birding and driving at the same time! There are some lay-bys, usually at bends in the road which provide a parking space that is visible from both directions. It takes about three hours to get to San Pedro, allowing for stops along the way, so gauge your time if you are planning to return to Lima before nightfall. A more modest undertaking would be to drive just as far as the village of Santa Eulalia, at about 1,000m and surrounded by some good habitat.

KEY SPECIES

Resident		
Torrent Duck, Andean Condor, Black-chested Buzzard-Eagle, Variable Hawk, Peruvian Pygmy Owl (Pacific slope form), Bare-faced Ground Dove, White-tipped Dove, Scarlet-fronted Parakeet, Mountain Parakeet, Andean Swift, Bronze-tailed Comet, Oasis Hummingbird, Peruvian	Sheartail, Purple-collared Woodstar, Black-necked Woodpecker, Bar-winged Cinclodes, Pied-crested Tit-Tyrant, Yellow-billed Tit-Tyrant, Dusky-capped Flycatcher, Spot-billed Ground Tyrant, Streak-necked Bush Tyrant, White-capped Dipper, House Wren, Long-tailed Mockingbird, Chiguanco Thrush, Cinereous	Conebill, Blue-and-yellow Tanager, Mourning Sierra Finch, Band-tailed Sierra Finch, Collared Warbling Finch, Rufous-breasted Warbling Finch, Band-tailed Seedeater, Rusty-bellied Brush Finch, Golden-bellied Grosbeak, Scrub Blackbird, Great Inca Finch, Hooded Siskin

6 Pucusana Harbour (60km)

Pucusana is a fishing port on the coast south of Lima, which is a notable site for many of the seabirds typical of the Peruvian coast. There are nearby beaches and it is a popular recreation resort, so getting there is quite easy if using public transport. The town is located in a natural harbour surrounded by rocks and cliff-faces teeming with seabirds, as well as the endemic Peruvian Seaside Cinclodes. You will probably be approached by boatmen to take a trip around the harbour and if you are averse to a full-blown pelagic trip, this perhaps offers a more sedate approach while still offering many of the seabird species as well as superb opportunities for photography. Just outside the harbour you can often find more pelagic birds quite near to shore including Waved Albatrosses and storm-petrels.

Even if you don't have a car, you can comfortably get to Pucusana by bus from Lima. There is a regular service from the corner of Jirón Montevideo and Jirón Ayacucho in Central Lima, which takes about two hours.

KEY SPECIES	
Resident Humboldt Penguin, Peruvian Pelican, Peruvian Booby, Blue-footed Booby (scarce), Guanay Cormorant, Red-legged Cormorant, Blackish Oystercatcher, Gray Gull, Band-tailed Gull, Grey-	headed Gull, Peruvian Tern, Inca Tern, Peruvian Seaside Cinclodes **Oct–Mar** Surfbird, Sanderling, Ruddy Turnstone

7 Pantanos de Puerto Viejo (70km)

The marshes at Puerto Viejo are just south of Pucusana and are worth travelling the extra few kilometres south as they hold a good selection of aquatic birds and at migration time could potentially turn up something unusual. There is a boardwalk through the reedbeds and an observation tower making it an easy place to survey. Check any hedges and fields which have been left fallow: these are good for seedeaters including Collared Warbling-finches.

Take the **Avenida Panamericana Sur** south for about 70km, past the turn-off for Pucusana. The entrance to the marshes is about 8km south of the Pucusana turn-off on the right-hand (west) side of the highway.

KEY SPECIES		
Resident Great Grebe, White-tufted Grebe, Guanay Cormorant, Striated Heron, Green Heron, Puna Ibis, Cinnamon Teal, Andean Duck, Harris's Hawk, Turkey Vulture, Slate-colored Coot, Peruvian Thick-knee, Black-necked Stilt, American Oystercatcher,	Blackish Oystercatcher, Band-tailed Gull, Kelp Gull, Peruvian Tern, Common Miner, Wren-like Rushbird, Many-colored Rush Tyrant, Grassland Yellow Finch, Peruvian Meadowlark, Collared Warbling Finch, Drab Seedeater, Chestnut-throated Seedeater, Parrot-billed Seedeater, Hooded Siskin	**Oct–Mar** Osprey, Spotted Sandpiper, Sanderling, Ruddy Turnstone, Lesser Yellowlegs, Greater Yellowlegs, Wilson's Phalarope, Franklin's Gull, Black Skimmer

8 Callao Pelagic trips

The coast of Peru is legendary for its abundance of marine life. The combination of the cold Humboldt Current and deep oceanic trenches about 60km offshore, with their nutrient-rich upwellings, create the essential ingredients for a very productive food-chain. Additionally, its location in the southern tropics means that Peruvian waters host both northern and southern hemisphere migrants at different times of year. These birds augment the local seabird populations of Peru and the Galapagos Islands, all of which feed in the rich waters of the Humboldt Current. This means that, unlike many pelagic trip centres, there is an interesting selection of birds year-round off Peru.

Regular seabird and whale-watching trips operate from **Callao**, just outside Lima, and if your visit coincides with one of these excursions, it would present the opportunity of a lifetime to see some of the world's most 'difficult to get' seabirds.

KEY SPECIES

Resident
Humboldt Penguin, Waved Albatross, Peruvian Diving-petrel, Ringed Storm-petrel, Markham's Storm-petrel, Wilson's Storm-petrel, White-vented Storm-petrel, Wedge-rumped Storm-petrel, White-chinned Petrel, Sooty Shearwater, Pink-footed

Shearwater, Peruvian Booby, Red-legged Cormorant, Guanay Cormorant, Blackish Oystercatcher, Band-tailed Gull, Inca Tern, Peruvian Seaside Cinclodes

May–August
Black-browed Albatross, Gray-headed Albatross, Cape Petrel, Blue-footed

Booby, Chilean Skua, Gray Gull, Grey-headed Gull, South American Tern

Oct-Mar
Black Storm-petrel, Westland Petrel, Buller's Shearwater, Red Phalarope, Red-necked Phalarope, Parasitic Jaeger, Pomarine Jaeger, Sabine's Gull, Swallow-tailed Gull, Arctic Tern

Useful contacts and websites

Birding Peru
www.birdingperu.com

Aves de Lima (Spanish)
www.avesdelima.com
Information about birding, bird distribution and habitats in Lima area.

Kolibri Expeditions (English)
www.kolibriexpeditions.com
Lima-based bird tour and pelagic company.

Ultimate Voyages (English)
www.ultimatevoyages.com
Birding trips and pelagics in Peru.

Neotropical Bird Club (English)
www.neotropicalbirdclub.org
c/o The Lodge, Sandy, Bedfordshire, SG19 2DL, United Kingdom.

Books and publications

A Field Guide to the Birds of Peru
James F Clements & Noam Shany (2001)
Lynx Edicions, ISBN 0934797188

Birds of Northern South America: An Identification Guide Vols 1 & 2. Robin Restall et al. (2006) Christopher Helm, ISBN 0713660260

Birds of Chile Alvaro Jaramillo et al. (2003)
Christopher Helm, ISBN 0713646888

The Birds of the Department of Lima Peru, 2nd edition, Maria Koepcke (1983) Harrowood Books, Pennsylvania, ISBN 0915180111

Where to Watch Birds in Peru Thomas Valqui (2004) Privately published, ISBN 9972330923

Where to Watch Birds in South America Nigel Wheatley (1994) Christopher Helm ISBN 0713639091

TYPICAL SPECIES

Resident
Brown Booby, Neotropic Cormorant, Magnificent Frigatebird, Snowy Egret, Great Egret, Cattle Egret, Black-crowned Night Heron, Cocoi Heron, Roadside Hawk, Southern Caracara, Black Vulture, American Kestrel, Southern Lapwing, Kelp Gull, Ruddy Ground Dove, Plain Parakeet, Monk Parakeet, Smooth-billed Ani, Guira Cuckoo, Swallow-tailed

Rio de Janeiro

As an internationally-renowned tourist centre, Rio de Janeiro is not normally associated with ecotourism or considered a birding destination, but this stunning city has much to offer the visiting birder. Apart from the potential offered by its coastline, the city also holds some pockets of Atlantic

Hummingbird, Ringed Kingfisher, Rufous Hornero, Blue-and-white Swallow, Southern Rough-winged Swallow, House Wren, Masked Water Tyrant, Great Kiskadee, Social Flycatcher, Cattle Tyrant, Rufous-bellied Thrush, Pale-breasted Thrush, Chalk-browed Mockingbird, Red-eyed Vireo, Bananaquit, Sayaca Tanager, Shiny Cowbird, House Sparrow, Rufous-collared Sparrow

Nov–Mar
Peregrine Falcon, Common Tern, Ashy-tailed Swift, Barn Swallow, Tropical Kingbird

Magnificent Frigatebirds patrol the beachfront at Rio

Forest (Mata Atlântica) which is now one of the most endangered ecosystems in the world, its present extent having been reduced to less than 5 per cent of the original area. This rich ecosystem supports over 100 endemic species, some of which are threatened with extinction. The best areas of forest can be found on the Serra do Mar mountain range which runs parallel to the coast approximately 50 to 100km inland of Rio de Janeiro. Additionally, the foothills of the mountains (Serra da Carioca) penetrate the city as far south as the coast at Tijuca, and the forested hills of Tijuca National Park provide some of the best urban birding in the world: you can access the rainforest without even leaving the city!

Despite its coastal location and having several coastal lagoons and artificial canals close by, the city lacks sufficient mudflats to attract large numbers of estuarine waders or concentrations of gulls and terns. Even the extensive beach front along the south side of the city offers few seawatching opportunities.

Like all southern hemisphere cities, Rio's large and diverse range of resident species is augmented from November to March by migrants from the northern hemisphere, and from April to October by a number of austral winter migrants, as well as by some post-breeding dispersers from the interior.

Apart from Tijuca, Rio has some other forest areas nearby which are accessible for a (longish) day's birding. However, for those who might like to venture a little further, the legendary Itatiaia National Park could also be visited in the course of a long weekend.

GETTING AROUND

Getting to the birding spots in the Rio de Janeiro area by taxi or public transport is either very easy, if you plan to remain within the city limits, or else rather complicated and will require careful planning, if you want to visit the more distant forested areas. For most practical purposes, the city is divided between north (Zona Norte) and south zones (Zona Sul) by the Serra da Carioca mountain ridge of the Tijuca National Park. The main city birding spots and most of the tourist and business hotels are in the Zona Sul, and you may find that a short taxi ride is all that is needed to reach one of the sites. The current layout of the metro does not serve any of the birding sites directly, but there are a number of feeder bus services from both Botafogo and Saens Peña stations which can bring you to the Botanic Gardens and Tijuca, respectively. In general, the bus services operate throughout the city and are frequent. However, in the absence of any

published route maps or schedules, you are probably better off using a taxi unless you are familiar with the route in question.

If you plan to go further afield and visit the forested areas of the Serra do Mar, you are best advised to go by car. There are two main routes north from Rio: Federal Highway 116 (BR116), which goes north hugging the shore of Baia de Guanabara, and Federal Highway 101 (BR 101), which is accessed by the Rio/Niteroi Bridge and passes northeast towards the mountains. If you are staying in the Zona Sul, getting to these key routes out of the city means taking the Túnel Rebouças at Lagoa or Humaitá and driving north through the Serra da Carioca into the city centre. This route carries very heavy traffic but avoids having to negotiate the one-way street system and the congested streets.

Without a car, you could contemplate taking a bus to Teresópolis from Rio's main bus terminal at Rodoviária Novo Rio, just north of the city centre, and from there a local bus or taxi to the Serra dos Órgãos National Park nearby, but this will be time-consuming and difficult to coordinate if you are trying to make it there and back in a day. Bear in mind that the national parks do not open until 08.00: however the approach roads to the park entrances can offer really good birding and would still repay an early morning visit.

Locations

1 Ilha do Governador (16km)   Bus Nos. 322/324/326/328/M-92

The Baia de Guanabara is not particularly good for birds, but it contains several islands, the largest of which is Ilha do Governador. This is the island on which the international airport is located and is connected by two bridges to the city. The beaches on Ilha do Governador can hold gulls, terns and waders, and are within 1km of the airport if you have some time to kill. The best, or rather, most accessible location is at the Praia do Dendê along the northern shore of the island.

Another island worth a visit is the Ilha de Paquetá which is near the eastern shore of the bay. You can get a ferry from Praca 15 de Novembro in the city centre, which takes about an hour and a half, and then rent a bicycle on the island: this trip also gives you a good chance to see South American Terns.

Take either the **Linha Vermelha** expressway or **Avenida Brasil** to reach Ilha do Governador. You can also take the airport bus or get a taxi on the island.

KEY SPECIES

Resident
Striated Heron, Little Blue Heron, White-faced Whistling Duck, Yellow-

headed Caracara, Cayenne Tern, White-tipped Dove, Glittering-throated Emerald, Yellowish Pipit,

Gray-breasted Martin, Long-billed Wren, Southern Beardless Tyrannulet, Common Tody-Flycatcher, Yellow-bellied Elaenia, Short-crested Flycatcher, Creamy-bellied Thrush, Tropical Parula, Chestnut-vented Conebill, Palm Tanager, Azure-shouldered Tanager, Grassland Sparrow

Apr–Oct
South American Tern, Royal Tern, Wedge-tailed Grass Finch

Nov–Mar
Semipalmated Sandpiper, White-

rumped Sandpiper, Solitary Sandpiper, Spotted Sandpiper, Greater Yellowlegs, Lesser Yellowlegs, Ruddy Turnstone

---

2 Morro do Leme (10km)   *Cardeal Arcoverde Metro Station*

Morro do Leme is a prominent headland located just at the east end of Copacabana Beach and although limited in its avifauna compared to the larger forested areas, its location makes it very convenient to the tourist areas of Rio. Recent efforts at reforestation have created much more habitat and over 90 species have been recorded here.

The best approach to Morro do Leme is from the east end of Avenida Atlântica which runs the complete length of Copacabana. Take the road up to the hilltop, passing a military base at the foot of the hill and climb up to the old fort (Forte do Vigia). This road ascends through some good secondary growth forest, and leads to several other paths. The Cardeal Arcoverde Metro Station is about 1km west of Morro do Leme.

KEY SPECIES

Resident
Yellow-headed Caracara, Slaty-breasted Wood Rail, White-tipped Dove, Scaly-headed Parrot, Glittering-throated Emerald, Glittering-bellied Emerald, White-barred Piculet, Yellow-eared Woodpecker, Sooretama

Slaty Antshrike, Chestnut-backed Antshrike, Yellow-bellied Elaenia, Yellow-lored Tody-Flycatcher, Common Tody-Flycatcher, Short-crested Flycatcher, Gray-breasted Martin, Lemon-chested Greenlet, Tropical Parula, Golden-crowned

Warbler, Chestnut-vented Conebill, Blue Dacnis, Violaceous Euphonia, Purple-throated Euphonia, Brazilian Tanager, Flame-crested Tanager, Ruby-crowned Tanager, Green-winged Saltator

---

3 Botanic Gardens (10km)   *Bus Nos. 170/571/572/594*

The Botanic Gardens are at the foot of the Corcovado Mountain and are contiguous with Tijuca National Park. This is actually one of the best sites in the Rio area for a birder not yet acquainted with rainforest birding, as the birds are relatively easy to see. It is also close to the tourist area of Ipanema, both safe and secure, and a short taxi ride from many hotels. Most of the planted trees and shrubs are non-native, but still maintain some attraction for up to 140 bird species, and the ponds within the gardens provide freshwater habitats. The gardens are open from 08.00 and there is a small entrance charge. The gardens (Jardim Botânico) are on **Rua Jardim Botânico**, about 2km north of Ipanema Beach. The area is well served by buses from the city centre, and from the Zona Sul.

Whilst in the area, it might also be worth checking the Lagoa Rodrigo de Freitas. This large shallow lagoon is right in the heart of the Zona Sul, just north of Ipanema Beach and is a well-known landmark in the city. All of the surrounding areas are residential, but the lake itself supports some aquatic species, and where the shoreline has not been paved or built-over, the scrubby areas provide sufficient cover to support some passerines. Avenida Epitácio Pessoa follows the circumference of the lagoon, and any of the bus routes serving the Botanic Gardens either from the city centre or from the beachfront will leave you near the lagoon.

4 Tijuca National Park (12km)   Bus Nos. 221/233/234

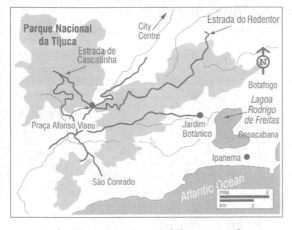

This area is reputed to be the largest urban forest in the world and is also a remnant of Atlantic Forest. Being effectively surrounded by the city and cut off from its natural hinterland, the range of species is more limited than the interior forests. It is still, however, a superb natural resource for the city of Rio and well worth a visit, even if you intend to travel to the interior. The park comprises three separate sectors which encompass several discreet peaks, up to an altitude of over 1,000m.

It is most likely that you will visit the east sector (Sector B) around the Corcovado, where the Statue of Christ is located and is a huge tourist draw. There is a small-gauge mountain railway (US$7) which accesses the statue, and you could then follow a trail down the southern face of the mountain to the Botanic Gardens. North of the Corcovado within Sector B, is the Estrada do Redentor road, running east to west along the spine of the mountain ridge: you may explore/drive along this road.

The northwest sector (Sector A) is the most distant from the city centre and the least visited by tourists, so for a birder it affords a less disturbed habitat.However, it is less well patrolled by the police and you might be better advised to visit accompanied by a guide. The highest point in the park (Pico da Tijuca) is here and this is best for the higher altitude species. If you continue west on the **Estrada do Redentor**, you exit Sector B at the Praça Afonso Viseu, on the opposite side of which is the entrance to Sector A. Go through the gate at **Praça Afonso Viseu** on **Estrada da Cascatinha** for about 2km to the car park at Bom Retiro. From here there is a trail up to Pico da Tijuca.

The Corcovado Cog Railway departs from Rua Cosme Velho (first train at 08.30) and there are several bus routes serving Sectors A and B. The Saens Peña Subway Station is the best place to take a bus (Bus Nos. 233, 234, 221 to Barra): the bus stop is in front of the subway station in Avenida Conde de Bonfim. Take the bus to Praça Afonso Viseu on Avenida Edson Passos. You may also find taxi drivers at Cosme Velho train station who are willing to take you to Estrada do Redentor and after a day's birding you could get another taxi or take the train back from Corcovado.

KEY SPECIES

Resident
Short-tailed Hawk, Slaty-breasted Wood Rail, Gray-fronted Dove, Ruddy Quail-Dove, Maroon-bellied Parakeet, Scaly-headed Parrot, Dusky-throated Hermit, Violet-capped Woodnymph, Channel-billed Toucan, White-barred Piculet, Yellow-eared Woodpecker, Gray-hooded Flycatcher, Eye-ringed Tody-Tyrant, Yellow-lored Tody-Flycatcher, Yellow-olive Flycatcher, White-crested Tyrannulet, Variable Antshrike, Plain Antvireo, Star-throated Antwren, Streak-capped Antwren, Scaled Antbird, Mouse-coloured Tapaculo, White-breasted Tapaculo, White-eyed Foliage-gleaner, Ochre-breasted Foliage-gleaner, Black-capped Foliage-gleaner, Olivaceous Woodcreeper, Swallow-tailed Manakin, Tropical Parula, Golden-crowned Warbler, White-thighed Swallow, Long-billed Wren, Flame-crested Tanager, Ruby-crowned Tanager, Green-headed Tanager, Red-necked Tanager, Blue Dacnis, Rufous-headed Tanager, Red-crowned Ant Tanager, Green-winged Saltator

5 Parque Ecológico de Marapendi (25km)

There are a number of coastal lagoons on the south side of Rio de Janeiro connected to the sea by narrow channels. They have been subjected to many environmental pressures due to the more recent growth of the city and pollution, but steps have been made to restore their ecosystems. An ecological park has been designated around the Lagoa de Marapendi which is a long shallow lagoon separated from the sea by a narrow

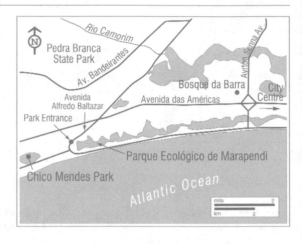

spit (a local geographical feature known as a *restinga*) which has been completely built over. However, the water quality of the lagoon has been much improved and its aquatic vegetation restored. There are now several nature trails around the lagoon, providing an opportunity to view the wildlife which includes caimans.

Marapendi can be reached by taking the main coastal route southwest from Rio de Janeiro and diverting inland on **Avenida das Américas**, which passes the Lagoa de Marapendi on the left-hand (south) side. Continue along Avenida das Américas for about 10km and then turn left (south) onto **Avenida Alfredo Balthazar da Silveira**. The entrance to the reserve is about 1km on the left-hand side of this road. However, you will need to continue to the next interchange in order to make a U-turn to get to the right side of the road.

If you stay on Avenida Alfredo Balthazar da Silveira and follow the road south to the coast, turn right onto **Avenida Lúcio Costa** and go west for a further kilometre. Then turn north on **Avenida Albert Sabin** and you will come to **Parque Ecológico Chico Mendes**. This park also has a smaller lagoon and is good for aquatic species.

Finally, when returning to Rio via Avenida das Américas, check out the Bosque da Barra park (also known as **Parque Arruda Câmara** on some maps) which is located at the intersection of **Avenida das Américas** and **Avenida Ayrton Senna**. This is another small park but a very good birding spot with forest, swamp and restinga habitat.

KEY SPECIES

Resident
Anhinga, Stripe-backed Bittern, Pinnated Bittern, Brazilian Teal, Cayenne Tern, Gray-breasted Martin, Long-billed Wren, Southern Beardless Tyrannulet, Tawny-crowned Pygmy Tyrant, Yellow-bellied Elaenia,

Short-crested Flycatcher, Creamy-bellied Thrush, Tropical Parula, Chestnut-vented Conebill, Palm Tanager, Grassland Sparrow

Nov–Mar
Osprey, Sanderling, Spotted

Sandpiper, Greater Yellowlegs, Lesser Yellowlegs

Apr–Oct
South American Tern, Royal Tern

6 Pedra Branca State Park (25km)

This area of Atlantic Forest is larger than that within Tijuca National Park, it lies further west of the city and its highest peak (Pedra Branca) is marginally higher than Pico da Tijuca. It therefore supports a wider range of species, including several endemics. There are a number of trails in the area, but visitors are guided through the park by officials for safety reasons. There are several routes into the park, but the most accessible is the Camorim entrance which leads to the Visitors' Centre, and from there up to the Camorim Reservoir in the mountains.

Access is from **Estrada dos Bandeirantes** which is the main road around the base of the mountains. Go north on **Estrada de Camorim** which is at the eastern end of Estrada dos Bandeirantes. This road follows the course of the **Camorim River** as it ascends into the mountains and passes through some excellent forest habitat. The park entrance is about 2km from the junction with dos Bandeirantes.

KEY SPECIES

Resident
White-necked Hawk, Black Hawk-Eagle, Barred Forest Falcon, Rusty-margined Guan, Slaty-breasted Wood Rail, Ruddy Quail-Dove, Maroon-bellied Parakeet, Scaly-headed Parrot, Brown-backed Parrotlet, Dusky-throated Hermit, Violet-capped Woodnymph, Spot-billed Toucanet,

Channel-billed Toucan, White-barred Piculet, Yellow-eared Woodpecker, Yellow-lored Tody-Flycatcher, Common Tody-Flycatcher, Yellow-olive Flycatcher, White-crested Tyrannulet, Tufted Antshrike, Star-throated Antwren, Sooretama Slaty Antshrike, Scaled Antbird, White-eyed Foliage-gleaner, Black-capped

Foliage-gleaner, Olivaceous Woodcreeper, Swallow-tailed Manakin, Golden-crowned Warbler, Flame-crested Tanager, Green-headed Tanager, Blue Dacnis, Rufous-headed Tanager, Red-crowned Ant Tanager, Buff-throated Saltator

Further Away

7 Serra dos Órgãos National Park (92km)

This national park is one of the ornithological world's most outstanding centres of endemism. Within the park boundaries, a fine selection of the endemics unique to the Brazilian Atlantic Forest zone can be found. Logistically it presents some difficulties for the short-term visiting birder, as the park does not open until 08.00, and it would take a pre-dawn start to climb to the summit and back in one day in order to see one of the park's specialities, the Grey-winged Cotinga: a species more often heard than seen. This is a steep climb of over 2,000m which takes nine hours: five hours uphill and four hours down. There is a campsite within the park so that you can stay overnight within the park and set off for the summit before dawn. However, even without a dawn ascent, the lower elevations of the mountain support a rich diversity of endemic and non-endemic species.

Take the **Rio–Bahia highway (BR116)** north from Rio de Janeiro, towards **Teresópolis**. The entrance to the main sector of the park is on the southern outskirts of Teresópolis, on the left-hand side of the road. You can drive into the park up to an altitude of about 1,000m where there is a car park beside the dam, and you can walk from there along the **Pedro do Sinho** trail to the mountaintop (extra fee payable). There are two campsites near the mountaintop and it is here that dedicated listers can camp in order to be on-site at the right elevation at dawn to find the extremely local and rare Grey-winged Cotinga. Even if you fail to see the cotinga there is a wealth of birdlife to be seen on this trail between the car park and the campsites.

Also worth checking is the sub-sector of the park, whose entrance is about 18km before Teresópolis on the right-hand (east) side of BR116 and several hundred metres lower down, some of the species here, for example, the Yellow-backed Tanager, are not found in the main sector of the park.

Yet another good spot, located about 7km before Teresópolis, is the turn-off at **Garrafão**: this is the site where the Kinglet Calyptura was rediscovered in 1998 after 100 years without a sighting, and is an excellent spot for forest birds. The turn-off is just north of the petrol station at Garrafão on the right-hand (east) side of the road.

KEY SPECIES

Resident
Barred Forest Falcon, Plumbeous Pigeon, Gray-fronted Dove, Maroon-bellied Parakeet, Brown-backed Parrotlet, Red-capped Parrot, Scaly-headed Parrot, Squirrel Cuckoo, Black Jacobin, Scale-throated Hermit, Black-breasted Plovercrest, Violet-capped Woodnymph, Brazilian Ruby, Surucua Trogon, Spot-billed Toucanet, White-barred Piculet, Yellow-browed Woodpecker, Blond-crested Woodpecker, Yellow-eared Woodpecker, Sepia-capped Flycatcher, Yellow-lored Tody-Flycatcher, Yellow-olive Flycatcher, White-throated Spadebill, Gray-hooded Attila, Giant Antshrike, Plain Antvireo, Star-throated Antwren, Ferruginous Antbird, Rufous-tailed Antbird, Scaled Antbird, White-bibbed Antbird, Mouse-coloured Tapaculo, Rufous Gnateater, White-eyed Foliage-gleaner, White-browed Foliage-gleaner, White-collared Foliage-gleaner, Sharp-billed Treehunter, Plain Xenops, Streaked Xenops, White-throated Woodcreeper, Olivaceous Woodcreeper, Black-billed Scythebill, Sharp-tailed Streamcreeper, Pale-browed Treehunter, Pin-tailed Manakin, Swallow-tailed Manakin, Swallow-tailed Cotinga, Hooded Berryeater, Black-and-gold Cotinga, Grey-winged Cotinga, Sharpbill, Golden-crowned Warbler, White-rimmed Warbler, Red-rumped Warbling Finch, Flame-crested Tanager, Golden-chevroned Tanager, Diademed Tanager, Red-necked Tanager, Brassy-breasted Tanager, Blue Dacnis, Rufous-headed Tanager, Yellow-backed Tanager, Red-crowned Ant Tanager, Violaceous Euphonia, Green-chinned Euphonia, Uniform Finch, Black-throated Grosbeak, Green-winged Saltator

Useful contacts and websites

PROAVES Associação Brasileira para Conservação das Aves (Portuguese)
www.proaves.org.br
Brazilian Association for Bird Conservation.

CBRO Comitê Brasileiro de Registros Ornitológicos (English and Portuguese)
www.cbro.org.br
Brazilian Ornithological Records Committee, Caixa Postal 64532, 05402-970 - São Paulo, SP.

Observação de Aves no Estado do Rio de Janeiro (Portuguese)
ricardo-gagliardi.sites.uol.com.br
Information about the birds of Rio de Janeiro and contact details for guided tours. Tel. (5521) 2288 9786. Email ricardo-gagliardi@uol.com.br

Aves do Brasil (Portuguese)
www.avesdobrasil.com.br
Information about Brazilian birds and habitats.

Atualidades Ornitológicas (English and Portuguese) www.ao.com.br
Atualidades Ornitológicas, PO Box 238 – 86870-000 Ivaiporã-PR – Brazil.

Arthur Grosset's Birds (English)
www.arthurgrosset.com
Personal website packed with useful information about birding in Brazil

Neotropical Bird Club (English)
www.neotropicalbirdclub.org
c/o The Lodge, Sandy, Bedfordshire, SG19 2DL, United Kingdom.

Jardim Botânico do Rio de Janeiro (Portuguese) www.jbrj.gov.br/institu.htm
Rio de Janeiro Botanic Gardens website.

Parque Nacional da Tijuca (Portuguese)
www.terrabrasil.org.br/pn_tijuca/pnt_1.htm
Tijuca National Park website, with information and trailmaps.

IEF (Portuguese)
www.ief.rj.gov.br
State parks information website.

IBAMA (Portuguese)
www.ibama.gov.br
National parks information website.

Sectran (Portuguese)
www.sectran.rj.gov.br
Public transport in Rio de Janeiro area.

Weather forecast and tidal conditions at Rio de Janeiro
www.myforecast.com

Books and publications

Birding Brazil Bruce C Forrester (1993)
Privately published, ISBN 0952156709

Birds in Brazil Helmut Sick (1993) Princeton
University Press, ISBN 0691085692

All the Birds of Brazil
Deodato Souza (2006, 2nd edition)
Subbuteo Natural History Books, ISBN 1905268017

Where to Watch Birds in South America
Nigel Wheatley (1994) Christopher Helm
ISBN 0713639091

● ●

TYPICAL SPECIES

Resident
Neotropic Cormorant, Snowy Egret, Great Egret, Cattle Egret, Black-crowned Night Heron, White-faced Whistling Duck, Roadside Hawk, American Kestrel, Crested Caracara, American Black Vulture, Common Moorhen, Southern Lapwing, Ruddy Ground Dove, Squirrel Cuckoo, Smooth-billed Ani, White-collared Swift, Ringed Kingfisher, Blue-winged Parrotlet, Plain Parakeet, Swallow-tailed Hummingbird, White-throated Hummingbird, Violet-capped Woodnymph, Sapphire-spangled Emerald, Campo Flicker, White-spotted Woodpecker, Rufous Hornero, Chicli Spinetail, Rufous-capped Spinetail, Southern Beardless Tyrannulet, Yellow-bellied Elaenia, Common Tody-Flycatcher, Cattle Tyrant, Great Kiskadee, Social Flycatcher, Blue-and-white Swallow, Brown-chested Martin, House Wren, Rufous-browed Peppershrike, Yellow-legged Thrush, Pale-breasted Thrush, Rufous-bellied Thrush, Chalk-browed Mockingbird, Red-eyed Vireo, Bananaquit, Orange-headed Tanager, Burnished-buff Tanager, Fawn-breasted Tanager, Sayaca Tanager, Palm Tanager, Rufous-collared Sparrow, Saffron Finch, Double-collared Seedeater, Shiny Cowbird, House Sparrow, Common Waxbill (I)

Nov-Mar
Pied-billed Grebe, Ashy-tailed Swift, Fork-tailed Flycatcher, Streaked Flycatcher, Tropical Kingbird

Great Kiskadee at the Interlagos Grand Prix racing circuit

São Paulo

São Paulo is one of the largest cities in the world, located right on the Tropic of Capricorn. The city occupies a plateau at 800m, between the Serra de Mar and the Serra da Cantareira mountain ranges in southeast Brazil. This huge urban sprawl spreads out over the plateau and encroaches up the sides of the mountains. The area was once covered by dense Atlantic Coastal Forest but this habitat has been depleted to such an extent that it is now one of the most threatened ecosystems on the planet. This region is one of the great centres of endemism in the world with many endemic species clinging on to the last refuges of Atlantic Forest that remain.

Amazingly, there are still some pockets of forest left quite close to the city. Although this is mostly secondary growth forest, which grew back after coffee plantations were abandoned, it supports a superb variety of forest species including many endemics. Even the urban greenbelts provide a good selection of birds, and a visitor from any other continent will be fascinated by the rich selection of hummingbirds, flycatchers and tanagers in even a modest city park.

The coast is only about 70km away, although it does require crossing the mountain ridge south of the city and then descending from the height of the escarpment to sea level. However this can easily be accomplished in a day and is well worth the trip, as the mangroves near the coastal city of Cubatão support a fantastic array of aquatic species and give an insight to what this stretch of coastline was like before settlement and urbanisation.

GETTING AROUND

Getting around one of the world's true megacities at any time of day or night is no easy matter. Traffic is extremely heavy and the street system makes it difficult to navigate across town. Luckily, the combination of the underground metro and the metropolitan train system is quite comprehensive and reaches several of the sites with relative ease, so even without a car you can visit two or more sites in a day. The four metro lines are numbered 1–5 (number 4 hasn't been constructed yet) and the six train

lines are lettered A–F. All converge at the two main railway stations: Estação da Luz and Estação do Brás, and there are several other interchange points on the network, so it is easy to switch from one to the other.

If you do have a car, you can get to the coast and back quite easily in a day, but a trip to one of the famous birding hotspots such as Ubatuba or Itatiaia would involve a two- to three-day trip in order to get any reasonable amount of birding time on-site. There are two major state highways south from São Paulo to the coast: SP 150 (Via Anchieta) and SP 160 (Rodovia dos Imigrantes). Unfortunately neither is linked to the system of expressways which interconnect all of the major highways to the north and west of the city. So if you are staying on the north side of the city, you have to drive across the congested city to get to these highways, and at weekends they can be really jam-packed with beachgoers.

As most of the birding round São Paulo takes place in parks, you are restricted to their opening hours, usually 08.00–17.00, although some open as early as 06.00. Being situated at an elevation of 800m, São Paulo is noticeably cooler than the adjacent coast and can experience frosty mornings during 'winter' months (June to August), so an extra layer of clothing would be needed.

Locations

1 Ibarapuera Park (City Centre)   Bus Nos. 5100/5175

This is a large urban park just south of the city centre and very conveniently located in relation to the subway and many central hotels. There are groves of mature trees, open parkland resembling the savanna typical of southern Brazil and two large lakes, which are good for aquatic species. Although it is hemmed in all round by the 'concrete jungle' it nevertheless supports a good range of species, and if you are new to the Neotropics, this is a good place to acquaint yourself with the more familiar species in the easy conditions of a city park. It does, however, get crowded during the day, especially at weekends and an early morning visit is recommended.

The park is about 1km south of Brigadeiro Metro Station and several bus routes serve the park from the metro station. It is open from 06.00.

KEY SPECIES

Resident		
Fulvous Whistling Duck, White-tipped Dove, White-eyed Parakeet, Maroon-bellied Parakeet, Yellow-chevroned Parakeet, Black Jacobin, Scale-throated Hermit, Amazon Kingfisher,	Green Kingfisher, White-barred Piculet, Blond-crested Woodpecker, White-crested Tyrannulet, Grey Monjita, Masked Water Tyrant, Yellow-lored Tody-Flycatcher, Grey-hooded Attila, Gray-breasted Martin,	Tropical Parula, Golden-crowned Warbler, White-rimmed Warbler, Blue Dacnis, Ruby-crowned Tanager, Uniform Finch, Green-winged Saltator

2 Serra da Cantareira (21 km)

The Serra da Cantareira is a ridge of low mountains (approximately 1,000m in altitude) just north of São Paulo and overlooking the city. They are covered by dense vegetation which constitutes one of the largest urban forests in the world. Although, mostly secondary growth, these forests support a good selection of Atlantic Coastal Forest species. The forest is also quite close to São Paulo Airport, so you could realistically hope to do some birding here if you had just half a day or so to spare prior to a flight.

The Serra da Cantareira State Park occupies a large area of the mountains and the southern tip of the park reaches into the suburbs of São Paulo. The main entrance is at the southeast corner, adjacent to the Horto Florestal Park. Several nature trails commence from this area. The Trilha de Bica and the Trilha das Figueiras both traverse some typical Atlantic Forest habitat and are each worth an hour or so. If you have more time available, and feel a little more energetic, the Caminho da Pedra Grande is a round trip of about 10km and ascends to the peak of Pedra Grande (Núcleo Pedra Grande) at over 1,000m. The forest here supports some of the higher altitude species.

If you are more pressed for time, then the Arboretum at Horto Florestal is a good alternative. This area of secondary-growth Atlantic Forest supports a good selection of birds typical of the São Paulo area with an easy walking trail for access. Unlike the state park, it is open all week from 06.00.

Take **Caetano Álvares Avenue** north from the city centre and after about 5km turn north on **Santa Inés Avenue**. At the first roundabout, turn right (east) at **José de Rocha Viana Avenue** which leads to the park entrance. The state park is open only at weekends from 08.30 to 17.00: at other times access is restricted to permit holders.

Resident
Solitary Tinamou, Tataupa Tinamou, Black Hawk-Eagle, Barred Forest Falcon, Yellow-headed Caracara, Spot-winged Wood Quail, Slaty-breasted Wood Rail, Plumbeous Pigeon, White-tipped Dove, Ruddy Quail Dove, Maroon-bellied Parakeet, Scaly-headed Parrot, Striped Cuckoo, Grey-rumped Swift, Swallow-tailed Hummingbird, Black Jacobin, Scale-throated Hermit, Brazilian Ruby, Surucua Trogon, Red-breasted Toucan, White-barred Piculet, Blond-crested Woodpecker, White Woodpecker, Olivaceous Woodcreeper, Giant Antshrike, Variable Antshrike, Plain Antvireo, Planalto Tyrannulet, White-crested Tyrannulet, Southern Bristle Tyrant, Mottle-cheeked Tyrannulet, Yellow-olive Flycatcher, Eye-ringed Tody-Tyrant, Gray-breasted Martin, Creamy-bellied Thrush, Tropical Parula, Golden-crowned Warbler, Blue Dacnis, Ruby-crowned Tanager, Brassy-breasted Tanager, Violaceous Euphonia, Uniform Finch, Blue-black Grassquit, Green-winged Saltator, Red-rumped Cacique, Chopi Blackbird, Chestnut-capped Blackbird

3 Tietê Ecological Park (18km)

Goulart Railway Station – Line F

The Rio Tietê flows westwards through the centre of the city, and several stretches of the river bank within the precincts of São Paulo have been set aside for nature conservation. One of the most important areas is at the Tietê Ecological Park in the east of the city. The park is on a broad bend in the river on the south bank and includes a number of lagoons and backwaters, lined by dense riparian woodland and aquatic vegetation. This mosaic of aquatic habitats permits a number of species which would not normally be seen so close to the city to occur here. A loop trail of approximately 4km encircles all the best areas. The west end of the park near the entrance is taken up mostly by sports facilities but the east end is less disturbed and easily accessed from the loop trail.

Take the main expressway east out of the city along the south bank of the river (**Rodovia Ayrton Senna**). There is a sign after 17km indicating the park. Turn right on to the slip-road which brings you to a car park, with access to the park via a footbridge. The railway station at Eng. Goulart is a short walk from the car park. The park is open from 06.00, although it is popular with joggers at this time.

Resident
Striated Heron, Whistling Heron, Fulvous Whistling Duck, White-cheeked Pintail, Brazilian Teal, Muscovy Duck, Snail Kite, Limpkin, Black-necked Stilt, Black Skimmer, Picazuro Pigeon, White-tipped Dove, White-eyed Parakeet, Sapphire-spangled Emerald, Amazon Kingfisher, White-barred Piculet, Green-barred Woodpecker, Yellow-chinned Spinetail, Masked Water-tyrant, Black-backed Water Tyrant, White-headed Marsh Tyrant, Yellowish Pipit, Tropical Parula, Masked Yellowthroat, Hooded Tanager, Ruby-crowned Tanager, Blue-black Grassquit, Red-cowled Cardinal, Green-winged Saltator, Chestnut-capped Blackbird

4 Ipiranga State Park (11km)

Jabaquara Metro Station – Line 1

This large park on the south side of São Paulo is the location for the zoo and Botanic Gardens. It was selected as a site for the Botanic Gardens because it possesses some remnants of the Atlantic Forest, the natural vegetation of the area, being one of the best preserved examples of this forest in the city area. Where the forest has been cleared, the terrain is now more open parkland resembling the savanna typical of the inland plateau of southern Brazil. Taken together with the artificial lakes and the riparian vegetation along the Rio Ipiranga, the park offers a good mix of habitats for the birder. It is also quite close to the National Airport (Congonhas), so is a convenient stopping-off point if travelling to or from the airport. The park is easy to get to by public transport. There is a zoo shuttle bus from Jabaquara Metro Station.

KEY SPECIES

Resident
Striated Heron, Brazilian Teal, Slaty-breasted Wood Rail, White-tipped Dove, Maroon-bellied Parakeet, Amazon Kingfisher, Ochre-collared Piculet, Variable Antshrike, White-crested Tyrannulet, Brown-crested Flycatcher, Yellow-olive Flycatcher, Euler's Flycatcher, Masked Yellowthroat, Golden-crowned Warbler, Purple-throated Euphonia, Violaceous Euphonia, Ruby-crowned Tanager, Blue Dacnis, Green-winged Saltator, Red-rumped Cacique

5 Guarapiranga Park (20km)

Socorro Railway Station – Line C

There are several large reservoirs south of São Paulo which supply the city with water. The one at Guarapiranga is accessible on its western shore, and a wooded park with nature trail has been created there. The park is at the northwest corner of the reservoir adjacent to the dam. It supports a number of different habitats including some secondary-growth Atlantic Forest near the dam, as well as riparian and more open cerrado-type woodland. However, as the area was formerly coffee plantations, much of the land has now been replanted with eucalyptus trees which are less attractive to the native wildlife.

Take the **Rio Pinheiros Avenue** south from the west end of the city for about 15km until it becomes **Guido Caloi Avenue** and continue for another 2km before turning right (west) on **Guarapiranga Avenue**. After about 500m Guarapiranga forks off to the left (towards the south) and the park entrance is about 500m on the left-hand side. Several bus routes serve Guarapiranga Avenue from **Socorro Railway Station**. The park is open from 06.00 to 18.00.

KEY SPECIES

Resident
Blue-winged Parrotlet, Ochre-collared Piculet, Blond-crested Woodpecker, Lineated Woodpecker, Pallid Spinetail, Boat-billed Flycatcher, Red-ruffed Fruitcrow, Creamy-bellied Thrush, Rufous-headed Tanager, Black-goggled Tanager, Blue-naped Chlorophonia

6 University of São Paulo Campus (15km)

Universitária Railway Station – Line C

The USP campus is spread out over 4km on the west bank of the Rio Pinheiros, and consists of parkland, with some secondary-growth Atlantic Forest as well as a number of waterways including the banks of the Rio Pinheiros. It adjoins the Villa Lobos Park on the east bank of the Pinheiros and this combination provides a continuum of habitats for grassland, savanna and forest species. The lakes and the river provide habitats for aquatic species and this makes the area one of the best birding spots in urban São Paulo.

The campus is on the west side of the city, just south of the intersection of **Jaguaré Avenue** and **Billings Avenue**, and a short walk from the Villa Lobos/Jaguaré and Universitária Railway Stations.

KEY SPECIES

Resident
Striated Heron, White-tailed Kite, Yellow-headed Caracara, Picazuro Pigeon, White-tipped Dove, Maroon-bellied Parakeet, Scaly-headed Parrot, Black Jacobin, Scale-throated Hermit, Red-breasted Toucan, Lineated Woodpecker, Blond-crested Woodpecker, White Woodpecker,

Variable Antshrike, Firewood-gatherer, Sharp-tailed Streamcreeper, White-crested Tyrannulet, Vermilion Flycatcher, Bran-colored Flycatcher,	Euler's Flycatcher, Crested Black-Tyrant, Grey-hooded Attila, Boat-billed Flycatcher, White-winged Becard, Southern Rough-winged Swallow,	Tropical Parula, Purple-throated Euphonia, Chestnut-backed Tanager, Blue Dacnis, Uniform Finch, Green-winged Saltator

Further Away

7 Cubatão Mangroves (60km)

Cubatão, an industrial city on the coast of São Paulo state, was once considered the most polluted city on Earth. Despite this rather dismal reputation, the city lies at the confluence of several rivers which flow into the sea at Baía de Santos and there are still large stretches of Manguezal left undisturbed. This is an area of mangrove swamp, mudflats and riparian forest. The only way to visit the Manguezal is by boat from the marina at Ilha Caraguatá near Cubatão. The most practical way to do this is by joining one of the tourist trips or angling trips on the river, which are usual-

ly organised and originate from São Paulo. If you are a photographer it will afford you the opportunity of photographing what must surely be the most photogenic bird in the world – the Scarlet Ibis. The mudflats are also one of the best places in southeastern Brazil for wintering waders from the northern hemisphere.

Cubatão is south of São Paulo and can be reached by taking either of the two main highways south from the city, the **SP 150** (Anchieta) or the **SP 160** (Imigrantes). If you take either Imigrantes or Anchieta, after about 55km you will come to a rather complex interchange which connects the two highways (**Exit 58** on Anchieta). At the interchange follow the signs for Ilha Caraguatá which is located just west of Imigrantes. Try to time your visit for low tide, when large areas of mud are exposed.

KEY SPECIES

Resident
Magnificent Frigatebird, Roseate Spoonbill, Little Blue Heron, Striated Heron, Yellow-crowned Night Heron, White-faced Ibis, Scarlet Ibis, Limpkin, Anhinga, Fulvous Whistling Duck, Brazilian Teal, Silver Teal, Southern Pochard, Snail Kite, Turkey

Vulture, Black Vulture, White-tailed Kite, Rufous Crab Hawk, Yellow-headed Caracara, Chimango Caracara, Purple Gallinule, Blackish Rail, Slaty-breasted Wood Rail, Collared Plover, South American Snipe, White-backed Stilt, Kelp Gull, Gull-billed Tern, Large-billed Tern,

Yellow-billed Tern, Black Skimmer, Picazuro Pigeon, Scaled Dove, Amazon Kingfisher, Green Kingfisher, Ochre-collared Piculet, Blond-crested Woodpecker, Lineated Woodpecker, Grey-rumped Swift, Glittering-throated Emerald, Black-backed Water Tyrant, Masked Water Tyrant,

Yellow-browed Tyrant, Gray-breasted Martin, Masked Yellowthroat, Tropical Parula, Creamy-bellied Thrush, Blue Dacnis, Ruby-crowned Tanager, Green-winged Saltator, Blue-black Grassquit, Red-rumped Cacique, Unicolored Blackbird, Chestnut-capped Blackbird, White-browed Blackbird

Apr–Oct
White-cheeked Pintail, South American Tern, Snowy-crowned Tern, Royal Tern, Cayenne Tern

Nov–Mar
Osprey, Black-bellied Plover, Semipalmated Plover, White-rumped Sandpiper, Semipalmated Sandpiper,

Spotted Sandpiper, Solitary Sandpiper, Lesser Yellowlegs, Greater Yellowlegs, Willet, Hudsonian Godwit, Purple Martin, Barn Swallow

Useful contacts and websites

Centro de Estudos Ornitológicos (Portuguese and English)
www.ib.usp.br/ceo/
Caixa Postal 64532, 05402-970 - São Paulo, SP
Information about birds in Brazil and São Paulo city.

PROAVES Associação Brasileira para Conservação das Aves (Portuguese)
www.proaves.org.br
Brazilian Association for Bird Conservation.

CBRO Comitê Brasileiro de Registros Ornitológicos (English and Portuguese)
www.ib.usp.br/cbro
Brazilian Ornithological Records Committee, Caixa Postal 64532, 05402-970 - São Paulo, SP.

Avesfoto (Portuguese and English)
www.avesfoto.com.br
Local tour guide with access to Serra da Cantareira State Park outside normal opening hours.

Aves do Brasil (Portuguese)
www.avesdobrasil.com.br
Information about Brazilian birds.

Atualidades Ornitológicos (English and Portuguese)
www.ao.com.br
PO Box 238 - 86870-000 Ivaiporã-PR – Brazil.
Brazilian ornithological website and online magazine.

Programa Ambiental: A Última Arca de Noé (English and Portuguese)
www.aultimaarcadenoe.com.br
Information about birding near São Paulo.

Náutica da Ilha (Portuguese)
www.nauticadailha.com.br
Boat trips to the Manguezal at Cubatão.

Arthur Grosset's Birds (English)
www.arthurgrosset.com
Personal website packed with useful information about birding in Brazil.

Neotropical Bird Club (English)
www.neotropicalbirdclub.org
c/o The Lodge, Sandy, Bedfordshire, SG19 2DL, United Kingdom.

Companhia do Metropolitano de São Paulo – Metrô (English)
www.metro.sp.gov.br
Metro service for São Paulo

Companhia Paulista de Trens Metropolitanos (Portuguese)
www.cptm.sp.gov.br
Suburban train service in São Paulo area.

Books and publications

Guia de Campo Aves da Grande São Paulo
Pedro F. Develey & Edson Endrigo

Birding Brazil Bruce C Forrester (1993) BC Forrester, ISBN 0952156709

Birds in Brazil Helmut Sick (1993)
Privately published, ISBN 0691085692

All the Birds of Brazil
Deodato Souza (2006, 2nd edition)
Subbuteo Natural History Books, ISBN 1905268017

Where to Watch Birds in South America
Nigel Wheatley (1994) Christopher Helm
ISBN 0713639091

Aves do Estado de São Paulo Edwin O. Willis & Yoshika Oniki (2003) Rio Claro: Divisa
ISBN 8590267628

Index of Species

Due to lack of space the following widespread species have been omitted from the index: Pied Avocet; Eurasian Blackbird; Blackcap; Brambling; Bullfinch; Corn Bunting; Reed Bunting; Common Buzzard; Chaffinch; Common Chiffchaff; Zitting Cisticola; Common (Eurasian) Coot; American Coot; Double-crested Cormorant; Great Cormorant; Carrion Crow; Common Cuckoo; Eurasian Curlew; Collared Dove; Stock Dove; (Eurasian) Turtle Dove; Ring-necked Duck; Ruddy Duck; Tufted Duck; Dunlin; Dunnock; Cattle Egret; Great (White) Egret; Little Egret; Snowy Egret; Fieldfare; Pied Flycatcher; Spotted Flycatcher; Gadwall; Garganey; Bar-tailed Godwit; Black-tailed Godwit; Goldcrest; Common Goldeneye; European Goldfinch; Greylag Goose; Canada Goose; Great Crested Grebe; Little Grebe; European Greenfinch; Common Greenshank; Black-headed Gull; Common Gull; Herring Gull; Black-crowned Night Heron; Grey Heron; Purple Heron; Western Jackdaw; Eurasian Jay; Common Kestrel; Common Kingfisher; Black Kite; Northern Lapwing; Linnet; Eurasian Magpie; Mallard; House Martin, Sand Martin, Common Moorhen; Common Myna; Common Nightingale; Eurasian Nuthatch; (Eurasian) Golden Oriole; Osprey; Common Pheasant; Wood Pigeon; Northern Pintail; Meadow Pipit; Tree Pipit; Black-bellied/Grey Plover; Little Ringed Plover; Ringed Plover; Common Pochard; Common Redshank; Common Redstart; Redwing; Robin; Rook; Ruff; Sanderling; Common Sandpiper; Curlew Sandpiper; Green Sandpiper; Spotted Sandpiper; Wood Sandpiper; Northern Shoveler; Eurasian Siskin; Eurasian Skylark; Common Snipe; House Sparrow; (Eurasian) Tree Sparrow; Eurasian Sparrowhawk; Common/European Starling; Black-winged Stilt; Little Stint; Common Stonechat; White Stork; Barn Swallow; Mute Swan; Common Swift; Blue-winged Teal; Eurasian Teal; Green-winged Teal; Black Tern; Caspian Tern; Common Tern; Little Tern; Whiskered Tern; White-winged Black Tern; Mistle Thrush; Song Thrush; Blue Tit; Coal Tit; Great Tit; Long-tailed Tit; Ruddy Turnstone; Grey Wagtail; White Wagtail; Yellow Wagtail; Garden Warbler; Great Reed Warbler; Marsh Warbler; Reed Warbler; Sedge Warbler; Willow Warbler; Wood Warbler; Northern Wheatear; Whimbrel; Whinchat; Common Whitethroat; Lesser Whitethroat; Eurasian Wigeon; Great Spotted Woodpecker; Green Woodpecker; (Winter) Wren; Yellowhammer; Greater Yellowlegs; Lesser Yellowlegs.